Nuclear Physics in a Nutshell

Nuclear Physics in a Nutshell

Carlos A. Bertulani

PRINCETON UNIVERSITY PRESS · PRINCETON AND OXFORD

Copyright © 2007 by Princeton University Press

Requests for permission to reproduce material from this work should be sent to Permissions, Princeton University Press

Published by Princeton University Press, 41 William Street, Princeton, New Jersey 08540

In the United Kingdom: Princeton University Press, 3 Market Place, Woodstock, Oxfordshire OX20 1SY

Library of Congress Control Number 2006940615

ISBN-13: 978-0-691-12505-3

ISBN-10: 0-691-12505-8

British Library Cataloging-in-Publication Data is available

This book has been composed in Scala and Scala sans

Printed on acid-free paper. ∞

press.princeton.edu

Printed in the United States of America

10 9 8 7 6 5 4 3 2

To Eliete, Henrique, and Daniel

Contents

Introduction

0.1 What is Nuclear Physics?

The most accepted theory for the origin of the universe assumes that it resulted from a great explosion, soon after which the primordial matter was extremely dense, compressed and hot. This matter was mainly composed of elementary particles, such as quarks and electrons. As it expanded and cooled down, the quarks united to form heavier particles, called hadrons, which contain 3 quarks (baryons) or 2 quarks (mesons). The protons and neutrons (which are baryons) formed nuclei, and the electrons were captured in orbits around the nuclei forming atoms.

The larger and heavier nuclei were created inside stars, which were formed by the collection of large amounts of the primordial matter. Some of those stars ejected parts of their mass to the interstellar space, leading to the formation of smaller stars, planets, nebulae, etc. The chemical substances were created by the union of atoms in molecules and, finally, by the grouping of several types of molecules in complex structures.

The evolution of the universe is the object of study of cosmology and astrophysics; nuclear astrophysics studies the synthesis of heavy nuclei starting from lighter ones, in temperature and pressure conditions existing in the stars. Nuclear physics studies the behavior of nuclei under normal conditions or in excited states, as well as the reactions among them. Chemistry studies the structure of the atomic molecules and the reactions among them. Finally, biology studies the formation and development of the great molecular agglomerates that compose living beings. In any of these sciences, the objective is to understand complex structures starting from simpler structures and from the interactions among them.

In nuclear physics the simpler structures are the nucleons, the generic name given to protons and neutrons. The nucleon-nucleon interaction, responsible for maintaining the nucleus bound, can be deduced from the analysis of scattering experiments, that is, from the collisions between nucleons. Knowledge about this interaction is in general quite good, which in principle should allow a description of the structure of an atomic nucleus with precision. But this is not exactly the case, and the reason is that in a many-particle system several structures appear, which most of the time do not depend on the interaction between them. For example, a molecule of benzene possesses a structure in the form of a hexagonal ring. We know that the interaction responsible for the formation of molecules is of a Coulombic nature. However, an attempt to describe the symmetry of a molecule solely as due to the properties of the Coulomb interaction will likely fail. Therefore, even if we knew exactly the form of the nucleon-nucleon interaction, that would be insufficient to describe the details of the structure of nuclei with precision. This is a general characteristic of a many-body system. In fact, it is not always clear that good knowledge of the nucleon-nucleon interaction is necessary to determine certain nuclear characteristics. Several interactions with different properties can lead to the same characteristics.

The known atomic nuclei possess at most about 280 nucleons. This number is not so large as to justify the description of the nucleus by macroscopic quantities such as pressure, temperature, elasticity coefficient, and so on, as we do with gases, liquids, and solids in thermodynamic balance. On the other hand, a nucleus with few nucleons is not so simple to describe either: the problem of three interacting particles already possesses a large enough degree of complexity not to allow an exact solution. This situation makes atomic nuclei ideal "laboratories" for the study of the effects of correlations that are developed in a many-body system.

Nuclear physics is intimately linked to other disciplines. In field theory, for example, both the weak interaction and the strong interaction were studied first in atomic nuclei. In fact, the atomic nucleus serves as a micro-laboratory for the study of these interactions in a many-body system. The most celebrated example in this sense is the experiments that demonstrated that the weak interaction is not symmetrical under space reflection.

Similarly, nuclear physics possesses a traditional connection with atomic physics. The interaction of nuclei with their atomic neighbors creates the hyperfine structures in the atomic spectrum. This is important not just in atomic physics, but also in solid state physics. In addition, radioactive nuclei are used as probes for the study of the electromagnetic fields in atomic bonds in crystals.

Nuclear physics is of essential importance for astrophysics. The "burning" of nuclei in the stars can only be studied through experiments of nuclear reactions accomplished in laboratories. This allows understanding of a star's temporal evolution, finally leading to the formation of neutron stars, good examples of the existence of macroscopic nuclear matter (this will be discussed in more detail in chapter 12). The burning of nuclei in stars leads to the creation of heavy elements in nature. In this way, the results of studies in nuclear physics are the basis of the "cosmic" chemistry, which studies the creation and distribution of the elements in the universe.

In a similar way, the methods of nuclear dating, as well as the micro-analytical methods (for example, activation induced by neutrons), are important applications in geology and archaeology.

Nowadays, in the medical as well as technological areas, one cannot neglect the use of nuclear methods. Examples of applications happen in the diagnosis and therapy in medicine, in the study of new materials, and elsewhere.

0.2 This Book

The general idea of the book is to present basic information on the atomic nucleus and the simple theories that try to explain it. Although there is reference to experiments or measurements when I find it necessary, there is no attempt to describe the equipment and methods of experimental nuclear physics in a systematic and consistent way. In the same way, practical applications of nuclear physics are mentioned sporadically, but there is no commitment to giving a general panorama of what exists in this area.

In the ordering of the subjects, I chose to begin with a study of the basic components of the nuclei, the protons and neutrons, and of other particles that compose the scenario of nuclear processes. Pions and quarks play an essential role here, and a summary of their properties is presented.

In chapter 1 the properties of hadrons are summarized. Chapters 2 and 3 treat the system of two nucleons, the deuteron and the nucleon-nucleon interaction, while in the next chapter the properties of nuclei with any number of nucleons is introduced. The nuclear models that have been developed in an effort to explain these properties are described in chapter 5.

Chapters 6 to 9 work with nuclear transformations, starting with a general study of radioactive properties followed by the description of alpha, beta, and gamma decay.

Chapters 10 and 11 embrace the second great block of study in nuclear physics, nuclear collisions, and chapter 12 treats the role of nuclear physics for stellar evolution in several contexts of astrophysics.

Chapter 13 discusses the rapidly growing field of rare nuclear isotopes, short-lived nuclei far from the valley of stability.

An adequate level for a complete understanding of this book corresponds to a student studying at the end of a first degree in physics, including, besides basic physics, a course in modern physics and a first course in quantum mechanics. Students of other exact sciences, and of technology in general, can profit in good part from the subjects presented in this book.

1 | Hadrons

1.1 Nucleons

The scattering experiments made by Rutherford in 1911 [Ru11] led him to propose an atomic model in which almost all the mass of the atom was contained in a small region around its center called the *nucleus*. The nucleus should contain all the positive charge of the atom, the rest of the atomic space being filled by the negative electron charges.

Rutherford could, in 1919 [Ru19], by means of the nuclear reaction

$$
{}_2^4\text{He} + {}_7^{14}\text{N} \; \rightarrow \; {}_8^{17}\text{O} + \text{p}, \tag{1.1}
$$

detect the positive charge particles that compose the nucleus called *protons*. The proton, with symbol p, is the nucleus of the hydrogen atom; it has charge $+e$ of the same absolute value as that of the electron, and mass

$$
m_p = 938.271998(38) \;\; \text{MeV/c}^2, \tag{1.2}
$$

where the values in parentheses are the errors in the last two digits.

From study of the hydrogen molecule one can infer that the protons in the molecule can be aligned in two different ways. The spins of the two protons can be parallel, as in *orthohydrogen*, or antiparallel, as in *parahydrogen*. Each proton has two possible orientations relative to the spin of the other proton, and like the electron the proton has spin $\frac{1}{2}$.

In orthohydrogen the wavefunction is symmetric with respect to the interchange of the spins of the two protons, since they have the same direction, and experiments show that the wavefunction is antisymmetric with respect to the interchange of the spatial coordinates of the protons. This justifies the wavefunction being antisymmetric with respect to the complete interchange of the protons. In parahydrogen the wavefunction is also antisymmetric with respect to the complete interchange of the two protons, being antisymmetric with respect to the interchange of the spins of the protons and symmetric with respect to the interchange of their spatial coordinates. This shows that the protons obey Fermi-Dirac statistics; they are *fermions* and the Pauli exclusion principle is applicable to them. At most one proton can exist in a given quantum state.

The *neutron*, with symbol n, has charge zero, spin $\frac{1}{2}$, and mass

$$m_n = 939.565330(38) \ \text{MeV}/c^2. \tag{1.3}$$

In 1930, Bothe and Becker [BB30] discovered that a very penetrating radiation was released when boron, beryllium, or lithium was bombarded with α-particles. At that time it was thought that this penetrating radiation was γ-rays (high-energy photons). In 1932, Curie and Joliot [CJ32] figured out that the radiation was able to pull out protons from a hydrogen-rich material. They suggested that this was due to Compton scattering, that is, the protons recoiled after scattering the γ-rays. This hypothesis, however, meant that the radiation consisted of extremely energetic γ-rays, and no explanation could be given for the origin of such high energies. Also in 1932, Chadwick [Ch32] showed, by means of an experiment conducted at the Cavendish laboratory in Cambridge, that the protons ejected from the hydrogen-rich material had collided with neutral particles with mass close to the mass of the proton. These were neutrons, the neutral particles that composed the penetrating radiation discovered by Bothe and Becker. The reaction that occurred when beryllium was bombarded with α-particles was

$$^{4}_{2}\text{He} + {}^{9}_{4}\text{Be} \ \rightarrow \ {}^{12}_{6}\text{C} + {}^{1}_{0}\text{n}. \tag{1.4}$$

The existence of the neutron was also necessary to explain some features of the molecular spectrum showing that the wavefunctions of nitrogen molecules were symmetric with respect to interchange of the two ^{14}N nuclei. As a consequence, the ^{14}N nuclei were *bosons*. This could not be explained if the ^{14}N nucleus were composed only of protons and electrons, since 14 protons and 7 electrons are needed for that, which means an odd number of fermions. A system made up of an odd number of fermions is a fermion, since the interchange of two systems of this type can be made by the interchange of each of their fermions, and each change of two fermions changes the sign of the total wavefunction. In the same way, we can say that a system composed of an even number of fermions is a boson. This shows that if the ^{14}N nucleus is formed by 7 protons and 7 neutrons it is a boson, assuming that the neutron is a fermion. In this way, the study of the N_2 molecule led Heitler and Hertzberg [HH29] to conclude that atomic nuclei are composed of protons and neutrons and not of protons and electrons.

Several other studies established that neutrons obey the Pauli principle and thus are fermions, having spin $\frac{1}{2}$. We recall that particles with fractional spin $(2n+1)/2$ are fermions, and that particles with integer spin are bosons. Protons and neutrons have similar properties in several aspects, and it is convenient to utilize the generic name *nucleon* for both.

1.2 Nuclear Forces

The origin of the Coulomb force between charged particles is the exchange of photons between them. This is represented by the *Feynman diagram* (a) of figure 1.1. In this diagram

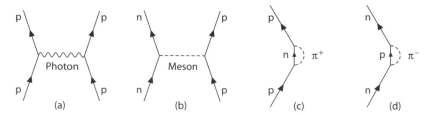

Figure 1.1 Diagrams that represent (a) the electromagnetic interaction, which occurs by exchange of photons, and (b) the nuclear interaction, due to meson exchange. (c and d) Virtual dissociation of the nucleons, giving rise to the anomalous magnetic moment.

lines oriented up represent the direction in which time increases. At some instant of time the particles exchange a photon, which gives rise to attraction or repulsion between them. The photon has zero mass and the Coulomb force is a long-range force.

The force that keeps the nucleus bound is the *nuclear force*. It acts between two nucleons of any type and, in contrast to the Coulomb force, it is of short range. In 1935 Yukawa [Yu35] suggested that the nuclear force has its origin in the exchange of particles with finite rest mass between the nucleons. These particles are called *mesons*, and this situation is described by the Feynman diagram of figure 1.1(b). In the emission of a meson with rest mass M, the total energy of the nucleon-nucleon system is not conserved by the amount $\Delta E = Mc^2$. From Heisenberg's uncertainty principle, $\Delta E \, \Delta t \simeq \hbar$, the exchanged meson can exist during a time Δt (in which violation of energy conservation is allowed), such that

$$\Delta t \simeq \frac{\hbar}{\Delta E} = \frac{\hbar}{Mc^2}. \tag{1.5}$$

During this time the exchanged meson can travel at most a distance

$$R = c\Delta t \simeq \frac{\hbar}{Mc}, \tag{1.6}$$

since the velocity of light c, is the maximum velocity. Then, if the nuclear force can be described by meson exchange, the mesons would exist "virtually" during a time permitted by the uncertainty principle. The nuclear force range would be approximately $\hbar/Mc$. Experimentally one finds that the nuclear force range is $R \simeq 10^{-13}$ cm. Thus, an estimate for the meson mass is

$$M \simeq \frac{\hbar}{Rc} \simeq 0.35 \times 10^{-24} \text{ g} \simeq 200 \text{ MeV}, \tag{1.7}$$

where 1 MeV/c^2 = 1.782×10^{-27} g (for brevity, one normally omits c^2).

In 1936, Anderson and Neddermeyer [AN36] observed cosmic rays in a bubble chamber and found a particle with mass approximately equal to that predicted by Yukawa. These particles were investigated during the next ten years but, because their interaction with nucleons was extremely weak, they could not be the Yukawa meson. This puzzle was solved by Lattes, Muirhead, Powell, and Ochialini [La47]. They discovered that there are two types of mesons: the μ-mesons and the π-mesons. The π-meson interacts strongly with nucleons, but has a very short lifetime and decays into a μ-meson, the particle identified previously by Anderson. The *muon*, as it is known today, has a longer lifetime and does not interact strongly with other particles. The muon does not enter into the description of the nuclear force and is classified among the *leptons*, the family of light particles to which the electron belongs.

The π-meson, known as the *pion*, is the particle predicted by Yukawa. Pions were produced in the laboratory for the first time by Gardner and Lattes in 1948 [GL48], using 340 MeV α-particles from the University of California synchrocyclotron.

1.3 Pions

The pion exists in three charge states, π^+, π^0, and π^-. The π^+ and π^- have the same mass, 139.56995(35) MeV, and the same mean lifetime, $\tau = 2.6 \times 10^{-8}$ s, and decay almost exclusively by the process

$$\pi^+ \rightarrow \mu^+ + \nu_\mu, \quad \pi^- \rightarrow \mu^- + \bar{\nu}_\mu, \tag{1.8}$$

where μ^+, μ^- are the positive and negative muons, ν_μ is the *muonic neutrino*, and $\bar{\nu}_\mu$ is the corresponding antineutrino. Only a small fraction, 1.2×10^{-4}, of pions decay by

$$\pi^+ \rightarrow e^+ + \nu_e, \quad \pi^- \rightarrow e^- + \bar{\nu}_e, \tag{1.9}$$

yielding, respectively, a positron (electron) and an *electron neutrino* (*antineutrino*). (Neutrinos are particles with zero charge and very small mass. Electron neutrinos have a significant role in the β-decay theory; see chapter 8.)

The decay fraction in a given mode is called the *branching ratio*. Charged pions can also decay as

$$\pi^+ \rightarrow \mu^+ + \nu_\mu + \gamma, \quad \pi^- \rightarrow \mu^- + \bar{\nu}_\mu + \gamma, \tag{1.10}$$

also with a branching ratio of 1.2×10^{-4}.

The mass of the neutral pion π^0 is 134.9764(6) MeV, a value 4.6 MeV smaller than that of the charged pions. π^0 decays as

$$\pi^0 \rightarrow \gamma + \gamma, \tag{1.11}$$

with a branching ratio of 98.8 %, by

$$\pi^0 \rightarrow e^+ + e^- + \gamma, \tag{1.12}$$

with a branching ratio of 1.2 %, and by other much less probable processes. The π^0 total lifetime is $(8.4 \pm 0.6) \times 10^{-17}$ s.

The simplest way to produce pions involves collisions between nucleons:

$$p + p \rightarrow p + p + \pi^0, \quad p + p \rightarrow p + n + \pi^+,$$
$$p + n \rightarrow p + p + \pi^-, \quad p + n \rightarrow p + n + \pi^0. \tag{1.13}$$

Pion properties can also be investigated by reactions induced by them, like elastic scattering,

$$\pi^- + p \rightarrow \pi^- + p, \tag{1.14}$$

inelastic scattering,

$$\pi^- + p \to \pi^0 + \pi^- + p, \tag{1.15}$$

or charge exchange reactions,

$$\pi^- + p \to \pi^0 + n. \tag{1.16}$$

The analysis of pion-nucleon and pion-deuteron reactions led to the conclusion that the pion spin is zero. The pions are bosons and obey *Bose statistics*, required for the treatment of particles with integer spin.

1.4 Antiparticles

For each particle in nature there is a corresponding antiparticle, with the same mass, and with charge of the same magnitude and opposite sign. This concept was established around 1930 with the development of relativistic quantum mechanics by Dirac and had its first experimental confirmation with the discovery of the positron (*antielectron*) by Anderson [An32] in 1932. The proton antiparticle (*antiproton*) was detected by Chamberlain and collaborators in 1955 [Ch55], using the 6 GeV bevatron at the University of California.

The first studies of the reaction $\bar{p} + p$, where $\bar{p}$ represents the antiproton, have shown that in the great majority of cases this reaction leads to the annihilation of the $p\bar{p}$ pair with production of pions, but in 0.3% of cases it is able to form the $n\bar{n}$ pair, where $\bar{n}$ is the antiparticle of the neutron, or *antineutron*. It was in this way that, in 1956, Cork and collaborators [Co56] first detected the antineutron, using antiprotons emerging from a beryllium target bombarded with 6.2 GeV protons.

Antiprotons and antineutrons are *antinucleons*. The magnitude of every quantity associated to some particle is identical to that of the corresponding antiparticle, but, as we shall see soon, there are, besides charge, other quantities for which the values for particles and antiparticles have opposite signs.

The mesons π^+ and π^- are antiparticles of each other. In this case it is not important to define which is the particle and which is the antiparticle, since mesons are not normal constituents of matter. In the case of π^0, particle and antiparticle coincide, since charge and magnetic moment are zero.

1.5 Inversion and Parity

Apart from rotational invariance (discussed in Appendix A), a system can have another important spatial symmetry, namely, that with respect to the inversion of coordinates. Let the quantum state of a particle be described by the wavefunction $\Psi(\mathbf{r})$. The *parity* of this state is connected to the properties of the wavefunction by an inversion of coordinates

$$\mathbf{r} \to -\mathbf{r}. \tag{1.17}$$

If

$$\Psi(-\mathbf{r}) = +\Psi(\mathbf{r}) \tag{1.18}$$

we say that the state has *positive parity*, and if

$$\Psi(-\mathbf{r}) = -\Psi(\mathbf{r}) \tag{1.19}$$

we say that the state has *negative parity*. An inversion of coordinates about the origin is represented in quantum mechanics by the operator Π, where

$$\Pi\Psi(\mathbf{r}) = \Psi(-\mathbf{r}). \tag{1.20}$$

Π is called the *parity operator*. The eigenvalues of Π are ± 1 (since $\Pi^2 = 1$):

$$\Pi\Psi(\mathbf{r}) = \pm\Psi(\mathbf{r}). \tag{1.21}$$

If the potential that acts on the set of particles is an even function, that is, $V(\mathbf{r}) = V(-\mathbf{r})$, the parity operator commutes with the Hamiltonian and the parity remains constant in time, that is, it is *conserved*.

From the analysis of a system of two particles 1 and 2 that do not interact, described by the product wavefunction

$$\Psi(\mathbf{r_1})\,\Psi(\mathbf{r_2}), \tag{1.22}$$

the parity of the system is the product of the parities of each particle. That is, the parity is a multiplicative quantum number.

Besides the parity connected to its spatial state, a particle can also have an intrinsic parity. In this case the total parity is the product of the intrinsic and spatial parities. In processes where no particle is created or destroyed, the intrinsic parities of the particles are irrelevant. In reactions where particles are created and destroyed, the application of parity conservation must include the intrinsic parities of the particles.

Since $\Pi = \Pi^{-1}$, there is no distinction between the active and passive viewpoints (see Appendix A). In Cartesian coordinates the inversion transformation means $(x, y, z) \rightarrow (-x, -y, -z)$, whereas in spherical polar coordinates

$$(r, \theta, \varphi) \rightarrow (r, \pi - \theta, \varphi + \pi). \tag{1.23}$$

Therefore $\sin\theta$ does not change, $\cos\theta$ changes sign, and the function $e^{-im\varphi}$ acquires the factor $(-)^m$.

Rotations commute with inversion, as can be easily understood from the geometrical picture and checked formally. This implies that if a state belonging to a rotational multiplet has a certain parity, this quantum number should be the same for all members of the multiplet. For the spherical function Y_{ll} given by equations (A.7) and (A.89) of Appendix A, we find parity $(-)^l$. Therefore we conclude that

$$\Pi Y_{lm}(\mathbf{n}) = Y_{lm}(-\mathbf{n}) = (-)^l Y_{lm}(\mathbf{n}). \tag{1.24}$$

If the particle has a positive intrinsic parity, the total parity will be $(-1)^l$. If the particle has a negative intrinsic parity, the total parity will be $(-1)^{l+1}$. The same result is clear for P_l which are polynomials of order l in $\cos\theta$. In particular, for the backward direction

$$P_l(-1) = (-)^l. \tag{1.25}$$

The operators, such as **r** or **p**, acting on a state with a certain parity, change this value to the opposite one. They can be called Π-odd operators.

For a particle in a spherically symmetric field, stationary wavefunctions have a certain value of the orbital momentum l,

$$\psi(\mathbf{r}) = R_l(r)Y_{lm}(\mathbf{n}), \tag{1.26}$$

where $R_l(r)$ is a radial function. We see that parity for single-particle motion is uniquely determined by the orbital momentum. This is not the case in many-body systems, where total momentum and parity are independent in general.

From an analysis of pion and nucleon reactions, one concludes that the intrinsic parity of the former is $\Pi_\pi = -1$ and for the nucleons $\Pi_n = \Pi_p = +1$.

1.6 Isospin and Baryonic Number

The elementary particles exist in groups of approximately the same mass, but with different charges. The mass of the neutron, for example, is about the same as that of the proton, and the mass of the neutral pion, π^0, is approximately equal to that of the charged pions, π^+ and π^-. In 1932, Heisenberg [He32] suggested that the proton and the neutron could be seen as two *charge states* of a single particle, using the name nucleon to identify this particle.

In the theory of atomic spectra, a state that has *multiplicity* $2s + 1$ has spin s[1]. It is common, however, to refer to the *spin quantum number s* simply as *spin s* (for more details on multiplets, see Appendix A). This is also true for the orbital and total angular momentum. One example is the Zeeman effect, which is the energy splitting among the $2s + 1$ states of an atom in a magnetic field. The spin s is identified with the angular momentum of the system, and operators for the components of this angular momentum, s_x, s_y, s_z, can be defined. The quantum commutation rules are defined as

$$[s_x, s_y] = i\hbar s_z, \quad [s_y, s_z] = i\hbar s_x, \quad [s_z, s_x] = i\hbar s_y. \tag{1.27}$$

The nucleon has, relative to its charge, a multiplicity $2 \times \frac{1}{2} + 1 = 2$. By analogy with the theory of atomic spectra, we create a quantity called *isospin*, $t = \frac{1}{2}$, to obtain the multiplicity $2t + 1 = 2$. Isospin cannot be identified with an angular momentum, and does not have any connection with the spatial properties of the nucleon. Nevertheless, we can introduce an isospin space, or charge space, where the isospin can be treated as a set of three components t_x, t_y, t_z, satisfying the same commutation rules as the spin

$$[t_x, t_y] = i\hbar t_z, \quad [t_y, t_z] = i\hbar t_x, \quad [t_z, t_x] = i\hbar t_y. \tag{1.28}$$

Thus we can deal with the isospin in the same way we deal with the angular momentum. Since the square of the spin, $\mathbf{s}^2$, has eigenvalues $s(s + 1)$, the square of the isospin, $\mathbf{t}^2$, has eigenvalues $t(t + 1)$. Addition of the isospins of several particles can be treated with the

---

[1] The spin is a vector with modulus $\hbar\sqrt{s(s + 1)}$.

vector model for addition, used in atomic spectra theory. For example, when we add two spins $\frac{1}{2}$, the total spin can be 0 or 1. This is also the case of the sum of isospins of two nucleons (each one has spin $\frac{1}{2}$).

The $2s + 1$ states of a system with spin s are denoted by the $2s + 1$ distinct values of the z component of $\mathbf{s}$ (in units of $\hbar$):

$$s_z = -s, -s + 1, \ldots, s - 1, s. \tag{1.29}$$

By analogy, the $2t + 1$ states of a system with isospin t are denoted by the $2t + 1$ distinct values of the t_z component,

$$t_z = -t, -t + 1, \ldots, t - 1, t. \tag{1.30}$$

The direction of the third axis in charge space is chosen in such a way that $t_z = +\frac{1}{2}$ for the proton and $t_z = -\frac{1}{2}$ for the neutron.

The nucleons can be represented by a "two-level" system with two basis states

$$p = \begin{pmatrix} 1 \\ 0 \end{pmatrix}, \quad n = \begin{pmatrix} 0 \\ 1 \end{pmatrix}. \tag{1.31}$$

The states (1.31) have a certain electric charge $z_p = 1$ and $z_n = 0$ in units of e, that is, they are the eigenstates of the charge operator Q. Referring to the basis (1.31), in analogy to the usual spin (A.86), as z-representation, we can say that this quantization axis is related to the interaction with the electromagnetic field that allows one to distinguish between the two charge states of the nucleon.

Let us call the spinor space represented by the basis vectors (1.31) *charge space*. All operators in this space are 2×2 matrices as in spinor states of a spin $\frac{1}{2}$ particle. We can construct the full set of matrices acting in this two-dimensional space of the unit matrix and matrices $\tau_{1,2,3}$ defined exactly as the Pauli matrices (A.82) in spin space. Evidently, the charge operator is

$$Q = \begin{pmatrix} 1 & 0 \\ 0 & 0 \end{pmatrix} = \frac{1}{2}(1 + \tau_z). \tag{1.32}$$

We can also introduce the off-diagonal operators inducing transitions between the charge states of the nucleon. The operator raising the charge is

$$\tau_+ = \begin{pmatrix} 0 & 1 \\ 0 & 0 \end{pmatrix}, \quad \tau_+ p = 0, \quad \tau_+ n = p; \tag{1.33}$$

the lowering operator is

$$\tau_- = \begin{pmatrix} 0 & 0 \\ 1 & 0 \end{pmatrix} = (\tau_+)^\dagger, \quad \tau_- p = n, \quad \tau_- n = 0. \tag{1.34}$$

The operators $\tau_\pm$ are built of the Pauli matrices

$$\tau_\pm = \tau_x \pm i\tau_y, \tag{1.35}$$

in the same way as the raising and lowering operators $J_\pm = J_x \pm iJ_y$ of any angular momentum operator, as explained in Appendix A. We can combine the matrices $\tau_{x,y,z}$ into a matrix "vector" $\boldsymbol{\tau}$ that is completely analogous to the vector $\boldsymbol{\sigma}$ of spin Pauli matrices.

Continuing this analogy, we speak about the *isospin* of the nucleon

$$t = \frac{1}{2}\tau,$$ (1.36)

which acts in 2×2 "isospace" with the basis (1.31), where the basis states have a certain charge

$$Z = \frac{1}{2} + t_z.$$ (1.37)

The correct full spelling of the "isospin" is "isobaric" spin, which unifies the *isobars*, these being states with the same mass number, as in, for instance, the proton and the neutron or nuclei with the same sum $A = Z + N$ of the proton and neutron numbers that coincides with the total baryonic charge B of a nucleus, but not "isotopic," which would relate the *isotopes* having the same electric charge Z at different masses A. In isospin language, the proton and neutron are the states with different projection of the isospin onto the z-axis of isospace,

$$t_3 p = \frac{1}{2}p, \quad t_3 n = -\frac{1}{2}n.$$ (1.38)

Here we use the convention accepted in particle physics. In nuclear physics the isospin projections are traditionally assigned in the opposite way, $-\frac{1}{2}$ for the proton, and $+\frac{1}{2}$ for the neutron. Then for stable nuclei, having as a rule more neutrons than protons, the total projection of the isospin would be positive; the charge operator (1.37) is redefined correspondingly.

The pion has three charge states and thus has isospin $t = 1$; the three pions form a charge *multiplet*, or isospin multiplet, with multiplicity $2t + 1 = 3$. The state $t_z = +1$ is attributed to the π^+, $t_z = 0$ to the π^0, and $t_z = -1$ to the π^-. This is connected to the convention that was adopted for the nucleons and is necessary for the validity of (1.39) below.

The isospin magnitude is an invariant quantity in a system governed by the strong interaction. In electromagnetic interactions this quantity is not necessarily conserved, and we shall verify below that this is the only conservation law that has different behavior in relation to these two forces.

The number of nucleons before and after a reaction is always the same (see, for example, eqs. (1.13) and (1.14)–(1.16). This suggests the introduction of a new quantity, B, called *baryonic number*, that is always conserved in reactions. We attribute to the proton and to the neutron the baryonic number $B = 1$, and to the antiproton and antineutron $B = -1$. To the pions we ascribe $B = 0$ (also for electrons, neutrinos, muons, and photons). In this way the conservation of baryonic number is extended to all reactions. This principle is extended to the leptons, defining a *leptonic number*, which is also conserved in reactions.

From the isospin and baryonic number definition we can write the charge q, in units of e, as

$$Z = t_z + \frac{B}{2}.$$ (1.39)

Since the antiparticle of a particle of charge q and baryonic number B has charge $-q$ and baryonic number $-B$, it must have also a third isospin component $-t_z$, where t_z is the isospin z-component of the corresponding particle.

1.7 Isospin Invariance

The *mirror symmetry* of strong interaction implies invariance under the charge symmetry transformation $p \leftrightarrow n$. By virtue of this symmetry the proton and neutron states are degenerate. Since their electromagnetic properties are different, this is equivalent to the statement that their mass difference comes exclusively from electromagnetic interactions, supposedly on the quark level. The charge symmetry transformation is a particular case of $SU(2)$ transformations (see Appendix A). Moreover, *isospin invariance* assumes that the strong Hamiltonian is invariant under all elements of the isospin group. In this case we have full rotational invariance in isospin space, and the stationary states can be labeled by the conserved quantum number total isospin, T:

$$\mathbf{T} = \sum_a \mathbf{t}_a, \tag{1.40}$$

which is the analog of the total angular momentum in isospace related to the eigenvalues of the isospin "length," $\mathbf{T}^2 = T(T+1)$. $\mathbf{t}_a$ represents the isospin of nucleon a.

Since the algebraic properties of spin and isospin are identical, the allowed values of T are quantized to be integer (half-integer) in a system of an even (odd) number of nucleons. They give rise to degenerate *isomultiplets* with given T that contain $2T+1$ states with projections $T_z = -T, \ldots, +T$ or, equivalently, with the charge [see (1.37)]

$$Z = \sum_a \left(\frac{1}{2} + t_{za} \right) = \frac{A}{2} + T_z, \tag{1.41}$$

where t_{za} is the z-component of $\mathbf{t}_a$.

The isospin invariance of the strong interaction Hamiltonian H_s can be written as the conservation law

$$[\mathbf{T}, H_s] = 0. \tag{1.42}$$

In the case of stationary states, all $2T+1$ states of a multiplet would have the same energy in the limit of exact isospin invariance. Let us emphasize that the states within a given isomultiplet belong to different nuclei (the same A but different Z). They are frequently called *isobaric analog* states (IAS). The conservation law (1.42) is certainly exact for the component T_z related to the electric charge, eq. (1.41). If we forget for a moment about electromagnetic interactions that single out the z-axis and violate the isotropy of isospace, we can classify all nuclear states by isomultiplets. All states in a given nucleus (the "vertical" scale) have the same

$$T_z = \frac{1}{2}(Z - N) = Z - \frac{A}{2}. \tag{1.43}$$

They belong to various isomultiplets (the "horizontal" scale). The allowed values of the magnitude T of the isospin of the nucleus (Z, A) cannot be less than the value of projection T_z, (1.43), but they cannot exceed the maximum value $A/2$,

$$\frac{1}{2}(Z - N) \leq T \leq \frac{1}{2}(Z + N). \tag{1.44}$$

We have already mentioned that the introduction of isospin does not increase the number of nuclear degrees of freedom, or the number of possible states. This is just a convenient classification associated with the invariance (1.42) of strong interactions. This classification is actually related to the permutational symmetry of the many-body wavefunction in the "normal" coordinate and spin variables. If the effects violating isospin invariance, in particular, due to electromagnetic interactions, can be neglected, we have an approximate isospin symmetry.

The concept of isospin SU(2) invariance is generalized to *higher symmetries* in quantum chromodynamics (see below), where it is related to the fact that the two lightest quarks, u (up) and d (down), have similar masses and interactions. In the approximation that neglects the difference of the u, d quarks from the s (strange) quark, the carrier of the strangeness, we have already three fundamental objects so that the corresponding invariance group is SU(3), or SU(6) if the interactions do not depend on spins.

1.8 Magnetic Moment of the Nucleons

A charged particle rotating around an axis can be visualized as a system equivalent to a small ring carrying an electric current. To this current is associated a magnetic dipole moment μ that is related to the particle angular momentum $\mathbf{L}$ through $\mu = e\mathbf{L}/2mc$, where e is the charge and m the mass of the particle. It is common to write

$$\mu_L = \frac{eg_L}{2mc}\mathbf{L}, \tag{1.45}$$

where the factor g_L is introduced. It is called *orbital g factor*, equal to 1 for protons and 0 for neutrons.

However, as we have already discussed, a particle can have an intrinsic angular momentum $\mathbf{s}$. Thus it is fair to admit that an intrinsic magnetic moment can also be associated to a particle, given by

$$\mu_S = \frac{eg_S}{2mc}\mathbf{s}, \tag{1.46}$$

where the constant g_S, the *spin g-factor*, does not necessarily have the same value g_L adequate to classical variables, since $\mathbf{s}$ and μ_S have pure quantum origin. In fact, from a relativistic treatment of quantum mechanics using the Dirac equation, a value $g_S = 2$ for spin $\frac{1}{2}$ charged particles is obtained [Di30]

The universal constant (using m as the proton mass)

$$\mu_N = \frac{e\hbar}{2mc} = 5.05 \times 10^{-27} \frac{\text{Joule m}^2}{\text{Weber}} \tag{1.47}$$

is known as the *nuclear magneton*, by analogy with the Bohr magneton defined for the electron. The nuclear magneton is used as a unit for magnetic moments and it is convenient to note that, owing to the proton mass in the denominator, its value is about 1800 times less than its electronic equivalent. With it, (1.45) and (1.46) can be rewritten as

$$\boldsymbol{\mu}_L = \mu_N g_L \frac{\mathbf{L}}{\hbar} \quad \text{and} \quad \boldsymbol{\mu}_S = \mu_N g_S \frac{\mathbf{S}}{\hbar}. \tag{1.48}$$

The prediction $g_S = 2$ obtained by Dirac works very well for the electron. For the proton and the neutron the values found experimentally are

$$g_S = 5.5856 \text{ (proton)} \quad \text{and} \quad g_S = -3.8262 \text{ (neutron)}, \tag{1.49}$$

when the expected values from Dirac's theory would be 2 and 0. The discrepancy above can be explained in part by the *virtual dissociation* of the nucleons. As we have already established, the uncertainty principle allows a nucleon to emit and reabsorb a pion during a time interval $\Delta t \sim \hbar/m_\pi c^2$, as described by the diagrams of figure 1.1b,c. One proton can be dissociated into a neutron and a π^+. The π^+ has spin zero and does not have intrinsic magnetic moment, but it can contribute to the proton magnetic moment due to its orbiting around the neutron. Supposing that this process conserves angular momentum and parity, one can show that the π^+ orbiting is in the same direction as the initial intrinsic spin of the proton. The effect of the virtual production of a π^+ is therefore to increase the proton magnetic moment, as is experimentally observed.

In the same way the π^- contributes to the neutron magnetic moment. There is a small contribution to the magnetic moment due to the proton orbit and spin but the largest contribution comes from the π^- orbiting, due to its small mass; the orbital magnetic moment depends on the inverse of the mass of the particle in orbit (1.47), and $m_\pi \ll m_p$. The contribution is negative due to the charge of the π^- and because it orbits in the same direction as the neutron spin. So, we expect that the neutron intrinsic magnetic moment is less than zero, as is experimentally verified. This analysis is only qualitative. A more exact explanation of the magnetic moments of the nucleons is still the object of theoretical studies.

The antiproton, being a particle with negative charge, has magnetic dipole moment in a direction opposite to that of the angular momentum and thus $g_L = -1$. The values given by (1.49) also have opposite sign for the antiparticles. Then, the antiproton magnetic moment points contrary to the spin and that of the antineutron stays aligned to the spin.

1.9 Strangeness and Hypercharge

In 1947, particles with properties different from those known at that time were found in cosmic rays. They were afterward (1953) observed in the laboratory. In cloud chambers, where their trajectories were detected and photographed, they appeared as a pair of tracks in form of a V, making clear that two particles were created simultaneously. These *strange particles* or *V-particles*, as they were initially known, form two distinct groups. One of them consists of particles heavier than the nucleons and decaying into them, called *hyperons*. The symbols Λ, Σ, Ξ, and Ω are utilized for the several hyperons. Because they decay into nucleons, hyperons are baryons and have baryonic number 1. They also have spin $\frac{1}{2}$, and they are fermions. The other group of strange particles are bosons with spin 0 and are called *K-mesons*, or *kaons*.

Typical reactions involving strange particles are

$$\Lambda^0 \to p + \pi^-, \qquad \Lambda^0 \to n + \pi^0,$$
$$\Xi^- \to \Lambda^0 + \pi^-, \qquad \Xi^0 \to \Lambda^0 + \pi^0,$$
$$K^0 \to \pi^+ + \pi^-, \qquad \Lambda^0 \to \pi^- + p,$$
$$\pi^- + p \to K^0 + \Lambda^0. \tag{1.50}$$

The *interaction time* for reactions involving nucleons and pions is obtained approximately by the time in which a pion, with velocity near that of light, travels a distance equal to the nuclear force range. This time is about $r/c \simeq 10^{-23}$ s, which is much less than the mean lifetime of the Λ^0 ($\tau = 2.5 \times 10^{-10}$ s), or of other strange particles. On the other side, it was found experimentally that the rate at which or other strange particles are produced is consistent with an interaction time on the order of 10^{-23} s. To explain why strange particles were produced so fast but decayed so slowly, Pais [Pa52] suggested that strong interactions (those acting between nucleons, or between pions and nucleons) are responsible for the production of strange particles. The reactions in which only one strange particle takes part, as in its decay, would proceed through weak interactions, similar to β-decay or the decay of muons or charged pions.

In 1953, Gell-Mann [Ge53] and Nishijima and Nakano [NN53] showed that the production of strange particles could be explained by the introduction of a new quantum number, *strangeness*, and postulating that strangeness is conserved in strong interactions. For example, two strange particles, but with opposite strangeness, could be produced by means of the strong interaction in a collision between a pion and a nucleon. Strangeness, however, is not conserved in the decay of a strange particle, and this decay is attributed to the weak interaction.

The great number of hadrons, as shown in table 1.1, and their apparently complex distribution led several investigators to question if these particles might be complex structures composed by the union of simpler entities. Models were proposed for those structures and, after some unsuccessful attempts, a model independently created by M. Gell-Mann [Ge64] and G. Zweig [Zw64], in 1964 imposed itself and has gained credibility over time. The inspiration for that model came from the symmetries observed when one put mesons and baryons in plots of strangeness versus the t_z-component of isospin, as shown in figure 1.2. The type of observed symmetry is a characteristic of the group called $SU(3)$, where three basic elements can generate singlets (the mesons η' and ϕ), octets (the other eight mesons of the figures above and the eight spin $\frac{1}{2}$ baryons) and decuplets (the spin $\frac{3}{2}$ baryons). These three basic elements, initially conceived only as mathematical entities able to generate the necessary symmetries, ended up acquiring the status of real elementary particles, for which Gell-Mann coined the name *quarks*. To obtain hadronic properties, these three quarks, presented in the *flavors up, down,* and *strange*, must have the characteristic values shown in table 2.

Table 1.1 Attributes of particles that interact strongly.

n	B	S	t	t_z	s	m (MeV/c^2)
p	+1	0	1/2	+1/2	1/2	938.272
n	+1	0	1/2	−1/2	1/2	939.565
$\bar{p}$	−1	0	1/2	−1/2	1/2	938.272
$\bar{n}$	−1	0	1/2	+1/2	1/2	939.565
Λ	+1	−1	0	0	1/2	1115.68
Σ^+	+1	−1	1	+1	1/2	1189.4
Σ^0	+1	−1	1	0	1/2	1192.6
Σ^-	+1	−1	1	−1	1/2	1197.4
$\bar{\Lambda}$	−1	+1	0	0	1/2	1115.68
$\bar{\Sigma}^+$	−1	+1	1	−1	1/2	1189.4
$\bar{\Sigma}^0$	−1	+1	1	0	1/2	1192.6
$\bar{\Sigma}^-$	−1	+1	1	+1	1/2	1197.4
Ξ^0	+1	−2	1/2	+1/2	1/2	1315
Ξ^-	+1	−2	1/2	−1/2	1/2	1321
$\bar{\Xi}^0$	−1	+2	1/2	−1/2	1/2	1315
$\bar{\Xi}^-$	−1	+2	1/2	+1/2	1/2	1321
Ω^-	+1	−3	0	0	3/2	1672
π^0	0	0	1	0	0	134.976
π^+	0	0	1	+1	0	139.567
π^-	0	0	1	−1	0	139.567
K^+	0	+1	1/2	+1/2	0	493.7
K^-	0	−1	1/2	−1/2	0	493.7
K^0	0	+1	1/2	−1/2	0	497.7
$\bar{K}^0$	0	−1	1/2	+1/2	0	497.7

Notes: The baryons are the particles with baryonic number $B \neq 0$; the mesons have $B = 0$. S is strangeness, t the isotopic spin and t_z its projection; s is the particle spin and m its mass. Baryons have positive intrinsic parity; mesons have negative ones.

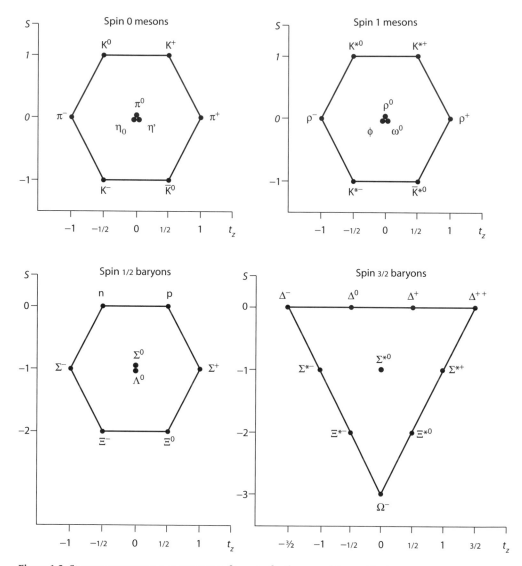

Figure 1.2 Strangeness versus t_z-component of isospin for the several types of hadrons.

The most striking fact is that, for the first time, the existence of particles with fractional charge (a fraction of the electron charge) is admitted. We can in this way construct a nucleon by composing three quarks (neutron = udd), and it is natural to attribute to quarks a baryonic number $B = \frac{1}{3}$. The pions, in turn, are obtained by the conjunction of a quark and an antiquark, ($\pi^+ = u\bar{d}$), ($\pi^0 = d\bar{d}$), ($\pi^- = d\bar{u}$), where the properties of the antiparticle for the quarks are obtained in the conventional way.

To reproduce the other baryons and mesons, the strange quarks have to play a role, and a hyperon like Σ^0, for example, has the constitution ($\Sigma^0 = uds$), while a meson has the constitution ($K^+ = u\bar{s}$). It is convenient to say at this point that a certain combination of quarks does not necessarily lead to only one particle. In the case of the combination above, we also have the possibility to build the hyperon ($\Sigma^{*0} = uds$). The reason for this is

Table 1.2 Quark characteristic quantum numbers.

Flavor	Charge	Spin	Strangeness
up	+2/3	1/2	0
down	−1/3	1/2	0
strange	−1/3	1/2	−1

that, besides other quantum numbers that will be discussed later, a combination of three fermions can give rise to particles with different spin. If we consider as zero the quarks' total orbital angular momentum, which is true for all particles we discussed, the total spin of the three quarks can be $\frac{1}{2}$ or $\frac{3}{2}$. The hyperon Σ^0 corresponds to the first case and the hyperon Σ^{*0} to the second.

A first difficulty in the theory appears when we examine the particles (Δ^{++} = uuu), (Δ^- = ddd), (Ω^- = sss). Since the three quarks in each case are fermions with $l = 0$, it is clear that at least two of them would be in the same quantum state, which violates the Pauli principle. To overcome this difficulty, a new quantum number was introduced, *color*: the quarks, besides the flavors up, down, or strange, would have a color, red (R), green (G), or blue (B), or *anticolor*, $\bar{R}$, $\bar{G}$, or $\bar{B}$. It is clear that, in the same way as flavor, color has nothing to do with the usual notion we have of that property. The introduction of this new quantum number solves the above difficulty; since now a baryon like Δ^{++} is written $\Delta^{++} = u_R u_G u_B$, the problems with the Pauli principle are eliminated. The addition of three new quantum numbers increases enormously the possibility of construction of hadrons, but a new rule comes to play, limiting the possible of color combinations: *all the possible states of hadrons are colorless*, where colorless in this context means absence of color or white color. White is obtained when, in a baryon, one adds three quarks, one of each color. In this sense the analogy with the common colors works, since the addition of red, green, and blue gives white. In a meson, absence of color results from the combination of a color and the respective anticolor. Another way to present this property is to understand the anticolor as the complementary color. In this case, the analogy with the common colors also works and the pair color-anticolor also results in white.

The concept of color is not only useful to solve the problem with the Pauli principle. It has a fundamental role in quark interaction processes. The accepted theory for this interaction establishes that the force between quarks works by the exchange of massless particles, with spin 1, called *gluons*. These gluons always carry one color and one different anticolor, and in the mediation process they interchange the respective colors; one example is seen in figure 1.3. One can also see in figure 1.4b that the gluons themselves can emit gluons.

The fields around hadrons where exchange forces act by means of colors are denominated *color fields*, and the gluons, the exchanged particles, turn out to be the field particles of the strong interaction. In this task they replace the pions that, in the new scheme, are composite particles. The fact that the gluons have colors and can interact mutually makes the study of color fields (quantum chromodynamics) particularly complex.

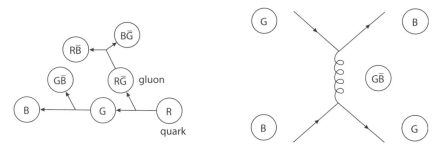

Figure 1.3 (a) Forces between quarks mediated by gluon exchange. (b) Diagram showing how a quark B changes to a quark G, and vice-versa, by the exchange of a gluon GB̄.

Despite the success of the *quark model*, new difficulties arose and, in 1970, with the purpose of explaining some decay times in disagreement with the predictions of the model, S. L. Glashow, J. Iliopoulos, and L. Maiani [Gl70] proposed the existence of a fourth quark whose flavor has received the designation *charm* (c). This c quark has a charge of $+\frac{2}{3}$; it has strangeness zero but has a new quantum number, the charm C, with an attributed value $C = 1$. The prediction of quark c received experimental confirmation in 1974, when two independent laboratories detected a new particle, called Ψ by the Stanford linear accelerator group (SLAC) and J by the Brookhaven National Laboratory team. The particle J/Ψ, as it is commonly designated, is interpreted as a cc̄ state, called *charmonium*, by analogy with positronium eē. The existence of particles with charm introduces some complications in the symmetries of figure 1.2: an axis with the new quantum number is added and the new symmetries have to be sought in three-dimensional space.

In 1977, a group of resonances in 10 GeV proton-proton reactions pointed to the existence of a new meson that received the name ϒ and that led to the proposition of a new quark. This quark, b (from *bottom* or *beauty*), has charge $-\frac{1}{3}$ and a new quantum number, *beauty*, B^*. The quark b has $B^* = -1$.

Theoretical reasons imply that quarks exist in pairs, and this led to a sixth flavor, which corresponds to quark t (from *top* or *true*), with charge $+\frac{2}{3}$. This quark was identified in experiments conducted at Fermilab in 1993 [Ab94].

The theory of quarks, with its colors and flavors, has created a scheme in which a great number of experimental facts can be explained. High energy electron beams have indeed detected an internal structure in nucleons with all the features of quarks [Fr91]. However, one can never pull out a quark from a hadron and study its properties separately. To eliminate this possibility, a theory of *asymptotic freedom* was developed *confining* quarks permanently to the hadrons. One consequence is that their mass cannot be directly determined, since it depends on the binding energies, which are also unknown. The quark model enjoys high prestige in the theory of elementary particles, and there is a substantial reduction in the number of elementary particles, that is, point particles without an internal structure. These are the quarks, the leptons, and the field bosons. An outline of the properties of these particles is shown in table 1.3.

Table 1.3 Properties of the elementary particles.

Quarks	Charge	Spin	Strangeness	Charm	Beauty	Truth
u	+2/3	1/2	0	0	0	0
d	−1/3	1/2	0	0	0	0
s	−1/3	1/2	−1	0	0	0
c	+2/3	1/2	0	1	0	0
b	−1/3	1/2	0	0	−1	0
t	+2/3	1/2	0	0	0	1

Leptons	Mass (MeV/c^2)	Charge	Spin	Half-life (s)
e^-	0.511	−1	1/2	∞
ν_e	0	0	1/2	∞
μ^-	105.66	−1	1/2	2.2×10^{-6}
ν_μ	0	0	1/2	∞
τ^-	1784	−1	1/2	3.4×10^{-13}
ν_τ	0	0	1/2	∞

Field particles	Mass (GeV/c^2)	Charge	Spin
Photon	0	0	1
$W^\pm$	81	1	1
Z^0	93	0	1
Gluons	0	0	1
Graviton	0	0	2

Notes: In the upper table each quark can appear in three colors, R, G, and B. Only one member of the particle-antiparticle pair appears in the table

1.10 Quantum Chromodynamics

It is well known that *quantum chromodynamics* (QCD) is the fundamental theory for strongly interacting particles. In this section we give a brief description of QCD. The formalism is best described in terms of the Lagrangian formalism. Within this approach, the *Euler-Lagrangian equations* yield the equations of motion for the fundamental particles. If the reader is not familiar with relativistic quantum mechanics and the notation used in this section, it is recommended to read Appendix D.

Strong interaction is indeed the strongest force of nature. It is responsible for over 80% of the baryon masses, and thus for most of the mass of everything on Earth. Strong interactions bind nucleons in nuclei, which, being then dressed with electrons and bound into molecules by the much weaker electromagnetic force, give rise to the variety of the physical world.

Quantum chromodynamics is the theory of strong interactions. The fundamental degrees of freedom of QCD, quarks and gluons, are already well established even though they cannot be observed as free particles, but only in color neutral bound states (*confinement*). Today, QCD has firmly occupied its place as part of the *standard model of particle physics* (for a good introduction to the field, see [HM84]). However, understanding the physical world does not only mean understanding its fundamental constituents; it means mostly understanding how these constituents interact and bring into existence the entire variety of physical objects composing the universe. Here, we try to explain why high energy nuclear physics offers us unique tools to study QCD.

QCD emerges when the naïve quark model is combined with local SU(3) gauge invariance. We will just summarize some of the main accomplishments in QCD. For an introduction to the concepts discussed here, we refer to, e.g., [PS95]. One can define a quark-state "vector" with three components (in this section I use $\hbar = c = 1$),

$$
q(x) = \begin{pmatrix} q^{\text{red}}(x) \\ q^{\text{green}}(x) \\ q^{\text{blue}}(x) \end{pmatrix}, \tag{1.51}
$$

where $q^{\text{color}}(x)$ are field quantities that depend on the space-time coordinate $x = (t, \mathbf{r})$. The transition from quark model to QCD is made when one decides to treat color similarly to the electric charge in electrodynamics. The entire structure of electrodynamics emerges from the requirement of local gauge invariance, that is, invariance with respect to the phase rotation of the electron field, $\exp(i\alpha(x))$, where the phase α depends on the space-time coordinate. One can demand similar invariance for the quark fields, keeping in mind that while there is only one electric charge in quantum electrodynamics (QED), there are three color charges in QCD.

To implement this program, one requires the free quark Lagrangian,

$$
\mathcal{L}_{\text{free}} = \sum_{q=u,d,s,\ldots} \sum_{\text{colors}} \bar{q}(x) \left(i\gamma_\mu \frac{\partial}{\partial x_\mu} - m_q \right) q(x), \tag{1.52}
$$

to be invariant under rotations of the quark fields in color space,

$$
U : q^j(x) \rightarrow U_{jk}(x) q^k(x), \tag{1.53}
$$

with $j, k \in \{1, 2, 3\}$ (we always sum over repeated indices). Since the theory we build in this way is invariant with respect to these "gauge" transformations, all physically meaningful quantities must be gauge invariant. In (1.52), $\partial/\partial x_\mu = (\partial/\partial t, \mathbf{V})$, and $\gamma_\mu = (\gamma_0, \gamma_i)$ are the 4×4 Dirac matrices, defined in Appendix D.

The indices (i, j, k) can only have values 1, 2, 3 (or, equivalently, x, y, z), and the index 0 means the time component of the matrix γ_μ. The γ-matrices are functions of the Pauli matrices σ (see Appendix D).

The contraction of two four-vectors is defined as $A^\mu B_\mu = A^0 B^0 - \mathbf{A} \cdot \mathbf{B}$. In (1.52), $\bar{q}(x)$ is a matrix multiplication of the complex conjugate of the transpose of (1.51) and the Dirac matrix γ_0, that is, $\bar{q}(x) = q^\dagger(x)\gamma_0$ (see Appendix D).

In electrodynamics, there is only one electric charge, and gauge transformation involves a single phase factor, $U = \exp(i\alpha(x))$. In QCD, one has three different colors, and U becomes a (complex-valued) unitary 3×3 matrix, that is, $U^\dagger U = UU^\dagger = 1$, with determinant Det $U = 1$. These matrices form the fundamental representation of the group SU(3), where 3 is the number of colors, $N_c = 3$. The matrix U has $N_c^2 - 1 = 8$ independent elements and can therefore be parameterized in terms of the 8 generators T_{kj}^a, $a \in \{1, \ldots, 8\}$ of the fundamental representation of SU(3),

$$U(x) = \exp\left(-i\phi_a(x)T^a\right). \tag{1.54}$$

By considering a transformation U that is infinitesimally close to the **1** element of the group, it is easy to prove that the matrices T^a must be Hermitian ($T^a = T^{a\dagger}$) and traceless (tr $T^a = 0$). The T^a's do not commute; instead, one defines the SU(3) structure constants f_{abc} by the commutator

$$[T^a, T^b] = if_{abc}T^c. \tag{1.55}$$

These commutator terms have no analog in QED, which is based on the abelian gauge group U(1). QCD is based on a non-abelian gauge group SU(3) and is thus called a *non-abelian gauge theory*.

The generators T^a are normalized to

$$\text{tr } T^a T^b = \frac{1}{2}\delta_{ab}, \tag{1.56}$$

where δ_{ab} is the Kronecker symbol. Useful information about the algebra of color matrices, and their explicit representations, can be found in many textbooks (see, e.g., [Fie89]).

Since U is x-dependent, the free quark Lagrangian (1.52) is not invariant under the transformation (1.53). In order to preserve gauge invariance, one has to introduce, following the familiar case of electrodynamics, the *gauge* (or "gluon") *field* $A_{kj}^\mu(x)$ and replace the derivative in (1.52) with the so-called *covariant derivative*,

$$\partial^\mu q^j(x) \rightarrow D_{kj}^\mu q^j(x) \equiv \left\{\delta_{kj}\partial^\mu - iA_{kj}^\mu(x)\right\} q^j(x), \tag{1.57}$$

where ∂^μ is the four-dimensional derivative $\partial^\mu = (\partial/\partial t, \nabla)$.

Note that the gauge field $A_{kj}^\mu(x) = A_a^\mu T_{kj}^a(x)$ as well as the covariant derivative are 3×3 matrices in color space. Note also that (1.57) differs from the definition often given in textbooks, because we have absorbed the strong coupling constant in the field A^μ. With the replacement given by (1.57), all changes to the Lagrangian under gauge transformations cancel, provided A^μ transforms as

$$U: A^\mu(x) \rightarrow U(x)A^\mu(x)U^\dagger(x) + iU(x)\partial^\mu U^\dagger(x). \tag{1.58}$$

(From now on, we will often not write the color indices explicitly.)

Figure 1.4 Due to the non-abelian nature of QCD, gluons carry color charge and can therefore interact with each other via these vertices.

The QCD Lagrangian then reads

$$\mathcal{L}_{QCD} = \sum_q \bar{q}(x) \left(i\gamma_\mu D^\mu - m_q \right) q(x) - \frac{1}{4} \text{tr } G^{\mu\nu}(x) G_{\mu\nu}(x), \qquad (1.59)$$

where the first term describes the dynamics of quarks and their couplings to gluons, while the second term describes the dynamics of the gluon field. The strong coupling constant g is the QCD analog of the elementary electric charge e in QED. The gluon field strength tensor is given by

$$G^{\mu\nu}(x) \equiv i\left[D^\mu, D^\nu \right] = \partial^\mu A^\nu(x) - \partial^\nu A^\mu(x) - i\left[A^\mu(x), A^\nu(x) \right]. \qquad (1.60)$$

This can also be written in terms of the color components A_a^μ of the gauge field,

$$G_a^{\mu\nu}(x) = \partial^\mu A_a^\nu(x) - \partial^\nu A_a^\mu(x) + f_{abc} A_b^\mu(x) A_c^\nu(x). \qquad (1.61)$$

For a more complete presentation, see modern textbooks like [Fie89], [ESW96], [Mu87].

The crucial difference, as will become clear soon, between electrodynamics and QCD is the presence of the commutator on the right-hand side of (1.60). This commutator gives rise to the gluon-gluon interactions shown in figure 1.4 that make the QCD field equations nonlinear: the color fields do not simply add like in electrodynamics. These nonlinearities give rise to rich and nontrivial dynamics of strong interactions.

Let us now turn to the discussion of the dynamical properties of QCD. To understand the dynamics of a field theory, one necessarily has to understand how the coupling constant behaves as a function of distance. This behavior, in turn, is determined by the response of the vacuum to the presence of external charge. The vacuum is the ground state of the theory; however, quantum mechanics tells us that the "vacuum" is far from empty—the uncertainty principle allows particle-antiparticle pairs to be present in the vacuum for a period of time inversely proportional to their energy. In QED, the electron-positron pairs have the effect of screening the electric charge; see figure 1.5. Thus, the electromagnetic coupling constant increases toward shorter distances. The dependence of the charge on distance (*running coupling constant*) is given by [HM84]

$$e^2(r) = \frac{e^2(r_0)}{1 + \frac{2e^2(r_0)}{3\pi} \ln \frac{r}{r_0}}, \qquad (1.62)$$

which can be obtained by resuming (logarithmically divergent, and regularized at the distance r_0) electron-positron loops dressing the virtual photon propagator.

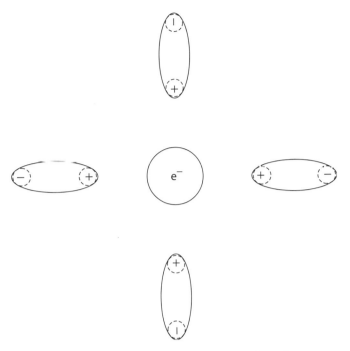

Figure 1.5 In QED, virtual electron-positron pairs from the vacuum screen the bare charge of the electron. The larger the distance, the more pairs are present to screen the bare charge and the electromagnetic coupling decreases. Conversely, the coupling is larger when probed at short distances.

The formula (1.62) has two surprising properties. First, at large distances r away from the charge which is localized at r_0, $r \gg r_0$, where one can neglect unity in the denominator, the "dressed" charge $e(r)$ becomes independent of the value of the "bare" charge $e(r_0)$—it does not matter what the value of the charge at short distances is. Second, in the local limit $r_0 \to 0$, if we require the bare charge $e(r_0)$ to be finite, the effective charge vanishes at any finite distance away from the bare charge! The screening of the charge in QED does not allow one to reconcile the presence of interactions with the local limit of the theory. This is a fundamental problem of QED, which shows that either i) it is not a truly fundamental theory, or ii) (1.62), based on perturbation theory, in the strong coupling regime gets replaced by some other expression with more acceptable behavior. The latter possibility is quite likely, since at short distances the electric charge becomes very large and its interactions with the electron-positron vacuum cannot be treated perturbatively.

Fortunately, because of the smallness of the physical coupling $\alpha_{em}(r) = e^2(r)/(4\pi) = 1/137$, this fundamental problem of the theory manifests itself only at very short distances $\sim \exp\left(-3/[8\alpha_{em}]\right)$. Such short distances will probably always remain beyond the reach of experiment, and one can safely apply QED as a truly effective theory.

In QCD, as we are now going to discuss, the situation is qualitatively different, and corresponds to *anti*-screening—the charge is small at short distances and grows at larger distances. This property of the theory is called *asymptotic freedom* [GW73], [Po73].

While the derivation of the running coupling is conventionally performed by using field theoretical perturbation theory, it is instructive to see how these results can be illustrated by using the methods of condensed matter physics. Indeed, let us consider the vacuum as a continuous medium with a dielectric constant ϵ. The dielectric constant is linked to the magnetic permeability μ and the speed of light c by the relation

$$\epsilon\,\mu = \frac{1}{c^2} = 1. \tag{1.63}$$

Thus, a screening medium ($\epsilon > 1$) will be diamagnetic ($\mu < 1$), and conversely a paramagnetic medium ($\mu > 1$) will exhibit antiscreening, which leads to asymptotic freedom. In order to calculate the running coupling constant, one has to calculate the magnetic permeability of the vacuum. In QED one has [Fie89], [ESW96], [Mu87]

$$\epsilon_{\mathrm{QED}} = 1 + \frac{2e^2(r_0)}{3\pi} \ln \frac{r}{r_0} > 1. \tag{1.64}$$

So why is the QCD vacuum paramagnetic while the QED vacuum is diamagnetic? The energy density of a medium in the presence of an external magnetic field $\mathbf{B}$ is given by

$$u = -\frac{1}{2} 4\pi \chi \, \mathbf{B}^2, \tag{1.65}$$

where the magnetic susceptibility χ is defined by the relation

$$\mu = 1 + 4\pi \chi. \tag{1.66}$$

When electrons move in an external magnetic field, two competing effects determine the sign of magnetic susceptibility:

- The electrons in the magnetic field move along quantized orbits, referred to as *Landau levels*. The current originating from this movement produces a magnetic field with opposite direction to the external field. This is the diamagnetic response, $\chi < 0$.

- The electron spins align along the direction of the external $\mathbf{B}$-field, leading to a paramagnetic response ($\chi > 0$).

In QED, the diamagnetic effect is stronger, so the vacuum is screening the bare charges. In QCD, however, gluons carry color charge. Since they have a larger spin (spin 1) than quarks (or electrons), the paramagnetic effect dominates and the vacuum is antiscreening.

Based on the considerations given above, the energy density of the QCD vacuum in the presence of an external color-magnetic field can be calculated by using the standard formulas of quantum mechanics, see, for example, [LL65], by summing over Landau levels and taking account of the fact that gluons and quarks give contributions of different sign. Note that a summation over all Landau levels would lead to an infinite result for the energy density. In order to avoid this divergence, one has to introduce a cutoff Λ with dimension of mass. Only field modes with wavelength $\lambda \gtrsim 1/\Lambda$ are taken into account. The upper limit for λ is given by the radius of the largest Landau orbit, $r_0 \sim 1/\sqrt{gB}$, which is the only dimensional scale in the problem; the summation thus is made over the wave lengths satisfying

$$\frac{1}{\sqrt{|gB|}} \gtrsim \lambda \gtrsim \frac{1}{\Lambda}.$$ (1.67)

The result is [Nie81]

$$u_{\text{vac}}^{\text{QCD}} = -\frac{1}{2}B^2 \frac{11N_c - 2N_f}{48\pi^2} g^2 \ln \frac{\Lambda^2}{|gB|},$$ (1.68)

where N_f is the number of quark flavors, and $N_c = 3$ is the number of colors. Comparing this with (1.65) and (1.66), one can read off the magnetic permeability of the QCD vacuum,

$$\mu_{\text{vac}}^{\text{QCD}}(B) = 1 + \frac{11N_c - 2N_f}{48\pi^2} g^2 \ln \frac{\Lambda^2}{|gB|} > 1.$$ (1.69)

The first term in the denominator $(11N_c)$ is the gluon contribution to the magnetic permeability. This term dominates over the quark contribution $(2N_f)$ as long as the number of flavors N_f is less than 17 and is responsible for asymptotic freedom.

The dielectric constant as a function of distance r is then given by

$$\epsilon_{\text{vac}}^{\text{QCD}}(r) = \frac{1}{\mu_{\text{vac}}^{\text{QCD}}(B)} \Bigg|_{\sqrt{|gB|} \to 1/r}.$$ (1.70)

The replacement $\sqrt{|gB|} \to 1/r$ follows from the fact that ϵ and μ in (1.70) should be calculated from the same field modes: the dielectric constant $\epsilon(r)$ could be calculated by computing the vacuum energy in the presence of two static colored test particles located at a distance r from each other. In this case, the maximum wavelength of field modes that can contribute is of order r, so that

$$r \gtrsim \lambda \gtrsim \frac{1}{\Lambda}.$$ (1.71)

Combining eqs. (1.67) and (1.71), we identify $r = 1/\sqrt{|gB|}$ and find

$$\epsilon_{\text{vac}}^{\text{QCD}}(r) = \frac{1}{1 + \dfrac{11N_c - 2N_f}{24\pi^2} g^2 \ln(r\Lambda)} < 1.$$ (1.72)

With $\alpha_s(r_1)/\alpha_s(r_2) = \epsilon_{\text{vac}}^{\text{QCD}}(r_2)/\epsilon_{\text{vac}}^{\text{QCD}}(r_1)$ one finds to lowest order in α_s (the strong interaction coupling constant)

$$\alpha_s(r_1) = \frac{\alpha_s(r_2)}{1 + \dfrac{11N_c - 2N_f}{6\pi} \alpha_s(r_2) \ln\left(\dfrac{r_2}{r_1}\right)}.$$ (1.73)

Apparently, if $r_1 < r_2$ then $\alpha_s(r_1) < \alpha_s(r_2)$. The running of the coupling constant is shown in figure 1.6, where $Q \sim 1/r$ is the *momentum transfer*. The intuitive derivation given above illustrates the original field-theoretical result of [GW73].

At high momentum transfer, corresponding to short distances, the coupling constant thus becomes small and one can apply perturbation theory, see figure 1.6. There are a

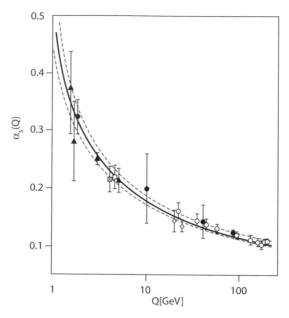

Figure 1.6 The running coupling constant $\alpha_s(Q^2)$ as a function of momentum transfer Q^2 determined from a variety of processes. The figure is from [Bet00].

variety of processes that involve high momentum scales, for example, deep inelastic scattering, Drell-Yan dilepton production, e^+e^--annihilation into hadrons, production of heavy quarks/quarkonia, high p_T hadron production, ... QCD correctly predicts the Q^2-dependence of these so-called "hard" processes, which is a great success of the theory [HM84].

While asymptotic freedom implies that the theory becomes simple and treatable at short distances, it also tells us that at large distances the coupling becomes very strong. In this regime we have no reason to believe in perturbation theory. In QED, as we have discussed above, the strong coupling regime starts at extremely short distances beyond the reach of current experiments—and this makes the "zero-charge" problem somewhat academic. In QCD, the entire physical world around us is defined by the properties of the theory in the strong coupling regime—and we have to construct accelerators to study it in the much simpler "QED–like," weak coupling limit.

We do not have to look far to find striking differences between the properties of QCD at short and large distances: the elementary building blocks of QCD—the "fundamental" fields appearing in the Lagrangian (1.59), quarks and gluons—do not exist in the physical spectrum as asymptotic states. For some reason still unknown to us, all physical states with finite energy appear to be color-singlet combinations of quarks and gluons, which are thus always "confined" at rather short distances on the order of 1 fm. This prevents us, at least in principle, from using well-developed formal S-matrix approaches based on analyticity and unitarity to describe quark and gluon interactions.

1.11 Exercises

1. a) Using the relativistic expression for the momentum-energy relation, find the de Broglie wavelength $\lambda = h/p$, for protons with kinetic energy 500 keV and 900 MeV. b) Repeat the calculation using a nonrelativistic expression for the momentum. c) Repeat (a) and (b) for electrons with the same energies.

2. For which kinetic energy does the proton have velocity equal to half that of light? Compare with the result for the electron.

3. From the uncertainty principle $\Delta p \Delta x \sim \hbar$, and the fact that a nucleon is confined within the nucleus, what can be concluded about the energies of nucleons within the nucleus?

4. What is the minimum photon energy required to dissociate the deuteron? Take the binding energy to be 2.224589 MeV.

5. Because pions had not been discovered in 1936 when Yukawa proposed the meson theory of the nuclear force, it was suggested that the muon was Yukawa's particle. What would the range of the nuclear force be if this were true?

6. Using relativistic expressions for momentum and energy conservation, show that a proton must have energy greater than 5.6 GeV to produce a proton-antiproton pair in a collision with another proton at rest.

7. Using the mass-energy relation, find the kinetic energy released in the decays in (1.8).

8. Find the threshold energies for the following reactions in the laboratory system, assuming that the initial proton is at rest:

(a) $p + p \rightarrow p + p + \pi^0$

(b) $p + p \rightarrow p + n + \pi^+$

(c) $p + p \rightarrow p + p + \pi^+ + \pi^-$

(d) $\pi^- + p \rightarrow p + \bar{p} + n$

9. Which of the following processes are absolutely forbidden?

(a) $\pi^0 + n \rightarrow \pi^- + p$

(b) $p + e^- \rightarrow \gamma + \gamma$

(c) $n \rightarrow p + e^- + \bar{\nu}_e$

(d) $n \rightarrow p + e^+ + \nu_e$

(e) $\gamma + p \rightarrow \bar{n} + \pi^+$

10. Verify expression (1.39) for p, n, $\bar{p}$, $\bar{n}$, π^+, π^-, and π^0.

11. Assuming that the virtual pions of figure 1.1 (c and d) describe a semicircle with diameter 1 fm, find, using (1.45), the extra contribution for the magnetic dipole moment of the proton and the neutron caused by the emission of a virtual pion (note that this is a very crude model to explain the anomalous magnetic moment of protons and neutrons).

12. Which of the following processes cannot occur through the strong interaction?

(a) $K^- \rightarrow \pi^- + \pi^0$
(b) $K^- + p \rightarrow \bar{K}^0 + n$
(c) $\Xi^0 + n \rightarrow \Sigma^- + p$
(d) $\Lambda^0 + n \rightarrow \Sigma^- + p$
(e) $K^- + p \rightarrow \Lambda^0 + n$
(f) $\pi^+ + n \rightarrow K^+ + \Sigma^0$

13. Write a reaction involving the proton and the kaons (K^+, K^-, and K^0) that obeys the conservation laws and that leads to the creation of the hyperon Ω^-.

14. Use table 1.1 and show that the hyperon Ω^- decay could not have any mode governed by the strong interaction (that conserves S) that does not violate some conservation law. For example, the decay $\Omega^- \rightarrow p + 2K^- + \bar{K}^0$ conserves S but is energetically forbidden. The particle Ω^- decays, in fact, only by means of the weak force (which does not conserve S), through the branching $\Omega^- \rightarrow \Lambda + K^-$ (69%), $\Omega^- \rightarrow \Xi^0 + \pi^-$ (23%), and $\Omega^- \rightarrow \Xi^- + \pi^0$ (8%).

15. Using relation (1.39), show that the quarks up and down are members of an isospin doublet $t_z = \pm 1/2$.

16. Find the quark composition of particles in table 1.1.

2 | The Two-Nucleon System

2.1 Introduction

The study of the hydrogen atom is relatively simple due the fact that the Coulomb force between the proton and the electron is very well known. The solution of this quantum problem resulted in the determination of a group of states of energy allowed for the system, permitting direct comparison with the measured values of the electromagnetic transitions between those states. Ever since, there has been great progress in understanding the hydrogen atom and atoms with many electrons. Nowadays, there are only small discrepancies between quantum theory and experimental data.

Nuclear systems are much more complex than atomic ones. Already the simplest case, the system of two nucleons, has its theoretical treatment hindered by the fact that the form of the force acting between them is not well known. In spite of that, quantum theory has been used with success in several areas of nuclear physics. In this chapter we will make a simple application of it to the system of two nucleons and an expression will be presented for their interaction potential.

Two groups of experimental data exist for the system of two nucleons. A first group arises from the study of the only bound system of that kind, the deuteron, composed of a proton and a neutron. Unlike the hydrogen atom, the deuteron has only one bound state, the ground state. Therefore, the theories for the neutron-proton interaction in the deuteron can only be tested by comparing their predictions with the experimental values of the energy, angular momentum, parity, magnetic dipole moment, and electric quadrupole moment of the ground state of the deuteron.

The second group of experimental data come from the study of nucleon-nucleon scattering. As it is difficult to produce a neutron beam for that goal (the neutrons have zero charge and cannot be accelerated by means of an electric field), the experiments are limited to collisions between protons and to proton-deuteron scattering, this last supplying indirect information on proton-neutron interaction. Comparison of the experimental data for those collisions and the properties of the ground state of the deuteron has been useful for the semi-phenomenological description of the interaction between two nucleons, as we shall see below.

2.2 Electrostatic Multipoles

Electromagnetic multipoles give one of the most important examples of tensor operators. They appear in classical field theory as a result of the *multipole expansion* of the fields created by a finite system of charges and currents. We start with the system of point-like classical particles with electric charges e_a located at the points $\mathbf{r}_a$. If you are not used to angular momentum algebra, read Appendices A and B first.

The electrostatic potential of this system measured at the point $\mathbf{r}$ is given by the Coulomb law,

$$\phi(\mathbf{r}) = \sum_a \frac{e_a}{|\mathbf{r} - \mathbf{r}_a|}. \tag{2.1}$$

The function

$$\frac{1}{|\mathbf{r} - \mathbf{r}'|} = \frac{1}{\sqrt{r^2 + r'^2 - 2rr' \cos \gamma}} \tag{2.2}$$

depends on the lengths r, r' of two vectors and the angle γ between them rather than on the angles of the vectors $\mathbf{r}$ and $\mathbf{r}'$ separately. If $\mathbf{r} \neq \mathbf{r}'$, this function has no singularities and can be expressed with the aid of the expansion over the infinite set of Legendre polynomials with coefficients depending on r and r',

$$\frac{1}{|\mathbf{r} - \mathbf{r}'|} = \sum_{l=0}^{\infty} P_l(\cos \gamma) f_l(r, r'). \tag{2.3}$$

Using the notation $r_<$ and $r_>$ for the smaller and greater r and r', one can show that the expansion (2.3) takes the form

$$\frac{1}{|\mathbf{r} - \mathbf{r}'|} = \sum_l \frac{r_<^l}{r_>^{l+1}} P_l(\cos \gamma). \tag{2.4}$$

The applications of the multipole expansion usually consider the potential (2.1) *outside the system*, that is, at the point $\mathbf{r}$ with $r > r_a$. Then we can use the expansion (2.4) and the *addition theorem* (A.112) to get

$$\phi(\mathbf{r}) = \sum_{lm} \frac{4\pi}{2l+1} \frac{1}{r^{l+1}} Y_{lm}^*(\mathbf{n}) \mathcal{M}(El, m). \tag{2.5}$$

Here the *electric multipole moment* of rank l, $l = 0, 1, \ldots$, is defined for a system of point-like charges $a = 1, 2, \ldots, A$ as a set of $2l + 1$ quantities,

$$\mathcal{M}(El, m) = \sum_a e_a r_a^l Y_{lm}(\mathbf{n}_a), \quad m = -l, -l+1, \ldots, +l, \tag{2.6}$$

where the sum runs over all charges e_a located at $\mathbf{r}_a = (r_a, \theta_a, \phi_a) \equiv (r_a, \mathbf{n}_a)$. Exactly in the same way one can define, instead of the charge distribution, multipole moments for any other property of the particles, for example, for the mass distribution, $e_a \to m_a$.

In quantum theory, multipole moments are to be considered as operators acting on the variables of the particles. Containing explicitly the spherical functions, the operator $\mathcal{M}(El, m)$ has the necessary features of the tensor operator of rank l. Introducing the *charge density* operator

$$\rho(\mathbf{r}) = \sum_a e_a \delta(\mathbf{r} - \mathbf{r}_a), \tag{2.7}$$

we come to a more general form of the multipole moment,

$$\mathcal{M}(El, m) = \int d^3r\, \rho(\mathbf{r}) r^l Y_{lm}(\mathbf{n}), \quad \mathbf{n} = \frac{\mathbf{r}}{r}. \tag{2.8}$$

In this form we do not even need to make an assumption of existence of point-like constituents in the system; for example, in the nucleus charged pions and other mediators of nuclear forces are included along with the nucleons if $\rho(\mathbf{r})$ is the total operator of electric charge density. As expected, we can separate the geometry of multipole operators from their dynamical origin. From any underlying structure $\rho(\mathbf{r})$, the operator (2.8) extracts the irreducible tensor of rank l, that is, the part with specific rotational properties.

The lowest multipole moment $l = 0$ is the *monopole* one. It determines the scalar part, the total electric charge Ze,

$$\mathcal{M}(E0, 0) = \frac{1}{\sqrt{4\pi}} \sum_a e_a = \frac{1}{\sqrt{4\pi}} \int d^3r\, \rho(\mathbf{r}) = \frac{1}{\sqrt{4\pi}} Ze. \tag{2.9}$$

The next term, $l = 1$, defines the vector of the *dipole* moment,

$$\mathbf{d} = \sum_a e_a \mathbf{r}_a = \int d^3r\, \rho(\mathbf{r})\mathbf{r}. \tag{2.10}$$

Taking into account the relation (B.8) between the vectors and the spherical functions of rank $l = 1$, we obtain

$$\mathcal{M}(E1, m) = \sqrt{\frac{3}{4\pi}} \sum_a e_a r_a (n_a)_m = \sqrt{\frac{3}{4\pi}} d_m. \tag{2.11}$$

Subsequent terms of the multipole expansion determine the quadrupole ($l = 2$), *octupole* ($l = 3$), *hexadecapole* ($l = 4$), and higher moments. Physical properties of the quadrupole tensor, recall (B.12), play an important role in molecular and nuclear structure.

In a similar way one can define *magnetic multipoles* $\mathcal{M}(Ml, m)$ related to the distribution of currents. The convection current due to orbital motion and the magnetization current generated by the spin magnetic moments determine corresponding contributions to the magnetic multipole moment of rank l,

$$\mathcal{M}(Ml, m) = \sum_a \left(g_a^s \mathbf{s}_a + \frac{2}{l+1} g_a^l \mathbf{l}_a \right) \cdot \nabla \left(r_a^l Y_{lm}(\mathbf{n}_a) \right). \tag{2.12}$$

Here $\mathbf{s}_a$ and $\mathbf{l}_a$ stand for the spin and orbital angular momentum of a particle a, respectively; g_a^s and g_a^l are corresponding *gyromagnetic ratios*. (Everywhere we measure all angular momenta in units of $\hbar$ and the gyromagnetic ratios in the magnetons $e\hbar/(2m_ac)$.) The expression (2.12) vanishes for $l = 0$ demonstrating the absence of magnetic monopoles. At $l = 1$, we come to the spherical components μ_m, (B.8), of the *magnetic moment* μ,

$$\mathcal{M}(M1, m) = \sqrt{\frac{3}{4\pi}}\mu_m, \tag{2.13}$$

$$\mu = \sum_a (g_a^s s_a + g_a^l l_a). \tag{2.14}$$

Higher terms determine magnetic quadrupole, $l = 2$, magnetic octupole, $l = 3$, and so on.

2.3 Magnetic Moment with Spin-orbit Coupling

The vector model described in Appendix B will be used to study the magnetic moment with *spin-orbit coupling*. Let a particle with spin s move in a central field where its orbit is characterized by an orbital momentum l. The energy of the orbit depends in general on the mutual orientation of the quantum vectors $\mathbf{l}$ and $\mathbf{s}$. This spin-orbit interaction is rather weak $(\sim v^2/c^2)$ for electrons in atoms, although it becomes increasingly important in heavy atoms with a large nuclear charge Z, when the electrons reach relativistic velocities, $v/c \sim \alpha Z$, where $\alpha = e^2/\hbar c \approx 1/137$ is the *fine structure constant*. In light atoms one can use the ls coupling scheme when, analogously to the states $|j_1m_1; j_2m_2\rangle$ in (B.41), the electron state $|ll_z; ss_z\rangle$ is described separately by the constituent angular momenta $\mathbf{l}$ and $\mathbf{s}$. For nucleons in nuclei, the spin-orbit coupling is strong, and one has to introduce the total angular momentum of the nucleon

$$\mathbf{j} = \mathbf{l} + \mathbf{s} \tag{2.15}$$

and use the corresponding basis of states $|(ls)jj_z = m\rangle$.

Since the lengths of all vectors in (2.15) are fixed,

$$\mathbf{j}^2 = j(j+1), \quad \mathbf{l}^2 = l(l+1), \quad \mathbf{s}^2 = s(s+1), \tag{2.16}$$

(Equation 2.15) enables us to find the average mutual orientations

$$(\mathbf{j} \cdot \mathbf{l}) = \frac{j(j+1) + l(l+1) - s(s+1)}{2}, \quad (\mathbf{j} \cdot \mathbf{s}) = \frac{j(j+1) + s(s+1) - l(l+1)}{2}. \tag{2.17}$$

Of course, the scalar quantities (2.17) are the same for all the states $|(ls)jm\rangle$ with different m.

The operator of magnetic moment (2.14) of a particle in a central field is given (in units of the corresponding magneton)

$$\mu = g^s\mathbf{s} + g^l\mathbf{l}. \tag{2.18}$$

Using the vector model (B.51), (B.52) together with the scalar quantities (2.17), we obtain the *effective operator* of the magnetic moment within the multiplet of states $|jm\rangle$,

$$\mu = g(j, l, s)\mathbf{j}, \tag{2.19}$$

where the gyromagnetic ratio (*Landé factor*) is

$$g(j, l, s) = g^l \frac{\langle(\mathbf{j} \cdot \mathbf{l})\rangle}{j(j+1)} + g^s \frac{\langle(\mathbf{j} \cdot \mathbf{s})\rangle}{j(j+1)}$$
$$= \frac{1}{2j(j+1)} \{(g^l + g^s)j(j+1) + (g^l - g^s)[l(l+1) - s(s+1)]\}. \tag{2.20}$$

As we mentioned, the tabular value corresponds to the state with $j_z = j$, that is, the magnetic moment is equal to $\mu = gj$.

For a free nucleon at rest $l = 0$ and $j = s = \frac{1}{2}$. Therefore spin gyromagnetic ratios are determined by the empirical magnetic moments μ_p and μ_n,

$$g_p^s = 2\mu_p = 5.58, \quad g_n^s = 2\mu_n = -3.82, \tag{2.21}$$

(in nuclear magnetons). The spin gyromagnetic ratio is predicted by the relativistic Dirac equation (see Appendix D) to be equal to $g^s = 2$ for a structureless particle of spin $\frac{1}{2}$ and charge e (in units of the corresponding magneton). This would lead to the spin magnetic moment of a free particle being exactly one magneton. This is the case for the electron (or positron) with small QED corrections of order 10^{-3} due to vacuum polarization by virtual electron-positron pairs. For the nucleons we see a large difference between the actual values (2.21) and the Dirac values, $g_p^s = 2$, $g_n^s = 0$. This difference (*anomalous* magnetic moments) is generated by strong QCD interactions responsible for the intrinsic structure of the nucleon.

Combining the rotational properties of the tensor operators and their properties with respect to spatial inversion (see Appendix B), we can come to important conclusions concerning multipole moments as physical observables.

The electric charge (2.9) is a scalar invariant under inversion. The electric dipole (2.10) changes sign as the radius vector, or any "normal" (*polar*) vector. The momentum $\mathbf{p}$ is also a polar vector. However, the orbital angular momentum (A.3) is an *axial* vector; its components do not change sign under inversion. As seen from the geometrical picture of rotation (it does not change sense in the inverted frame), any angular momentum including spin should be an axial vector. The scalar product of an axial vector by a polar vector is a *pseudoscalar*. Similar to scalars, pseudoscalars are invariant under rotations, but change sign under inversion. An important example of a pseudoscalar is given by the *helicity* of a particle,

$$h = \left(\mathbf{s} \cdot \frac{\mathbf{p}}{|\mathbf{p}|}\right), \tag{2.22}$$

that is, a spin component along the direction of motion.

Thus, in addition to the tensor properties under transformations from the rotation group, we can classify the operators O by their behavior under spatial inversion $\mathcal{P}$, that is, by their parity $\Pi(O)$, defined by the operator transformation $O' = \mathcal{P}O\mathcal{P} = \pm O$. Acting between the states $|i\rangle$ and $|f\rangle$ with parity Π_i and Π_f, respectively, an operator O has the additional selection rule

$$\Pi_f = \Pi(O)\Pi_i \quad \text{or} \quad \Delta\Pi = \Pi_O. \tag{2.23}$$

Table 2.1 Allowed (+) electromagnetic multipoles for quantal systems with different values of angular momenta. The entries in parentheses are allowed by rotational symmetry but forbidden by parity.

Spin	E0	M0	E1	M1	E2	M2	E3	M3
0	+	−	−	−	−	−	−	−
1/2	+	−	(−)	+	−	−	−	−
1	+	−	(−)	+	+	(−)	−	−
3/2	+	−	(−)	+	+	(−)	(−)	+

It is easy to see that the parity selection rules for electric and magnetic multipoles are complementary (the electric multipoles sometimes are said to have *natural parity*),

$$E\lambda : \quad \Delta\Pi = (-)^{\lambda}, \quad M\lambda : \quad \Delta\Pi = (-)^{\lambda+1}. \tag{2.24}$$

Therefore, the expectation values (diagonal matrix elements, $f = i$) are forbidden for odd electric and even magnetic multipoles if the state has definite parity. In particular, any system in a state of certain parity cannot have a nonzero electric dipole moment.

Table 2.1 summarizes the allowed (+) electromagnetic multipoles for quantal systems with different values of angular momenta (spins). The entries in parentheses are allowed by rotational symmetry but forbidden by parity. The nucleons can have electric charge and magnetic moment; the electric dipole moment can be allowed if parity is not conserved and the stationary states do not have certain parity. Higher multipoles are strictly forbidden by rotational symmetry.

Parity conservation in strong and electromagnetic interactions means that the corresponding Hamiltonian is invariant under inversion (genuine scalar). Then its eigenstates can always be chosen in such a way that they have definite parity. However, this choice is not mandatory. If some states with opposite parities have the same energy (are degenerate), any linear combination of them is also stationary but with no definite parity. For example, the photon circular polarization is its *helicity*. The photon with the left circular polarization has no definite parity. Under inversion this state is transformed into the state with the right circular polarization and the same energy. If the radiation from an unpolarized system contains the left- and right-polarized photons with different probabilities, this means that parity is not conserved in the transition. Similar conclusions may be drawn from experiments with longitudinally (along the momentum) polarized particles.

2.4 Experimental Data for the Deuteron

(a) *Binding energy.* The deuteron has a smaller mass than the sum of the masses of the proton and the neutron. That "missing mass" is emitted in the form of gamma radiation when the proton and the neutron join to form the deuteron, and it should be returned in the form of energy if we want to separate the deuteron again in its constituents. This *binding energy* exists, with different values, for any nucleus; it is in fact a general property imposed by the theory of the relativity to all bound systems.

An indirect method to measure the binding energy of the deuteron is through measurement of its atomic mass, comparing the result with the sum of the masses of the proton and the neutron. Another method, more direct, consists of an experimental measurement of the gamma ray energy emitted when the neutron and the proton combine to form a bound state (n-p capture). One can also measure the inverse process, that is, the energy of the gamma ray necessary to break the binding between the proton and the neutron (photo-disintegration). The first two methods supply the more precise results and one can extract the value for the binding energy of the deuteron,

$$E_B = (2.22464 \pm 0.00005) \text{ MeV}. \tag{2.25}$$

(b) *Angular momentum and parity*. The angular momentum of the deuteron was determined as being $J = 1$, coming from results of optical, radio frequency, and microwave methods.

The parity of a nuclear state cannot be measured directly. It is obtained by analyzing the conservation of parity of certain nuclear disintegrations. Those studies show that a wavefunction with even parity supplies the more appropriate theoretical description for the deuteron.

(c) *Magnetic dipole moment*. The magnetic moment of the deuteron can be obtained as a function of the magnetic moment of the proton using the method of magnetic resonance for a molecular beam. This method measures the frequency, or quantum energy, necessary to redirect through $180°$ the magnetic moment of a nucleus in a periodic magnetic field. The result of the measurements gives the value

$$\mu_d/\mu_p = 0.30701218 \pm 0.00000002. \tag{2.26}$$

That is,

$$\mu_d - (0.857393 \pm 0.000001)\mu_N, \tag{2.27}$$

where $\mu_N = e\hbar/(2m_N c)$ is the nuclear magneton.

(d) *Electric quadrupole moment*. Rabbi and collaborators [Ra33] have shown that the deuteron also has an electric quadrupole moment that makes it look like a prolate spheroid along its spin axis, with the mean square values of the coordinates z and r of the proton obeying the ratio

$$\frac{\langle z^2 \rangle}{\langle r^2 \rangle} = \frac{1.14}{3}, \tag{2.28}$$

instead of $\frac{1}{3}$, as it should be for a spherically symmetric distribution of charge, since $\langle r^2 \rangle = \langle x^2 \rangle + \langle y^2 \rangle + \langle z^2 \rangle$. The quadrupole moment that corresponds to that deformation has the experimental value $Q_d = 0.00282$ barns (1 barn $= 10^{-28} \text{m}^2$).

(e) *The radius of the deuteron*. Hofstadter and collaborators [Ho62] have performed precise measurements of the radius of the proton, deuteron, and complex nuclei, through electron scattering. They obtained the value 2.1 fm for the mean square radius of the deuteron. The same measurement for the proton resulted in the value of 0.8 fm.

2.5 A Square-well Model for the Deuteron

In a quantum description of the deuteron, it is reasonable to suppose that the ground state is an S state that is, a state with zero angular momentum. With $l = 0$, Ψ is spherically symmetric and the angular momentum of the nucleus is entirely due to its spin. As the spins of the proton and neutron are $\frac{1}{2}$, and that of the deuteron is 1, this means that their spins are parallel. In such situation, the magnetic moments would be due to a simple sum:

$$\mu_d \cong \mu_p + \mu_n = 0.8797 \, \mu_N. \tag{2.29}$$

Comparing this result with the value given in (2.27) we see that the difference is 0.0223 μ_N. Therefore, the magnetic moment value of the deuteron almost agrees with the sum of the moments of the proton and of the neutron. The reason for the small difference will be discussed later.

Similarly to that for the hydrogen atom, the Schrödinger equation for the deuteron can be solved reducing the two-body problem to a one-body problem that has the reduced mass of the system and whose distance from the origin is the distance between the two bodies. In the center of mass system of the deuteron, the Schrödinger equation is

$$-\frac{\hbar^2}{2M} \nabla^2 \Psi + V(\mathbf{r})\Psi = E\Psi, \tag{2.30}$$

where

$$M = \frac{M_n M_p}{M_n + M_p} \tag{2.31}$$

is the reduced mass and $V(\mathbf{r})$ is the potential that describes the force between the proton (mass M_p) and the neutron (mass M_n). The difference from the hydrogen atom is that the nuclear potential $V(\mathbf{r})$ is not well known, but, for a first approach, we shall use a very simple form for that potential. We shall suppose initially that V is spherically symmetric, that is, that it depends only on the separation distance between the proton and the neutron, or, $V(\mathbf{r}) = V(r)$. In this case, the wavefunctions Ψ, solutions of (2.30), can be separated into radial and angular parts,

$$\Psi = \sum_{l,m} \frac{u_l(r)}{r} Y_l^m(\theta, \phi), \tag{2.32}$$

where the indexes l and m admit the values

$$l = 0, 1, 2, \ldots, \quad m = -l, -l+1, \ldots, l. \tag{2.33}$$

$Y_l^m(\theta, \phi)$ is the spherical harmonic function (see Appendix A), and u_l the solution of the radial equation

$$\frac{d^2 u_l}{dr^2} + \frac{2M}{\hbar^2} \left[E - V(r) - \frac{l(l+1)}{2Mr^2} \right] u_l = 0. \tag{2.34}$$

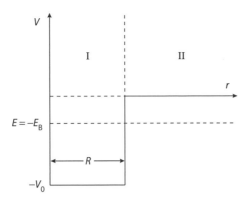

Figure 2.1 Potential well proposed for the deuteron.

The last term inside the brackets is known as the *centrifugal potential*. When $V(r)$ is negative (attractive potential), the centrifugal potential acts to decrease the attraction, making the system less tightly bound. In those circumstances it is easy to see that when $l = 0$ we have a situation where the system is more bound, that is, the energy is the lowest possible; the ground state of a spherically symmetric system is always a state with $l = 0$.

The simplest potential we can imagine for the deuteron is the "tri-dimensional square-well," illustrated in figure 2.1. The radius R and depth V_0 should be adjusted in a way to reproduce the experimental data. In region I, for $l = 0$, we have, omitting the index of u, that

$$\frac{d^2u}{dr^2} + \frac{2M}{\hbar^2}[V_0 - E_B]u = 0, \tag{2.35}$$

where the well-known experimental fact $E = -E_B = -2.225$ MeV is used, with V_0 and E_B positive numbers. The solution of this equation, which satisfies the boundary condition $u = 0$ at $r = 0$, is

$$u_I = A \sin(Kr), \tag{2.36}$$

where A is a normalization constant and

$$K = \frac{1}{\hbar}\sqrt{2M(V_0 - E_B)}. \tag{2.37}$$

In region II the radial equation assumes the form

$$\frac{d^2u}{dr^2} - \frac{2M}{\hbar^2}E_B u = 0, \tag{2.38}$$

whose solution, which satisfies the boundary condition $u = 0$ at $r = \infty$, is

$$u_{II} = Be^{-kr}, \tag{2.39}$$

where B is a normalization constant and

$$k = \frac{1}{\hbar}\sqrt{2ME_B}. \tag{2.40}$$

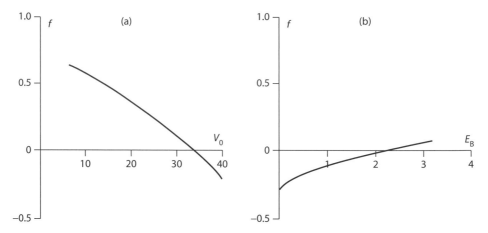

Figure 2.2 One way of solving an equation like (2.43) is to find graphically the roots of the function $f = K \cot(KR) + k$. In (a) f is plotted as a function of the depth V_0 of the deuteron potential, using $E_B = 2.225$ MeV. The function has a root for $V_0 \cong 34$ MeV. Using this value of V_0 in (b), the same function is plotted against the binding energy E_B, showing that there is no other solution besides the previous one.

The solutions (2.36) and (2.39) should match at $r = R$ so that both the function u and its derivative are continuous at that point. That is also the only way to satisfy the Schrödinger equation at that point. Those conditions imply that

$$AK \cos(KR) = -kBe^{-kR},$$ (2.41)

$$A \sin(KR) = Be^{-kR}.$$ (2.42)

Dividing (2.41) by (2.42), we find

$$K \cot(KR) = -k.$$ (2.43)

Equation (2.43) implicitly relates the binding energy E_B to the width R and depth V_0 of the potential. As E_B is measured experimentally, (2.43) gives us a relationship between the unknown parameters V_0 and R. If we use $R = 2.1$ fm, the "electromagnetic radius" of the deuteron obtained in the measurement of Hofstadter with electron scattering, (2.43) for V_0 can be solved numerically or graphically (figure 2.2a), yielding a value $V_0 \cong 34$ MeV for the depth of the well for the deuteron. If, inversely, we use the value of V_0 in 2.43 we can see (figure 2.2b) that it does not have a solution for any other value of E_B. We conclude that there are no excited states for $l = 0$. For $l > 0$, the centrifugal potential inhibits even more the formation of a bound state. The nuclear potential would have to be deeper so that the binding is not broken by the centrifugal force.

An important experimental fact is that the total angular momentum of the deuteron is $J = 1$. If the assumption of a $l = 0$ state is correct, it is certain that the intrinsic spin of the deuteron is $S = 1$ (the proton and the neutron with parallel spin vectors). In spectroscopic terminology, this is known as a *triplet state*, denoted by $^3S_1$, where the upper index is the same as $2S + 1$ and the lower index is the value of J. If the nuclear forces were independent of the spin, we would observe a *singlet state*, $^1S_0$, with the same energy, and that is not the case. In fact, no state with $J = 0$ for the deuteron has been found experimentally, indicating that the neutron-proton force is stronger when the spins are parallel than when they are antiparallel. These forces will be studied in detail later.

2.6 The Deuteron Wavefunction

The fact that the quadrupole moment of the deuteron is different from zero and that the magnetic moment of the deuteron is not the sum of the magnetic moments of the proton and of the neutron indicates that the deuteron cannot have a spherically symmetric wavefunction as in the case of a $^3S_1$ state.

We can discover the nature of the ground state wavefunction of the deuteron using the information that it has a definite parity, and that $J = 1$. The possible states of orbital angular momentum l and spin S are

$$l = 0 \quad S = 1 \;\rightarrow\; {}^3S_1,$$
$$l = 1 \quad S = 0 \;\rightarrow\; {}^1P_1,$$
$$l = 1 \quad S = 1 \;\rightarrow\; {}^3P_1,$$
$$l = 2 \quad S = 1 \;\rightarrow\; {}^3D_1.$$

Therefore, the deuteron wavefunction can only be a combination of the $^3S_1$ and $^3D_1$ states (even parity) or a combination of the $^1P_1$ and $^3P_1$ states (odd parity). We should verify which of those combinations represents better the ground state wavefunction of the deuteron with the observed properties.

To build the wavefunction it will be necessary to consider that it is an eigenfunction of the total angular momentum of the system, which is obtained by the addition (coupling) of three vectors, the spins of each nucleon and the orbital angular momentum. At this point it is good to recall how one obtains the wavefunction resulting from the coupling of two angular momenta. For more details, see Appendix B.

2.6.1 Angular momentum coupling

Let us imagine two independent systems, without interaction, where the wavefunctions $|j_1 m_1\rangle$ and $|j_2 m_2\rangle$ describe the behavior of each system, with respective angular momenta $\mathbf{j}_1$ and $\mathbf{j}_2$. In this case,

$$j_1^2 \,|j_1 m_1\rangle = j_1(j_1 + 1)\,|j_1 m_1\rangle, \quad j_{1z}\,|j_1 m_1\rangle = m_1\,|j_1 m_1\rangle, \tag{2.44}$$

with identical expressions for the ket 2.[1]

---

[1] Using the Dirac notation, $\Psi^+ \equiv \langle\Psi|$ is a *bra* and $\Psi \equiv |\Psi\rangle$ is a *ket*.

Little will change if we include the kets $|j_1 m_1\rangle$ and $|j_2 m_2\rangle$ in a single ket

$$|j_1 j_2 m_1 m_2\rangle \equiv |j_1 m_1\rangle |j_2 m_2\rangle, \tag{2.45}$$

for which the relationships

$$j_1^2 |j_1 j_2 m_1 m_2\rangle = j_1(j_1 + 1) |j_1 j_2 m_1 m_2\rangle, \text{ etc.} \tag{2.46}$$

are valid.

Let us now place the two systems in interaction. $\mathbf{j}_1$ and $\mathbf{j}_2$ will not be constants of motion any more, but will precess around the total angular momentum $\mathbf{j} = \mathbf{j}_1 + \mathbf{j}_2$. The possible values of j will be $|j_1 - j_2|, |j_1 - j_2| + 1, \ldots, j_1 + j_2$. The projections m_1 and m_2 will not be good quantum numbers any more, but the projection $m = m_1 + m_2$ will be, so that the wavefunction adequate to the description of the system can be represented by the ket $|j_1 j_2 jm\rangle$.

The wavefunction (2.46) that describes the two joint systems without interaction is no longer valid in the present case. However, it does form a basis in which the wavefunction with interaction can be expanded. Thus,

$$|j_1 j_2 jm\rangle = \sum_{m_1, m_2} \langle j_1 j_2 m_1 m_2 | jm \rangle |j_1 j_2 m_1 m_2\rangle \tag{2.47}$$

is the proper expansion, where the sum on m_1 and m_2 reduces to a single sum, if we take in consideration that $m_1 + m_2 = m$. The quantities

$$\langle j_1 j_2 m_1 m_2 | jm \rangle \tag{2.48}$$

are denominated *Clebsh-Gordan coefficients*, and they can be obtained from the algebra of angular momenta. The coefficients are also found in tables where the input data are the six values that appear in (2.48).

2.6.2 Two particles of spin $\frac{1}{2}$

Before proceeding to general theory let us consider a simple important example of two particles of spin $\frac{1}{2}$, $s_1 = s_2 = \frac{1}{2}$. The single-particle spinors χ_m, $m = \pm\frac{1}{2}$, are studied in Appendix A. Two states of each particle give rise to four states $\chi_m(1)\chi_{m'}(2)$ of the representation (B.25). Now we can explicitly proceed along the line sketched in the preceding subsection.

According to our rules, the vector coupling of two spins defines two multiplets, *triplet* and *singlet*, with the values of the total spin $\mathbf{S} = \mathbf{s}_1 + \mathbf{s}_2$ equal to 1 (three states, $S_z = \pm 1, 0$) and 0 (one state $S_z = 0$), respectively. The highest, $M = 1$ and the lowest, $M = -1$, states (B.27) belonging to the triplet are constructed uniquely from the corners of the diagram,

$$\left|\frac{1}{2}, \frac{1}{2}; 11\right\rangle = \chi_+(1)\chi_+(2), \quad \left|\frac{1}{2}, \frac{1}{2}; 1-1\right\rangle = \chi_-(1)\chi_-(2). \tag{2.49}$$

Two states with $S_z = 0$ should be combined into correct linear combinations with $S = 1$ and $S = 0$. As shown in Appendix B, for $j_1 = j_2 = \frac{1}{2}$ and $J = S = 1$, we come to the analog of (B.34), the triplet combination with $S_z = 0$,

$$|10\rangle = \frac{1}{\sqrt{2}}(\chi_+(1)\chi_-(2) + \chi_-(1)\chi_+(2)). \tag{2.50}$$

The orthogonal combination with $S_z = 0$, (B.35),

$$|00\rangle = \frac{1}{\sqrt{2}}(\chi_+(1)\chi_-(2) - \chi_-(1)\chi_+(2)), \tag{2.51}$$

belongs to the singlet $S = 0$.

Note that all three triplet states, (2.49) and (2.50), are *symmetric* with respect to interchange of spins $1 \leftrightarrow 2$, whereas the singlet state (2.51) is *antisymmetric*. We show in Appendix B, in relation to parity and in decomposition of the second rank tensor, that the intrinsic symmetry given by an operation commuting with rotations should be the same for all members of the multiplet. Let the *spin exchange* operator $\mathcal{P}^\sigma$ interchange the spin variables of the pair. Then it can be expressed via the total spin S of the pair,

$$\mathcal{P}^\sigma = (-)^{S+1}. \tag{2.52}$$

The alternative expression can be derived in terms of the Pauli spin operators (A.82), $\sigma = 2\mathbf{s}$. Using

$$(\sigma_1 \cdot \sigma_2) = 4(\mathbf{s}_1 \cdot \mathbf{s}_2) = 4\frac{\mathbf{S}^2 - \mathbf{s}_1^2 - \mathbf{s}_2^2}{2}, \tag{2.53}$$

and replacing the angular momentum squares by their eigenvalues, we get

$$(\sigma_1 \cdot \sigma_2) = 2S(S+1) - 3 = \begin{cases} -3 \ (S = 0, \text{ singlet}), \\ +1 \ (S = 1, \text{ triplet}). \end{cases} \tag{2.54}$$

Therefore the exchange operator (2.52) can be written as

$$\mathcal{P}^\sigma = \frac{1 + (\sigma_1 \cdot \sigma_2)}{2}. \tag{2.55}$$

The spin wavefunctions, $\chi_S^{m_S}$, should now be coupled to the eigenfunctions of orbital angular momentum, Y_l^m, to obtain the angular part $\mathcal{Y}$ of the total wavefunction. Based on (2.47) we can write

$$\mathcal{Y}_{lSJ}^M = \sum_{m_S, m_l} \langle lSm_lm_S \mid JM\rangle \chi_S^{m_S} Y_l^{m_l}. \tag{2.56}$$

2.6.3 Total wavefunction

The ground state of the deuteron has $J = 1$. Thus, only the four states calculated below will be considered:

$^3S_1 : J = 1, M = 1, l = 0,$ and $S = 1$

$$\mathcal{Y}_{011}^1 = \langle 0101 \mid 11 \rangle Y_0^0 \alpha(1)\alpha(2) = Y_0^0 \alpha(1)\alpha(2); \tag{2.57}$$

where we define

$$\chi_{1/2}^{+1/2} = \alpha \quad \text{and} \quad \chi_{1/2}^{-1/2} = \beta . \tag{2.58}$$

$^3D_1 : J = 1, M = 1, l = 2,$ and $S = 1$

$$\mathcal{Y}_{211}^1 = \sum_{m_S} \langle 21(1 - m_S)m_S \mid 11 \rangle Y_2^{1-m_S} \chi_1^{m_S}$$

$$= \sqrt{3/5} \, Y_2^2 \beta(1)\beta(2) - \sqrt{3/10} \, Y_2^1 \frac{1}{\sqrt{2}} [\alpha(1)\beta(2) + \beta(1)\alpha(2)]$$

$$+ \sqrt{1/10} \, Y_2^0 \alpha(1)\alpha(2);$$

$^1P_1 : J = 1, M = 1, l = 1,$ and $S = 0$

$$\mathcal{Y}_{101}^1 = \langle 1010 \mid 11 \rangle Y_1^1 \chi_0^0 = Y_1^1 \frac{1}{\sqrt{2}} [\alpha(1)\beta(2) - \alpha(2)\beta(1)]; \tag{2.59}$$

$^3P_1 : J = 1, M = 1, l = 1,$ and $S = 1$

$$\mathcal{Y}_{111}^1 = \sum_{m_S} \langle 11(1 - m_S)m_S \mid 11 \rangle Y_1^{1-m_S} \chi_1^{m_S}$$

$$= \frac{1}{\sqrt{2}} \, Y_1^1 \frac{1}{\sqrt{2}} [\alpha(1)\beta(2) + \beta(1)\alpha(2)] - \frac{1}{\sqrt{2}} \, Y_1^0 \alpha(1)\beta(2).$$

For the deuteron, let us make the index 1 in α and β to correspond to the proton and the index 2 to the neutron. We can write the wavefunction $\Psi_{lSJ}^M(\mathbf{r}, \chi)$ for the four cases discussed in terms of the radial wavefunctions $u_l(r)$, where l refers to the orbital angular momentum:

$$^3S_1 : \Psi_S \equiv \Psi_{011}^1 = \frac{u_0(r)}{r} \mathcal{Y}_{011}^1 = \frac{u_0(r)}{r} Y_0^0 \chi_1^1; \tag{2.60}$$

$$^3D_1 : \Psi_D \equiv \Psi_{211}^1 = \frac{u_2(r)}{r} \mathcal{Y}_{211}^1 = \frac{u_2(r)}{r} \left[\sqrt{\frac{3}{5}} \, Y_2^2 \chi_1^{-1} - \sqrt{\frac{3}{10}} \, Y_2^1 \chi_1^0 + \sqrt{\frac{1}{10}} \, Y_2^0 \chi_1^1 \right]; \tag{2.61}$$

$$^1P_1 : \Psi_{1P} \equiv \Psi_{101}^1 = \frac{u_1(r)}{r} \mathcal{Y}_{101}^1 = \frac{u_1(r)}{r} Y_1^1 \chi_0^0; \tag{2.62}$$

$$^3P_1 : \Psi_{3P} \equiv \Psi_{111}^1 = \frac{u_1(r)}{r} \mathcal{Y}_{111}^1 = \frac{u_1(r)}{r} \left[\frac{1}{\sqrt{2}} \, Y_1^1 \chi_1^0 - \frac{1}{\sqrt{2}} \, Y_1^0 \chi_1^1 \right]. \tag{2.63}$$

Noticing that in the center of mass system the orbital angular momentum associated to the proton is half of the relative orbital momentum, $l_p = \frac{1}{2}l$, gives z-component of the magnetic moment of the deuteron:

$$\mu_z = \frac{1}{2}l_z + g_p S_p^z + g_n S_n^z, \tag{2.64}$$

in units of μ_N, the nuclear magneton, with $g_p = 5.58$ and $g_n = -3.82$.

The magnetic moment of the deuteron is defined as

$$\mu = \left\langle \Psi_{lSJ}^{J} \left| \mu_z \right| \Psi_{lSJ}^{J} \right\rangle = \left\langle \frac{u_l}{r} \mathcal{Y}_{lSJ}^{J} \left| \mu_z \right| \frac{u_l}{r} \mathcal{Y}_{lSJ}^{J} \right\rangle. \tag{2.65}$$

If we take into consideration that (in units of $\hbar$)

$$l_z Y_l^m = m Y_l^m, \tag{2.66}$$

and that

$$S_p^z \chi_1^1 = \frac{1}{2} \chi_1^1, \qquad S_n^z \chi_1^1 = \frac{1}{2} \chi_1^1,$$

$$S_p^z \chi_1^0 = \frac{1}{2} \chi_0^0, \qquad S_n^z \chi_1^0 = -\frac{1}{2} \chi_0^0,$$

$$S_p^z \chi_1^{-1} = -\frac{1}{2} \chi_1^{-1}, \qquad S_n^z \chi_1^{-1} = -\frac{1}{2} \chi_1^{-1},$$

$$S_p^z \chi_0^0 = \frac{1}{2} \chi_1^0, \qquad S_n^z \chi_0^0 = -\frac{1}{2} \chi_1^0,$$

we obtain

$$\mu_z \mathcal{Y}_{011}^1 = \frac{1}{2} (g_p + g_n) \mathcal{Y}_{011}^1 = \frac{1}{2} (g_p + g_n) Y_0^0 \chi_1^1; \tag{2.67}$$

$$\mu_z \mathcal{Y}_{211}^1 = \sqrt{\frac{3}{5}} \left(1 - \frac{g_p + g_n}{2} \right) Y_2^2 \chi_1^{-1} - \frac{1}{2} \sqrt{\frac{3}{10}} Y_2^1 [\chi_1^0 + (g_p - g_n)\chi_0^0] + \sqrt{\frac{1}{10}} \frac{g_p + g_n}{2} Y_2^0 \chi_1^1; \tag{2.68}$$

$$\mu_z \mathcal{Y}_{101}^1 = \frac{1}{2} Y_1^1 [\chi_0^0 + (g_p - g_n)\chi_1^0]; \tag{2.69}$$

$$\mu_z \mathcal{Y}_{111}^1 = \frac{1}{2\sqrt{2}} \left[Y_1^1 \chi_1^0 + g_p(Y_1^1 \chi_0^0 - Y_1^0 \chi_1^1) - g_n(Y_1^1 \chi_0^0 + Y_1^0 \chi_1^1) \right]. \tag{2.70}$$

Using the orthonormality of Y_l^m and of the $\chi_S^{m_s}$, we obtain the expected values

$${}^3S_1 : \langle \Psi_{011}^1 \mid \mu_z \mid \Psi_{011}^1 \rangle = \frac{g_p + g_n}{2} \int_0^\infty u_0^2(r)\, dr = \frac{g_p + g_n}{2} = 0.88 \mu_N; \tag{2.71}$$

$${}^3D_1 : \langle \Psi_{211}^1 \mid \mu_z \mid \Psi_{211}^1 \rangle = \left[\frac{3}{4} - \frac{1}{4}(g_p + g_n) \right] \int_0^\infty u_2^2(r)\, dr$$
$$= \frac{3}{4} - \frac{1}{4}(g_p + g_n) = 0.31 \mu_N; \tag{2.72}$$

$${}^1P_1 : \langle \Psi_{101}^1 \mid \mu_z \mid \Psi_{101}^1 \rangle = \frac{1}{2} \int_0^\infty u_1^2(r)\, dr = 0.5 \mu_N; \tag{2.73}$$

$${}^3P_1 : \langle \Psi_{111}^1 \mid \mu_z \mid \Psi_{111}^1 \rangle = \left[\frac{1}{4} + \frac{1}{4}(g_p + g_n) \right] \int_0^\infty u_1^2(r)\, dr$$
$$= \frac{1}{4} + \frac{1}{4}(g_p + g_n) = 0.69 \mu_N. \tag{2.74}$$

The experimental value of the magnetic moment of the deuteron is $0.8573\ \mu_N$, different from all values found above. Thus, the deuteron cannot be found exclusively in any of the states mentioned previously. In order for the magnetic moment to be equal to the experimental value, one is forced to conclude that the ground state of the deuteron should be a mixture of these states. Therefore, we construct the possible mixtures of states of same parity,

$$\Psi = C_S\Psi_S + C_D\Psi_D \tag{2.75}$$

or

$$\Psi = C_{1P}\Psi_{1P} + C_{3P}\Psi_{3P}, \tag{2.76}$$

and the normalization condition $\langle \Psi \mid \Psi \rangle = 1$ gives

$$C_S^2 + C_D^2 = 1 \tag{2.77}$$

or

$$C_{1P}^2 + C_{3P}^2 = 1. \tag{2.78}$$

From (2.71), (2.72), and (2.75) we find

$$\langle \Psi \mid \mu_z \mid \Psi \rangle \frac{1}{2}(g_p + g_n)C_S^2 + \left[\frac{3}{4} - \frac{1}{4}(g_p + g_n)\right]C_D^2 = (0.88C_S^2 + 0.31C_D^2)\mu_N. \tag{2.79}$$

If we make $\langle \Psi \mid \mu_z \mid \Psi \rangle$ equal to the experimental value $0.85573\,\mu_N$, we obtain $C_S^2 = 0.96$ and $C_D^2 = 0.04$. On the other hand, a mixture of $^1P_1$ and $^3P_1$ states cannot take us to the experimental value of μ since in both states the magnetic moment is smaller than μ_d. We should conclude that the ground state of the deuteron is basically (96%) a $^3S_1$ state, with a small (4%) contribution of the $^3D_1$ state. Therefore, the wavefunction of the deuteron can be written as

$$\Psi_d = C_S\frac{u_0}{r}\mathcal{Y}_{011}^1 + C_D\frac{u_2}{r}\mathcal{Y}_{211}^1. \tag{2.80}$$

Identical results can be obtained by analysis of the electric quadrupole moment.

2.7 Particles in the Continuum: Scattering

When a wave of any type hits a small obstacle, secondary waves (circular or spherical) are produced and move away from it, going to infinity. In the same way, a monoenergetic beam of particles, which can be represented by a plane wave, undergoes scattering when it finds a region in which there exists a potential $V(\mathbf{r})$ created by a nucleus (figure 2.3). Unlike the quantum bound state problem, where one searches for the possible values of the energy of the system, the solution of scattering problems consists in finding the angular distribution of the scattered particles, where the total energy of the system target+projectile can have any positive value.

The angular distribution is determined by the probability of finding the scattered particles as a function of the scattering direction, and this probability is directly connected to the wavefunctions. Thus, given an incident plane wave, whose stationary part can be represented by

$$\Psi i(\mathbf{r}) = e^{i\mathbf{k}\cdot\mathbf{r}} = e^{ikz}, \tag{2.81}$$

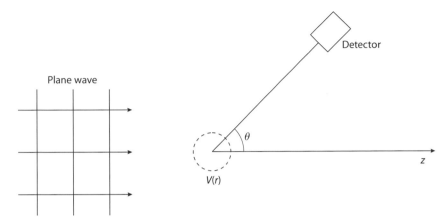

Figure 2.3 Scattering of a plane wave by a potential $V(r)$ limited to a small region of space.

and a scattering potential $V(\mathbf{r})$, our problem reduces to finding the wavefunction of the scattered particles, or the scattered wavefunction.

In problems of atomic and nuclear physics the detectors lie far away from the scattering centers compared to their dimensions, that is, they are in a region where the particles no longer feel the action of the potential. Thus, our interest will be limited to the asymptotic part of the scattered wavefunction, namely, its form when $r \to \infty$. When a short range potential $V(r)$, supposed for simplicity to be spherically symmetric, acts on the particles of an incident beam, a detector placed in the asymptotic region will register not only the presence of the plane wave but also the particles scattered by the potential. That is, to the plane wave will be added an outgoing spherical wave created by the scattering center, and we can write the wavefunction far from this center as

$$\Psi \sim e^{ikz} + f(\theta)\frac{e^{ikr}}{r}. \tag{2.82}$$

where the symbol $\sim$ means asymptotic value. The presence of the function $f(\theta)$ expresses the fact that the scattering directions do not have not the same probability. This function is called *scattering amplitude* and has, as we will see next, an essential role in the theory for the process.

The probability current,

$$\mathbf{j} = \frac{\hbar}{m}\,\mathrm{Im}\,(\Psi^*\nabla\Psi), \tag{2.83}$$

will be now employed in the definition of a function that measures the angular distribution of the particles scattered by $V(r)$. The current for the incident plane wave is

$$j_i = \frac{\hbar}{m}\mathrm{Im}\left(e^{-ikz}\frac{d}{dz}e^{ikz}\right) = \frac{\hbar k}{m} = v, \tag{2.84}$$

and for the outgoing spherical wave

$$j_r \sim \frac{\hbar}{m}\mathrm{Im}\left\{f^*(\theta)\frac{e^{-ikr}}{r}\frac{\partial}{\partial r}\left[f(\theta)\frac{e^{ikr}}{r}\right]\right\} = \frac{v}{r^2}|f(\theta)|^2. \tag{2.85}$$

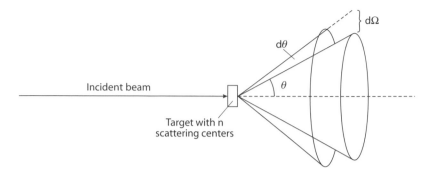

Figure 2.4 Quantities used in the definition of the differential cross section.

We define the *differential cross section*, a function of the angle θ (see figure 2.4), by

$$\frac{d\sigma}{d\Omega} = \frac{dN/d\Omega}{n\Phi},$$
(2.86)

dN being the number of observed events in $d\Omega$ per unit time, n the number of target scattering centers comprised by the beam, and Φ the incident flux (number of incident particles per unit area and per unit time). $d\Omega = 2\pi\sin\theta\,d\theta$ is the solid angle located between the cones defined by the directions θ and $\theta + d\theta$. If our assumption of spherical symmetry for the scatter potential is not valid, the solid angle is the one defined by the direction θ, ϕ, namely, $d\Omega = \sin\theta\,d\theta\,d\phi$.

Definition (2.86) is a general one, and the *observed events*, in the present case, are particles scattered by the potential $V(r)$. $d\sigma/d\Omega$ has the dimension of area, and its value is obtained from

$$\frac{d\sigma}{d\Omega} = \frac{j_r r^2}{j_i}$$
(2.87)

by the fact that the number of particles that cross a given area per unit time is measured by the probability current flux through that area. With (2.84) and (2.85) it is clear that

$$\frac{d\sigma}{d\Omega} = |f(\theta)|^2,$$
(2.88)

being thus the determination of the angular distribution reduced to the evaluation of the scattering amplitude $f(\theta)$.

The *total cross section* is obtained by integrating (2.88):

$$\sigma = \int \frac{d\sigma}{d\Omega}\,d\Omega = 2\pi \int_{-1}^{+1} |f(\theta)|^2\,d(\cos\theta),$$
(2.89)

and its meaning is obvious from (2.86): the total cross section measures the number of events per target nucleus per unit time divided by the incident flux defined above. It must include, in this way, events for which we cannot define a differential cross section, such as the absorption of particles from the incident beam by the nucleus.

2.8 Partial Wave Expansion

When we study interactions governed by a central potential $V(r)$, solutions of the Schrödinger equation

$$\nabla^2 \Psi + \frac{2m}{\hbar^2}\left[E - V(r)\right]\Psi = 0 \tag{2.90}$$

can be written as linear combinations of products of solutions separated into radial and angular parts:

$$\Psi = \sum_{l,m} a_{lm} \frac{u_l(r)}{r} Y_l^m(\theta, \phi), \tag{2.91}$$

where $u_l(r)$ obeys the radial equation

$$\frac{d^2u}{dr^2} + \frac{2m}{\hbar^2}\left[E - V(r) - \frac{\hbar^2}{2m}\frac{l(l+1)}{r^2}\right]u = 0 \tag{2.92}$$

and the boundary condition

$$u_l(0) = 0. \tag{2.93}$$

The axial symmetry of our problem allows to eliminate the dependence in ϕ of (2.91) so that

$$\Psi = \sum_l a_l P_l(\cos\theta)\frac{u_l(r)}{kr}, \tag{2.94}$$

where the constant $k = \sqrt{2mE}/\hbar$ was introduced to make easier later applications of the expansion.

The terms of (2.94) can be understood as *partial waves* from which the general solution Ψ can be constructed. An expression like (2.94) is convenient: if $V(r)$ is spherically symmetric, the angular momentum is a constant of motion, and states of different values of the angular momentum contribute in an independent way to the scattering. Thus, it is also useful to present the plane wave by an expansion in Legendre polynomials

$$e^{ikz} = e^{ikr\cos\theta} = \sum_{l=0}^{\infty}(2l+1)i^l j_l(kr)P_l(\cos\theta), \tag{2.95}$$

where $j_l(x)$ are spherical Bessel functions and $P_l(\cos\theta)$ the Legendre polynomials.

Expression (2.95) has the form of (2.94). This means that the plane wave $e^{i\mathbf{k}\cdot\mathbf{r}}$ can be understood as the sum of a set of partial waves, each one with orbital angular momentum $\sqrt{l(l+1)}\,\hbar$. The terms $j_l(kr)P_l(\cos\theta)$ specify the radial and angular dependence of the partial wave l, the weight of the contribution of each term being given by the amplitude $2l+1$ and the phase factor i^l.

Using classical arguments we can give an interpretation to this amplitude value. Let us consider a surface normal to the plane wave propagation direction, and imagine a set of circles of radius $b_l = l\bar{\lambda}$, with the wavelength $\bar{\lambda} = \lambda/2\pi = 1/k$, centered at the point where the z-axis crosses the surface (see figure 2.5). If the beam of particles moves along

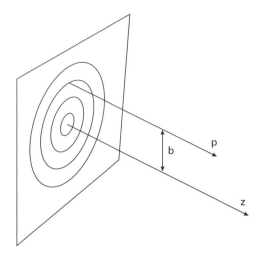

Figure 2.5 Classical representation of a plane wave: to each partial wave
l corresponds an impact parameter b.

the z-axis, the classical angular momentum of a particle about the origin of the coordinate system is the product of the impact parameter b and the linear momentum $p = \hbar k$. Hence, all particles that pass by a ring of internal radius b_l and external radius b_{l+1} will have orbital angular momentum between $l\bar{\lambda}\hbar k = l\hbar$ and $(l+1)\bar{\lambda}\hbar k = (l+1)\hbar$. In the classical limit l is large and $l + 1 \cong l$. So we can say that all the particles that pass through the ring have orbital angular momentum $l\hbar$. However, still using classical reasoning, a particle belonging to a uniform beam can have any impact parameter, and its probability of passing one of the rings is proportional to the area A of that ring:

$$A = \pi(b_{l+1}^2 - b_l^2) = \pi\bar{\lambda}^2\left[(l+1)^2 - l^2\right] = \pi\bar{\lambda}^2(2l+1). \tag{2.96}$$

We see that $2l + 1$ is the relative probability that a particle in a uniform beam has an orbital angular momentum $l\hbar$, which is the classical limit for the orbital angular momentum $\sqrt{l(l+1)}$ associated to the partial wave l.

At large distances from the origin the spherical Bessel functions reduce to the simple expression

$$j_l(kr) \sim \frac{\sin\left(kr - \frac{l\pi}{2}\right)}{kr} = \frac{e^{i(kr - \frac{l\pi}{2})} - e^{-i(kr - \frac{l\pi}{2})}}{2ikr}. \tag{2.97}$$

Using (2.97) in (2.95) results in

$$e^{ikr\cos\theta} \sim \frac{1}{2i}\sum_{l=0}^{\infty}(2l+1)i^l P_l(\cos\theta)\left[\frac{e^{i(kr - \frac{l\pi}{2})} - e^{-i(kr - \frac{l\pi}{2})}}{kr}\right], \tag{2.98}$$

which represents the asymptotic form of a plane wave.

In (2.98) the first term in the brackets corresponds to an outgoing spherical wave and the second to an incoming spherical wave. Thus, each partial wave in (2.98) is, at large distances from the origin, a superposition of two spherical waves, namely, the incoming and the outgoing components. The total radial flux for the wavefunction $\Psi_i = e^{ikr\cos\theta}$ vanishes, since the number of free particles that enter a region is the same as the number that exit. This can be easily shown using (2.98) in (2.83) (exercise 9 proposes this demonstration for the more general expression 2.99).

Let us now understand Ψ in (2.94) as a solution of a scattering problem, the scattering being caused by a potential $V(r)$. The asymptotic form of Ψ can be obtained if we observe that the presence of the potential has the effect of causing a perturbation in the outgoing part of the plane wave, and such a perturbation can be represented by a unitary module function, $S_l(k)$.

From (2.98), this leads to

$$\Psi \sim \frac{1}{2i} \sum_{l=0}^{\infty} (2l+1) i^l P_l(\cos\theta) \frac{S_l(k) e^{i(kr - \frac{l\pi}{2})} - e^{-i(kr - \frac{l\pi}{2})}}{kr}, \tag{2.99}$$

where the function $S_l(k)$ can be represented by

$$S_l(k) = e^{2i\delta_l}. \tag{2.100}$$

When we write the form (2.100) we admit that the scattering is elastic. The unitary module of $S_l(k)$ keeps the same value for the probability current and does not allow that the presence of the potential removes or add particles to the elastic channel k. From a comparison of (2.99) and (2.94) we can obtain the expressions for a_l and for the asymptotic form of $u_l(r)$:

$$a_l = i^l (2l+1) e^{i\delta_l} \tag{2.101}$$

and

$$u_l(r) \sim \sin\left(kr - \frac{l\pi}{2} + \delta_l\right). \tag{2.102}$$

$u_l(r)$ differs from the asymptotic form of the radial function of a free particle by the presence of the *phase shifts* δ_l; the presence of the scattering potential creates in each partial wave a phase shift δ_l, and the scattering problem would be solved with the determination of these phase shifts for a given potential $V(r)$. In fact, the use of (2.99) and (2.98) in (2.82) results in

$$f(\theta) = \frac{1}{k} \sum_{l=0}^{\infty} (2l+1) e^{i\delta_l} \sin\delta_l P_l(\cos\theta), \tag{2.103}$$

and the differential cross section (2.88) is obtained from the knowledge of the phase shifts δ_l.

The phase shifts are evaluated by solving (2.92) for each l and comparing the phase of $u_l(r)$, for some large r, with the phase of $j_l(kr)$ for the same value of r. This is depicted in figure 2.6, for a generic value of l and three different situations of the potential $V(r)$.

The first curve shows $u_l(r)$ for the case in which $V(r) = 0$ for all r. In this case, $u_l(r) = j_l(kr)$ and we do not have phase shifts for any l. The middle curve shows $u_l(r)$ when one introduces a small attractive potential acting inside a certain radius r_0, that is, $V(r) < 0$ for $r < r_0$, and $V(r) = 0$ for $r > r_0$. From (2.92) we see that, with this attractive potential, $|E - V(r)| > E$ in the potential region and the quantity d^2u_l/dr^2 will be greater in that

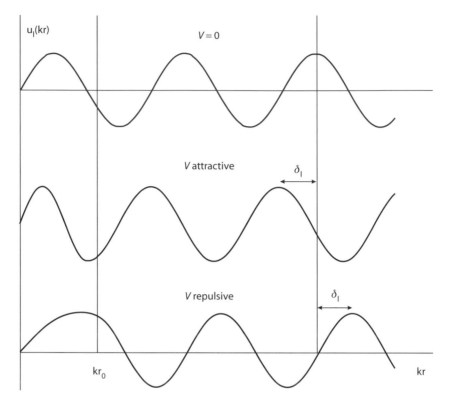

Figure 2.6 Radial part of the wavefunction for three different potentials, showing how the phase shift sign is determined by the function behavior in the region $r < r_0$ where the potential acts.

region than in the region of zero potential. Thus, $u_l(r)$ will oscillate rapidly for $r < r_0$. For $r > r_0$, the behavior is the same as that in the case $V(r) = 0$, except that the phase is displaced. In this way we see that with a small attractive potential, $u_l(r)$ is "pulled in," which in turn advances its phase and makes the phase shift positive. The last curve shows $u_l(r)$ for the case of a small repulsive potential, that is, $V(r) > 0$ for $r < r_0$, and $V(r) = 0$ for $r > r_0$. In this case, $| E - V(r) | < E$ in the potential region and the quantity $d^2 u_l / dr^2$ will be smaller in that region than in the region of zero potential. The result is that, for a repulsive potential, $u_l(r)$ is "pulled out," its phase is retarded, and the phase shift is negative.

The total cross section, in turn, has the expression

$$\sigma = \frac{4\pi}{k^2} \sum_l (2l + 1) \sin^2 \delta_l, \tag{2.104}$$

obtained by the integration (2.89).

From (2.103) and (2.104) one extracts an important relation. For this, it is enough to observe in (2.103) that

$$\text{Im } f(0) = \frac{1}{k} \sum_{l=0}^{\infty} (2l + 1) \sin^2 \delta_l \tag{2.105}$$

and to compare this result with (2.104) to obtain

$$\sigma = \frac{4\pi}{k} \, \text{Im} f(0). \tag{2.106}$$

This relation is known as the *optical theorem*. It connects the total cross section with the scattering amplitude at angle zero. This is physically understandable: the cross section measures the removal of particles from the incident beam that arises, in turn, from the destructive interference between the incident and scattered beams at zero angle [Sc54]. We will see in chapter 10 that the optical theorem is not restricted to elastic scattering, being also valid for inelastic processes.

In the sum (2.104) each angular momentum contributes at most a cross section

$$(\sigma)_{\text{max}} = \frac{4\pi}{k^2} (2l + 1), \tag{2.107}$$

a value of the same order of magnitude as the maximum classical cross section per unit $\hbar$ of angular momentum. Actually, if we use the estimate $b = l/k$ for the impact parameter, the contribution of an interval $\Delta l = 1$, or $\Delta b = 1/k$, for the total cross section will be

$$\sigma_l = 2\pi \, b \, \Delta b = 2\pi \, \frac{l}{k^2}. \tag{2.108}$$

For large l this agrees with $(\sigma)_{\text{max}}$, except for a factor 4. The difference is due to the unavoidable presence of diffraction effects for which the wave nature of matter is responsible.

The partial wave analysis (2.103) gives an exact procedure to solve the scattering problem at all energies. For a given potential $V(r)$, (2.92) should be solved and its asymptotic solutions (2.102) used to find the phase shifts δ_l. The infinite number of terms of (2.103) is not a problem in practice, since

$$\lim_{l \to \infty} \delta_l = 0. \tag{2.109}$$

This result can be verified by examining (2.92): for large l the centrifugal potential term, proportional to $l(l + 1)$, is totally dominant, making irrelevant the phase shifts generated by $V(r)$. However, at high energies the sum (2.103) will have the contribution of many terms, since, in this case, $kr_0 \gg 1$, and for all l up to $l_{\text{max}} \simeq kr_0$ there will be appreciable phase shifts. The partial wave analysis is of great utility, especially in the low energy case, which will be treated in the next section.

2.9 Low Energy Scattering

We have seen that the partial wave expansion is useful only at low energies since in this case the number of terms of (2.103) that we have to deal with is small. If the energy is low enough, the sum (2.103) reduces to the term with $l = 0$. We have, in this case,

$$f(\theta) = \frac{1}{k} e^{i\delta_0} \sin \delta_0 \tag{2.110}$$

and

$$\sigma = \frac{4\pi}{k^2} \sin^2 \delta_0. \tag{2.111}$$

The differential cross section that results from (2.110) is independent of θ: the scattering is isotropic. This is easily understandable since at low energies the incident particle wavelength is much greater than the dimension of the target nucleus; during its passage all points in the nucleus are with the same phase at each time, and it is impossible to identify the direction of incidence.

In the extreme case $E \to 0$ the scattering amplitude (2.110) remains finite only if $\delta_0 \to 0$ together with the energy. In this case the phase difference is no more the main scattering parameter. A better parameter is the *scattering length a*, defined as the limit

$$\lim_{E \to 0} f(\theta) = \lim_{k \to 0} \frac{\delta_0}{k} = -a, \tag{2.112}$$

yielding the equation

$$\sigma = 4\pi a^2 \tag{2.113}$$

as the expression for the total cross section at the zero energy limit.

The physical meaning of the scattering length can be obtained observing that for $l = 0$, and in the limit $E \to 0$, (2.92) in the region outside the potential reduces to its first term. Hence, if $d^2 u/dr^2 = 0$, we see that the wavefunction u tends to a straight line and the abscissa at the point where this line crosses the r-axis is the scattering length a. This can be easily seen if we rewrite (2.102) in the limit $E \to 0$:

$$u_0(r) \cong kr + \delta_0 = k(r - a). \tag{2.114}$$

This property will be used later to determine if a state of the system is or not bound.

As an application let us evaluate the cross section of scattering of low energy neutrons by protons. The attractive nuclear potential between a proton and a neutron is put in the simple form

$$V(r) = \begin{cases} -V_0, & \text{if } r < r_0, \\ 0, & \text{if } r < r_0, \end{cases} \tag{2.115}$$

where r_0 represents the range of the nuclear force. We know that this problem can be reduced to a single particle problem, a particle carrying the reduced mass of the system and subject to the same potential V_0. E turns to be the total system energy in the center of mass frame. For a neutron with energy E_n incident on a proton at rest in the laboratory, E is very close to $E_n/2$.

If $l = 0$ is the only partial wave to contribute, (2.92) assumes the form

$$\frac{d^2 u}{dr^2} + \frac{2m}{\hbar^2}(E + V_0)u = 0 \quad (r < r_0),$$

(2.116)

$$\frac{d^2 u}{dr^2} + \frac{2m}{\hbar^2} E u = 0 \quad (r > r_0),$$

(2.117)

with the boundary conditions $u = 0$ in $r = 0$ and u and du/dr continuous at $r = r_0$. From this results

$$u = A \sin{(Kr)} \quad (r < r_0),$$

(2.118)

with

$$K = \frac{\sqrt{2m(E + V_0)}}{\hbar},$$

(2.119)

and

$$u = \sin{(kr + \delta_0)} \quad (r > r_0),$$

(2.120)

where

$$k = \frac{\sqrt{2mE}}{\hbar}.$$

(2.121)

Note that both solutions are of sinusoidal type because $E > 0$. The continuities of the function and its derivative at $r = r_0$ can be expressed by the continuity of $(du/dr)/u$:

$$K \cot{(Kr_0)} = k \cot{(kr_0 + \delta_0)}$$

(2.122)

or

$$k \tan{(Kr_0)} = K \tan{(kr_0 + \delta_0)},$$

(2.123)

which in the limit $k \to 0$ gives

$$\delta_0 \approx kr_0 \left[\frac{\tan{(K_0 r_0)}}{K_0 r_0} - 1 \right],$$

(2.124)

where $K_0 = \sqrt{2mV_0}/\hbar$. The total cross section calculated from (2.124) using (2.111) shows singularities at the points where $K_0 r_0$ have values $\pi/2$, $3\pi/2$, etc., since the tangent in the numerator of (2.124) makes the cross section diverge. This corresponds physically to the appearance of a bound state at that depth. As we are dealing with very small E, there is a resonance whenever an increase in the well depth gives rise to a new level at zero energy. But, in fact, (2.124) is not valid for the $K_0 r_0$ values above, which violates the approximation from which it was deduced. The exact equation (2.123) shows that, when $Kr_0 = \pi/2$, $3\pi/2$, etc., the phase shift is $\delta_0 = Kr_0$, and from (2.111) we see that, for these values of δ_0, the cross section has very large but finite values, given by

$$\sigma = \frac{4\pi}{k^2}.$$

(2.125)

Note that (2.125) can be written as $\sigma = [4/(kr_0)^2] \pi r_0^2$. Since $kr_0 \ll 1$, σ is much greater than πr_0^2, which is the geometrical cross section of the scattering potential. When σ has its maximum possible value for an s-wave scattering, one says that the cross section is in an

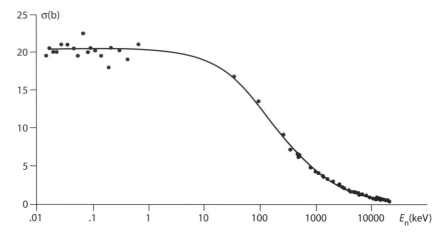

Figure 2.7 Neutron-proton cross sections at low energies. E_n is the energy of the incident neutron. The experimental points were obtained from [Ad50] and [Ho71]. The curve was calculated using (2.144).

s-wave resonance. Resonances in other partial waves occur if E is large enough to create large phase shifts for $l > 0$. For example, when $\delta_1 = \pi/2, 3\pi/2$, etc., the cross section is especially large and one says that we have a p-wave resonance.

Another interesting fact to see in (2.124) is that, if $\tan(K_0 r_0) = K_0 r_0$, the phase shift and the scattering cross section vanish. Hence, for certain values of the well depth there will be no s-wave scattering. This is known as the *Ramsauer effect*, owing to the discovery by C. Ramsauer, in 1921, that the effective cross section for electron scattering in inert gas atoms is very low at energies close to 0.7 eV. The quantum theory gives a simple explanation of this effect, which cannot be explained by the classical theory. The atomic field of inert gases decreases faster with distance than the fields of other atoms; as a first approximation, we can replace this field by a rectangular well and use (2.111) and (2.124) to evaluate the cross section of the low energy electrons. For an electron energy of approximately 0.7 eV, we get $\sigma \sim 0$, if we use r_0 equal to the atomic dimensions.

Let us make an estimate of the neutron-proton cross section. Let us use $V_0 = 34$ MeV for the approximate depth of the deuteron potential. We have

$$K_0 = \frac{\sqrt{2mV_0}}{\hbar} = \frac{\sqrt{2mc^2 V_0}}{\hbar c} = \frac{\sqrt{938.93 \times 34}}{197.33} = 0.91 \text{ fm}^{-1}, \tag{2.126}$$

where m is the reduced mass, equal to half a nucleon mass. Still using $r_0 \cong 2.1$ fm as the deuteron radius, we have

$$a = -\frac{\delta_0}{k} = r_0 \left[1 - \frac{\tan(K_0 r_0)}{K_0 r_0}\right] = 2.1 \times 10^{-13} \left[1 - \frac{\tan(0.91 \times 2.1)}{0.91 \times 2.1}\right] = 5.2 \times 10^{-13} \text{ cm}. \tag{2.127}$$

Hence,

$$\sigma = 4\pi (5.2 \times 10^{-13})^2 \text{cm}^2 \cong 3.5 \text{ b}. \tag{2.128}$$

Figure 2.7 shows experimental values of the neutron-proton cross sections up to 20 MeV. In the zero energy limit the cross section has the value $\sigma = (20.43 \pm 0.02)$ b, six times

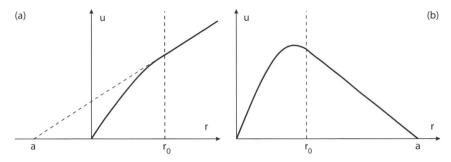

Figure 2.8 Radial part of the scattering wavefunction u at zero energy for two values of the well depth. In (a) the depth is not large enough for the existence of a bound state and the scattering length is negative. In (b) one sees the existence of a bound state, the function u has a maximum inside the potential range and a has a positive value.

greater than the value obtained in (2.128). The reason for this discrepancy was explained by Wigner, proposing that the nuclear force depends on the spin, being different when nucleons collide with parallel (triplet) spins or with antiparallel spins (singlet). As there are three triplet states and only one singlet, an experiment where the nucleons are not polarized will register three times more events of the first type than of the second, resulting for the cross section the combination

$$\sigma = \frac{3}{4}\sigma_t + \frac{1}{4}\sigma_s. \tag{2.129}$$

If then $\sigma = 20.4$ b and $\sigma_t = 3.4$ b, one gets for σ_s the value 71 b! The explanation for this high value is found in the fact that the singlet potential is shallower than the triplet one, being within the threshold for the appearance of the first bound state. This gives rise to a resonance when the incident particle has very low energy, as happens in the present case.

There are ways to find whether the singlet potential has a bound state with a very low negative energy or if the resonance occurs at a very low yet positive energy. For that let us initially study the scattering amplitude variation as a function of the well depth in the case of very low incident energy E. When $V_0 = 0$, (2.92) has the simple form $u'' = 0$, and the trivial solution for u is a straight line crossing the origin. When V_0 is small and forms the well, the form of the wavefunction looks like figure 2.8a. There is no bound state yet, and the scattering length a is negative (see 2.114). When V_0 is deep enough to allow the existence of the first bound state, the form of the wavefunction is as in figure 2.8b, with a maximum inside the well. This is an expected behavior: the internal part of the wavefunction is not sensitive to the fact that E is a little positive or negative, and a bound state wavefunction should have in r_0 a negative derivative to match to the exponential decay of the external part. The essential result is that the sign of a is, in this case, positive. Thus, the sign of the scattering length can show us if a resonance occurs with a negative (bound) state or a positive (virtual) energy.

The combination of (2.113) and (2.129) gives

$$\sigma = \pi(3a_t^2 + a_s^2), \tag{2.130}$$

where a_t and a_s are the scattering length for the triplet and singlet potentials, with respective cross sections σ_t and σ_s given by expressions such as (2.113). For an equation like (2.130) it does not matter what the sign of the scattering amplitude is, since it expresses an *incoherent* combination of singlet and triplet scattering. The cross section is proportional to the square of the amplitudes, in the same way as the intensity of a light beam is proportional to the square of the magnitude of the electric (or magnetic) field. One form of *coherent* scattering is achieved when the wavelength of the incident particles is of the order of the distance between the nuclei inside a molecule. In the case of neutrons incident on H_2, the distance between the protons in the molecule is about 0.8×10^{-8} cm and the coherent scattering is reached with a neutron energy near 2×10^{-3} eV. The scattering of these very slow neutrons by a hydrogen molecule produces an interference phenomenon similar to that occurring with light waves that emerge from the two slits of a Young experiment. An additional ingredient in the present case is that the H_2 molecule can be in two states, one with the spins of the protons forming a triplet (orthohydrogen) and the other where the spins form a singlet (parahydrogen). When a neutron interacts with an orthohydrogen molecule the two scattering amplitudes are of the same type; when the interaction is with parahydrogen they are of different types. Schwinger and Teller [ST37] have found expressions for the scattering cross sections of slow neutrons with ortho- and parahydrogen:

$$\sigma = c_1(a_t - a_s)^2 + c_2(3a_t + a_s)^2, \tag{2.131}$$

where c_1 and c_2 are numerical coefficients, with different values for the two types of hydrogen and also depending upon the incident neutron energy and the gas temperature. This last dependence is natural since the incident neutrons have very low velocities and the thermal molecular motion, in this case, has a non-negligible influence. As an example, at a temperature of 20.4 K and with 0.0045 eV neutrons the coefficients are, for orthohydrogen, $c_1 = 13.762$ and $c_2 = 6.089$, and for parahydrogen, $c_1 = 0$ and $c_2 = 6.081$. Note that in (2.131), contrary to (2.130), the signs of a_t and a_s are important for the calculation of σ.

From (2.131) it is possible to obtain the scattering lengths a_t and a_s if the experimental cross sections are known. Measurements of these cross sections using gaseous hydrogen have been done since 1940 [SS55], improving former work done with liquid H_2, where the effects due to the intermolecular forces are difficult to separate. The cross sections were measured from room temperature, where the proportion is 75% orthohydrogen and 25% parahydrogen, down to 20 K. At the lowest temperature only parahydrogen exists since it has a greater binding energy than orthohydrogen and can be formed by the decay of orthohydrogen, provided the process is accelerated by a catalyzer (a substance with paramagnetic atoms that induces the spin change of one of the protons of the H_2 molecule).

The results found for the scattering lengths in these and other experimental works using different methods are not free of some systematic errors. Houk [Ho71] recommended the values

$$a_t = (5.423 \pm 0.005) \text{ fm}, \quad a_s = (-23.71 \pm 0.01) \text{ fm} \tag{2.132}$$

for the scattering lengths. Note that these values also satisfy (2.130). The negative sign of the singlet scattering length is the answer to our question related to figure 2.8: the proton-neutron system has no bound state except the deuteron ground state. The resonance in the low energy scattering of neutrons by protons is due to a state of the system with a small positive energy.

2.10 Effective Range Theory

In the previous section we developed the theory of elastic scattering at energies close to zero, where the cross sections are expressed by (2.113). When the incident neutron energy goes beyond this limit we have two problems to face. First, the limit (2.112) is no longer applicable and the scattering length alone cannot determine the cross section. Second, the series (2.103) cannot be truncated in $l = 0$, since the terms $l = 1, l = 2, \ldots$ begin to have a significant contribution. In fact, this problem begins to be important only at energies of tens of MeV. The results that we shall obtain show that we can safely proceed up to 20 MeV using only the partial wave $l = 0$. The title of this section has thus the meaning that the zero energy approximation will no longer be valid but we will remain restricted to the $l = 0$ component of the angular momentum.

Our purpose is to investigate the behavior of the cross section (2.111) when we move away from the zero energy limit. The first results in this direction were established by Schwinger using a variational principle, but were reproduced afterward with a simpler method that is based only on the properties of the wavefunction. We will follow closely the work of H. Bethe [Bet49], where this method is explicated with clarity.

Let us consider the incident neutron with energy E_1 and wavenumber k_1. If we write (2.92) for $l = 0$ and use (2.121), the radial wavefunction satisfies

$$\frac{d^2 u_1}{dr^2} + k_1^2 u_1 - \frac{2m}{\hbar^2} V(r) u_1 = 0. \tag{2.133}$$

For another energy E_2, we have

$$\frac{d^2 u_2}{dr^2} + k_2^2 u_2 - \frac{2m}{\hbar^2} V(r) u_2 = 0. \tag{2.134}$$

Multiplying (2.133) by u_2 and (2.134) by u_1, subtracting and integrating, we arrive at

$$u_2 u_1' - u_1 u_2'|_0^R = (k_2^2 - k_1^2) \int_0^R u_1 u_2 \, dr, \tag{2.135}$$

where the limit R is arbitrary.

Let us define the function Ψ as the asymptotic form of u, but valid for every point in space:

$$\Psi_1 = \frac{\sin(k_1 r + \delta_1)}{\sin \delta_1} \tag{2.136}$$

where the normalization was chosen to make $\Psi = 1$ at the origin; this also determines the normalization of u. Note that the sub-index of the phase shift δ refers to the energy and not to the angular momentum.

A relation analogous to (2.135) is valid for Ψ:

$$\Psi_2 \Psi_1' - \Psi_1 \Psi_2' |_0^R = (k_2^2 - k_1^2) \int_0^R \Psi_1 \Psi_2 \, dr. \tag{2.137}$$

Let us subtract (2.135) from (2.137). If R is chosen beyond the range of the nuclear forces, where Ψ and u coincide, the contribution of the left side in the limit R will be zero. At the lower limit $u_1 = u_2 = 0$ and thus this term does not contribute. Extending the integration limit to infinity, we obtain:

$$\Psi_1(0)\Psi_2'(0) - \Psi_2(0)\Psi_1'(0) = (k_2^2 - k_1^2) \int_0^\infty (\Psi_1 \Psi_2 - u_1 u_2) \, dr. \tag{2.138}$$

The derivatives of (2.138) are obtained from (2.136), resulting in

$$k_2 \cot \delta_2 - k_1 \cot \delta_1 = (k_2^2 - k_1^2) \int_0^\infty (\Psi_1 \Psi_2 - u_1 u_2) \, dr. \tag{2.139}$$

Let us now apply (2.139) to the special case $k_1 = 0$. In this case, $k_1 \cot \delta_1 = -1/a$, where a is the scattering length. Ignoring the lower index 2, (2.139) can be rewritten as

$$k \cot \delta = -\frac{1}{a} + k^2 \int_0^\infty (\Psi_0 \Psi - u_0 u) \, dr. \tag{2.140}$$

No approximation has been done up to now and (2.140) is exact. But, looking at the integrand of (2.140), we see that it is different from zero inside the range of the nuclear forces; in this situation, Ψ and u depend very weakly on the energy, since E is supposedly small compared to $V(r)$ (see 2.92). A reasonable approximation is therefore to replace Ψ and u by Ψ_0 and u_0, respectively, and to write

$$k \cot \delta = -\frac{1}{a} + \frac{1}{2} k^2 r_{\text{eff}}, \tag{2.141}$$

with

$$r_{\text{eff}} = 2 \int_0^\infty (\Psi_0^2 - u_0^2) \, dr. \tag{2.142}$$

The quantity r_{eff}, which has the dimension of length, does not depend on the energy and is called *effective range*. The factor 2 was inserted in its definition to make it resemble the potential range. Gathering (2.111) and (2.141), we can express the cross section by

$$\sigma = \frac{4\pi}{k^2} \frac{1}{1 + \cot^2 \delta_0} = \frac{4\pi a^2}{a^2 k^2 + (1 - \frac{1}{2} a r_{\text{eff}} k^2)^2}, \tag{2.143}$$

where the influence of the potential is represented by two parameters, the effective range r_{eff} and the scattering length a. Thus the cross section is not affected by the details of the form of the potential since with any other reasonable form it will be always possible to adjust the depth and the range in such a way as to reproduce the values of a and r_{eff}. The result is that a study of low energy scattering does not lead to information about the form of the nucleon-nucleon potential. The theory of effective range is therefore sometimes called the *form independent approximation*.

For the application of the cross section (2.143) we have to remember that there are in fact two potentials, one for the singlet and another for the triplet state, and (2.143) should be, using (2.129), more appropriately written

$$\sigma = \frac{3}{4} \frac{4\pi a_t^2}{a_t^2 k^2 + (1 - \frac{1}{2} a_t r_t k^2)^2} + \frac{1}{4} \frac{4\pi a_s^2}{a_s^2 k^2 + (1 - \frac{1}{2} a_s r_s k^2)^2} , \qquad (2.144)$$

implying the existence of four parameters to be determined: a_t, a_s, r_t, r_s, namely, the scattering lengths in the singlet and triplet states and the respective effective ranges. For the first two we have the established values (2.132). The effective range in the triplet state, r_t, can be obtained from well-known experimental information, the deuteron binding energy. For that goal it is enough to see that there is no restriction to employing the above theory to negative energies, namely, bound states. So let us use for u the deuteron radial wavefunction; in this case Ψ is the decreasing exponential (2.39),

$$\Psi = e^{-\sqrt{2mE_B}\, r/\hbar}, \qquad (2.145)$$

already properly normalized. E_B is the deuteron binding energy and m its reduced mass. Using (2.138), with $\Psi_2 = \Psi$ and $\Psi_1 = \Psi_0$, we have

$$-\frac{\sqrt{2mE_B}}{\hbar} + \frac{1}{a_t} = -\frac{2mE_B}{\hbar^2} \frac{r_t}{2}, \qquad (2.146)$$

being an expression that allows us to deduce r_t from a_t and E_B. The sign of the right-hand side of (2.146) was introduced because in the case of negative energy the sign of the second term of (2.134) should be changed. Using the values of (2.25) and (2.132), we get the effective range for the triplet potential:

$$r_t = 1.76 \text{ fm.} \qquad (2.147)$$

The value of r_s cannot be obtained as a direct result of an experiment and is normally used in (2.144) as the parameter that best reproduces experimental values of the cross section. The value

$$r_s = 2.56 \text{ fm} \qquad (2.148)$$

produces cross sections with (2.144) that are in very good agreement with experiment, as we can see in figure 2.7.

2.11 Proton-Proton Scattering

This type of scattering is more difficult to deal with than that of the neutron-proton case. Aiming to pinpoint the origin of some difficulties, we shall initially describe the essential differences between the two types of scattering.

1) In p-p scattering there is a repulsive *Coulomb force* between the protons in addition to the nuclear force. The Coulomb forces are of long range: the differential cross section (Rutherford formula) diverges for small angles and the total cross section is infinite.

2) When we deal with identical particles, the Pauli principle puts restrictions on the spatial and spin wavefunctions. In particular, at low energies ($l = 0$), the spatial part is

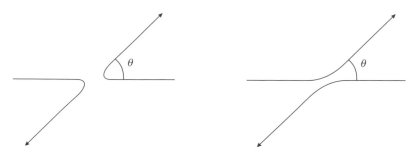

Figure 2.9 Two possible and indistinguishable ways to detect a proton in the angle θ after p-p scattering

symmetric and the spin part is, as a consequence, antisymmetric. In this way only the singlet state contributes to the cross section.

3) The indistinguishability between the protons implies that it is not possible to discriminate the two situations shown in figure 2.9. The wavefunction and the scattering amplitude should have contributions of θ and $\pi - \theta$. In the calculation of cross sections (square of the scattering amplitude) interference terms between the two parts show up. This is a purely quantum phenomenon, with no analog in classical physics.

4) The two independent forces, the nuclear and the Coulomb, contribute their own terms to the cross section. But, the nuclear scattering acts coherently with the Coulomb one and an interference term between the two effects also appears in the cross section.

To take care of all these questions the differential cross section (2.86) is composed of several parts. Its expression for the s-wave ($l = 0$), which we present without demonstration (see [BD04]), can be written as the sum

$$\frac{d\sigma}{d\Omega} = \left[\left(\frac{d\sigma}{d\Omega} \right)_c + \left(\frac{d\sigma}{d\Omega} \right)_n + \left(\frac{d\sigma}{d\Omega} \right)_{cn} \right]. \tag{2.149}$$

The first two terms of (2.149) are due to the Coulomb and nuclear potential, respectively, and the third is the interference term between them. Explicitly (see [BD04]),

$$\left(\frac{d\sigma}{d\Omega} \right)_c = \left(\frac{e^2}{2E_p} \right)^2 \left\{ \frac{1}{\sin^4(\theta/2)} + \frac{1}{\cos^4(\theta/2)} - \frac{\cos\left\{ \eta \ln\left[\tan^2(\theta/2) \right] \right\}}{\sin^2(\theta/2) \cos^2(\theta/2)} \right\}, \tag{2.150}$$

where $e^2 = 1.44$ MeV $\cdot$ fm, E_p is the kinetic energy of the incident proton in the laboratory system, assuming the second proton to be at rest, θ is the scattering angle in the center of mass system, and $\eta = e^2/(\hbar v)$, v is the relative velocity between the protons. The first term in (2.150) refers to the usual Rutherford scattering; the second is due to the necessary existence of a term in $\pi - \theta$ explained in item 3 above. The last term is the interference term between the two previous contributions, namely, the Coulomb scattering in θ and in $\pi - \theta$. This term was first studied by Mott [Mo30] and the full expression (2.150) is known as *Mott scattering*.

The term in (2.149) due to the nuclear potential has the expected form

$$\left(\frac{d\sigma}{d\Omega} \right)_n = \frac{\sin^2 \delta_0}{k^2} \tag{2.151}$$

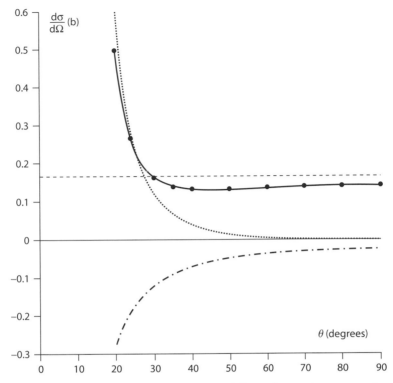

Figure 2.10 Composition of the proton-proton differential scattering cross section (solid line) by the sum of Coulomb (dotted line) and nuclear (dashed line) parts and the interference term (dashed-dotted line). The energy of the incident protons is 3.037 MeV. The experimental points were taken from [Kn66] and the best fit to (2.149) is obtained with $\delta_0 = 50.952°$. Due to the symmetry about 90°, only values up to this angle are shown.

and is written as a function of a pure nuclear phase shift δ_0. Since the nuclear scattering is isotropic, integration of (2.151) yields trivially the result (2.111). It remains to explain the interference term between the Coulomb and nuclear scattering:

$$\left(\frac{d\sigma}{d\Omega}\right)_{cn} = -\frac{1}{2}\left(\frac{e^2}{E_p}\right)^2\frac{\sin\delta_0}{\eta}\left\{\frac{\cos\left[\delta_0 + \eta\ln\sin^2(\theta/2)\right]}{\sin^2(\theta/2)} + \frac{\cos\left[\delta_0 + \eta\ln\cos^2(\theta/2)\right]}{\cos^2(\theta/2)}\right\}. \quad (2.152)$$

The nuclear phase shift δ_0 of (2.151) and (2.152) should be found by a best fit of (2.149) to the experimental points at each energy. In the example of figure 2.10 one obtains $\delta_0 = 50.952°$ for $E_p = 3.037$ MeV. We also see in the figure that the interference yields a total cross section that can be smaller than the purely Coulomb or nuclear part.

When the above procedure is repeated for several energies one obtains the graph in figure 2.11. Note that the interference term allows one to obtain the sign of δ_0 from the experimental cross sections, which is not possible with low energy neutron-proton scattering (see 2.151).

In the study of neutron-proton scattering we developed the effective range theory, where the phase shifts have their values linked to only two parameters connected to the potential,

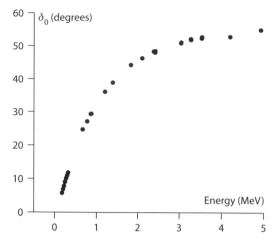

Figure 2.11 Phase shift variation as a function of the incident proton energy for proton-proton collision. The experimental points are from reference [JB50].

not depending on details about the form of the potential. The application of a form-independent theory to the case of proton-proton scattering demands some care, since the Coulomb potential has infinite range and even in the zero energy limit the approximation (2.141) is no longer valid. Despite this, a theory for the process was developed [JB50], resulting in parameters comparable to the scattering length and to the effective range of the neutron-proton singlet scattering. For proton-proton scattering we have

$$a_s = (-7.82 \pm 0.01) \text{ fm}, \quad r_s = (2.79 \pm 0.02) \text{ fm}. \tag{2.153}$$

The much lower absolute value of a_s now found is not of great significance in comparative terms because the Coulomb force adds to the nuclear one. There are means, however, to approximately subtract the effect of this force [JB50] and to evaluate proton-proton scattering amplitudes as if there were only the nuclear force. This new value is $a_s \cong -17$ fm, closer to but yet different from the value corresponding to neutron-proton scattering ($a_s = -23.71$ fm). It is still an open question whether this difference is real or caused by deficiency in the methods, but, anyway, an examination of figure 2.8 show us that a difference like this has little influence on the wavefunction. Similar values for the parameters of neutron-proton and proton-proton collisions would support the assumption of charge independence of the nuclear force, and with the available results we can say that at least approximately this assumption is true.

2.12 Neutron-Neutron Scattering

Low energy neutron-neutron scattering does not present any additional theoretical difficulty as compared to neutron-proton scattering, since in both cases the nuclear force is the only agent. The problem here is of experimental character, since a neutron target is not available and the study of the interaction should be done indirectly.

One of the methods employed consists in analyzing the energy spectrum of the protons resulting from the reaction

$$n + d \rightarrow p + n + n. \tag{2.154}$$

It is a continuous spectrum but presents a peak near the maximum energy. This indicates a resonance related to the formation of the *di-neutron* virtual state, and the peak width can give information about the scattering length. When one collects this and other results from reactions with two neutron formation, one can extract the average values

$$a_s = (-17.6 \pm 1.5) \text{ fm}, \quad r_s = (3.2 \pm 1.6) \text{ fm} \tag{2.155}$$

for the scattering length and effective range of the neutron-neutron scattering, respectively. These values are closer to the proton-proton nuclear scattering than to the neutron-proton scattering. Hence, since they are stronger than the charge independence, there is the indication of a *charge symmetry* of the nuclear force.

2.13 High Energy Scattering

When the energy of the incident nucleon reaches some tenths of an MeV, new modifications in the elastic scattering treatment are necessary: waves with $l > 0$ begin to be important and the differential cross section, according to (2.88) and (2.103), will be determined by the interference of different l values. If, for example, $l = 0$ and $l = 1$ waves are present in the scattering, then

$$\frac{d\sigma}{d\Omega} = \frac{1}{k^2} \left[\sin^2 \delta_0 + 6 \sin \delta_0 \sin \delta_1 \cos (\delta_0 - \delta_1) \cos \theta + 9 \sin^2 \delta_1 \cos^2 \theta \right]. \tag{2.156}$$

As a consequence, the interference between s and p scattered waves leads to breaking the scattering symmetry about the angle $\theta = 90°$ that would exist if each wave scattered independently.

Up to an energy close to 280 MeV, elastic scattering is the only process to occur in a nucleon-nucleon collision. The nucleons do not have low energy excited states and the weak interaction (chapter 8) is very slow to manifest itself in a scattering process. Figure 2.12 sketches proton-proton total cross section behavior as a function of energy. The smooth decrease in energy is interrupted at 280 MeV (135 MeV in the center of mass), value that sets the threshold of pion creation. With the beginning of the contribution of these inelastic processes, the total cross section separates from the elastic one.

2.14 Laboratory and Center of Mass Systems

With the aim of completing the elastic scattering study, we will see in this section how the change in the reference frame affects the quantities related to the scattering, specially angular distributions.

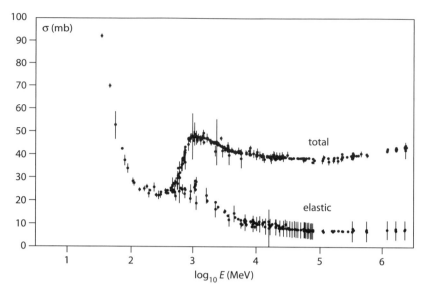

Figure 2.12 Total and elastic proton-proton cross sections as a function of the laboratory energy [Ba96].

The center of mass of a system of particles is defined by the vector

$$\mathbf{r}_c = \frac{\sum m_i \mathbf{r}_i}{\sum m_i}, \tag{2.157}$$

which moves with the velocity

$$\mathbf{v}_c = \frac{\sum m_i \mathbf{v}_i}{\sum m_i}. \tag{2.158}$$

In the special case of two particles with the second at rest, the center of mass velocity has the simple expression

$$V_c = \frac{m_1 V}{m_1 + m_2} = V \frac{m_R}{m_2}, \tag{2.159}$$

where V is the velocity of particle 1 and m_R the system reduced mass, defined by

$$m_R = \frac{m_1 m_2}{m_1 + m_2}. \tag{2.160}$$

Figure 2.13 shows this collision as seen by an observer located at the center of mass. One sees that the total linear momentum is always zero when computed at the center of mass. This property can be used to make it easier to evaluate the energy balance of a reaction. From the laboratory point of view, one should add the center of mass velocity to the velocities of figure 2.13. The result is seen in figure 2.14: θ and ϕ are the emerging angles of particles 1 and 2 (the latter being initially at rest), respectively. The triangle ABC

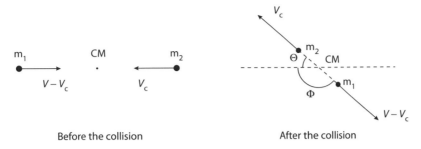

Before the collision   After the collision

Figure 2.13 Collision seen by an observer located at the center of mass.

is isosceles; hence $\Phi - \phi = \phi$ and thus

$$\phi = \frac{1}{2}\Phi = \frac{1}{2}(\pi - \Theta), \tag{2.161}$$

relations that are independent of m_1 and m_2. In the triangle CDE,

$$\cot\theta = \frac{CD}{DE} = \frac{V_c + (V - V_c)\cos\Theta}{(V - V_c)\sin\Theta} = \frac{V_c/(V - V_c) + \cos\Theta}{\sin\Theta} = \frac{m_1/m_2 + \cos\Theta}{\sin\Theta}; \tag{2.162}$$

thus

$$\cot\theta = \frac{m_1}{m_2}\operatorname{cosec}\Theta + \cotan\Theta. \tag{2.163}$$

Relations (2.161) and (2.163) define how the angles change in the passage from one system to the other.

The differential cross sections $\sigma(\theta)$ and $\sigma(\Theta)$ can also be related. For that, we have to keep in mind that if ω and Ω are the solid angles associated to θ and Θ, respectively, then $\sigma(\theta)d\omega = \sigma(\Theta)d\Omega$, since the detected particles are the same in both cases (see definition

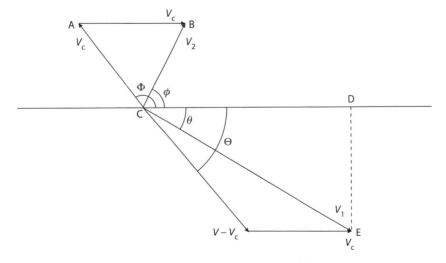

Figure 2.14 Transformation of center of mass velocities to the laboratory.

2.86). In this way

$$\frac{\sigma(\theta)}{\sigma(\Theta)} = \frac{d\Omega}{d\omega} = \frac{2\pi \sin \Theta \, d\Theta}{2\pi \sin \theta \, d\theta}. \tag{2.164}$$

Using now (2.163) to obtain $d\Theta/d\theta$, we get

$$\frac{\sigma(\theta)}{\sigma(\Theta)} = \frac{\sin^3 \Theta}{\sin^3 \theta} \left(\frac{1}{1 + m_1 \cos \Theta / m_2} \right), \tag{2.165}$$

which is the relation between the cross sections in the two systems that we were searching for.

2.15 Exercises

1. Use the listed values for the masses of the proton, neutron, and deuteron and deduce the value (2.25) for the binding energy of the deuteron.

2. Calculate the percent loss of mass due to the binding energy for the systems: a) earth-moon; b) hydrogen atom; c) deuteron. Verify that only in the last case is this effect important.

3. Using (2.36) and (2.39) as the wavefunction of the deuteron, calculate: a) the fraction of time that the neutron and the proton spend out of range of the force between them; b) the mean square radius of the deuteron.

4. A common generalization of the potential of figure 2.1 is the addition of a "hard core," that is, $V = +\infty$ for $r < c$, c being the radius of the core. Show that the presence of the core modifies the wavefunctions but does not alter the relationship between E_B, R, and V_0 given by (2.43).

5. Suppose that the interaction potential between the neutron and the proton is exponential, of the form $V = V_0 e^{-r/2r_n}$, where V_0 and r_n are, respectively, the depth and range of the nuclear potential. a) Write the Schrödinger equation (in the center of mass system) for the ground state of the deuteron, of angular momentum $l = 0$. b) Use the definition $x = e^{-2r/r_n}$ and $\psi(r) = u(r)/r$. Show that the Schrödinger equation has a Bessel function as solution. Write the general solution of this equation. c) Applying the boundary conditions (ψ finite for $r = 0$ and $r = \infty$), determine the relationship between V_0 and r_n.

6. For a system of two nucleons, show that $L + S + T$ should be odd, where L, S, and T are, respectively, the quantum numbers of orbital momentum, spin, and isospin of the system.

7. The deuteron has spin 1. What are the possible states of total spin and total angular momentum of two deuterons in a state with orbital angular momentum L?

8. Suppose that the meson π^- (spin 0 and negative parity) is captured from the orbit P in a pionic atom, giving rise to the reaction

$$\pi^- + d \longrightarrow 2n. \tag{2.166}$$

Show that the two neutrons should be in a singlet state.

9. Show that if $S_l(k)$ has the form (2.100), the wavefunction (2.99) describes an elastic scattering. Suggestion: show that the flux of the probability current vector through a sphere that involves the scattering center is zero.

10. Demonstrate the relation (2.103), using (2.82), (2.98), and (2.99).

11. Demonstrate the relation (2.104), using (2.103) and

$$\int d\Omega \, P_l(\cos\theta) \, P_{l'}(\cos\theta) = \frac{4\pi}{2l+1} \, \delta_{ll'} \, . \tag{2.167}$$

12. Find the cross section for low energy particles incident in a "hard sphere" potential

$$V(r) = \infty \quad (r < R), \tag{2.168}$$

$$V(r) = 0 \quad (r > R). \tag{2.169}$$

13. Low energy neutrons are scattered by protons. Let θ and ϕ be the emerging angles of neutrons and protons, respectively. a) Show that, for a given event, $\theta + \phi = 90°$. b) The scattering is isotropic in the center of mass and (2.165) shows that the neutron angular distribution in the laboratory system is given by $\sigma(\theta) = 4\cos\theta \, \sigma(\Theta)$. Show that for protons there is the relationship $\sigma(\phi) = 4\cos\phi \, \sigma(\Phi)$. c) Since $\sigma(\Theta)$ and $\sigma(\Phi)$ are constants, the functions $\sigma(\theta)$ and $\sigma(\phi)$ have maxima in $0°$. How does this result harmonize with the result of item (a)?

14. Say why it is not possible for a proton at rest to scatter another proton of low energy if both spins have the same direction.

15. A neutron of kinetic energy E_1 is elastically scattered by a nucleus of mass M, remaining with a final kinetic energy E_2. a) With Θ the scattering angle in the center of mass, show that

$$\frac{E_1}{E_2} = \frac{1}{2} \left[(1+\alpha) - (1-\alpha)\cos\Theta \right], \tag{2.170}$$

where $\alpha = [(M-1)/(M+1)]^2$. b) What is the maximum loss of kinetic energy as a function of E_1 and M? For which angle Θ does it occur? Which angle θ in the laboratory system corresponds to this value of Θ?

16. Assume that, in the scattering of a particle by a central potential, all phase shifts δ_l except for δ_0 and δ_1 are negligibly small.

(a) Find the differential and total (integrated over angles) cross sections.

(b) For $\delta_0 = 20°$ and $\delta_1 = 2°$ calculate the relative contribution of the p-wave to the total cross section and the ratio of cross sections in the forward, $\theta \to 0$, and in the backward, $\theta \to \pi$, directions.

17. Consider the *Born approximation* scattering amplitude (see [BD04])

$$f(\mathbf{k'}, \mathbf{k}) = -\frac{m}{2\pi\hbar^2} \int d^3r\, e^{-i(\mathbf{k'}\cdot\mathbf{r})} U(\mathbf{r}) e^{i(\mathbf{k}\cdot\mathbf{r})} \tag{2.171}$$

for the spherically symmetric potential $U(r)$.

(a) Using in (2.171) the expansion of the plane wave over spherical waves and the addition theorem for spherical harmonics (see Appendix A), present the scattering amplitude as an expansion over Legendre polynomials of the angle θ between $\mathbf{k}$ and $\mathbf{k'}$ and find the partial amplitudes f_l.

(b) For the short-range potential of radius R, $kR \ll 1$, and typical magnitude $\bar{U}$, estimate the energy dependence of the phase shifts δ_l.

18. Which of the following quantities are conserved? Energy E, components of the momentum $\mathbf{p}$, components of the orbital momentum $\mathbf{l}$, its square $\mathbf{l}^2$, parity $\mathcal{P}$, when a particle is moving

(a) with no external fields (free motion);

(b) in the static uniform field along the z direction;

(c) in the static central field $U(r)$;

(d) in the field $U = f(\rho)$ where ρ is the radius in the x, y-plane;

(e) in the uniform field along the x-direction with time-dependent amplitude?

19. The nucleus of the deuterium atom (heavy hydrogen isotope) is the deuteron, the only existing bound state of a neutron + proton system, with binding energy $E_B = 2.2$ MeV.

(a) Assuming that the orbital momentum of relative motion in the deuteron is $l = 0$, calculate the penetration length $1/\kappa$ of the deuteron wave function in the classically forbidden region outside the nuclear potential (see problem 3).

(b) Consider the neutron-proton potential as a square well of radius $R = 1.7$ fm (1 fm $= 10^{-13}$ cm). Calculate the critical depth for the appearance of a bound state in this well.

(c) For a bound state with a small but nonzero binding energy ϵ, the square potential has to be deeper compared to its critical depth U_0^{crit} by $\delta U_0 = U_0 - U_0^{crit}$. Considering the matching conditions for a well slightly deeper than the critical one, $\delta U_0 \ll U_0^{crit}$, derive the connection between δU_0 and ϵ and calculate the depth of the potential for the deuteron (the radius value is given in part b).

(d) Estimate the probability for the proton and the neutron in the deuteron to be outside the region $r < R$ of nuclear attraction.

3 | The Nucleon-Nucleon Interaction

3.1 Introduction

The starting point for any dynamical description of a physical system is knowledge of the relevant degrees of freedom and of the interaction. In the previous chapters we have seen that nucleons are the basic components of nuclei. Their degrees of freedom are determined by the position $\mathbf{r}_i$, momentum $\mathbf{p}_i$, spin $\mathbf{s}_i$ and isospin $\boldsymbol{\tau}_i$ of the ith nucleon. For the interaction one first takes the simplest assumption that it is a two-body interaction that can be described by a potential. A further extension of the model introduces three- and many-body interactions for a deeper understanding of the many-body system. For historical reasons we will first give with a phenomenological description and later the more fundamental meson-exchange theory.

In the following section we shall examine in what way the knowledge we already have about the deuteron can help us find a reasonable description of the nucleon-nucleon interaction. We should, at the beginning, consider a group of experimental facts that indicate that the nuclear force is independent of the charge of the nucleons. This means that the force between a neutron and a proton has the same form as the force between two neutrons and also between two protons, if we subtract the Coulomb part. It also means that there is a physical quantity involved for which there is a conservation law. Such a quantity is the isospin T, defined in chapter 1. In terms of the component T_z of that quantity we can express three possibilities to build a system of two nucleons: the *di-neutron* with $T_z = -1$, the *di-proton* with $T_z = +1$, and the deuteron with $T_z = 0$. T_z denotes the sum of the Z-component of the isospin of each nucleon. Since we only have two nucleons, T cannot be larger than 1. Therefore, for both systems, the di-neutron and di-proton, T has to be equal to 1. For the deuteron with $T_z = 0$, T can be 0 or 1.

The wavefunction of a system of two nucleons can be written as the product of a space function, a spin function, and one of isospin:

$$\Psi = \psi_{\text{spa}} \chi_s^{m_s} \phi_T^{T_z}. \tag{3.1}$$

Table 3.1 Isospin wavefunction for the two-nucleon system.

Isospin wavefunction	T	T_z	Symmetry by isospin exchange
$\phi_1^1 = \pi(1)\pi(2)$	1	1	Triplet (symmetric)
$\phi_1^0 = \frac{1}{\sqrt{2}}[\pi(1)\nu(2) + \pi(2)\nu(1)]$	1	0	
$\phi_1^{-1} = \nu(1)\nu(2)$	1	−1	
$\phi_0^0 = \frac{1}{\sqrt{2}}[\pi(1)\nu(2) - \pi(2)\nu(1)]$	0	0	Singlet (antisymmetric)

We denote π as a state of the proton and ν as a state of the neutron, so that $\pi(1)\,\nu(2)$ means that the first nucleon is a proton and the second is a neutron. We can build the isospin part $\phi_{T_z}^T$ of the wavefunction of the two-nucleon system in a similar way to the case of spin, as indicated in table 3.1.

3.2 Phenomenological Potentials

In the phenomenological method one uses the appropriate functional form for the potential with a sufficient amount of parameters. The parameters are chosen so that the potential describes as closely as possible the experimental data of the NN system. There are two classes of such potentials: *local* and *nonlocal* potentials.

3.3 Local Potentials

The following general ansatz is made for the potential as a function of the relevant degrees of freedom of both nucleons:

$$V(1,2) = V(\mathbf{r}_j, \mathbf{p}_j, \boldsymbol{\sigma}_j, \boldsymbol{\sigma}_j; j = 1, 2). \tag{3.2}$$

Symmetry and invariance properties of the Hamiltonian operator limit the general form of the interaction (see Appendix C). These properties are the requirement of invariance through translation, rotation, Galilean transformations, and particle exchange in connection with the Pauli principle, that is,

$$V(1,2) = V(2,1). \tag{3.3}$$

To account for these invariance properties one introduces the relative and center-of-mass coordinates and momenta, where the small mass difference between the neutron

Table 3.2 Tensors in two-nucleon space.

Type	Operator	Parity	Time reversal	Number
scalar	1	$+$	$+$	1
scalar	$\sigma_1 \cdot \sigma_2$	$+$	$+$	1
vector	$\sigma_1 \times \sigma_2$	$+$	$+$	3
vector	$\sigma_1 - \sigma_2$	$+$	$-$	3
vector	$\sigma_1 + \sigma_2$	$+$	$-$	3
tensor	$\left[\sigma_1^{[1]} \times \sigma_2^{[1]}\right]^{[2]}$	$+$	$+$	5
	Total number			16

and the proton is neglected:

$$\mathbf{r} = \mathbf{r}_1 - \mathbf{r}_2, \qquad \mathbf{R} = \frac{1}{2}(\mathbf{r}_1 + \mathbf{r}_2),$$

$$\mathbf{p} = \frac{1}{2}(\mathbf{p}_1 - \mathbf{p}_2), \qquad \mathbf{P} = \mathbf{p}_1 + \mathbf{p}_2. \tag{3.4}$$

The requirement of invariance under *translations* $\mathbf{r} \to \mathbf{r}_1 + \mathbf{a}$ leads to the condition

$$V(\mathbf{r}, \mathbf{p}, \mathbf{R}, \mathbf{P}) = V(\mathbf{r}, \mathbf{p}, \mathbf{R} + \mathbf{a}, \mathbf{P}),$$

and invariance under *Galilean transformation* $\mathbf{p}_j \to \mathbf{p}_j + \mathbf{p}_0$ implies

$$V(\mathbf{r}, \mathbf{p}, \mathbf{R}, \mathbf{P}) = V(\mathbf{r}, \mathbf{p}, \mathbf{R}, \mathbf{P} + 2\mathbf{p}_0).$$

Since $\mathbf{a}$ and $\mathbf{p}_0$ can take any values, these relations mean that V cannot depend on $\mathbf{R}$ and $\mathbf{P}$, so it can only possess the form

$$V(1, 2) = V(\mathbf{r}, \mathbf{p}, \sigma_j, \tau_j; j = 1, 2). \tag{3.5}$$

Next we study the *rotational invariance* property. This determines the structure of the spin degrees of freedom. Any function $f(\sigma_1, \sigma_2)$ represents a 4×4 matrix in the space of two-nucleon spin that can be spanned by a linear combination of 16 matrices. These can be classified by their tensor properties, as shown in table 3.2. The indices [1] and [2] refer the coupling scheme of two tensor operators $T_1^{[L_1]}$ and $T_2^{[L_2]}$ into a new operator

$$T_{[M]}^{[L]} = \left[T_1^{[L_1]} \times T_2^{[L_2]}\right]_{[M]}^{[L]} = \sum_{M_1 M_2} (L_1 M_1 L_2 M_2 | LM) T_{[M_1]}^{[L_1]} T_{[M_2]}^{[L_2]}.$$

The vector operator $\sigma_1 \times \sigma_2$ in the third row of table 3.2 does not carry the similar notation, $\left[\sigma_1^{[1]} \times \sigma_2^{[1]}\right]^{[1]}$, for the sake of simplicity. For more details on tensor operators, see Appendix B.

Table 3.3 Tensors build from r and p.

Type	Operator	Parity	Time reversal
scalar	r^2	+	+
scalar	p^2	+	+
scalar	$\mathbf{r} \cdot \mathbf{p}$	+	−
vector	$\mathbf{r}$	−	+
vector	$\mathbf{p}$	−	−
vector	$\mathbf{r} \times \mathbf{p}$	+	−
tensor	$\left[\mathbf{r}^{[1]} \times \mathbf{r}^{[1]}\right]^{[2]}$	+	+
tensor	$\left[\mathbf{p}^{[1]} \times \mathbf{p}^{[1]}\right]^{[2]}$	+	+
tensor	$\left[\mathbf{r}^{[1]} \times \mathbf{p}^{[1]}\right]^{[2]}$	+	−

When one constructs the potentials in terms of these linear combinations, one must be sure that the result is a scalar and that the symmetries under particle exchange, parity, and time reversal are observed. This means that one has to combine the vector and tensor symmetry operators with the corresponding vector and tensor operators obtained from $\mathbf{r}$ and $\mathbf{p}$. The possible operators obtained in this way are shown in table 3.3.

Due to consideration of symmetry and invariance properties, only the following vector-vector and tensor-tensor combinations are possible:

(a) Vector-vector: spin-orbit operator

$$\mathbf{L} \cdot \mathbf{S} = \frac{1}{2}(\mathbf{r} \times \mathbf{p}) \cdot (\boldsymbol{\sigma}_1 + \boldsymbol{\sigma}_2). \tag{3.6}$$

(b) Tensor-tensor:

$$\left[\mathbf{r}^{[1]} \times \mathbf{r}^{[1]}\right]^{[2]} \cdot \left[\boldsymbol{\sigma}_1^{[1]} \times \boldsymbol{\sigma}_2^{[1]}\right]^{[2]} = (\boldsymbol{\sigma}_1 \cdot \mathbf{r})(\boldsymbol{\sigma}_2 \cdot \mathbf{r}) - \frac{1}{3}\boldsymbol{\sigma}_1 \cdot \boldsymbol{\sigma}_2 \, r^2,$$

$$\left[\mathbf{p}^{[1]} \times \mathbf{p}^{[1]}\right]^{[2]} \cdot \left[\boldsymbol{\sigma}_1^{[1]} \times \boldsymbol{\sigma}_2^{[1]}\right]^{[2]} = (\boldsymbol{\sigma}_1 \cdot \mathbf{p})(\boldsymbol{\sigma}_2 \cdot \mathbf{p}) - \frac{1}{3}\boldsymbol{\sigma}_1 \cdot \boldsymbol{\sigma}_2 \, p^2,$$

$$\left[\boldsymbol{\sigma}_1^{[1]} \times \boldsymbol{\sigma}_2^{[1]}\right]^{[2]} \cdot \left[\mathbf{r}^{[1]} \times \mathbf{p}^{[1]}\right]^{[2]} \, \mathbf{r} \cdot \mathbf{p}. \tag{3.7}$$

Instead of the last tensor operator, one uses the equivalent square of the spin-orbit operator $(\mathbf{L} \cdot \mathbf{S})^2$.

From the above considerations we see that the nucleon-nucleon potential, which respects invariance under particle exchange, translation, Galilean transformation, rotation, parity, and time reversal, is given by

$$V(1,2) = V_C + V_S(\boldsymbol{\sigma}_1 \cdot \boldsymbol{\sigma}_2) + V_T S_{12}(\mathbf{r}) + V'_T S_{12}(\mathbf{p}) + V_{LS}\mathbf{L} \cdot \mathbf{S} + V_Q(\mathbf{L} \cdot \mathbf{S})^2, \tag{3.8}$$

where the operator S_{12} is given by

$$S_{12} = 3\left(\boldsymbol{\sigma}_1 \cdot \frac{\mathbf{r}}{r}\right)\left(\boldsymbol{\sigma}_2 \cdot \frac{\mathbf{r}}{r}\right) - (\boldsymbol{\sigma}_1 \cdot \boldsymbol{\sigma}_2). \tag{3.9}$$

In (3.8) the quantities V_α with $\alpha \in \{C, S, T, T', LS; Q\}$ are scalar functions of the remaining scalars r^2, p^2, and $(\mathbf{r} \cdot \mathbf{p})^2$. Due to the relation

$$(\mathbf{r} \cdot \mathbf{p})^2 = r^2 p^2 - L^2, \tag{3.10}$$

one chooses instead the variables r^2, p^2, and L^2 as the independent ones. One must also be sure that the total V is a Hermitian operator.

As a last point, we have to consider the isospin dependence of the interaction. The experimental data indicate that the NN interaction is approximately independent of the charge state of the nucleons, that is, of nn, pp, or np. In fact, the states ϕ_1^1 (di-proton) and ϕ_1^{-1} (di-neutron) discussed in the last chapter constitute, together with ϕ_1^0, a triplet in the isospin space. Now we want to know if some member of that triplet can be part of a bound state of the two particles. To show that this is not possible, let us examine the ground state of the deuteron. We saw that this state has $J = 1$, $S = 1$, and $l = 0$. The last value indicates that the space part is symmetrical and that $S = 1$ also corresponds to a symmetrical spin part. As Ψ in (3.1) should be antisymmetric, $\phi_T^{T_z}$ should also be antisymmetric for the ground state of the deuteron, and the isospin wavefunction of that state can only be ϕ_0^0. The function ϕ_1^0 is, therefore, the isospin wavefunction of an excited state of the deuteron. But we know experimentally that this state is not bound. As the nuclear force does not depend on the charge, the absence of a bound state for ϕ_1^0 should be extended to ϕ_1^1 and ϕ_1^{-1}. This last result exhibits that *the bound proton-proton or neutron-neutron system does not exist.*

But, how can states with $T = 1$ and $T = 0$ correspond to different energies if the nuclear forces are independent of the charge (isospin)? This is due to the dependence of the nuclear force on the spin. To each group of isospin states is associated a different orientation for the spins, so that to each group correspond different energies. The dependence of the nuclear force on the spin has a connection with the fact that there is no state for the deuteron other than the ground state (triplet spin). The force between the proton and the neutron when they have antiparallel spins (singlet) is smaller than when they have parallel spins (triplet), not strong enough to form a bound state. This force has a value just a little below that necessary to produce a bound state.

One can formally account for isospin independence by using the commutator property $[H, T_\pm] = 0$, where $\mathbf{T}$ is the total isospin operator $\mathbf{T} = \mathbf{t}_1 + \mathbf{t}_2$. Together with the charge conservation property $[H, T_z] = 0$, it follows that $[H, \mathbf{T}^2] = 0$, that is, an invariance under complete rotations in the isospin space. In other words, the interaction between a neutron and a proton cannot be different from that in any coherent superposition of both. Under these assumptions for isospin invariance, the functions V_α in (3.8) must be scalars in the isospin space in the form

$$V_\alpha = V_{\alpha 0} + V_{\alpha 1} \boldsymbol{\tau}_1 \cdot \boldsymbol{\tau}_2. \tag{3.11}$$

Sometimes it is convenient to describe the spin and isospin dependence of the NN interaction in terms of projection operators. We will show in the following that the terms of (3.8) can be derived in a more physically transparent way. For example, the spin part of the interaction can be written as

$$V_\sigma(r) \cdot \frac{1}{2}(1 + \boldsymbol{\sigma}_1 \cdot \boldsymbol{\sigma}_2) \equiv V_\sigma(r) P_\sigma, \tag{3.12}$$

where $V_\sigma(r)$ describes the radial dependence and the operator $P_\sigma = \frac{1}{2}(1 + \boldsymbol{\sigma}_1 \cdot \boldsymbol{\sigma}_2)$ has the expected values $+1$ for the triplet state and -1 for the singlet state. This can be shown starting from the vector $\mathbf{S} = \frac{\hbar}{2}(\boldsymbol{\sigma}_1 + \boldsymbol{\sigma}_2)$. Since $S^2 = \frac{\hbar^2}{4}(\sigma_1^2 + \sigma_2^2 + 2\boldsymbol{\sigma}_1 \cdot \boldsymbol{\sigma}_2)$, then

$$\boldsymbol{\sigma}_1 \cdot \boldsymbol{\sigma}_2 = \frac{1}{2}\left(-\sigma_1^2 - \sigma_2^2 + \frac{4S^2}{\hbar^2}\right); \tag{3.13}$$

the eigenvalues $\hbar^2 S(S+1)$ of S^2 are $+2\hbar^2$ for the triplet state ($S = 1$) and 0 for the singlet state. The eigenvalues of $\sigma^2 = \sigma_x^2 + \sigma_y^2 + \sigma_z^2$ are equal to 3, so that the eigenvalues of $\boldsymbol{\sigma}_1 \cdot \boldsymbol{\sigma}_2$ are equal to $+1$ for the triplet state and -3 for the singlet state, resulting in the expected values of P_σ predicted above.

P_σ is known as the *Bartlett potential* or *spin exchange potential* since, if we use for the spin functions similar to those given in table 3.1, we shall obtain that the operation of spin exchange is equivalent to multiplication by a factor $+1$ for the triplet state and a factor -1 for the singlet state.

If one assumes that the nuclear force depends on the parity of the wavefunction that describes the two particles, then a way of expressing that dependence is by means of a term of the form

$$V_r(r) P_r, \tag{3.14}$$

referred to as the *Majorana potential*, which contains the operator P_r that exchanges the space coordinates of the two particles. The eigenvalues of P_r are $+1$ and -1, if the wavefunction is even or odd, respectively.

The isospin dependence of the interaction can also be defined by the quantity

$$V_t(r) \cdot \frac{1}{2}(1 + \mathbf{t}_1 \cdot \mathbf{t}_2) \equiv V_t(r) P_t, \tag{3.15}$$

where the operator P_t changes the isospin of the two particles. The antisymmetry of the total wavefunction implies that P_t is not independent of P_σ and P_r, embodied in the relation

$$P_t = -P_\sigma P_r, \tag{3.16}$$

which can be verified easily by the application of both sides to (3.1). The operator $P_\sigma P_r$ is known as the *Heisenberg operator*.

Gathering the terms presented up to now, we can write the expression that represents the central part of the nucleon-nucleon potential:

$$V_C(r) = V_W(r) + V_r(r) P_r + V_\sigma(r) P_\sigma + V_t(r) P_t, \tag{3.17}$$

where the portion V_W, dependent only on r, is usually referred to as the *Wigner potential*.

The existence of tensor interactions becomes necessary to explain certain experimental results. One of them is the absence of a well-defined value for l, represented by the disturbance of a state $l = 2$ caused in the ground state $l = 0$ of the deuteron, (2.80). This forces us to think in terms of a noncentral nucleon-nucleon interaction potential, $V(\mathbf{r})$, since a central potential $V(r)$ conserves angular momentum and has l as a good quantum number. This explains the existence of the S_{12} term in (3.8).

It is common to describe the action of those noncentral forces by a function of the angles between the spin vectors of the neutron and of the proton and of the radial vector $\mathbf{r}$ that separates them. Such a potential is known as the *tensor potential*. The candidate functions to represent the tensor potential should, as a first requirement, be a scalar. Thus, with $\mathbf{u}_r$ the unitary vector in the direction $\mathbf{r}$, products of the type $\boldsymbol{\sigma}_1 \cdot \mathbf{u}_r, \boldsymbol{\sigma}_2 \cdot \mathbf{u}_r$, and $(\boldsymbol{\sigma}_1 \times \boldsymbol{\sigma}_2) \cdot \mathbf{u}_r$ must be rejected as pseudoscalars, that is, they change sign under reflection of the system of coordinates. Powers of those expressions are useless since we have, for example, $(\boldsymbol{\sigma} \cdot \mathbf{u}_r)^2 = 1$ and $(\boldsymbol{\sigma} \cdot \mathbf{u}_r)^3 = \boldsymbol{\sigma} \cdot \mathbf{u}_r$. In this situation, the simplest form of scalars we are looking for is $(\boldsymbol{\sigma}_1 \cdot \mathbf{u}_r)(\boldsymbol{\sigma}_2 \cdot \mathbf{u}_r)^1$. This expression is usually modified to satisfy the condition that the average value about all directions is zero. Since we know that the average of $(\mathbf{A} \cdot \mathbf{u}_r)(\mathbf{B} \cdot \mathbf{u}_r)$ is $\frac{1}{3}\mathbf{A} \cdot \mathbf{B}$, we define the potential tensor as in (3.9).

For the singlet state, $\boldsymbol{\sigma}_1 = -\boldsymbol{\sigma}_2$, where it follows that $(\boldsymbol{\sigma}_1 \cdot \boldsymbol{\sigma}_2) = -\sigma_1^2 = -3$, and $(\boldsymbol{\sigma}_1 \cdot \mathbf{u}_r)(\boldsymbol{\sigma}_2 \cdot \mathbf{u}_r) = -(\boldsymbol{\sigma}_1 \cdot \mathbf{u}_r)^2 = -1$. Thus, for the singlet state, $S_{12} = 0$, the tensor force is zero. This is an expected result since there is no preferential direction for the singlet state.

The terms described above are all characteristic of a *local potential*, an expression that denotes a potential that is perfectly defined at each point $\mathbf{r}$ of the space. Potentials dependent on momentum are, on the other hand, examples of potentials that do not only depend on one point and are called *nonlocal*. Among these, it is common to include in the nuclear potential a term of the form $V_{LS}(r)\mathbf{L} \cdot \mathbf{S}$. Thus, (3.8) is not completely local, as it is linear in $\mathbf{p}$, and is known as the *spin-orbit interaction*. This interaction can be observed, for example, in the scattering of polarized protons by a spinless target nucleus (figure 3.1). Depending on which direction the proton travels, the spin $\mathbf{S}$ and the angular momentum $\mathbf{L}$ can be parallel or antiparallel. Therefore, the term $V_{LS}(r)\mathbf{L} \cdot \mathbf{S}$ in the potential has a scalar product which is some times positive, other times negative. This leads to an asymmetry in the scattering cross section.

To establish the form of the unknown functions contained in (3.8), one adopts the approach that this potential describes correctly the experimental observations on nucleon-nucleon scattering, or the properties of certain nuclei, as for instance the deuteron. The values of these functions should be adjusted in such a way that they satisfy the approach above; we shall get a *phenomenological potential*. Phenomenological potentials are broadly employed, not only in the construction of nucleon-nucleon forces, but also in the interaction of complex nuclei, where the participation of the individual nucleons becomes extremely difficult to describe.

Phenomenological parameterizations for the nuclear potential possess attractive and repulsive components. At great distances they are reduced to the *one-pion exchange potential*

---

[1] The alternative $(\boldsymbol{\sigma}_1 \times \mathbf{u}_r)(\boldsymbol{\sigma}_2 \times \mathbf{u}_r)$ is not relevant, being a linear combination of $(\boldsymbol{\sigma}_1 \cdot \mathbf{u}_r)(\boldsymbol{\sigma}_2 \cdot \mathbf{u}_r)$ and $\boldsymbol{\sigma}_1 \cdot \boldsymbol{\sigma}_2$.

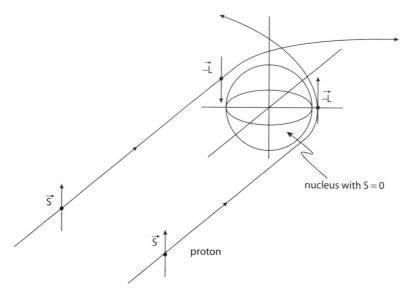

Figure 3.1 Scattering of polarized protons.

(OPEP), while at small distances they possess an extremely repulsive part. That repulsive part is usually referred to as the "hard core," with $V(r) \to \infty$ for $r < r_c \cong 0.4$ fm. Some authors use a repulsive potential that goes to infinity only for $r \to 0$. These potentials are known as "soft core" potentials. The most popular of these potentials is the *Reid soft-core* potential [Re68]. It has the form

$$V = V_c(\mu r) + V_{12}(\mu r)S_{12} + V_{LS}(\mu r)\mathbf{L} \cdot \mathbf{S}, \tag{3.18}$$

where

$$V_c(x) = \sum_{n=1}^{\infty} a_n \frac{e^{-nx}}{x}, \quad V_{LS}(x) = \sum_{n=1}^{\infty} c_n \frac{e^{-nx}}{x}, \tag{3.19}$$

and

$$V_{12}(x) = \frac{b_1}{x}\left[\left(\frac{1}{3} + \frac{1}{x} + \frac{1}{x^2}\right)e^{-x} - \left(\frac{b_0}{x} + \frac{1}{x^2}\right)e^{-b_0 x}\right] + \sum_{n=2}^{\infty} b_n \frac{e^{-nx}}{x}. \tag{3.20}$$

The constants are different for all values of T, S, and L. Only a_1, b_1, and c_1 are given in order to reproduce the OPEP potential at great distances. For $l > 2$, the Reid potential is replaced by the OPEP potential. The Reid potential is quite realistic and describes well, within its range of validity, the properties of a system of two nucleons.

3.3.1 Nonlocal potential

The most general form of a potential, including local and nonlocal characteristics, can be represented by an integral operation of the form

$$\langle \mathbf{r} | \widehat{V} | \psi \rangle = \int d^3 r' \langle \mathbf{r} | \widehat{V} | \mathbf{r}' \rangle \langle \mathbf{r}' | \psi \rangle = \int d^3 r' V(\mathbf{r}, \mathbf{r}') \psi(\mathbf{r}'). \tag{3.21}$$

This potential leads to an integro-differential Schrödinger equation. The special case of a local potential is represented by the diagonal form

$$V\left(\mathbf{r}, \mathbf{r}'\right) = V(\mathbf{r})\delta\left(\mathbf{r} - \mathbf{r}'\right). \tag{3.22}$$

In other words, a local potential is such that the relation

$$\langle \mathbf{r} | \widehat{V} | \psi \rangle = V(\mathbf{r})\psi(\mathbf{r}) \tag{3.23}$$

is valid. It means that the action of the interaction at the point $\mathbf{r}$ only depends on the value of $\psi(\mathbf{r})$ at that point. Thus, a momentum-dependent potential $V(\mathbf{r}, \mathbf{p})$ does not belong to the family of local potentials because the dependence on $\mathbf{p}$ implies a dependence of the potential on the neighborhood of $\mathbf{r}$. In fact, at the end of this section we will see that there is an equivalence between the dependence on momentum (or velocity) and nonlocal potentials.

For $V(\mathbf{r}, \mathbf{r}')$ one can again use a few symmetry invariance conditions to fix the form of the potential. We will not take this route here. Instead, we will describe a simpler class of potentials called *separable potentials*, which can be represented by

$$V\left(\mathbf{r}, \mathbf{r}'\right) = f^*(\mathbf{r})f\left(\mathbf{r}'\right). \tag{3.24}$$

One obtains for this class of potentials

$$\langle \mathbf{r} | \widehat{V} | \psi \rangle = f^*(\mathbf{r}) \int d^3 r' f\left(\mathbf{r}'\right) \psi\left(\mathbf{r}'\right), \tag{3.25}$$

which leads to a great simplification of the Schrödinger equation.

The main source of nonlocality of the NN interaction is the size of the nucleon: the fact that it has an internal structure and dynamics. The relativistic effects, e.g., retardation effects, also lead to nonlocality.

Let us calculate the matrix element of V between any states in the coordinate representation

$$\langle \phi | \widehat{V} | \psi \rangle = \int d^3 r' d^3 r'' \langle \phi | \mathbf{r}'' \rangle V\left(\mathbf{r}, \mathbf{r}'\right) \langle \mathbf{r}' | \psi \rangle. \tag{3.26}$$

With the help of the transformation of the integration variables

$$\mathbf{r} = \frac{1}{2}\left(\mathbf{r}'' + \mathbf{r}'\right), \quad \rho = \mathbf{r}'' - \mathbf{r}',$$

and the Taylor expansion, represented by the translation operator, $D(\mathbf{r})$, described by (A.23) of Appendix A,

$$\langle \mathbf{r}' | \psi \rangle = \left\langle \mathbf{r} - \frac{1}{2}\rho \middle| \psi \right\rangle = \left\langle \mathbf{r} \middle| D\left(\frac{1}{2}\rho\right) \middle| \psi \right\rangle, \tag{3.27}$$

one can reduce the matrix element to the form

$$\langle \phi | \widehat{V} | \psi \rangle = \int d^3 r \, \langle \phi | \mathbf{r} \rangle \, \widetilde{V}\left(\mathbf{r}, \mathbf{p}\right) \langle \mathbf{r} | \psi \rangle, \tag{3.28}$$

where

$$\tilde{V}\left(\mathbf{r}, \mathbf{p}\right) = \int d^3\rho \, e^{-\frac{i}{2}\rho \cdot \mathbf{p}} V\left(\mathbf{r} + \frac{1}{2}\rho, \mathbf{r} - \frac{1}{2}\rho\right) e^{-\frac{i}{2}\rho \cdot \mathbf{p}}$$

$$= \int d^3\rho \, V\left(\mathbf{r} + \frac{1}{2}\rho, \mathbf{r} - \frac{1}{2}\rho\right) + \frac{i}{2} \int d^3\rho \left\{\rho \cdot \mathbf{p}, \, V\left(\mathbf{r} + \frac{1}{2}\rho, \mathbf{r} - \frac{1}{2}\rho\right)\right\} + \cdots \qquad (3.29)$$

with $\mathbf{p} = -i\nabla$ and $\{ \}$ meaning the anticommutator. One alternative form is

$$\tilde{V}(\mathbf{r}, \mathbf{p}) = \frac{1}{2} \int d^3\rho \left[V(\mathbf{r}, \mathbf{r} - \rho) \, e^{-i\rho \cdot \mathbf{p}} + \text{h.c.}\right]. \qquad (3.30)$$

Now let us assume the opposite situation, in which a momentum-dependent potential $V(\mathbf{r}, \mathbf{p})$ is given. One can always bring it to the form

$$V(\mathbf{r}, \mathbf{p}) = v(\mathbf{r}, \mathbf{p}) + v^\dagger(\mathbf{r}, \mathbf{p}), \qquad (3.31)$$

where in $v(\mathbf{r}, \mathbf{p})$ all $\mathbf{p}$-terms are on the right of the $\mathbf{r}$ operators. For such matrix elements one gets

$$\langle \mathbf{r}' \,|v\,(\mathbf{r}, \mathbf{p})|\, \mathbf{r}'' \rangle = \int d^3 r'' \delta\left(\mathbf{r} - \mathbf{r}''\right) v\left(\mathbf{r}'', \mathbf{p}''\right) \delta\left(\mathbf{r}' - \mathbf{r}''\right)$$

$$= \frac{1}{(2\pi)^3} \int d^3 r'' \delta\left(\mathbf{r} - \mathbf{r}''\right) v\left(\mathbf{r}'', \mathbf{p}''\right) \int d^3 k \, e^{i\mathbf{k}\cdot(\mathbf{r}''-\mathbf{r}')}$$

$$= \frac{1}{(2\pi)^3} \int d^3 r'' \delta\left(\mathbf{r} - \mathbf{r}''\right) \int d^3 k \, v\left(\mathbf{r}'', \mathbf{k}\right) e^{i\mathbf{k}\cdot(\mathbf{r}''-\mathbf{r}')}$$

$$= \frac{1}{(2\pi)^3} \int d^3 k \, v\left(\mathbf{r}, \mathbf{k}\right) e^{i\mathbf{k}\cdot(\mathbf{r}-\mathbf{r}')}, \qquad (3.32)$$

and consequently nonlocal potential

$$\langle \mathbf{r}' \,|V|\, \mathbf{r}'' \rangle = \frac{1}{(2\pi)^3} \int d^3 k \left[v\left(\mathbf{r}, \mathbf{k}\right) e^{i\mathbf{k}\cdot(\mathbf{r}-\mathbf{r}')} + \text{h.c.}\right]. \qquad (3.33)$$

The construction of an equivalent $\mathbf{p}$-dependent potential V leads naturally back to the original potential, as one can easily verify by using (3.30).

3.4 Meson Exchange Potentials

3.4.1 Yukawa and Van der Waals potentials

Another way to attack the nuclear force problem is to be found in analyzing the meson exchange processes directly. The simplest exchange potential is due to the exchange of just one pion. But, only the long distance part of the potential can be explained in that way. Since the pion has spin zero, its wavefunction should be described by the Klein-Gordon equation

$$\left(\nabla^2 - \frac{m_\pi^2 c^2}{\hbar^2}\right)\Phi = \frac{1}{c^2}\frac{\partial^2 \Phi}{\partial t^2}. \tag{3.34}$$

Performing a separation of variables, we obtain a time-independent wave equation when the total energy of the pion is equal to 0 (binding energy equal to the rest mass),

$$(\nabla^2 - \mu^2)\phi = 0, \tag{3.35}$$

where

$$\mu = \frac{m_\pi c}{\hbar}. \tag{3.36}$$

An acceptable solution for (3.35) is

$$\phi = g\frac{e^{-\mu r}}{r}, \tag{3.37}$$

in which g is a constant that has the same role as the charge in the case of electrostatics, where the potential that results from the interaction between two equal charges is $qV = q^2/r$. That interaction is due to the continuous exchange of virtual photons between the charges.

We can assume that the potential between two nucleons is proportional to the wavefunction of the pion, that is, to the probability amplitude that the emitted pion finds itself close to the other nucleon. We thus find the *Yukawa potential*

$$V = g^2\frac{e^{-\mu r}}{r}, \tag{3.38}$$

where we have used the factor g^2, in analogy with electrostatics. The potential above decays exponentially, and its range can be estimated by

$$R \cong \frac{1}{\mu} = \frac{\hbar}{m_\pi c} \cong 0.7 \text{ fm}, \tag{3.39}$$

in agreement with the result obtained in section 1.2, where the uncertainty principle was used. The force field between two protons, or two neutrons, can only be produced by the exchange of neutral pions. Between a proton and a neutron the exchange can be done by means of charged pions.

It is a well-established experimental fact that the nuclear force is strongly repulsive at very short distances and the form of the central part of the nuclear potential should be given, schematically, as in figure 3.2. The potential well, that is, its attractive part of medium range, can be described by the exchange of two pions. It is interesting to observe that this part of the potential is created similarly to the Van der Waals force between two molecules (figure 3.4).

From QCD, the fundamental theory of strong interactions, we know that nucleons are colorless objects, that is, when they are looked upon from the outside, no net color charge is visible. The same is true for neutral nonpolar molecules that contain equal positive and negative electromagnetic charges distributed with no net shift, and hence no net charge or dipole moment. However, when two molecules approach one another, the charges become polarized and each molecule acquires a nonzero dipole moment. Then

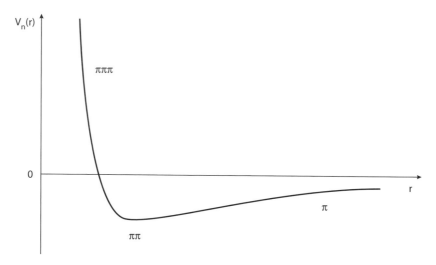

Figure 3.2 Sketch of the nucleon-nucleon potential. See section 3.4.3 for a fuller interpretation of the form of the potential.

the leading-order interaction energy between molecules equals $V(\mathbf{r}) = -2\mathbf{E}(\mathbf{r}) \cdot \mathbf{d}(\mathbf{r})$, where $\mathbf{E}(\mathbf{r})$ is the average electric field felt by one of the molecules when the second one is located at $\mathbf{r}$, and $\mathbf{d}(\mathbf{r})$ is its induced dipole moment. Assuming that the induced dipole moment $\mathbf{d}(\mathbf{r})$ depends linearly on the electric field, and knowing that the electric field created by a dipole decreases as $1/r^3$, we obtain immediately that $V(\mathbf{r}) \sim -1/r^6$, which gives the well-known *Van der Waals potential*. At intermediate and small distances, polarization effects become stronger, and higher induced multipole moments begin to be active; however, we can model these effects by a phenomenological term that is equal to the square of the Van der Waals term. One thus obtains the Lennard-Jones potential,

$$V_{LJ}(r) = 4E_{p,0}\left[\left(\frac{\sigma}{r}\right)^{12} - \left(\frac{\sigma}{r}\right)^{6}\right], \tag{3.40}$$

where $E_{p,0}$ and σ are parameters fitted to data.

Figure 3.3 shows a comparison of the NN Argonne v18 potential in the $^1S_0$ channel, with the Lennard-Jones potential between two O_2 molecules ($E_{p,O} = 10$ meV and $\sigma = 0.358$ nm). The *Argonne v18* [WSS95] is a phenomenological potential, very successful in describing NN scattering properties. In figure 3.3 both the v18 and the molecular potential are drawn in the same figure with two abscissas (the lower one for O_2-O_2, the upper one for the NN potential) and two ordinates (the left one for O_2-O_2, the right one for NN). Scales on the abscissas were fixed so as to put the minima of potentials at the same point, and differ by a factor of about 0.5×10^8, while scales on the ordinates differ by a factor of 10^{10}.

Despite the tremendous differences in scales, the two both potentials are qualitatively very similar. Amazingly, the electromagnetic molecule-molecule potential is stiffer at the minimum than the neutron-neutron "strong" potential. In this respect, it is fully justified to put the word "strong" in quotation marks—this potential is not strong at all! Both potentials exhibit a very strong repulsion at short distances—the so-called hard core (the

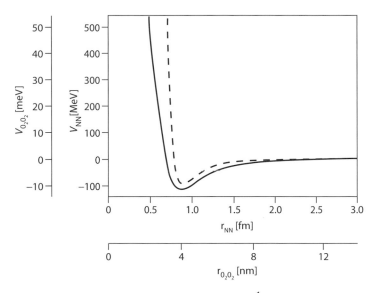

Figure 3.3 The NN potential in MeV (solid curve) in the $^1S_0$ channel as a function of the NN distance in fm compared to the O_2-O_2 molecular potential in meV (dashed curve) as a function of the distance of separation in nanometers (outer axes).

O_2-O_2 repulsion is stronger!). At large distances, there appears a weak attraction (the NN attraction vanishes more slowly despite the exponential form of the OPEP potential). Neither of the potentials is strong enough to bind the constituents into a composite object.

The analogy between the "strong" NN force and the electromagnetic molecule-molecule force is extremely instructive. First of all, we can demystify the OPEP potential in the sense that the exchange of real particles (pions) is, in fact, not its essential element. The OPEP potential is a remnant of a tool (quantum field theory) that one uses to derive it, but on a deeper level it is an effect of the color force between color-polarized composite particles. After all, nobody wants to interpret the dipole-dipole intermolecular O_2-O_2 force by an exchange of a "particle." This force can be understood in terms of a more fundamental interaction—the Coulomb force. Second, although the asymptotic large-distance leading-order behavior of both potentials can fairly easily be derived, at intermediate and small distances the interaction becomes very complicated. This is not a reflection of complications on the level of fundamental forces (color or electromagnetic), but a reflection of the complicated polarization effects that take place when composite objects are put close to one another. Moreover, these polarization effects have per se quantum character, because the fermionic constituents do not like being put close to one another—the Pauli exclusion principle creates additional polarization and repulsion effects. And third, it is obvious that at small distances effects must appear that are of a three-body character. Namely, when three O_2 molecules approach each other (e.g., in liquid oxygen), the basic assumption that they polarize one another only in pairs does not hold. There are certainly polarization effects that depend on explicit positions of all three of them. Similarly, when three nucleons

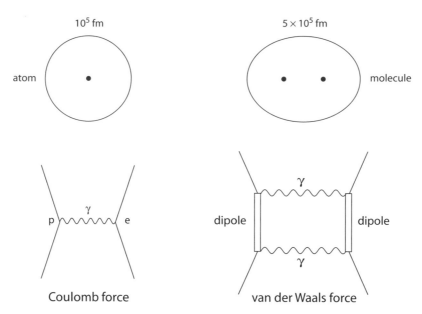

Figure 3.4 The Coulomb force between protons and electrons in an atom can be described in quantum electrodynamics by the exchange of one photon, and the Van der Waals force between atoms and molecules by the exchange of more than one photon.

approach each other within the nucleus, their quark-gluon magma becomes polarized in a fairly complicated way, which on the level of potential energy (total-interaction energy) reveals additional terms depending on the three positions simultaneously; this gives the *three-body NNN force*.

3.4.2 Field theory picture

If we stick to the quantum field theory picture in which forces are described by the exchange of particles, the molecular binding can be described by the exchange of two photons. The first photon, emitted by molecule 1, induces an electric dipole in molecule 2, and this dipole emits a virtual photon that induces another electric dipole in molecule 1. The interaction between the two dipoles gives rise to the Van der Waals force. The pions take the place of the photons in the case of the nuclear forces. The production of an electric dipole is similar to the excitation of a nucleon to a Δ-resonance. In this way, the nuclei are bound by a type of Van der Waals force (figure 3.4).

A more detailed comparison between the theory of pion (or other types of mesons) exchange and the electrostatic Coulomb field is necessary to account for the spin (and isospin) dependence of the interaction. Since the Coulomb field is represented by a static scalar field ϕ_γ, a massive static pion field can be represented by a static (pseudoscalar) isovector field $\boldsymbol{\phi}_\pi$. The relationships between the equations applicable to the static fields ϕ_γ and $\boldsymbol{\phi}_\pi$ are shown in table 3.4. Following this table, the static field generated by one

Table 3.4 Comparison between the equations for the static Coulomb and meson fields.

	Photon exchange	Meson exchange
Scalar field	ϕ_γ	ϕ_π
Static field eq.	$\Delta\phi_\gamma(\mathbf{x}) = -4\pi\rho_\gamma(\mathbf{x}, \mathbf{r}_1)$	$(\Delta - m_\pi^2)\,\boldsymbol{\phi}_\pi(\mathbf{x}) = -\rho_\pi(\mathbf{x}, \mathbf{r}_1)$
Point charge	$\rho_\gamma(\mathbf{x}, \mathbf{r}_1) = q_1\delta(\mathbf{x} - \mathbf{r}_1)$	$\rho_\pi(\mathbf{x}, \mathbf{r}_1) = \frac{f_\pi}{m_\pi}\boldsymbol{\tau}_1\boldsymbol{\sigma}_1 \cdot \nabla_{\mathbf{r}_1}\delta(\mathbf{x} - \mathbf{r}_1)$
Solution	$\phi_\gamma(\mathbf{x}, \mathbf{r}_1) = \frac{q_1}{\lvert\mathbf{r}_1 - \mathbf{r}_2\rvert}$	$\boldsymbol{\phi}_\pi(\mathbf{x}, \mathbf{r}_1) = \frac{f_\pi}{4\pi m_\pi}\boldsymbol{\tau}_1\boldsymbol{\sigma}_1 \cdot \nabla_{\mathbf{r}_1}\frac{\exp[-m_\pi\lvert\mathbf{r}_1 - \mathbf{r}_2\rvert]}{\lvert\mathbf{r}_1 - \mathbf{r}_2\rvert}$

particle leads to a potential energy due to the interaction with a second particle given by

$$V_\gamma(\mathbf{r}_1, \mathbf{r}_2) = \int d^3x\, \rho(\mathbf{x}, \mathbf{r}_2)\, \phi_\gamma(\mathbf{x}, \mathbf{r}_1) = \frac{q_1 q_2}{\lvert\mathbf{r}_1 - \mathbf{r}_2\rvert},$$

$$V_\pi(\mathbf{r}_1, \mathbf{r}_2) = \int d^3x\, \rho_\pi(\mathbf{x}, \mathbf{r}_2) \cdot \phi_\pi(\mathbf{x}, \mathbf{r}_1)$$

$$= -\frac{f_\pi^2}{4\pi m_\pi^2}\boldsymbol{\tau}_1 \cdot \boldsymbol{\tau}_2\boldsymbol{\sigma}_1 \cdot \nabla_1\boldsymbol{\sigma}_2 \cdot \nabla_2 \frac{\exp[-m_\pi\lvert\mathbf{r}_1 - \mathbf{r}_2\rvert]}{\lvert\mathbf{r}_1 - \mathbf{r}_2\rvert}. \tag{3.41}$$

One thus obtains in the case of the static pion field

$$V_\pi(\mathbf{r}) = \frac{f_\pi^2}{4\pi m_\pi^2}\boldsymbol{\tau}_1 \cdot \boldsymbol{\tau}_2\boldsymbol{\sigma}_1 \cdot \nabla_1\boldsymbol{\sigma}_2 \cdot \nabla_2 \frac{\exp[-m_\pi r]}{r}, \tag{3.42}$$

which is the so-called one-pion exchange potential (OPEP) with coupling constant $f_\pi^2/4\pi$. This coupling has an empirical value given by $f_\pi^2/4\pi \simeq 0.08$. The spin operator can be written in terms of a scalar and a tensor operator by means of

$$\boldsymbol{\sigma}_1 \cdot \mathbf{a}_1\boldsymbol{\sigma}_2 \cdot \mathbf{a}_2 = \frac{1}{3}\boldsymbol{\sigma}_1 \cdot \boldsymbol{\sigma}_2\, \mathbf{a}^2 + \sqrt{5}\left[\left[\boldsymbol{\sigma}_1^{[1]} \times \boldsymbol{\sigma}_2^{[1]}\right]^{[2]} \cdot \left[\mathbf{a}^{[1]} \times \mathbf{a}^{[1]}\right]^{[2]}\right]^{[0]}$$

$$= \frac{1}{3}\left(\boldsymbol{\sigma}_1 \cdot \boldsymbol{\sigma}_2\, \mathbf{a}^2 + S_{12}(\mathbf{a})\right),$$

where

$$S_{12}(\mathbf{a}) = 3\boldsymbol{\sigma}_1 \cdot \mathbf{a}\boldsymbol{\sigma}_2 \cdot \mathbf{a} - \boldsymbol{\sigma}_1 \cdot \boldsymbol{\sigma}_2\mathbf{a}^2$$

is the tensor operator (see equation 3.9). Replacing $\mathbf{a}$ by ∇ one obtains finally

$$V_{\mathrm{OPEP}} = \frac{f_\pi^2}{12\pi}\boldsymbol{\tau}_1 \cdot \boldsymbol{\tau}_2\left[\boldsymbol{\sigma}_1 \cdot \boldsymbol{\sigma}_2\left(\frac{\exp[-m_\pi r]}{r} - \frac{4\pi}{m_\pi^2}\delta(\mathbf{r})\right)\right.$$

$$\left. + S_{12}(\mathbf{r})\left(1 + \frac{3}{m_\pi r} + \frac{3}{m_\pi^2 r^2}\right)\frac{\exp[-m_\pi r]}{r}\right]. \tag{3.43}$$

The exchange of a pseudoscalar meson leads to a spin-dependent central potential as well as a tensor part. As we have seen in (3.38), the exchange of a scalar meson, that is, $\rho(\mathbf{x}, \mathbf{r}) = g\delta(\mathbf{x} - \mathbf{r})$, leads to a central potential with no spin dependence. A more detailed derivation of the equations above can be found in many textbooks, for example, [BD64].

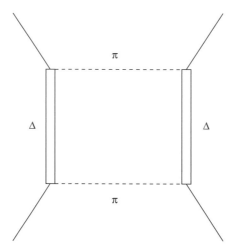

Figure 3.5 Nuclear force due to the exchange of two pions. The Δ-particle is a kind of polarized nucleon in the pion field generated by the other nucleon.

The OPEP potential describes well the nucleon-nucleon scattering for angular momenta $l \geq 6$. The high value of this limit shows that the OPEP potential describes the nuclear force in a reasonable way at great distances ($r \geq 2$ fm).

3.4.3 Short range part of the NN interaction

The short range part of the nucleon-nucleon potential represented in figure 3.2 is due to the exchange of three pions or more. The essential part of this process can be described by the effective exchange of a resonance of three pions, known as the ω-meson with spin 1 and m_ω= 783.8 MeV. The ω exchange is important for two properties of the nuclear force: the repulsive part of the potential and the spin-orbit interaction.

Both properties also have analogies with the electromagnetic case. In the case of electromagnetism, the exchange of a photon also gives rise to the repulsive force between charges of the same sign. In the case of the nuclear force, due to the large ω-meson mass, the repulsive force is of short range. Starting from this argument we can also conclude that the strongly repulsive potential becomes a strongly attractive potential for a nucleon-antinucleon pair at short distances.

At intermediate distances the nucleon-nucleon potential is, as we have already emphasized, adequately described in terms of the exchange of two pions. Another way to describe this part of the potential is with the exchange of a single particle, the ρ-meson, of mass (768.1 ± 0.5) MeV/c^2. It is believed that this meson is built of two pions, thus the equivalence with the exchange of two pions follows.

The potentials derived from the hypothesis of π, ρ, and ω exchange consist of combinations of central, tensorial, and spin-orbit parts and terms of higher order. The radial functions that accompany these terms have a total of up to 50 parameters, which are adjusted to the experimental data of the deuteron and to the nucleon-nucleon scattering.

3.4.4 Chiral symmetry

A meson is a complicated solution of the QCD quark and gluon fields that involves a real quark-antiquark pair. However, without ever being able to find this solution, we can identify basic features of the meson that result from the underlying QCD structure.

Let us concentrate on a small piece of the QCD Lagrangian density (1.59), that is, on the up-and-down quark components of the first term,

$$\mathcal{L}_\chi = -\bar{u}\gamma^\mu D_\mu u - \bar{d}\gamma^\mu D_\mu d = -\bar{q}\gamma^\mu D_\mu q.$$

The gluon fields and the color $SU(3)$ matrices are not essential now, so we have hidden all that in the $SU(3)$ covariant derivative: $D_\mu = \partial_\mu - igA^\alpha_\mu t_\alpha$. On the other hand, we have explicitly indicated the up-and-down quark fields, u and d, and moreover, we have combined the fields into the quark iso-spinor,

$$q = \begin{pmatrix} u \\ d \end{pmatrix}. \tag{3.44}$$

To be specific, q contains 24 components, i.e., two quarks, each in three colors, and each built as a four-component Dirac spinor. However, the Dirac and color structure is again not essential, so in the present section we may think about q as a two-component spinor. For a moment we have also disregarded the quark mass terms—we reinsert them to some degree below.

What is essential now are the symmetry properties of $\mathcal{L}_\chi$. This piece of the Lagrangian density looks like a scalar in the two-component field q, i.e., it is manifestly invariant with respect to unitary mixing of up-and-down quarks. We formalize this observation by introducing the isospin Pauli matrices, τ_1, τ_2, and τ_3, which are equal to the matrices defined in Appendix A (A.82), and we introduce unitary mixing of up-and-down quarks in the language of rotations in the abstract isospin space. This is exactly the same iso-space that we know very well from chapter 1, where the upper and lower components are the neutron and proton.

$\mathcal{L}_\chi$ is also invariant with respect to multiplying the quark fields by the γ_5 Dirac matrix (see Appendix D). This property results immediately from the commutation properties of the γ matrices (remember that $\bar{q} = q^\dagger \gamma_0$). So in fact we have altogether six symmetry generators of $\mathcal{L}_\chi$, namely,

$$\mathbf{t} = \tfrac{1}{2}\boldsymbol{\tau} \quad \text{and} \quad \mathbf{x} = \gamma_5 \mathbf{t}, \tag{3.45}$$

where the boldface denote vectors in the isospace.

It is now easy to identify the symmetry group of $\mathcal{L}_\chi$. We introduce the left-handed $\mathbf{t}_L$ and right-handed $\mathbf{t}_R$ generators,

$$\mathbf{t}_L = \tfrac{1}{2}(1 + \gamma_5)\mathbf{t} = \tfrac{1}{2}(\mathbf{t} + \mathbf{x}) \quad \text{and} \quad \mathbf{t}_R = \tfrac{1}{2}(1 - \gamma_5)\mathbf{t} = \tfrac{1}{2}(\mathbf{t} - \mathbf{x}). \tag{3.46}$$

Since $(\gamma_5)^2 = 1$, they fulfill the following commutation relations:

$$\left[t_{Li}, t_{Lj}\right] = i\epsilon_{ijk}t_{Lk}, \quad \left[t_{Ri}, t_{Rj}\right] = i\epsilon_{ijk}t_{Rk}, \quad \left[t_{Li}, t_{Rj}\right] = 0; \tag{3.47}$$

that is, $\mathbf{t}_L$ generates the SU(2) group, $\mathbf{t}_R$ generates another SU(2) group, and since they commute with one another, the complete symmetry group is SU(2)×SU(2). We call this group *chiral*.

This result is in disagreement with experiment. On the one hand, particles appear in iso-multiplets. For example, there are two nucleons, a neutron and a proton, that can be considered as upper and lower components of an isospinor, and there are three pions, π_+, π_0, and π_-, that can be grouped into an isovector. So there is no doubt that there is an isospin SU(2) symmetry in nature, but, what about the second SU(2) group? In the Lorentz group, the γ_5 Dirac matrix changes the parity of the field, so if γ_5 were really a symmetry then particles should appear in pairs of species having opposite parities. This is not so in our world. Nucleons have positive intrinsic parity, and their negative-parity brothers or sisters are nowhere to be seen. Parity of pions is negative, and again, the positive-parity mirror particles do not exist with any being nearly of the same mass.

So nature tells us that the SU(2)×SU(2) symmetry of the QCD Lagrangian must be *dynamically broken*. It means that the Lagrangian has this symmetry, while the physical solutions do not. We have already learned that these physical solutions are very complicated, and we are unable to find them and check their symmetries. But we do not really need that—experiment tells us that chiral symmetry must be broken, and hence we can build theories that incorporate this feature on a higher level of description.

Let us now reinsert the quark-mass terms into the discussed piece of the Lagrangian:

$$\mathcal{L}'_\chi = -\bar{u}\gamma^\mu D_\mu u - \bar{d}\gamma^\mu D_\mu d - m_u \bar{u}u - m_d \bar{d}d . \tag{3.48}$$

Neither of the two mass terms, nor any linear combination thereof, is invariant with respect to the chiral group SU(2)×SU(2). For certain, had the quark masses been equal, the two combined mass terms would have constituted an isoscalar (an invariant with respect to the isospin group), but even then they would not be chiral scalars (invariants with respect to the chiral group). So, the nonzero quark masses break the chiral symmetry. What the values of these masses are needs to be taken from experiment, and indeed, the up-and-down quark masses are neither zero nor equal to one another. The chiral symmetry is therefore broken in two ways: explicitly, by the presence of a symmetry breaking term in the Lagrangian, and dynamically, as discussed above. Without going into details, we just mention that the nonzero mass of the π-mesons results from the nonzero quark masses; see [Wei99], chap. 19. For more quark flavors, when taken into account, the dimensionality of the chiral group increases; for example, when three quarks u, d, and s are considered the chiral group is SU(3)×SU(3).

3.4.5 Generalized boson exchange

So far we have used a purely classical approach for the nucleon-nucleon potential. The quantum mechanical treatment delivers the same result, if one relies on the lowest order perturbation theory in the static case. The starting point is a system of coupled nucleon and meson fields

$$H = H_N^0 + H_m^0 + H_{mN}, \tag{3.49}$$

whereby the free fields are described by H_N^0 and H_m^0. The meson-nucleon coupling H_{mN} depends on the meson type. The most important are the following three types (ψ denotes the nucleon field):

(1) *Scalar meson:* $\phi^{(s)}$ with coupling constants g_s,

$$H_{mN}^{(s)} = g_s \overline{\psi} \psi \phi^{(s)}, \tag{3.50}$$

(2) *Pseudoscalar mesons:* $\phi^{(ps)}$. Here one distinguishes pseudoscalar (ps) and pseudovector (pv) couplings,

$$H_{mN}^{(ps)} = i g_{ps} \overline{\psi} \gamma_5 \psi \phi^{(ps)}, \tag{3.51}$$

$$H_{mN}^{(pv)} = i \frac{f_{ps}}{m_{ps}} \overline{\psi} \gamma_5 \gamma^\mu \psi \partial_\mu \phi^{(ps)} \tag{3.52}$$

with corresponding coupling constants g_{ps} and f_{ps}.

(3) *Vector mesons:* $\phi_\mu^{(V)}$. Here also one has two possibilities, vector coupling (γ^μ) with coupling constant g_V and tensor coupling with coupling constant f_V.

$$H_{mN}^{(V)} = g_V \overline{\psi} \gamma^\mu \psi \phi_\mu^{(V)} - \frac{f_V}{4M} \overline{\psi} \sigma^{\mu\nu} \psi \left(\partial_\mu \phi_\nu^{(V)} - \partial_\nu \phi_\mu^{(V)} \right) \tag{3.53}$$

with M = nucleon mass.

In these equations, the meson field is denoted by ϕ or ϕ_μ. Furthermore, the mesons carry isoscalar and isovector properties. In the latter case, the couplings $\overline{\psi} \cdots \psi \phi$ are to be replaced by $\overline{\psi} \cdots \tau \psi \cdot \phi$. For pseudoscalar mesons coupled to nucleons on the mass shell, the following relation holds;

$$\frac{f_{ps}}{m_{ps}} = \frac{g_{ps}}{2M}. \tag{3.54}$$

This relation is, however, no longer valid for off-shell situations. In particular, the antiparticle contributions to the ps-coupling are large, while they are strongly suppressed for the pv-coupling. At this point it is also important to recall that in processes for which the chiral symmetry is of importance, the pv-coupling is preferred, since in contrast to the ps-coupling the chiral symmetry holds for the pv-coupling.

In second order perturbation theory the *Feynman diagram* represented in figure 3.6 has an amplitude given by

$$\overline{u}_1(\mathbf{q}') \Gamma_1 \overline{u}_1(\mathbf{q}) G_m \left(q' - q \right) \overline{u}_2(-\mathbf{q}') \Gamma_2 \overline{u}_2(-\mathbf{q}), \tag{3.55}$$

where $u(\mathbf{p})$ is the Dirac spinor of a nucleons, Γ_k is the vertex function arising from H_{mN}, and G_m is the free meson propagator. The vertex function Γ_k and propagator G_m depend

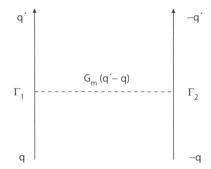

Figure 3.6 Feynman diagram for one-boson exchange.

on the meson type. For example, in the case of scalar and pseudoscalar mesons

$$G_m \left(q' - q \right) = \frac{1}{\left(q' - q \right)^2 - m^2} \,. \tag{3.56}$$

For vector mesons one must also take into account the spin structure. In the above relation $q \equiv (q_0, \mathbf{q})$ is a 4 vector.

The range of the potential described by a meson exchange is given by its Compton wavelength m^{-1}. An overview of the related mesons is given in table 3.5. One recognizes that the π-meson describes the long range of the interaction due to its low mass, while η, ρ, and ω are the responsible for the short range part of the interaction. Realistic potentials need, however, another meson of middle range ($\sim$500 MeV), to produce the necessary attraction at middle range (σ-meson). The existence of this meson is, however, disputed.

As an example of a realistic potential we discuss the *Bonn potential* given in coordinate space [Mac01]. The mesons included in this potential are shown together with their corresponding coupling constant in table 3.6. In each spin-isospin channel the potential

Table 3.5 Mesons related to the meson exchange problem with spin J, parity P, isospin I, G-parity G, and mass in MeV. G-parity is a combination of charge conjugation and a 180° rotation around the 2nd axis of isospin space (see, e.g., [Wei99]).

Meson	J^P	I^G	Mass
$\pi^\pm$	0^-	1^-	139.57
π^0	0^-	1^-	134.97
η	0^-	0^+	548.8
ρ	1^-	1^+	769
ω	1^-	0^-	769

Table 3.6 Mesons included in the Bonn potential, with their coupling constants and the cutoff parameters Λ ($n = 1$; see (3.58). Also used are $f_\rho/m_\rho = 6.1$ and $f_\omega/m_\omega = 0$.

Meson	Mass [MeV]	$g^2/4\pi$	Λ [GeV]
π	138.03	14.9	1.3
η	548.8	2	1.5
ρ	769	1.2	1.2
ω	782.6	25	1.4
δ	983	2.742	2.0
σ	550	8.77171	2.0

is written in the form

$$V = V_C + V_T S_{12} + V_{LS} \, \mathbf{L} \cdot \mathbf{S}. \tag{3.57}$$

Vertex functions were also introduced in these potentials to account for the finite size of the nucleons and mesons. One often chooses the simple analytical form

$$F(q^2) = \left(\frac{\Lambda^2 - m^2}{\Lambda^2 - q^2} \right)^{n/2} \quad (n = 1, 2, \dots), \tag{3.58}$$

with a suitable choice of cutoff parameters that are also shown in table 3.6. The high momentum components are suppressed; in particular, the δ-function in the scalar part of the potential is eliminated. Other functional forms are also used, for example, Gaussian functions. In coordinate space, the introduction of form-factor functions is viewed as a weakening of the potential at short distances. In principle, this is a purely heuristic procedure based on our ignorance of the interaction at short distances. It parametrizes this part of the interaction in the simplest way.

With regard to the role of the different mesons, the following qualitative features arise:

(a) The long range part ($r > 2$ fm) is described exclusively through the pion. Also, modern phenomenological potentials account for this fact.
(b) The middle range part ($1 < r < 2$ fm) is dominated by attraction. That is described through two-pion exchange (TPE).
(c) The short range part is dominated by ρ and ω exchange, responsible for a strong repulsion.
(d) Finally, the very short range is described purely phenomenologically, either through a sharp cutoff radius ("hard core"), or in a soft form ("soft core").

3.4.6 Beyond boson exchange

One of the main problems of the OBE potentials is justification of the σ-exchange that is needed for description of the attraction at middle range. Within the set of physical mesons,

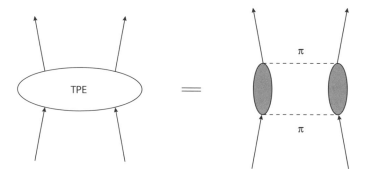

Figure 3.7 Two-pion exchange diagram.

there is no meson with the demanded characteristics. Investigations have shown that the underlying mechanism is a correlated two-pion exchange (TPE). Two methods are often used to attack this problem.

(1) Dispersion relation method. This is based on the assumption that the TPE contribution, shown diagrammatically in figure 3.7, can be cut in the middle and separated into two pion lines representing two disjunct pion-nucleon off-shell scatterings. The off-shell pion-nucleon scattering is separated into a sum of pure pion-nucleon scattering without $\pi\pi$ scattering and an additional contribution with $\pi\pi$ scattering. This is shown schematically in figure 3.8. The two-pion exchange amplitude is then written as a dispersion relation integral

$$T_{2\pi}(t) = \frac{1}{\pi} \int_{4m_\pi^2}^{\infty} dt' \, \frac{\rho_{2\pi}(t')}{t'-t}, \tag{3.59}$$

where $t = (p - p')^2$ is a *Mandelstan variable*. In this integral $\rho_{2\pi}(t)$ is called the *spectral function*. It characterizes both the strength and the range of the interaction and it is obtained by the "crossing" symmetry from the process $N\overline{N} \to 2\pi$.

One can show that with the dispersion relation (3.59) one can obtain $\rho_{2\pi}$ from $\pi\pi$ and πN scattering amplitudes [ChR79]. The contribution to the NN potential is then given by

$$\frac{1}{r} \int_{4m_\pi^2}^{\infty} dt' \, \rho_{2\pi}(t') \, e^{-\sqrt{t}r}. \tag{3.60}$$

The *Paris potential* is constructed with this method. The explicit form is further parametrized in terms of a sum of Gaussians [Lac80].

(2) Field theory approach. Here, as in the OBEP, one uses an explicit field theory model to obtain the TPE amplitude. The lowest order diagrams are represented in figure 3.9. Relevant for a realistic description are the couplings to resonances, most importantly the $\Delta(1232)$ resonance, together with pure $\pi\pi$ exchange. The nucleon resonances are herewith treated as elementary particles. One gets in this manner a detailed microscopic description of the $\pi\pi$ exchange. In contrast to the simple OBE potentials, the correlated $\pi\pi$ exchange is described here in terms of realistic ρ and σ with wide mass distribution.

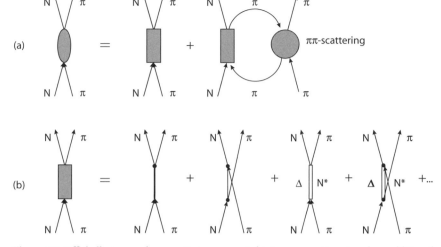

(a)

$\pi\pi$-scattering

(b)

Figure 3.8 Off-shell pion-nucleon scattering separated into a pure πN part and an additional part with $\pi\pi$ scattering [ChR79].

With regard to the meaning of 2π exchange one can raise the question how far in higher order diagrams one has to go and in which manner one can undertake a systematic development. That is represented schematically in figure 3.10, where the arrows indicate which diagrams must be considered simultaneously in order to reach convergence.

Parallel to these refinements that follow in the frame of conventional nuclear physics, the question arises in the development of QCD as a fundamental theory of the strong interaction as to how the NN interaction can be described in this framework. The basic difficulty to answering this question, given that the structure of the hadrons is already complicated by the nuclear many-body environment, lies in the so-called nonperturbative area of QCD, dominated by the phenomenon of *confinement* and described by nonlinear equations of motion.

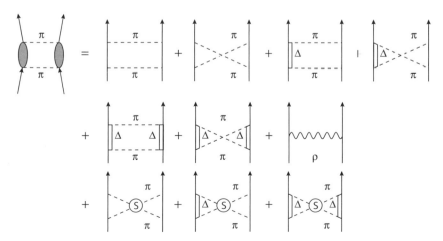

Figure 3.9 Description of two-pion exchange in the field theory approach.

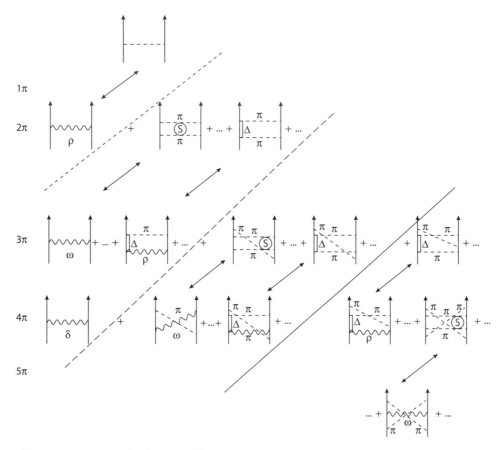

Figure 3.10 Systematic development of the field theory approach [Mac89].

For this reason hybrid models have been developed, in which the internal hadron structure is described with effective quark models. The interaction is described in these models in the long range part through conventional meson exchange. However, the mesons are now coupled directly to the quarks, so the extended structure of the meson-nucleon vertices

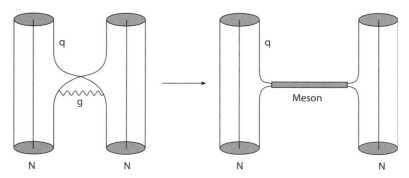

Figure 3.11 Quark-gluon exchange between two nucleons at small distances (left) and effective meson exchange at larger distances (right).

is obtained automatically by the quark wavefunctions. Only for very short distances does the interaction become dominated by explicit quark-gluon exchange (see figure 3.11).

3.5 Effective Field Theories

As we have discussed earlier, the Argonne v18 interaction uses the OPEP potential at large distances and the phenomenological interaction at intermediate and small distances. One can also follow the standard approach of quantum field theory and model the second piece by the exchange effects for heavier mesons. Larger meson masses mean shorter distances of the interaction, so we can understand why, by adding more mesons and using the corresponding Yukawa interactions, we can parametrize the NN force equally well.

Although this way of proceeding works very well in practice, it creates two conceptual problems. First, one has to include the scalar-isoscalar meson called σ, which has the quantum numbers of a pair of pions. It fulfills the role of an exchange of a pair of pions. However, such a meson does not exist in nature as a free particle, nor is its mass (which has to be used in the corresponding Yukawa term) close to the doubled pion mass. The exchange of such a virtual particle simply corresponds to higher order effects in the exchange of pions, which is a perfectly legitimate procedure, but it departs from the idea that real, physical particles mediate the NN interaction. Second, two other heavy mesons have to be included, the vector isovector meson ρ and the vector isoscalar meson ω. They are physical particles, with rest masses of about 800 MeV, and the corresponding ranges of the Yukawa potentials are very small, of the order of 0.25 fm. These small ranges allow modeling the NN interaction at very short distances, but at these distances nucleons really start to touch and overlap. Therefore, it is rather unphysical to think that nucleons can still interact as unchanged objects by exchanging physical particles. Within the image of the strong color polarization taking place at such a small distances, one would rather think that the internal quark-gluon structure of nucleons becomes strongly affected, which creates strong repulsion effects, predominantly through the Pauli blocking of overlapping quark states.

At present, we are probably not at all able to tell what happens with the nucleons when they are so near to one another. However, we do not really need such complete knowledge when describing low energy NN scattering and structure of nuclei. All we need is some kind of parametrization of the short range, high energy effects when we look at their influence on the long range, low energy observables. Such separation of scales is at the heart of *effective field theory* (EFT).

One can apply similar ideas to almost all physical systems, where our knowledge of the detailed structure is neither possible nor useful. The simplest example is the effect of the electromagnetic charge and current distributions inside a small object, when we shine at it an electromagnetic wave of a much longer length (the long wave length limit). It is well known that all we need are a few numbers—low multiplicity electric and magnetic moments. Of course, the best would be to be able to calculate these moments from the exact charge and current distributions, but once we know these numbers, we know everything.

On the other hand, if the internal structure is not known, we can fit these numbers to the measured long wave scattering and thus obtain the complete information needed to describe such a scattering process.

Examples of other such situations are plenty in physics. Interested students are invited to go through the very good introductory lecture notes by Lepage [Lep97], where nice instructive examples are presented in the framework of ordinary quantum mechanics. In particular, it is shown how a short range perturbation of the ordinary Coulomb potential influences the hydrogen atomic wavefunctions, and how such a perturbation (no matter what its physical origin) can be parametrized by a zero range, delta-like potential.

3.6 Exercises

1. Suppose that the interaction potential between the neutron and the proton is exponential, of the form $V = V_0 e^{-r/2r_n}$, where V_0 and r_n are, respectively, the depth and the range of the nuclear potential. a) Write the Schrödinger equation (in the center of mass system) for the ground state of the deuteron of angular momentum $l = 0$. b) Use the definition $x = e^{-2r/r_n}$ and $\psi(r) = u(r)/r$. Show that the Schrödinger equation has a Bessel function as a solution. Write the general solution of this equation. c) Applying the boundary conditions (ψ finite for $r = 0$ and $r = \infty$), determine the relationship between V_0 and r_n.

2. For a system of two nucleons show that $L + S + T$ should be odd, where L, S, and T are, respectively, the quantum numbers of orbital momentum, spin, and isospin of the system.

3. The deuteron has spin 1. What are the possible states of total spin and of total angular momentum of two deuterons in a state with orbital angular momentum L?

4. A particle with spin 1 moves in a central potential of the form

$$V(r) = V_1(r) + \mathbf{S} \cdot \mathbf{L} V_2(r) + (\mathbf{S} \cdot \mathbf{L})^2 V_3(r).$$

What are the values of $V(r)$ in the states with $J = L + 1$, L, and $L - 1$?

5. Suppose that the meson π^- (spin 0 and negative parity) is captured from the orbit P in a pionic atom, giving rise to the reaction

$$\pi^- + d \rightarrow 2n.$$

Show that the two neutrons should be in a singlet state.

6. Consider the operator S_{12} defined in (3.9). Show that, for the spin singlet and triplet state of the two particles, the following relations are valid:

$$S_{12}\chi_{\text{siinglet}} = 0, \quad (S_{12} - 2)(S_{12} + 4)\chi_{\text{triplet}} = 0.$$

7. Let s_1 and s_2 be the spin operators of two particles and $\mathbf{r}$ the radius vector that connects them. Show that any positive integer power of the operators

$$s_1 \cdot s_2 \quad \text{and} \quad \frac{3(s_1 \cdot \mathbf{r})(s_2 \cdot \mathbf{r})}{r^2} - (s_1 \cdot s_2)$$

can be written as a linear combination of these operators and the unit matrix.

8. Prove the relations

$$s_1 \times s_2 = \frac{2i}{\hbar}(s_1 \cdot s_2)s_1 - \frac{i\hbar}{2}s_2,$$

$$(s_1 \times \mathbf{r}) \cdot (s_2 \times \mathbf{r}) = R^2(s_1 \cdot s_2) - (s_1 \cdot \mathbf{r})(s_2 \cdot \mathbf{r}).$$

9. Show that the tensorial force S_{12} has a zero angular average; that is, show that $\int S_{12} d\Omega = 0$.

10. Find the functional form in coordinate space of a potential expressed in momentum space by equation (3.58).

11. Consider, in addition to nuclear np forces, the interaction of the neutron magnetic moment with the Coulomb field of the proton.

(a) Show that, for nonrelativistic relative np motion, this interaction leads to a new Hamiltonian term with the structure

$$H' = V(r)(\mathbf{l} \cdot s_n), \tag{3.61}$$

where $\mathbf{l}$ and s_n are the operators of the relative orbital momentum and of the neutron spin, respectively. Find the coordinate dependence of $V(r)$.

(b) Find the constants of motion in the np system in the presence of the additional interaction (3.61).

(c) Construct the matrix of interaction (3.61) in the basis of the unperturbed n-p wavefunctions with given values of l, total spin S, and total angular momentum J. Estimate the shift of the deuteron binding energy due to this interaction.

(d) Write down the Schrödinger equations for the radial wave functions outside the range of nuclear forces. Are there any new effects expected in the np scattering?

4 General Properties of Nuclei

4.1 Introduction

The basic properties of nucleons were presented in chapters 1, 2, and 3, together with the development of the deuteron theory. Our purpose in this and the following chapters is to study the physics of nuclei with any number A of nucleons, to establish the systematics of their properties, and to present the theories that aim to explain them. However, the approach we have followed for the deuteron is not applicable here. The Schrödinger equation is already not exactly soluble for a three-nucleon system, and to establish the properties of a heavy nucleus starting from the interaction of all its constituents is not a feasible task. The reasonable approach is the use of idealized models that incorporate part of the physics involved and explain a limited set of experimental data. This chapter presents the general characteristics of nuclei and introduces some basic ideas that will be employed in the elaboration of nuclear models. The detailed presentation of these models will be done in chapter 5.

4.2 Nuclear Radii

The radius of protons and neutrons that compose the nucleus is of the order of 1 fm. Suppose that a nucleus has A nucleons distributed inside a sphere of radius R. If the nucleons could be considered as small hard spheres of radius r in contact with each other, we could write

$$A \cong \frac{\frac{4}{3}\pi R^3}{\frac{4}{3}\pi r^3}$$

or, in another way,

$$R \cong r_0 A^{1/3}, \tag{4.1}$$

where we put r_0 in place of r to take into account that, even in this model of "packed"

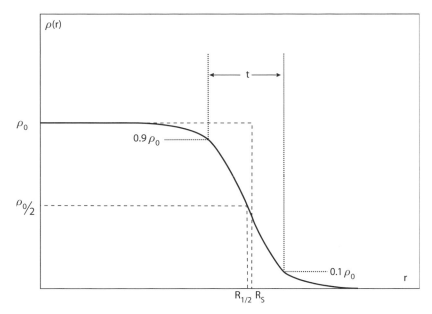

Figure 4.1 Nucleon density as a function of the distance to the center of the nucleus, obeying a typical Fermi distribution.

spheres, there are empty spaces among them, and the nuclear volume should be greater than the simple sum of volumes of each sphere. We expect, therefore, that r_0 is somewhat greater than 1 fm.

We can also infer the radius of a nucleus experimentally. The experiments that give the most precise results are the ones that use electron scattering. The electrons are accelerated and thrown against a target, interacting electromagnetically with the protons and bringing, on their way out, information on how these protons are distributed inside the nuclei. In other words, measurement of electron scattering allows us to deduce the charge distribution in nuclei. If we suppose that the neutron and proton densities have the same distribution shape, then the charge distribution in nuclei will be identical to the mass distribution.

The method used to measure the charge distribution in nuclei was developed mainly by R. Hofstadter and collaborators [Ha56] using the Stanford University linear accelerator. The results from several experiments show that the charge (or mass) density, that is, the number of charges (or mass) per unit volume, can be well described by

$$\rho(r) = \frac{\rho_0}{1 + e^{(r-R_{1/2})/a}}, \tag{4.2}$$

where ρ_0, $R_{1/2}$, and a are fitting parameters. The functional form of r above is known as the *Fermi distribution*. It falls to half its center value at $r = R_{1/2}$ (figure 4.1). Expression (4.2) tell us that the nucleon distribution inside the nucleus is not like a homogeneously occupied sphere with a well-defined radius.

The nuclei have a diffuse surface, with the density decreasing rapidly for $r \gtrsim R_{1/2}$. The quantity a gives the surface diffuseness width. The interval where the density decreases from 90% to 10% from the center value has width $t = 4.4a$. Figure 4.2 shows the charge

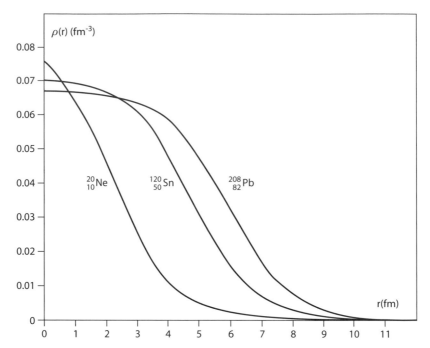

Figure 4.2 Charge distribution in three nuclei, representing light, medium, and heavy elements [Ho57].

distribution in several nuclei, obtained from the analysis of experimental results of Hofstadter and colleagues in 1957 [Ho57]. It gives a good idea of the behavior of this quantity for a large range of masses. An examination of these results shows that the charge distribution of nuclei with $A > 20$ is well described by (4.2), with

$$\rho_0 = 0.17 \frac{Z}{A} \text{ fm}^{-3}, \qquad a = 0.54 \text{ fm},$$

$$R_S = 1.128 \, A^{1/3} \text{ fm}, \qquad R_{1/2} = R_S - 0.89 A^{-1/3} \text{ fm}. \tag{4.3}$$

R_S is the radius of a homogeneously charged sphere, with constant charge density ρ_0 and total charge Ze. If we wish to use (4.2) to describe nucleon instead of charge density for $A > 20$, we can use the same values as in (4.3) but with $\rho_0 = 0.17$ nucleons/fm^3. For $A < 20$ the Fermi distribution, (4.2), is not adequate to describe the charge (or nucleon) distribution, since for nuclei with few nucleons the idea of defining a surface is less clear.

Besides electron scattering, other methods are used to determine the nuclear radius experimentally. One of them is to study the *muonic atom*. The muon is the nearest relative of the electron and has mass equal to $207m_e$. Muons can be captured by nuclei and form atoms, where they play the role of the electrons. The atomic levels in a muonic atom are analogous to those of a normal atom; the difference is in the energy of the levels and in the radii of the muonic orbits.

The energy of an atomic level is given (without relativistic corrections) by

$$E = -\frac{\mu Z^2 e^4}{2\hbar^2 n^2}, \tag{4.4}$$

where n is the principal quantum number and

$$\mu = \frac{m_e M}{m_e + M} \tag{4.5}$$

the reduced mass of the atom (M is the nuclear mass). In a muonic atom the value of μ is about 200 times greater and the levels will be more strongly bound. If this were the only difference, the transition energy between two equal levels for normal and muonic atoms (ΔE and $\Delta E'$, respectively) would be given by the ratio $\Delta E/\Delta E' \cong 1/200$. However, since the orbit radius (determined from the Bohr atom model) is given by

$$r = \frac{n^2 \hbar^2}{\mu Z e^2}, \tag{4.6}$$

the radius of a muonic atom is 200 times smaller than that of a normal atom. If the muon is in the lowest level (K shell) there will be a reasonable probability of finding it inside the nucleus. The atomic levels of a muonic atom will be modified due to the interaction of the part of the wavefunction of the muon that lies inside the nucleus. Thus, instead of $\Delta E'$, the transition energy to the lowest level will be $\Delta E' + \Delta E_{vol}$, with ΔE_{vol} being given by

$$\Delta E_{vol} = e \int_0^R \left[\psi_2^2(r) - \psi_1^2(r) \right] \left[U_V(r) - U(r) \right] 4\pi r^2 \, dr, \tag{4.7}$$

where $\psi_n^2(r)dV$, with $dV = 4\pi r^2 dr$, is the probability density of finding the n orbital muon in the volume dV inside the nucleus. $U(r) = Ze/r$ is the Coulomb potential that the muon would feel if the nucleus were pointlike, and $U_V(r)$ is the realistic potential at point r for a finite size nucleus. For a uniform charge distribution one can show that (R is the nuclear radius)

$$U_V(r) = \frac{Ze}{R} \left[\frac{3}{2} - \frac{1}{2} \left(\frac{r}{R} \right)^2 \right]. \tag{4.8}$$

The above relations point out that from a measurement of the transition energy between two levels of a muonic atom we can infer the radius R of the corresponding nucleus [En74].

The nuclear radii can also be obtained from a study of nuclear reactions or collisions induced by α-particles and other nuclei. The Rutherford experiments with α-particles around 1911 [Ru11] obtained the value (4.1) with $r_0 \cong 1.2$ fm.

4.3 Binding Energies

For every bound system, the mass of the system is smaller than the sum of the masses of its constituents, if measured separately. This property was presented earlier for the case of the deuteron and is an important attribute of the nucleus for each A value. In this respect nuclear physics is unique, since in other fields of physics the loss of mass corresponding to binding is negligible compared to the mass of the system itself.

The binding energy of a nucleus, which is conceptually the energy needed to separate all the nucleons in the nucleus, is easily calculated if we remember that it should be equal

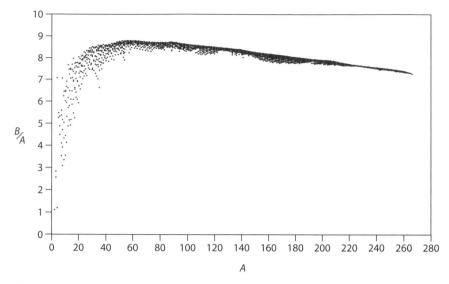

Figure 4.3 Binding energy per nucleon, B/A, as a function of the mass number A.

to the mass loss when the nucleus is formed. For a nucleus $^A_Z X$, with proton number Z and neutron number $N = A - Z$, it is given by

$$B(Z, N) = \{Zm_p + Nm_n - m(Z, N)\}c^2, \tag{4.9}$$

where m_p is the proton mass, m_n the neutron mass, and $m(Z, N)$ the mass of the nucleus. These masses can be measured by means of the *mass spectrograph*, an apparatus based on the trajectory that a charged particle describes under the action of an electric and a magnetic field. Since the neutron does not have charge, its mass has to be measured by other processes. We can, for example, measure the deuteron and proton masses and, knowing the deuteron binding energy by means of its dissociation by a photon, deduce the rest mass of the neutron.

The binding energy defined by (4.9) is always positive. In figure 4.3 we show the binding energy per nucleon, B/A, as a function of A, for all known nuclei. The average value of B/A increases quickly with A for light nuclei and decreases slowly from 8.5 MeV to 7.5 MeV beginning with $A \cong 60$, where it has a maximum. We can say that for with nuclei $A > 30$ the binding energy B is approximately proportional to A.

In the light nuclei region, four points are observed whose binding energy per nucleon is greater than the local average: $^4_2 He$, $^8_4 Be$, $^{12}_6 C$, and $^{16}_8 O$. The nuclei $^{20}_{10} Ne$ and $^{24}_{12} Mg$ also lie in the upper part of the graph. Notice that these nuclei have equal and even proton and neutron numbers.

The initial rise of the B/A curve indicates that the fusion of two light nuclei produces a nucleus with greater binding energy per nucleon, releasing energy. This is the origin of the energy production in the stars. The initial stage in the evolution of a star is the production of helium by means of hydrogen fusion; in later stages the production of heavier elements occurs by fusion of lighter nuclei. It is not difficult to conclude from figure 4.3 that if a star follows the normal course of its evolution without the occurrence of major incidents (gravitational, etc.) it will end as a cold cluster of nuclei with $A \cong 60$, since from that time on nuclear fusion is no longer energetically advantageous.

On the other side of the maximum binding energy per nucleon, for heavy nuclei the division into approximately equal parts (*nuclear fission*) releases energy. Figure 4.3 shows that in this case the energy gain is nearly 1 MeV per nucleon and thus about 200 MeV is gained in each event. The nuclear fission process is the basis of nuclear reactor operation, where neutrons strike heavy elements (normally uranium or plutonium), leading them to fission and to produce more neutrons, forming chain reactions. It is also the basis of war artifacts. The explosive devices of nuclear origin that receive the somewhat inappropriate denomination atomic bombs have this same character due to the fast but now uncontrolled chain reactions that release enormous amounts of energy in a small volume.

The fact that the binding energy per nucleon is approximately constant for $A > 30$ is due to the *saturation of the nuclear forces*. Each nucleon is bound to $A - 1$ other nucleons, in such a way that there are in total $A(A - 1)/2$ nucleon-nucleon bindings in a nucleus with mass number A. Thus, if the range of nucleon-nucleon forces were greater than the nuclear dimension, the binding energy B should be proportional to the number of bindings between them, that is, B should be proportional to A^2. Since this is not the case, one concludes that the nucleon-nucleon forces have a range much smaller than the nuclear radius.

The binding energy B is the energy necessary to separate all the protons and all the neutrons of a nucleus. Another quantity of interest is the *separation energy* of a nucleon from the nucleus. The separation energy of a neutron from a nucleus (Z, N) is given by

$$S_n(Z, N) = \{m(Z, N - 1) + m_n - m(Z, N)\}c^2 = B(Z, N) - B(Z, N - 1). \tag{4.10}$$

In the same way we can define the separation energy of a proton or an α-particle. The separation energy can vary from a few MeV to about 20 MeV and depends very much on the structure of the nucleus. One observes that S_n is greater for nuclei with an even number of neutrons. We can define a *pairing energy* as the difference between the separation energy of a nucleus with an even number of neutrons and that of a neighbor nucleus, that is,

$$\delta_n(Z, N) = S_n(Z, N) - S_n(Z, N - 1), \tag{4.11}$$

where N is even. One observes experimentally that both δ_n and δ_p are about 2 MeV.

When one plots the separation energy versus Z or N one sees that at the values 2, 8, 20, 50, 82, 126, 184[1] the separation energy changes abruptly. These values are known as *magic numbers*, and nuclei with magic Z (or N) have the last proton (or neutron) shell complete, similar to that which occurs with the closed shells of electrons in noble gases. This subject will be discussed in connection with the shell model for the nucleus.

---

[1] The last two values refer only to neutrons.

4.4 Total Angular Momentum of the Nucleus

The nucleus is a quantum system composed of A nucleons and the nucleons are fermions (half-spin particles): the laws of quantum mechanics for the addition of angular momenta establish that the total angular momentum, or the "spin" of the nucleus, is equal to $n\hbar$ (n integer) if A is even and $(n + \frac{1}{2})\hbar$ if A is odd. This is a direct consequence of the nucleon spin value $\frac{1}{2}\hbar$, since orbital angular momenta only contribute with integer values of $\hbar$.

The experimental determination of the spin of a nucleus can be done through, among other techniques, the magnetic perturbation caused in the atomic spectra by the nuclear spin (hyperfine structure). The results not only confirm the above scheme but also provide additional important information: the spin of even-even nuclei (Z and N even) in their ground state is always zero. We can summarize these results in a scheme:

$I = n\hbar$ for odd-odd nuclei,

$I = (n + \frac{1}{2})\hbar$ for odd nuclei (even-odd or odd-even),

$I = 0$ for even-even nuclei,

I being the total angular momentum quantum number and n an integer greater than or equal to zero.

The correct determination of the nuclear spin values is an important element in the broad or restricted acceptance of models that try to describe the nuclear properties.

4.5 Multipole Moments

Electrostatic moments were discussed in chapter 2. For clarity we follow here an alternative derivation and show how they describe the geometry of the nucleus. A given distribution of charges $\rho(\mathbf{r}')$ confined to a certain region produces at each point $\mathbf{r}$ of space an electrostatic potential

$$V = \int \frac{\rho(\mathbf{r}')}{|\mathbf{r} - \mathbf{r}'|} \, d\mathbf{r}'. \tag{4.12}$$

The factor $1/|\mathbf{r} - \mathbf{r}'| = [(x - x')^2 + (y - y')^2 + (z - z')^2]^{-1/2}$ in (4.12) can be expanded in a Taylor series for 3 variables,

$$f(t, u, v) = \sum_{n=0}^{\infty} \frac{1}{n!} \left[t\frac{\partial}{\partial t} + u\frac{\partial}{\partial u} + v\frac{\partial}{\partial v} \right]^n f(t, u, v), \tag{4.13}$$

the derivatives being calculated at the point $t = 0$, $u = 0$, $v = 0$. The convention for the power is such that $(t\,\partial/\partial t)^n = t^n\,\partial^n/\partial t^n$, etc. Using (4.13) for the source coordinates ($t = x'$, etc.), (4.12) becomes

$$V = \frac{\int \rho(\mathbf{r}') \, d\mathbf{r}'}{r} + \frac{x_i \int x_i' \rho(\mathbf{r}') \, d\mathbf{r}'}{r^3} + \frac{1}{2} \frac{x_i x_j \int (3x_i' x_j' - r'^2 \delta_{ij}) \rho(\mathbf{r}') \, d\mathbf{r}'}{r^5} + \cdots, \tag{4.14}$$

where $(x_1, x_2, x_3) \equiv (x, y, z)$. In (4.13) the *sum convention* was used, where the repetition of an index in the same term indicates a sum over the index, that is, $x_i x_i' \equiv \sum_{i=1}^{3} x_i x_i'$, etc.

The first term of (4.14) is identical to the potential of a charge $q = \int \rho(\mathbf{r}') d\mathbf{r}'$ (*monopole*) placed at the origin. The second has the form of the potential of a *dipole* (two charges of the same magnitude and opposite signs placed near each other), the integrals representing each component i of the *electric dipole moment* vector

$$\mathbf{p} = \int \mathbf{r}' \rho(\mathbf{r}') \, d\mathbf{r}'. \tag{4.15}$$

The third term of (4.14) represents the contribution of a *quadrupole*, the six integrals

$$Q_{ij} = \int (3x_i' x_j' - r'^2 \delta_{ij}) \rho(\mathbf{r}') \, d\mathbf{r}' \tag{4.16}$$

being the components of the *electric quadrupole moment* tensor. Using (4.15) and (4.16) we can rewrite (4.14) more compactly as

$$V = \frac{q}{r} + \frac{\mathbf{p} \cdot \mathbf{x}}{r^3} + \frac{1}{2} \frac{Q_{ij} x_i x_j}{r^5} + \cdots . \tag{4.17}$$

The increasing powers of the denominator make the contribution of higher order multipoles (octupole, hexadecapole, etc.) less and less important.

The expansion into multipoles can be employed for the Coulomb potential created by the protons in nuclei, but it is necessary that the above treatment be adapted to a quantum system. For this purpose, the charge density $\rho(\mathbf{r}')$ must be understood as Ze times the probability density $|\psi|^2 = \psi^*(\mathbf{r}')\psi(\mathbf{r}')$ of finding a proton at the point $\mathbf{r}'$. One consequence for the nuclear case is that the electric dipole moment (4.15) vanishes. In fact, the wavefunction $\psi(\mathbf{r})$ represents nuclear states of definite parity and $|\psi|^2$ must necessarily be an even function, which makes the integral in (4.15) identically zero over the entire space. The first important information about the charge distribution in the nucleus must therefore come from the quadrupole term, the third term of the expansion (4.14). The electric quadrupole moment (4.16) associated with it will be the subject of later study.

In the same way that a charge distribution gives place to an expansion of the type (4.17), a localized current distribution $\mathbf{J}(\mathbf{r}')$ produces a vector field, with the vector potential $\mathbf{A}$ expanded in a sum of multipole terms. In this case, however, the monopole term does not exist, and the other even terms (even powers of r) cancel each other in the nuclear case by considerations similar to those in the former case. Thus, our expansion reduces to a single important term [Ja75],

$$\mathbf{A}(\mathbf{r}) = \frac{\mu \times \mathbf{r}}{r^3} + \cdots , \tag{4.18}$$

with μ, the *magnetic dipole moment*, given by

$$\mu = \frac{1}{2c} \int \mathbf{r}' \times \mathbf{J}(\mathbf{r}') \, d^3\mathbf{r}'. \tag{4.19}$$

The information about the nuclear structure that can be extracted from the measurement of the magnetic dipole moment will be described in the next section. The quantum operator corresponding to this quantity will be obtained not from (4.16) but from the relationship between μ and the corresponding angular momentum, (1.45) and (1.46).

4.6 Magnetic Dipole Moment

Expression (4.19) shows the value of the magnetic dipole moment due to the presence of currents; in the nuclear case it would describe the magnetic moment due to the orbital motion of protons. However, this description is not complete. We saw that the nucleons have intrinsic magnetic moments, and these need to be taken into account in the total computation.

To relate the total magnetic moment of the nucleus with the moments of all its constituents is not, as we have seen, a feasible task. However, experimental information is of great relevance for our study: as for angular momentum, the magnetic moment of all even-even nuclei is zero, that is, the coupling among the moments of the nucleons of an even-even nuclei is such that the total magnetic moment vanishes. This allows us to consider, as a first hypothesis, that the magnetic moment of an odd nucleus is produced by the unpaired single nucleon, called, by analogy with atomic physics, the *valence nucleon*. To calculate its value it is necessary first to recall that, for a quantum system, the expected value of a quantity has to be obtained from the corresponding operator and the system wavefunction. The dipole magnetic moment is *defined* as the expectation value of the z-component of the corresponding operator, calculated in the state where the total angular momentum component is maximum. This definition was employed for the deuteron case in writing (2.65). What we propose now is to calculate this expression for an odd nucleus.

Let l_z and s_z be the spin and orbital angular momentum projections of the unpaired nucleon (proton or neutron), which, as we have seen, is responsible for the properties of the odd nucleus. The magnetic moment of this nucleon is then described, in nuclear magneton units (see 2.18), by

$$\mu_z = g^l l_z + g^s s_z \tag{4.20}$$

where we used $\hbar = 1$. The wavefunction ψ_{lsj}^j that appears in the definition (2.65) of μ is not an eigenfunction of l_z and s_z separately but only of their sum j_z. It is then more convenient to rewrite (4.20) as

$$\mu_z = g^l j_z + (g^s - g^l) s_z, \tag{4.21}$$

and now, using (2.65),

$$\mu = g^l j + (g^s - g^l)\langle s_z \rangle. \tag{4.22}$$

The expected value of s_z can be obtained if we consider that the vector $\mathbf{s}$, as well as $\mathbf{l}$, precesses around the vector $\mathbf{j}$, and that $\mathbf{s}$ can be decomposed into a component along $\mathbf{j}$ and in two other perpendicular components of average value zero. This allows us to write

$$\langle s_z \rangle \;=\; \langle (s_j)_z \rangle \;=\; \left\langle \frac{(\mathbf{j} \cdot \mathbf{s}) j_z}{j^2} \right\rangle \;=\; \frac{j}{2j(j+1)} [j(j+1) - l(l+1) + s(s+1)], \tag{4.23}$$

where the fact was used that $\mathbf{j} \cdot \mathbf{s} = (j^2 - l^2 + s^2)/2$. Thus, for the two possible values of j, $j = l + \frac{1}{2}$ and $j = l - \frac{1}{2}$, the magnetic moment calculated by (4.23) separates into two expressions (compare to 2.18 and 2.19)

$$\mu = g^l \left(j - \frac{1}{2} \right) + \frac{1}{2} g^s, \quad j = l + \frac{1}{2},$$

$$\mu = g^l \frac{j(j + \frac{3}{2})}{j + 1} - \frac{j g^s}{2(j + 1)}, \quad j = l - \frac{1}{2}. \tag{4.24}$$

Comparison of (4.24) with the experimental values can be seen in the so-called *Schmidt diagrams*, which are constructed for the case when the unpaired nucleon is a proton or a neutron. Figure 4.4 shows the two respective Schmidt diagrams, where μ calculated by (4.23) is compared with the experimental value of the magnetic moment for several *nuclides* (a nuclide is a nuclear specimen, that is, a nucleus with a given value of Z and A). Based on these diagrams we can make the following observations:

a) If the model of a single unpaired nucleon describing the nuclear properties were entirely satisfactory, the experimental values would fall on the solid lines, called *Schmidt lines*. Although this does not occur, it is possible to clearly distinguish groups of values, each one associated with one line. The identification of a nuclide of known total angular momentum j with one of the groups implies, from knowledge of the value l of its unpaired nucleon, that there is information that cannot be obtained directly from experiment. With the value of l one can determine the ground state parity of that nuclide, and this determination gives results that are correct in the majority of the cases.

b) The experimental values follows the inclination of the lines and are, with very few exceptions, placed in the region between them. This can be explained by considering that the orbital part of (4.24) (where g^l is the angular coefficient for $j = l + \frac{1}{2}$ and approximately also for $j = l - \frac{1}{2}$) is well described by the model and the absolute value of g^s is, in fact, smaller than the measured value for free neutrons and protons. We saw previously that these values are connected to the phenomenon of emission and absorption of virtual mesons, and it is probable that this process is disturbed by the presence of other nucleons inside the nucleus.

c) The idea that an odd nucleus has its properties dictated by the valence nucleon would be more reasonable if the "core" had a magic number of protons or neutrons. In this case it would be more "closed" and its interaction with the isolated nucleon would be smaller. However, this is not observed in practice if we take into account that the open circles in figure 4.4 are not, on the average, markedly closer to the Schmidt lines than the other values. The idea is that, to a certain extent, the isolated nucleon polarizes the core, adding a new contribution to the magnetic moment of the nucleus. These considerations also apply to the case of the nuclear electric quadrupole moment that we analyze next.

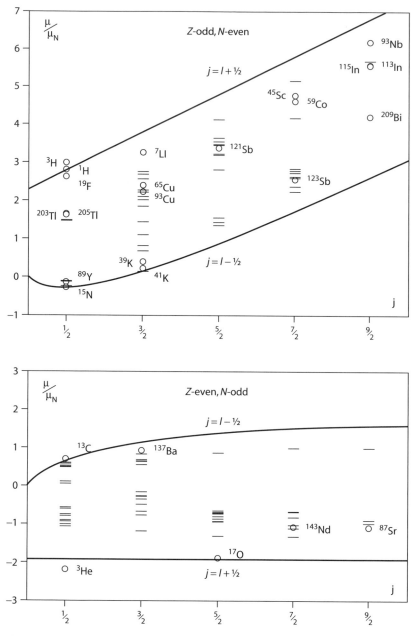

Figure 4.4 Schmidt diagrams for nuclei with: a) Z-odd, N-even and b) Z-even, N-odd. Circles represents nuclei with an extra or missing nucleon from a closed shell of 2, 8, 20, 28, 40, 50, 82, or 126 nucleons.

4.7 Electric Quadrupole Moment

The values

$$Q_{ij} = \int (3x_i' x_j' - r'^2 \delta_{ij}) \psi^*(\mathbf{r}') \psi(\mathbf{r}') \, d\mathbf{r}' \tag{4.25}$$

are the components of the quadrupole moment tensor of the nucleus. The charge Ze of the nucleus was excluded from the definition, (4.25), which gives for the quadrupole moment the dimension of area. The normally employed unit is the *barn*, corresponding to 10^{-24} cm^2 (or 10^2 fm^2).

In applying (4.25) we have to take into account that the existence of symmetries can greatly simplify the calculation of multipole moments. If the charge distribution has axial symmetry, with respect to the z-axis, for example, the six components of the quadrupole moment can be reduced to only one. This is due to the fact that the integrals over the three components with $i \neq j$ are zero, due to identical positive and negative contributions, and that the integrals in x'^2 and y'^2 are equal. Taking into account that $r^2 = x^2 + y^2 + z^2$, we obtain for the quadrupole part of the potential:

$$\begin{aligned} V_Q &= \frac{Ze}{2} \left\{ \frac{x^2 + y^2}{r^5} \int \frac{3(x'^2 + y'^2) - 2r'^2}{2} \psi^*(\mathbf{r}') \psi(\mathbf{r}') \, d\mathbf{r}' + \frac{z^2}{r^5} \int (3z'^2 - r'^2) \psi^*(\mathbf{r}') \psi(\mathbf{r}') \, d\mathbf{r}' \right\} \\ &= \frac{Ze}{2r^3} \int (3z'^2 - r'^2) \psi^*(\mathbf{r}') \psi(\mathbf{r}') \, d\mathbf{r}', \end{aligned} \tag{4.26}$$

which depends on a single integral over the source variables,

$$Q = \int (3z'^2 - r'^2) \psi^*(\mathbf{r}') \psi(\mathbf{r}') \, d\mathbf{r}', \tag{4.27}$$

understood as the electric quadrupole moment of the nucleus. Expression (4.27) indicates that a *prolate* (egg-shaped) nucleus has a positive quadrupole moment and an *oblate* (pancake-shaped) nucleus has a negative quadrupole moment. A spherical nucleus has in consequence a null quadrupole moment.

It is necessary at this point to include an additional consideration. The ground state of a spin I nucleus is composed of $2I + 1$ degenerate states, one for each possible projection of I. The wavefunction to be used in (4.27) is, *by definition*, one that has the maximum value of the total angular momentum component. This leads, for a single proton with wavefunction $\psi_{lsj}^{m_j}$,

$$Q = \left\langle \psi_{lsj}^j \left| 3z^2 - r^2 \right| \psi_{lsj}^j \right\rangle. \tag{4.28}$$

Now we can, by using (4.28) and based on the considerations presented in the last section, calculate the quadrupole moment of an odd-Z, even-N nucleus. There are again two possibilities: $j = l + \frac{1}{2}$ and $j = l - \frac{1}{2}$. For the first, the condition $m_j = j$ implies $m_l = l$ and $m_s = \frac{1}{2}$. The valence proton wavefunction can be written as

$$\psi = \frac{u_l(r)}{r} Y_l^l(\theta, \phi) \chi_{spin} \chi_{isospin}. \tag{4.29}$$

Observing that the spin and isospin functions are normalized and are not acted on by the quadrupole operator, we have

$$Q = \int u_l^2(r) \mid Y_l^l(\theta, \phi) \mid^2 r^2(3\cos^2\theta - 1)\sin\theta \, dr \, d\theta \, d\phi \tag{4.30}$$

or

$$Q = \langle r^2 \rangle \int \mid Y_l^l(\theta, \phi) \mid^2 (3\cos^2\theta - 1)\sin\theta \, d\theta \, d\phi. \tag{4.31}$$

Taking into account that

$$Y_l^l(\theta, \phi) = (-1)^l \sqrt{\frac{2l+1}{4\pi} \frac{1}{(2l)!}} e^{il\phi}(\sin\theta)^l \frac{(2l)!}{2^l l!}, \tag{4.32}$$

we find the quadrupole moment value

$$Q = -\langle r^2 \rangle \frac{2j-1}{2j+2}. \tag{4.33}$$

In simple applications it is common to use the fact that the extra nucleon is found essentially at the surface and write $\langle r^2 \rangle = R^2 = r_0^2 A^{2/3}$.

In the case of $j = l - \frac{1}{2}$, we have two possible sets of values for the components:

$$m_l = l - 1, \quad m_s = \frac{1}{2}$$

and

$$m_l = l, \quad m_s = -\frac{1}{2},$$

and Ψ is the coupling of the two cases:

$$\Psi = C_l \frac{u_l(r)}{r} Y_l^l \chi_{1/2}^{-1/2} \chi_{isospin} + C_{l-1} \frac{u_l(r)}{r} Y_l^{l-1} \chi_{1/2}^{1/2} \chi_{isospin}, \tag{4.34}$$

where C_l and C_{l-1} are the appropriate Clebsh-Gordan coefficients. Application of (4.34) in (4.28) results in the same value of Q as previously. Thus, (4.33) is in both cases, $j = l + \frac{1}{2}$ or $j = l - \frac{1}{2}$, the value of the electric quadrupole moment of a nucleus with odd Z and even N.

It is a known fact that the electric quadrupole moments of even-even nuclei with total angular momentum $j = 0$ are identically zero. Equation (4.33) shows that odd nuclei with $j = \frac{1}{2}$ also have zero quadrupole moment. These results are part of a general theorem that says that for a multipole moment 2^λ to exist it is necessary that the angular momentum be at least equal to $\lambda/2$.

Table 4.1 shows some values of Q calculated by (4.33) and a comparison with experimental results. We see that in the three first cases the calculation reproduces the sign, but the magnitude of Q is about three times smaller than the measured one. For the fourth element, ^{175}Lu, the discrepancy is significantly larger, and the last two cases show that even when the valence nucleon is a neutron the experimental value is different from zero, completely contradicting our basic hypothesis that the extra nucleon is responsible for the quadrupole moment of the odd nucleus. The reason for such discrepancies was mentioned at the end of the previous section. They are more pronounced in the present

Table 4.1 Comparison of calculated and measured values of the electric quadrupole moment for several nuclei.

Nucleus	$Q_{calc}(b)$	$Q_{exp}(b)$
$^{7}_{3}$Li	−0.013	−0.037
$^{35}_{17}$Cl	−0.037	−0.082
$^{93}_{41}$Nb	−0.13	−0.32
$^{175}_{71}$Lu	−0.18	5.68
$^{17}_{8}$O	0	−0.026
$^{73}_{32}$Ge	0	−0.17

case and ^{175}Lu is not an isolated case: it belongs to a region, the rare earths, where nuclei have a large deformation in their ground state, giving rise to large quadrupole moments. An overview of the situation is seen in figure 4.5. We observe here the alternation between positive values of Q (prolate nuclei) and negative ones (oblate nuclei), and the presence of the magic numbers of protons or neutrons in the transitions between these values.

The considerations made until now in this section refer to the *observed quadrupole moment*, relative to some z-axis in space. We can also think of an *intrinsic quadrupole moment*, relative to the symmetry axis of a deformed nucleus. Calling Q' this moment, one can, through a quantum calculation, show that the two moments are, in the ground state, related by

$$Q = \frac{I(2I - 1)}{(I + 1)(2I + 3)} Q',$$

(4.35)

where I is the spin of the nucleus. It is clear from (4.35) that $Q = 0$ for nuclei with spin $I = 0$ and $I = \frac{1}{2}$, as established previously, without the moment Q' being necessarily zero. As an example of this case, we mention the mass regions where nuclei have a permanent deformation in the ground state.

The values of Q' for even-even nuclei in this region are positive, indicating a deformation in a shape of an elongated ellipsoid. The corresponding values of Q are, however, zero, since for even-even nuclei, $I = 0$. Physically this result corresponds to the fact that a nucleus with zero spin does not have any reference that could characterize a direction in space.

4.8 Excited States of Nuclei

The set of A nucleons that form the nucleus possesses, like all quantum bound systems, a sequence of excited energy states, above the ground state (the most bound) to which one ascribes energy zero. The values of the energy of these states are normally presented in

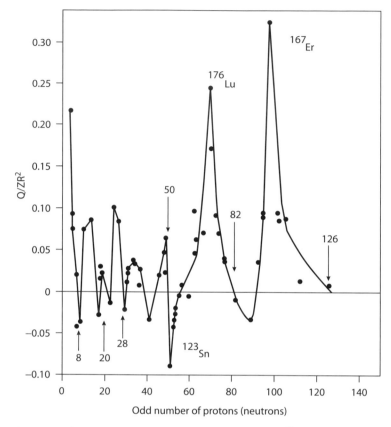

Figure 4.5 Observed quadrupole moments Q, in units of ZR^2. The abscissa measures the (odd) number of protons or neutrons. The positions of the magic numbers are indicated by arrows.

diagrams that also give, when known, the values of spin and parity corresponding to each state. Examples can bee seen in figure 4.6, which shows the level spectra of the elements $^{82}_{36}$Kr and $^{111}_{49}$In.

The distribution of states can vary enormously from nucleus to nucleus. The nucleus $^{4}_{2}$He, for example, has its first excited state around 19 MeV, while nuclei like $^{182}_{73}$Ta, $^{198}_{79}$Au, $^{223}_{88}$Ra, and $^{223}_{90}$Th have more than 50 states below 1 MeV. But it is a general rule that the density of states increases rapidly with the energy, forming practically a continuum for high energies.

The energies of the first excited states are also affected by the presence of a magic number of protons or neutrons in the nucleus. Figure 4.7 shows the variation of the average energy of the first excited state of stable even-even nuclei as a function of the number of neutrons. The maxima at the magic numbers 8, 20, 28, 50, 82, and 126 are evident.

The apparent mess that appears through the arrangement and spacing of the levels for the greater number of nuclei is due to the simultaneous presence of several excitation modes, each of which can have, in its turn, a complex structure. It is true, on the other side, that some types of excitation give place to a group of levels with a well-defined arrangement

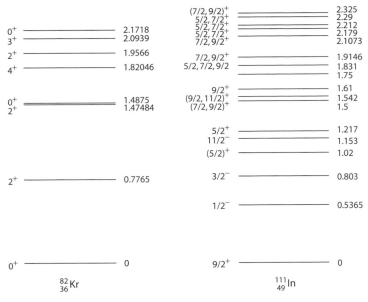

Figure 4.6 First excited states of an even-even and an odd nucleus. The numbers to the right of each level are the energies in MeV and the numbers to the left are the spin and parity of the state. Parentheses mark doubtful values.

that can be easily identified within the global set of states. It is a task of the several branches of nuclear spectroscopy to establish the existence of the various states and to measure the parameters associated with each of them. Thus it is possible, in a great number of cases, to identify the mode of excitation responsible for the presence of a given state in the energy

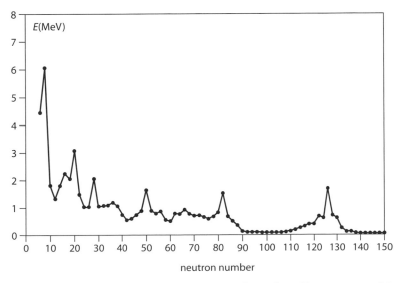

Figure 4.7 Energy of the first excited state, averaged over the stable even-even nuclei, as a function of the neutron number.

spectra of a nucleus. We will see in the next chapter that nuclear models based on simple ideas are, in fact, able to furnish a great quantity of information about the composition of the excited states of nuclei.

4.9 Nuclear Stability

The excited states discussed in the previous section are not stable: in common situations an excited state decays to a lower energy state of the same nucleus with the emission of a γ-ray. The sequence of these transitions leads normally to the ground state of that nucleus. However, the ground state itself may not be stable for many nuclides, which can decay into other nuclides by the spontaneous emission of one or more particles or fragments. The several options for the transformation of an unstable nucleus will be discussed in detail in later chapters and are described briefly in what follows:

a) β^- and β^+ decays. Stable light nuclei have their proton number Z similar to their neutron number N. In heavy nuclei a greater number of neutrons is necessary to compensate the Coulomb force between the protons. In both cases, when a nucleus has a value N greater than necessary for equilibrium it can decay by the emission of an electron and a antineutrino ($\beta^- decay$) in the form

$$^A_Z X_N \rightarrow \, ^A_{Z+1} Y_{N-1} + e^- + \bar{\nu}, \tag{4.36}$$

being then closer to a situation of greater equilibrium. If, on the other side, N is less than necessary, then $\beta^+ decay$ can occur:

$$^A_Z X_N \rightarrow \, ^A_{Z-1} Y_{N+1} + e^+ + \nu, \tag{4.37}$$

the emitted particles now being one positron and one neutrino. In both cases the product nucleus Y is not necessarily stable, and can also decay by the same way or by another form of disintegration.

b) *Electron capture.* This process consists in the capture of an atomic electron by the nucleus, giving place to a decrease of the proton number and an increase of the neutron number by 1. The effect is the same as in β^+ decay, and the electron capture can compete strongly with that in heavy nuclei.

c) α decay. In this disintegration mode an α-particle (^{4}He nucleus) is emitted, being the process energetically allowed for heavy nuclei. One example is the decay

$$^{238}_{92} U \rightarrow \, ^{234}_{90} Th + \alpha \tag{4.38}$$

with the emission of a 4.2 MeV α-particle. The α-disintegration mode is responsible for the nonexistence of stable elements with $Z > 83$.

d) *Light fragment emission.* Nuclei heavier than ^{4}He can also be emitted in the few cases where the process is energetically allowed. Emissions of ^{14}C, ^{24}Ne, ^{28}Mg, and ^{32}Si by heavy nuclei have already been experimentally observed [Pr89], but they are all rare and difficult to measure.

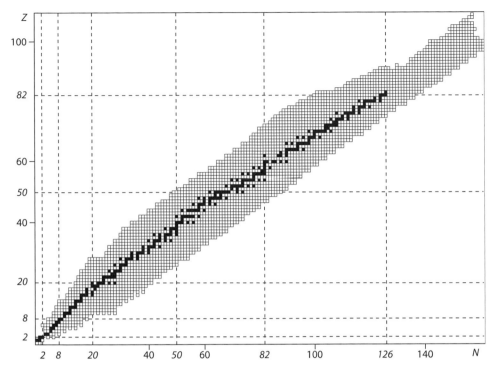

Figure 4.8 Distribution of stable nuclides (solid squares) and of the known unstable nuclides (open squares) as a function of the atomic number Z and X the neutron number N. The dashed lines mark the magic numbers on both axes.

e) *Fission*. The energetics of fission was discussed in section 4.3. It is necessary to remark that the occurrence of this process can take place spontaneously for very heavy nuclei. It is responsible, together with α emission, for the extremely short half-lives of $Z > 100$ nuclides.

The processes described above take place in unstable nuclei in their ground state. They can also proceed from an excited state, but only in special situations. The basic option for an excited state is to decay to other states of lower energy, emitting γ-radiation or ejecting an electron from the atomic shells (*internal conversion*). We can also mention that for states of high excitation energy, the emission of a nucleon is a process that can be energetically possible and compete with γ-emission.

Figure 4.8 (the so-called Segrè plot) shows the distribution in Z and N of the stable nuclides and the known unstable ones. Among the latter, the greater majority are artificially produced in the laboratory and only a few exist in nature in significant amounts. In this last group, $^{235}_{92}\text{U}$, $^{238}_{92}\text{U}$, and $^{232}_{90}\text{Th}$ are of great importance in nuclear engineering.

The stable nuclei define a band in figure 4.8 called the β *stability line*. The nuclides to the right of (or below) that line have an excess of neutrons and are unstable by β^--emission. Conversely, the nuclides to the left of (or above) the line have an excess of protons and tend to decay by β^+ emission. The more distant from the line of β stability the nuclide is, the more unstable it will be. Recently, nuclei with very unequal balances between proton and neutron numbers (called *exotic nuclei*), such as ^{11}Li and ^{8}He, have been studied. We will return to this subject in chapter 13.

It is worthwhile to establish at this point the nomenclature for nuclides that have some common characteristics. Thus we define:

Isotopes: nuclides that have the same Z, that is, the same number of protons. It is common to use the name isotope as being synonymous with nuclide, and in this case the term is also used in the singular.

Isotones: nuclides that have the same number N of neutrons.

Isobars: nuclides that have the same mass number $A = Z + N$.

The set of 284 stable nuclei, distributed in 83 elements, shows some striking features:

a) Light nuclei have approximately the same number of protons and neutrons, that is, $Z \cong N$. Heavy nuclei have $N > Z$.

b) Even-Z nuclei are much more numerous than odd-Z nuclei, and even-N nuclei are much more numerous than odd-N nuclei.

c) As a consequence of (b), even-A nuclides are much more numerous. Among them the even-even are much more common. In fact, there are only a few examples of stable odd-odd nuclides: ^{2}H, ^{6}Li, ^{10}B, and ^{14}N.

d) Of the 20 elements that have only one isotope, only ^{9}Be has Z even.

e) The element that has the greatest number of stable isotopes is $_{50}$Sn, with 10.

We will see that these features have immediate explanations in the nuclear models that will be presented in chapter 5.

4.10 Exercises

1. On a scale where a water droplet (radius = 1 mm) is increased to the size of the earth (radius = 6400 km), what would be the radius of a ^{238}U nucleus?

2. Find the approximate density of nuclear matter in g/cm^3.

3. Find the values of the constants in (4.3) for $^{120}_{50}$Sn. With these values, find the nucleon fraction that is outside the radius $R_{\frac{1}{2}}$. Use a reasonable approximation to avoid a numerical integration.

4. A 10 MeV proton beam hits a ^{208}Pb target. What is the smallest distance to the ^{208}Pb surface that the proton can reach?

5. Find the binding energy of the last neutron in ^{4}He and of the last proton in ^{16}O. How do these energies compare with B/A for these nuclei? What does this say about ^{4}He stability relative to ^{3}He, and ^{16}O relative to ^{15}N?

6. Find the energy in MeV released in the fusion of two deuterons.

7. Knowing that the binding energy of the electrons in an atom of atomic number Z is given by $B_e(Z) = 15.73 Z^{7/3}$ eV, find the error in the calculation of the binding energy of the nuclei a) $^{59}_{27}$Co, b) $^{156}_{54}$Gd, c) $^{238}_{92}$U, when one neglects the difference between the binding energy of the Z electrons in the atom and Z times the electron binding energy in the hydrogen atom (13.6 eV).

8. Evaluate the difference between the binding energy of a ^{12}C nucleus and the sum of the binding energies of three nuclei of ^{4}He (α-particles). Assuming that ^{12}C is composed of a triangular connection of three α-particles, what would the binding per α-connection be?

9. Evaluate the total binding energy of $^8_4$Be. According to this result, $^8_4$Be would be a stable nucleus, but in reality it is extremely unstable. Try to justify this discrepancy.

10. Derive (4.14) from (4.12) and (4.13).

11. The vector potential at a distance $\mathbf{r}$ from a system of localized charges, and with current density $\mathbf{J}(\mathbf{r}')$ per unit of volume, is given by

$$\mathbf{A}(\mathbf{r}) = \int \frac{\mathbf{J}(\mathbf{r}')}{|\mathbf{r} - \mathbf{r}'|} \, d^3 r' \ .$$

For $r > r'$, use a procedure similar to that of the previous exercise to demonstrate (4.18) and (4.19).

12. The deuteron has $J^\pi = 1^+$, and $S = 1$. What would be the value of its magnetic moment if the parity π were $(-)$, the other quantities remaining unchanged?

13. Assuming the proton to be a sphere of radius 1.5×10^{-13} cm, of uniform density, with angular momentum $\hbar\sqrt{\frac{1}{2}\left(\frac{1}{2}+1\right)}$, evaluate the angular velocity in revolutions/s and the tangential velocity in the proton equator. What is the proton magnetic dipole moment in this classical model?

14. Evaluate the magnetic dipole moment for the nuclides below and compare with the experimental values.

Nuclides	I^π	$\mu_{exp}\,(\mu_N)$
^{75}Ge	$\frac{1}{2}^-$	0.51
^{87}Sr	$\frac{9}{2}^+$	−1.093
^{91}Zr	$\frac{5}{2}^+$	−1.304
^{47}Sc	$\frac{7}{2}^-$	5.34
^{147}Eu	$\frac{1}{2}^-$	6.06

15. In a nuclear magnetic resonance equipment, electromagnetic radiation of frequency f releases energy to a proton, changing the direction of its spin from antiparallel to parallel to an external magnetic field B. For a field B of 1 T, find f.

16. Suppose that a nucleus has the form of an ellipsoid described by

$$\frac{x^2 + y^2}{a^2} + \frac{z^2}{b^2} = 1,$$

and that it is uniformly charged. Show that the electric quadrupole moment of this nucleus is given by

$$Q = \frac{2}{5}Z(b^2 - a^2).$$

Apply the above result to the deuteron and, comparing it with the experimental value of Q_d, find the ratio

$$\frac{\Delta R}{R} \simeq 2\frac{b - a}{b + a}$$

of the deuteron "elongation" (use $R = \langle r_d^2 \rangle^{1/2} = 2.1$ fm).

17. What is the quadrupole moment created by a proton located at the surface of a spherical nucleus of $A = 120$?

18. Show that $^{238}_{92}U$ is stable with respect to decay by p, n, e$^-$, e$^+$ emission, but is unstable with respect to the emission of an α-particle.

19. In figure 4.3 ^{6}Li has an average binding energy B/A smaller than that of ^{4}He. Why is ^{6}Li not an α-emitter?

5 | Nuclear Models

5.1 Introduction

In the previous chapters we have talked about the impossibility of obtaining the properties of a system of A nucleons starting from its constituents and their underlying interactions, and it was clearly evidenced that there is a need to use models that represent some aspects of the real problem.

The models are essentially of two classes. The first class of models assume that the nucleons interact strongly in the interior of the nucleus and that their mean free path is small. This is a situation identical to that of molecules of a liquid, and the liquid drop model belongs to this first class. These are called *collective models* and they study phenomena that involve the nucleus as a whole.

Apart from such an approach there exist a class of *independent particle models* that assume that the Pauli principle restricts the collisions of the nucleons inside the nuclear matter, leading to a larger mean free path. The several forms of shell models belong to this class.

Today we have a clear notion that the nucleus can exhibit both collective and independent particle phenomena and that each model finds its usefulness in the explanation of a specific group of nuclear properties.

5.2 The Liquid Drop Model

This model is based on the hypothesis that the nucleus has behavior identical to that of a liquid, due mainly to the fact of the occurrence, in both cases, of saturation of forces between its constituents. This idea is the starting point in obtaining an equation for the binding energy of the nucleus, introduced in chapter 4. In its simplest form this equation contains five contributory parts:

1) The main part of the binding energy is called *volume energy*. It is based on the experimental fact that the binding energy per nucleon is approximately constant (see figure 5.1);

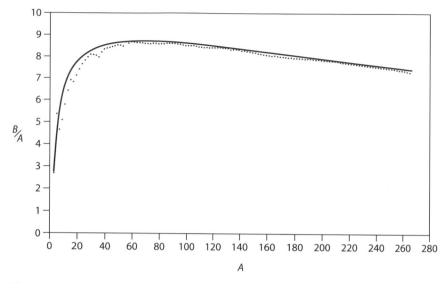

Figure 5.1 Average experimental values of B/A for A-odd nuclei, and the corresponding curve calculated by (5.7) and (5.11).

thus the total binding energy is proportional to A:

$$B_1 = a_V A. \tag{5.1}$$

If the nucleon-nucleon interaction were the same for all possible nucleon pairs, the total binding energy would be proportional to the total number of pairs, which is equal to $A(A-1)/2 \cong A^2/2$. Therefore, the binding energy per nucleon would be proportional to A. The fact that this energy is constant is due to the short range of the nuclear force, leading to the interaction of a nucleon with its neighbors. This property implies the saturation of the nuclear forces, as studied in section 4.3.

2) The surface nucleons contribute less to the binding energy since they only feel the nuclear force from the inner side of the nucleus. The number of nucleons in the surface should be proportional to the surface area, $4\pi R^2 = 4\pi r_0^2 A^{2/3}$. We should, therefore, correct (5.1) by adding the *surface energy*

$$B_2 = -a_S A^{2/3}. \tag{5.2}$$

3) The binding energy should also be smaller due to the Coulomb repulsion between the protons. The Coulomb energy of a charged sphere with homogeneous distribution and total charge Ze is given by $\frac{3}{5}(Ze)^2/R = \frac{3}{5}(e^2/r_0^2)(Z^2/A^{1/3})$. Thus, the *Coulomb energy* contributes negatively to the binding energy, given by

$$B_3 = -a_C Z^2 A^{-1/3}. \tag{5.3}$$

4) If the nucleus has a different number of protons and neutrons, its binding energy is smaller than for a symmetric nucleus. The reason for this term will be clear when we study

the Fermi gas model in the next section. This *asymmetry term* also contributes negatively and it is given by

$$B_4 = -a_A \frac{(Z - A/2)^2}{A}. \tag{5.4}$$

5) The binding energy is larger when the proton and neutron numbers are even (even-even nuclei); it is smaller when one of the numbers is odd (odd nuclei) and also when both are odd (odd-odd nuclei). Thus, we introduce a *pairing term*

$$B_5 = \begin{cases} +\delta & \text{for even-even nuclei,} \\ 0 & \text{for odd nuclei,} \\ -\delta & \text{for odd-odd nuclei.} \end{cases} \tag{5.5}$$

Empirically we find that

$$\delta \cong a_P A^{-1/2}. \tag{5.6}$$

Gathering the terms in the equations above we obtain

$$B(Z, A) = a_V A - a_S A^{2/3} - a_C Z^2 A^{-1/3} - a_A \frac{(Z - A/2)^2}{A} + \frac{(-1)^Z + (-1)^N}{2} a_P A^{-1/2}. \tag{5.7}$$

Substituting into (4.9) we obtain an equation for the mass of a nucleus,

$$m(Z, A) = Z m_p + (A - Z) m_n - a_V A + a_S A^{2/3}$$
$$+ a_C Z^2 A^{-1/3} + a_A \frac{(Z - A/2)^2}{A} - \frac{(-1)^Z + (-1)^N}{2} a_P A^{-1/2}. \tag{5.8}$$

Expression (5.8) is known as the *semi-empirical mass formula* or Weizsäcker formula [We35]. The constants appearing in (5.8) are determined empirically, that is, from the data analysis. A good adjustment is obtained using [Wa58]:

$$a_V = 15.85 \text{ McV/c}^2, \quad a_S - 18.34 \text{ MeV/c}^2, \quad u_C = 0.71 \text{ MeV/c}^2,$$
$$a_A = 92.86 \text{ MeV/c}^2, \quad \text{and} \quad a_P = 11.46 \text{ MeV/c}^2. \tag{5.9}$$

However, other good groups of parameters can also be found. Observe that a small variation in a_V or a_S leads to a large variation in the other parameters, due to a larger relevance of the corresponding terms in the mass formula.

Figure 5.1 exhibits a comparison of (5.7) to experimental data for odd nuclei. For $A < 20$ there is an absence of agreement with experience. This is expected, since light nuclei are not so similar to a liquid drop.

Equation (5.8) allows us to deduce important nuclear properties. Observe that this equation is quadratic in Z. For odd nuclei one has a parabola, as viewed in figure 5.2.

For A even, we obtain two parabolas due to the pairing energy $\pm \delta$. Nuclei with a given Z can decay into neighbors by β^+ (positron) or β^- (electron) particle emission. In figure 5.2 we see that for a nucleus with odd A there is only one stable isobar, while for even A countless stable isobars are possible.

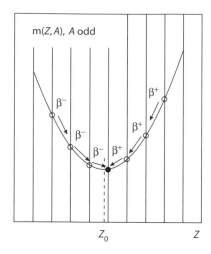 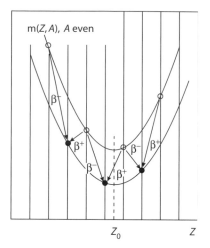

Figure 5.2 Mass of nuclei with a fixed A. The stable nuclei are represented by solid circles.

Fixing the value of A, the number of protons Z_0 for which $m(Z, A)$ is a minimum is obtained by

$$\left.\frac{\partial m(Z, A)}{\partial Z}\right|_{A=\text{const}} = 0. \tag{5.10}$$

From (5.8) we get

$$Z_0 = \frac{A}{2}\left(\frac{m_n - m_p + a_A}{a_C A^{2/3} + a_A}\right) = \frac{A}{1.98 + 0.015 A^{2/3}}. \tag{5.11}$$

We see from (5.11) that the stability is obtained with $Z_0 < A/2$, that is, with a number of neutrons larger than that of protons. We know that this in fact happens, and figure 5.3 exhibits that the stability line obtained with (5.11) accompanies perfectly the valley of stable nuclei.

From (5.8) we can also ask if a given nucleus is stable against the emission of an α-particle. For this it is necessary that

$$E_\alpha = [m(Z, A) - m(Z - 2, A - 4) - m_\alpha]c^2 > 0, \tag{5.12}$$

situation that happens for $A \gtrsim 150$. We can also verify the possibility of a heavy nucleus decaying by fission, that is, breaking in two pieces of approximately the same size. This will be possible if

$$E_f = \left[m(Z, A) - 2m(Z/2, A/2)\right]c^2 > 0, \tag{5.13}$$

a relation valid for $A \gtrsim 90$.

The liquid drop model provides a good description of the average behavior of the binding energy with mass number, but it has nothing to say about other effects such as, for example, the existence of a magic number of nucleons. In figure 5.4 a plot is provided of the difference between the experimental separation energy of a neutron (see 4.10) and that calculated using the liquid drop model for about 2000 nuclei. It is well recognized that

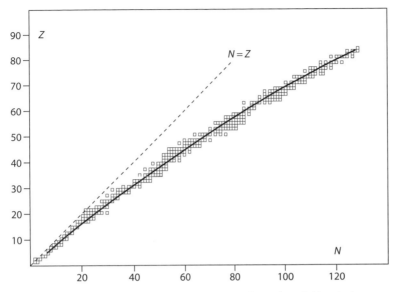

Figure 5.3 Location of the stable nuclei in the N, Z-plane. The solid line is the curve of Z_0 against $N = A - Z_0$, obtained from (5.11).

there exist values of N where the difference is positive and then drops abruptly, becoming negative. We see that these values are the magic numbers that appear in several other experiments of a different nature. The presence of these numbers is due to a structure of shells that cannot be obtained by the liquid drop model. We shall see in the following

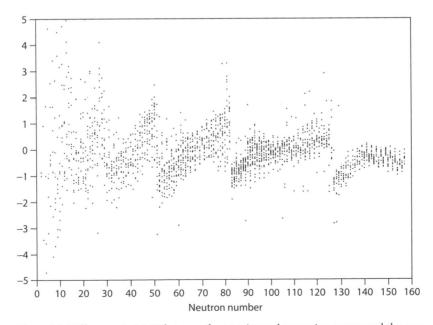

Figure 5.4 Difference (in MeV) between the experimental separation energy and the one calculated with the liquid drop model for about 2000 nuclei.

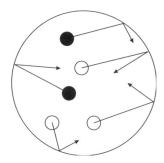

Figure 5.5 Representation of the nucleus by a Fermi gas.

sections that models that treat the nucleus as a quantum system are the only ones capable of giving a justification for the existence of these structures.

5.3 The Fermi Gas Model

This model, quite simple in its structure, is based on the fact that the nucleons move almost freely inside the nucleus due to the Pauli principle. Since two of them cannot occupy the same energy state, they do not scatter, as all possible final states that could be scattered to are already occupied by other nucleons. But when a nucleon approaches the surface and tries to fly off the nucleus, it suffers an attractive force by the nucleons that are left behind, forcing it to return toward the interior. Inside the nucleus it feels the attraction forces of all the nucleons that are around it, resulting in a net force approximately equal to zero. We can imagine the nucleus as a balloon, inside of which the nucleons move freely, but occupying states of different energy (figure 5.5).

The nucleons in the Fermi gas model obey the Schrödinger equation for a free particle,

$$-\frac{\hbar^2}{2m} \nabla^2 \Psi = E\Psi, \tag{5.14}$$

where m is the nucleon mass and E its energy. To simplify let us assume that, instead of a sphere, the region to which the nucleons are limited to is the interior of a cube. The final results of our calculation will be independent of this hypothesis. In this way, Ψ will have to satisfy the boundary conditions

$$\Psi(x, y, z) = 0 \tag{5.15}$$

for

$$x = 0, \ y = 0, \ z = 0 \quad \text{and} \quad x = a, \ y = a, \ z = a,$$

where a is the side of the cube. The solution of (5.14) and (5.15) is given by

$$\Psi(x, y, z) = A \sin(k_x x) \sin(k_y y) \sin(k_z z), \tag{5.16}$$

since

$$k_x a = n_x \pi, \quad k_y a = n_y \pi, \quad \text{and} \quad k_z a = n_z \pi, \tag{5.17}$$

where n_x, n_y, and n_z are positive integers and A is a normalization constant.

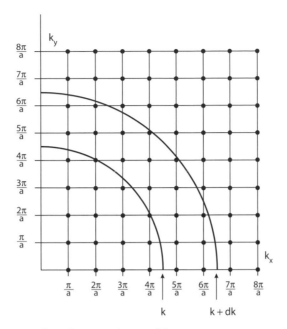

Figure 5.6 Allowed states in the part of the momentum space contained in the k_x, k_y-plane. Each state is represented by a point in the lattice.

For each group (n_x, n_y, n_z) we have an energy

$$E(n_x, n_y, n_z) = \frac{\hbar^2 k^2}{2m} = \frac{\hbar^2}{2m}(k_x^2 + k_y^2 + k_z^2) = \frac{\hbar^2 \pi^2}{2ma^2}\, n^2, \tag{5.18}$$

where $n^2 = n_x^2 + n_y^2 + n_z^2$.

Equations (5.17) and (5.18) represent the quantization of a particle in a box, where $\mathbf{k} \equiv (k_x, k_y, k_z)$ is the momentum (divided by $\hbar$) of the particle in the box. Due to the Pauli principle, a given momentum can only be occupied by at most four nucleons: two protons with opposite spins and two neutrons with opposite spins. Consider the space of vectors $\mathbf{k}$: by virtue of (5.18), for each cube of side length π/a in this space only one point exists that represents a possible solution of the relationship (5.16). The possible number of solutions (see figure 5.6) $n(k)$ with magnitude $\mathbf{k}$ between k and $k + dk$ is given by the ratio between the volume of the spherical slice displayed in the figure and the volume $(\pi/a)^3$ for each allowed solution in the $\mathbf{k}$-space:

$$dn(k) = \frac{1}{8} 4\pi k^2 dk \frac{1}{(\pi/a)^3}, \tag{5.19}$$

where $4\pi k^2 dk$ is the volume of a spherical box in the $\mathbf{k}$-space with radius between k and $k + dk$. Only $\frac{1}{8}$ of the shell is considered, since only positive values of k_x, k_y, and k_z are necessary for counting all the states with eigenfunctions defined by (5.16). With the aid of (5.18) we can make the energy appear explicitly in (5.19):

$$dn(E) = \frac{\sqrt{2}\, m^{3/2} a^3}{2\pi^2 \hbar^3} E^{1/2} dE. \tag{5.20}$$

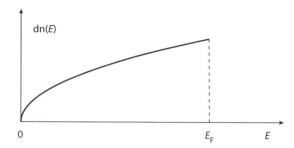

Figure 5.7 Fermi distribution for $T = 0$.

The total number of possible states of the nucleus is obtained by integrating (5.20) from 0 to the minimum value needed to include all the nucleons. This value, E_F, is called the *Fermi energy*. Thus, we obtain

$$n(E_F) = \frac{\sqrt{2}\, m^{3/2} a^3}{3\pi^2 \hbar^3} E_F^{3/2} = \frac{A}{4}, \tag{5.21}$$

where the last equality is due to the mentioned fact that a given state can be occupied by four nucleons. Inverting (5.21) we obtain

$$E_F = \frac{\hbar^2}{2m} \left(\frac{3\pi^2 \rho}{2} \right)^{2/3}, \tag{5.22}$$

where $\rho = A/a^3$. We assume that the maximum energy is the same for both nucleons, which means equal values for protons and neutrons. If that is not true, the Fermi energy for protons and neutrons will be different. If $\rho_p = Z/a^3$ and $\rho_n = N/a^3$ are the respective proton and neutron densities, we will have

$$E_F(p) = \frac{\hbar^2}{2m} (3\pi^2 \rho_p)^{2/3} \tag{5.23}$$

and

$$E_F(n) = \frac{\hbar^2}{2m} (3\pi^2 \rho_n)^{2/3}, \tag{5.24}$$

for the corresponding Fermi energies.

The number of nucleons with energy between E and $E + dE$, given by (5.20), is plotted in figure 5.7 as a function of E. This distribution of particles is referred to as the *Fermi distribution* for $T = 0$, which is characterized by the absence of any particle with $E > E_F$. This corresponds to the ground state of the nucleus. An excited state $(T > 0)$ can be obtained by the passage of a nucleon to a state above the Fermi level, leaving a vacancy (hole) in the energy state it previously occupied.

If we use $\rho = 1.72 \times 10^{38}$ nucleons/cm$^3 = 0.172$ nucleons/fm^3, which is the approximate density of all nuclei with $A \gtrsim 12$, we obtain

$$k_F = \frac{\sqrt{2mE_F}}{\hbar} = 1.36 \text{ fm}^{-1}, \tag{5.25}$$

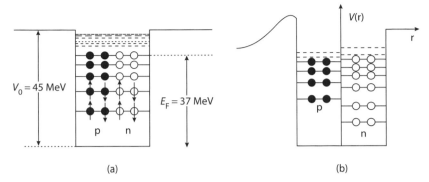

Figure 5.8 a) Potential well and states of a Fermi gas for $T = 0$. b) When one takes into account the Coulomb force. The potentials for protons and neutrons are different and we can imagine a well for each nucleon type. In two of the levels we show the spins associated to each nucleon.

which corresponds to

$$E_F = 37 \text{ MeV}. \tag{5.26}$$

We know that the separation energy of a nucleon is of the order of 8 MeV. Thus, the nucleons are not inside a well with infinite walls as we supposed, but in a well with depth $V_0 \cong (37 + 8) \text{ MeV} = 45 \text{ MeV}$ (figure 5.8a).

We are now prepared to explain the origin of the asymmetry term (5.4) of the liquid drop model. Let us imagine the nucleus as a mixture of a proton gas with Fermi energy $E_F(p)$ and a neutron gas with Fermi energy $E_F(n)$. Taking C as a constant, we can write

$$E_F(p) = C(Z/A)^{2/3}, \quad E_F(n) = C(N/A)^{2/3}. \tag{5.27}$$

If dn is the number of particles with energy between E and $E + dE$, then the total energy E_T of the gas is written as

$$E_T = \int_0^{E_F} E \, dn = \frac{3}{5} Z E_F(p) \quad \text{for protons,}$$

$$= \frac{3}{5} N E_F(n) \quad \text{for neutrons,} \tag{5.28}$$

where we used (5.20). Defining C' as a new constant, we can write the total energy as

$$E(Z, A) = C' A^{-2/3}(Z^{5/3} + N^{5/3}), \tag{5.29}$$

with the constraint $Z + N = A$. The minimum of the energy (5.29) happens when $Z = N = A/2$. Calling $E(Z, A)_{\min}$ that value, let us calculate

$$\Delta E = E(Z, A) - E(Z, A)_{\min} = C' A^{-2/3} \left[N^{5/3} + Z^{5/3} - 2(A/2)^{5/3} \right]. \tag{5.30}$$

Let us define $D = (N - Z)/2 = N - A/2 = A/2 - Z$; we have

$$\Delta E = C' A^{-2/3} \left[(A/2 + D)^{5/3} + (A/2 - D)^{5/3} - 2(A/2)^{5/3} \right]. \tag{5.31}$$

Expanding $(D + A/2)^{5/3}$ and $(-D + A/2)^{5/3}$ in a Taylor series

$$f(x + a) = f(a) + xf'(a) + \frac{x^2}{2}f''(a) + \cdots, \tag{5.32}$$

we obtain

$$\Delta E = \frac{10}{9}C'\frac{(Z - A/2)^2}{A} + \cdots. \tag{5.33}$$

We see that the imbalance between the proton and the neutron number increases the energy of the system (decreasing the binding energy) by the amount specified in (5.33). This justifies the existence of the asymmetry term

$$B_4 = -a_A\frac{(Z - A/2)^2}{A}$$

in the liquid drop model.

Despite its simplicity, the Fermi gas model is able to explain many of the nuclear properties discussed in the previous chapter. At the beginning, the occupation of the states indicates that in a light nucleus $Z \cong N$, since in this way the energy is lowered. For heavy nuclei the Coulomb force makes the proton well shallower than that for neutrons; as a consequence, the proton number is smaller than that of neutrons, being with agreement with the practical occurrence (see figure 5.8b). Another explained characteristic is the verified abundance of even-even nuclei contrasted to the almost nonexistence of stable odd-odd nuclei. It is easy to see why this happens: when we have a nucleon isolated in a level, the lower possible state of energy for a subsequent nucleon is in that same level. In other words, in an odd-odd nucleus we have one isolated proton and one isolated neutron, each in its potential well. But between those states there is, generally, a difference of energy, creating the possibility of passage of one of the nucleons to the well of the other through β-emission and, thus, the nucleus returns to stability.

Finally, we notice that we have described here an independent particle model, and the above results predict the success applying that idea to more sophisticated models, as for instance the shell model that we will study next.

5.4 The Shell Model

This model admits that the nucleons move within the nucleus independently of each other, in the same spirit as the Fermi gas model. The difference is that now the nucleons are not treated as free particles but are subject to a central potential, similar to the central potential that acts on electrons in the atom. At first sight the idea is a bit strange because we cannot, as in the atomic case, identify the origin of a such potential. This difficulty is resolved by assuming that each nucleon moves in an average potential created by the other nucleons, a potential that should be determined in a way to best reproduce the experimental results.

The first proposals of the model appeared at the end of the 1920s, motivated by the fluctuations in the relative abundance and masses of the nuclei along the periodic table.

However, the lack of an apparent theoretical basis and the low acceptance of the idea of independent motion of nucleons, together with poor initial results, meant that the model took a long time to succeed. Finally, the introduction of a spin-orbit term, in 1949, established in a definitive way the shell model as an important tool of vast use in nuclear physics.

We are going to describe the shell model idea in a more formal way. The exact Hamiltonian for a problem of A bodies can be written as

$$H = \sum_i^A T_i(\mathbf{r}_i) + V(\mathbf{r}_1, \ldots, \mathbf{r}_A), \tag{5.34}$$

where T is the kinetic energy operator and V the potential function.

If we restrict ourselves to two-body interactions (e.g., nucleon-nucleon interaction), (5.34) takes the form

$$H = \sum_i^A T_i(\mathbf{r}_i) + \frac{1}{2} \sum_{ji} V_{ij}(\mathbf{r}_i, \mathbf{r}_j). \tag{5.35}$$

In the model proposal, the nucleon i feels not the potential $\sum_j V_{ij}$, but a central potential $U(r_i)$, that depends only on the coordinates of nucleon i. This potential can be introduced in (5.35), with the result

$$H = \sum_i^A T_i(\mathbf{r}_i) + \sum_i^A U(r_i) + H_{\text{res}}, \tag{5.36}$$

$$H_{\text{res}} = \frac{1}{2} \sum_{ji} V_{ij}(\mathbf{r}_i, \mathbf{r}_j) - \sum_i U(r_i). \tag{5.37}$$

H_{res} refers to the *residual interactions*, that is, the part of potential V not embraced by the central potential U. The hope of the shell model is that the contribution of H_{res} is small or, alternatively, that the *shell model Hamiltonian,*

$$H_0 = \sum_{i=1}^A \left[T_i(r_i) + U(r_i) \right], \tag{5.38}$$

represents a good approximation for the exact expression of H. Later we shall see that part of the lost accuracy when we pass from (5.35) to (5.38) can be recovered by an approximated treatment of the effect of the residual interaction H_{res}.

The solutions $\Psi_1(\mathbf{r}_1), \Psi_2(\mathbf{r}_2), \ldots$, of the equation

$$H_0 \Psi = E \Psi, \tag{5.39}$$

with respective eigenvalues $E_1, E_2, \ldots$, are called *orbits* or *orbitals*. In the shell model prescription the A nucleons fill the orbitals of lower energy in a way compatible with the Pauli principle. Thus, if the sub-index 1 of Ψ_1, which represents the group of quantum numbers of the orbital 1, includes spin and isospin, we can say that the first nucleon is described by $\Psi_1(\mathbf{r}_1), \ldots,$ and the A-th by $\Psi_A(\mathbf{r}_A)$. Thus, the wavefunction

$$\Psi = \Psi_1(\mathbf{r}_1)\Psi_2(\mathbf{r}_2)\ldots\Psi_A(\mathbf{r}_A) \tag{5.40}$$

is a solution of (5.39) with eigenvalues

$$E = E_1 + E_2 + \cdots + E_A, \tag{5.41}$$

and it would be, in principle, the wavefunction of the nucleus, with energy E given by the shell model. We should have in mind, however, that we are treating a fermion system and that the total wavefunction should be antisymmetric for an exchange of coordinates of two nucleons. Such a wavefunction is obtained from (5.40) for the construction of the *Slater determinant*

$$\Psi = \frac{1}{\sqrt{A!}} \begin{vmatrix} \Psi_1(\mathbf{r}_1) & \Psi_1(\mathbf{r}_2) & \cdots & \Psi_1(\mathbf{r}_A) \\ \Psi_2(\mathbf{r}_1) & \Psi_2(\mathbf{r}_2) & \cdots & \Psi_2(\mathbf{r}_A) \\ \vdots & \vdots & \ddots & \vdots \\ \Psi_A(\mathbf{r}_1) & \Psi_A(\mathbf{r}_2) & \cdots & \Psi_A(\mathbf{r}_A) \end{vmatrix}, \tag{5.42}$$

where the change of coordinates (or of the quantum numbers) of two nucleons changes the sign of the determinant.

An inconvenience of the construction in (5.42) is that the function Ψ, by mixing well-defined angular momenta J and isotopic spin T, is no longer an eigenfunction of these operators. The solution to this difficulty involves the construction of linear combinations of Slater determinants that are eigenfunctions of J and T. The problem has a well-known solution, but it involves a great amount of calculation. It is important to make clear, however, that many of the properties of the nuclear states can be extracted from the shell model without knowledge of the wavefunction, as we will see next.

We will analyze what is obtained when one starts with potentials $U(r)$ (5.38) with well-known solutions. Let us initially examine the simple harmonic oscillator. Being a potential that always grows with distance, at first it would seem not to be adaptable to representation of the nuclear potential, which goes to zero when the nucleon is at a larger distance than the radius of the nucleus. It is expected, however, that this is not very important when we analyze just the bound states of the nucleus. The oscillator potential has the form

$$V(r) = \frac{1}{2}m\omega^2 r^2, \tag{5.43}$$

where the frequency ω should be adapted to the mass number A.

We will seek solutions of (5.43) of the type

$$\Psi(\mathbf{r}) = \frac{u(r)}{r}Y_l^m(\theta, \phi), \tag{5.44}$$

where the substitution of Ψ in the Schrödinger equation for a particle reduces the solution of (5.44) to the solution of an equation for u:

$$\frac{d^2u}{dr^2} + \left\{ \frac{2m}{\hbar^2}\left[E - V(r)\right] - \frac{l(l+1)}{r^2} \right\}u = 0. \tag{5.45}$$

The solution of (5.45) with the potential (5.43) is

$$u_{nl}(r) = N_{nl} \exp\left(-\frac{1}{2}vr^2\right) r^{l+1} \mathcal{V}_{nl}(r), \tag{5.46}$$

where $v = m\omega\hbar$ and $\mathcal{V}_{nl}(r)$ is the associated Laguerre polynomial

$$\mathcal{V}_{nl}(r) = L_{n+l-\frac{1}{2}}^{l+\frac{1}{2}}(vr^2) = \sum_{k=0}^{n-1}(-1)^k 2^k \binom{n-1}{k} \frac{(2l+1)!!}{(2l+2k+1)!!}(vr^2)^k, \tag{5.47}$$

where $L_k^\alpha(t)$ are solutions of the equation

$$t\frac{d^2 L}{dt^2} + (\alpha + 1 - t)\frac{dL}{dt} + kL = 0. \tag{5.48}$$

From the normalization condition

$$\int_0^\infty u_{nl}^2(r)dr = 1 \tag{5.49}$$

we obtain that

$$N_{nl}^2 = \frac{2^{l-n+3}(2l+2n-1)!!}{\sqrt{\pi}(n-1)![(2l+1)!!]^2}v^{l+\frac{3}{2}}. \tag{5.50}$$

The energy eigenvalues corresponding to the wavefunction $\Psi_{nlm}(\mathbf{r})$ are

$$E_{nl} = \hbar\omega\left(2n + l - \frac{1}{2}\right) = \hbar\omega\left(\Lambda + \frac{3}{2}\right) = E_\Lambda, \tag{5.51}$$

where

$$n = 1, 2, 3, \ldots, \quad l = 0, 1, 2, \ldots, \quad \text{and} \quad \Lambda = 2n + l - 2. \tag{5.52}$$

For each value of l there are $2(2l+1)$ states with the same energy (degenerate states). The factor 2 is due to two spin states. However, the eigenvalues that correspond to the same value of $2n + l$ (same value of Λ) are also degenerate. As $2n = \Lambda - l + 2 =$ even, a given value of Λ corresponds to the degenerate eigenstates

$$(n, l) = \left(\frac{\Lambda+2}{2}, 0\right)\left(\frac{\Lambda}{2}, 2\right), \ldots, (2, \Lambda - 2), (1, \Lambda) \tag{5.53}$$

for Λ even and

$$(n, l) = \left(\frac{\Lambda+1}{2}, 1\right)\left(\frac{\Lambda-1}{2}, 3\right), \ldots, (2, \Lambda - 2), (1, \Lambda) \tag{5.54}$$

for Λ odd.

We obtain then that the neutron or proton numbers with eigenvalues E_Λ are given by (we will use $l = 2k$ or $2k + 1$, in the case that Λ is even or odd)

$$N_\Lambda = \sum_{k=0}^{\Lambda/2} 2[2(2k)+1] \quad \text{for } \Lambda \text{ even}, \tag{5.55}$$

$$N_\Lambda = \sum_{k=0}^{\Lambda-1/2} 2[2(2k+1)+1] \quad \text{for } \Lambda \text{ odd}. \tag{5.56}$$

In both cases, the result is

$$N_\Lambda = (\Lambda + 1)(\Lambda + 2). \tag{5.57}$$

The quantum number Λ defines a *shell* and each shell can accommodate N_Λ protons and N_Λ neutrons.

The accumulated number of particles for all the levels up to Λ is

$$\sum_\Lambda N_\Lambda = \frac{1}{3}(\Lambda + 1)(\Lambda + 2)(\Lambda + 3). \tag{5.58}$$

Based on these results, we can make an estimate of the frequency ω of the harmonic oscillator applied to a nucleus with atomic number A. For a harmonic oscillator, the expectation value of the kinetic energy of a given state is equal to the expectation value of the potential energy. Thus, the sum of the energies of the occupied states in a nucleus of mass A is

$$E = m\omega^2 A \langle r^2 \rangle. \tag{5.59}$$

We can estimate $\langle r^2 \rangle$ using

$$\langle r^2 \rangle \cong \frac{3}{5} R^2, \tag{5.60}$$

with $R \cong 1.2 A^{\frac{1}{3}}$. Assuming that $N = Z$ and that all states up to an energy E_Λ are occupied, one gets

$$A = \sum_{\Lambda=0}^{\Lambda_0} 2N_\Lambda = \frac{2}{3}(\Lambda_0 + 1)(\Lambda_0 + 2)(\Lambda_0 + 3) \cong \frac{2}{3}(\Lambda_0 + 2)^2 + \text{terms of order}(\Lambda_0) \tag{5.61}$$

and

$$\frac{E}{\hbar\omega} = \sum_{\Lambda=0}^{\Lambda_0} 2N_\Lambda \left(\Lambda + \frac{3}{2} \right) \cong \frac{1}{2}(\Lambda_0 + 2)^4 - \frac{1}{3}(\Lambda_0 + 2)^3 + \cdots . \tag{5.62}$$

Eliminating $\Lambda_0 + 2$ from the equations above and keeping terms of larger order in $\Lambda_0 + 2$, one gets

$$\frac{E}{\hbar\omega} \cong \frac{1}{2} \left(\frac{3}{2} A \right)^{\frac{4}{3}}. \tag{5.63}$$

Using (5.59) and (5.60) gives

$$\hbar\omega \cong 41 A^{-\frac{1}{3}} \text{ MeV}. \tag{5.64}$$

The giant dipole resonances are excitations with $\Delta l = \pm 1$. The position of the peak varies with the mass of the nucleus as $A^{-\frac{1}{3}}$, being a good example of application of (5.64).

The levels predicted by the harmonic oscillator are given in table 5.1. We can observe that the closed shells appear in levels 2, 8, and 20, in agreement with the experimental facts, since the nuclei should close their shells (of protons and of neutrons) with a magic number. But, the same does not happen in closed shells for nucleon numbers larger than 20, which is in disagreement with experience.

Table 5.1 Nucleon distribution for the first shells of a simple harmonic oscillator. The last column indicates the total number of neutrons (or protons) accumulated up to that shell.

$\Lambda = 2n + l - 2$	$E/\hbar\omega$	l	States	N_Λ = number of neutrons (protons)	Total
0	3/2	0	1s	2	2
1	5/2	1	1p	6	8
2	7/2	0,2	2s,1d	12	20
3	9/2	1,3	2p,1f	20	40
4	11/2	0,2,4	3s,2d,1g	30	70
5	13/2	1,3,5	3p,2f,1h	42	112
6	15/2	0,2,4,6	4s,3d,2g,1i	56	168

For an infinite square well we have the same approximate situation. The solutions for that potential obey the equation

$$\frac{d^2u}{dr^2} + \left[\frac{2m}{\hbar^2}E - \frac{l(l+1)}{r^2}\right]u = 0, \tag{5.65}$$

whose solutions

$$u = A\,r\,j_l(kr) \tag{5.66}$$

involve spherical Bessel functions that obey the boundary condition

$$j_l(kR) = 0, \tag{5.67}$$

where R is the radius of the nucleus and $k = \sqrt{2mE}/\hbar$. From (5.66) and (5.67) we build the allowed states for that potential. Table 5.2 shows these states, the nucleon numbers admitted in each of them, and the nucleon numbers that close the shells. Again here the magic numbers are only reproduced in the initial shells.

The nuclear potential should actually have an intermediary form between the harmonic oscillator and the square well, not being as smooth as the first or as abrupt as the second. It is common to use the "Woods-Saxon" form $V = V_0/\{1 + \exp[(r - R)/a]\}$, where V_0, r, and a are adjustable parameters. A numerical solution of the Schrödinger equation with such a potential does not supply us, however, with the expected results.

A considerable improvement was obtained in 1949 by Maria Mayer [Ma49] and, independently, by Haxel, Jensen, and Suess [Ha49], with the introduction of a term of spin-orbit interaction in the form

$$f(r)\mathbf{l} \cdot \mathbf{s}, \tag{5.68}$$

Table 5.2 Proton (or neutron) distribution for the first shells of an infinite square well. The principal quantum number n indicates the order in that a zero appears for a given l in eq. (5.67). Notice that here there is no longer the degeneracy in l. The third column gives the proton and neutron numbers that can be fitted in each orbit.

orbit: nl	kR	$2(2l+1)$	Total
1s	3.142	2	2
1p	4.493	6	8
1d	5.763	10	18
2s	6.283	2	20
1f	6.988	14	34
2p	7.725	6	40
1g	8.183	18	58
2d	9.095	10	68
3s	9.425	2	70

where $f(r)$ is a radial function that should be obtained by comparison with experiments. However, we will soon see that its form is not important for the effect that we want.

A spin-orbit term already appears in atomic physics as a result of the interaction between the magnetic moment of the electrons and the magnetic field created by orbital motion. In nuclear physics this term has a different nature and is related to the quantum field properties of an assembly of nucleons.

We will see that the addition of such a term to the potential of (5.43) alters the energy values. The new values are given to first order by

$$E = \int \Psi^* H \Psi = \left(n + \frac{3}{2}\right)\hbar\omega + \alpha \int \Psi^* f(r) \mathbf{l} \cdot \mathbf{s}\, \Psi, \qquad (5.69)$$

where α is a proportionality constant. If we suppose now that the spin-orbit term is small and that it can be treated as a small perturbation, the wavefunctions in (5.69) are basically those of a central potential. Recalling that $\mathbf{l} \cdot \mathbf{s} = (j^2 - l^2 - s^2)/2$, we have

$$\int \Psi^* \mathbf{l} \cdot \mathbf{s}\, \Psi = \frac{l}{2} \quad \text{for } j = l + \frac{1}{2},$$
$$\int \Psi^* \mathbf{l} \cdot \mathbf{s}\, \Psi = -\frac{1}{2}(l+1) \quad \text{for } j = l - \frac{1}{2}. \qquad (5.70)$$

Thus, the spin-orbit interaction removes the degeneracy in j and, anticipating that the best experimental result will be obtained if the orbitals for larger j have the energy lowered, we admit a negative value for α, allowing us to write for the energy increment

$$\Delta E \mid_{j=l+1/2} = -\mid \alpha \mid \langle f(r) \rangle \frac{l}{2}, \tag{5.71}$$

$$\Delta E \mid_{j=l-1/2} = +\mid \alpha \mid \langle f(r) \rangle \frac{1}{2}(l+1). \tag{5.72}$$

Figure 5.9 exhibits the level scheme of a central potential with the introduction of the spin-orbit interaction. It is easy to see the effect of (5.71) and (5.72) in the energy distribution of the levels. Referring now to a shell as a group of levels of closed energy, not necessarily associated to only one principal quantum number of the oscillator, we obtain a perfect description of all the magic numbers.

We will use this level scheme[1] to establish what one refers to as the *single particle model* or *extreme shell model*. In this version, the model allows that an odd nucleus is composed of an inert even-even core plus an unpaired nucleon, and that this last nucleon determines the properties of the nucleus. This idea was discussed earlier; what we can now do is to determine, starting from figure 5.9, in which state we find the unpaired nucleon.

Let us take ^{17}O as an example. This nucleus has a shell closed with 8 protons but has a remaining neutron above the closed core of 8 neutrons. A quick examination of figure 5.9 indicates that this neutron finds itself in level $1d_{5/2}$. We can say that the neutron *configuration* of this nucleus is

$$\left(1s_{1/2}\right)^2 \left(1p_{3/2}\right)^4 \left(1p_{1/2}\right)^2 \left(1s_{1/2}\right)^2 \left(1d_{5/2}\right)^1,$$

it being evident that in this definition we list the filled levels for the neutrons, with the upper indices, equal to $2j+1$, indicating the number of particles in each of them. It is also common to restrict the configuration to the partially filled levels, the subshells being completely ignored. Thus, the configuration of neutrons in ^{17}O would be $(1d_{5/2})^1$. The prediction of the model is, in this case, that the spin of the ground state of ^{17}O is $\frac{5}{2}$ and the parity is positive ($l=2$). This prediction is in agreement with experiments. For similar reasons, the model predicts that the ground state of ^{17}F is a $\frac{5}{2}^+$ state and this is indeed the measured value.

The extreme shell model works well when we have a nucleon above a closed shell, as in the examples above. It also works well for a hole (absence of a nucleon) in a closed shell. Examples of this case are the nuclei ^{15}O and ^{15}N for which the model predicts correctly a $\frac{1}{2}^-$ ground state. There are situations, however, in which the model needs a certain adaptation. Such is the case, for example, for the stable nuclei ^{203}Tl and ^{205}Tl. They have 81 protons, with a resulting hole in $\frac{11}{2}$. Their ground state is, however, $\frac{1}{2}^+$ instead of $\frac{11}{2}^-$. In order to understand what happens it is necessary to recall that the model, in the simple form we are using, totally neglects the individual interactions of the nucleons; a correction of the model would be to take into account certain nucleon-nucleon interactions that we know are present and that are part of the residual interaction (5.37).

---

[1] Each level of figure 5.9, characterized by the quantum numbers n, l, j, contains $2j+1$ nucleons of a same type and is also referred to as a subshell.

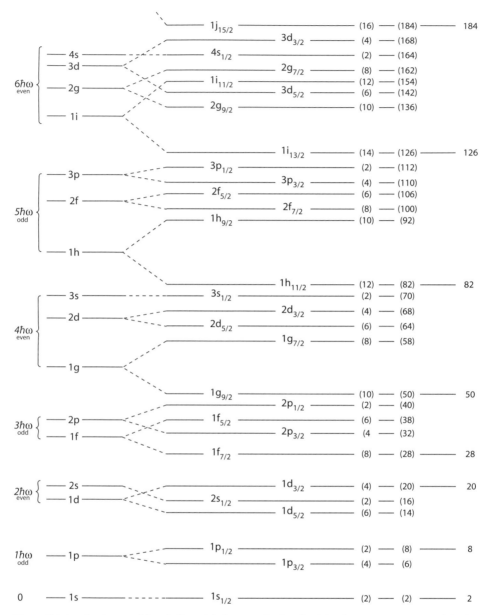

Figure 5.9 Level scheme of the shell model showing the break of the degeneracy in j caused by the spin-orbit interaction term and the emergence of the magic numbers in the shell closing. The values in the first set of parentheses indicate the number of nucleons of each type that the level admits and the values in the second set of parentheses provide the total number of nucleons of each type up to that level. Finally, the numbers outside parentheses indicate the total number of nucleons at shell-closure, reobtaining the magic numbers in their entirety. The ordering of the levels is not rigid, and there could be level inversions when changes occur in the form of the potential [MJ55].

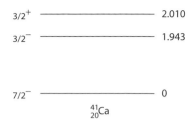

$3/2^+$ ——————————— 2.010

$3/2^-$ ——————————— 1.943

$7/2^-$ ——————————— 0

$^{41}_{20}$Ca

Figure 5.10 Level scheme for ^{41}Ca. The values to the right are the energies in MeV.

A class of interactions of special interest is the one that involves a proton (or a neutron) pair of equal orbits n, l, and j with symmetric values of m_j. A collision of these particles can take the pair to other orbits with the same quantum numbers n, l, j but with new projections m'_j and $-m'_j$. These collisions conserve energy (the $2j + 1$ states are degenerate), angular momentum, and parity, and we expect that there is a permanent alternation between the several possible values of the pair m_j, $-m_j$. The interaction between two nucleons in these circumstances is commonly called the *pairing force*. It leads to an increase of the binding energy of the nucleus; since the nucleons belong to the same orbit, their wavefunctions have the same space distribution and the average proximity between them is maximum. As the nuclear force is attractive, this leads to an increase in the binding energy. The pairing force is responsible for the pairing term (5.6) of the mass formula (5.7). It is the same type of force that, acting between the conduction electrons of a metal, in special circumstances and at low temperatures, yields the superconductivity phenomenon [Ba57].

The pairing force increases with the value of j, since the larger the angular momentum, the larger the location of the wavefunction of the nucleon around a classical orbit, and also the stronger the argument of the previous paragraph. This implies that it is sometimes more energetically advantageous that the isolated nucleon is not in the last level but below it, leaving the last level for a group of paired nucleons. This happens in our example of Tl; the hole is located not in $h_{\frac{11}{2}}$ but in $s_{\frac{1}{2}}$, leaving the orbital $h_{\frac{11}{2}}$ (of high j) occupied by a pair of protons. Another example is ^{207}Pb, for which the hole in the closed shell of 126 neutrons is not in $i_{\frac{13}{2}}$ but in $p_{\frac{1}{2}}$, resulting in the value $\frac{1}{2}^-$ for the spin of its ground state.

An example of another kind is ^{23}Na. This nucleus has the last 3 protons in the orbital $1d_{\frac{5}{2}}$. The value of its spin is, however, $\frac{3}{2}$. This is an example of a flaw in the predictions of the extreme shell model. Here, it is the coupling between the three nucleons that determines the value of the spin and not separately the value of j of each of them. This type of behavior will be analyzed in the following section.

Having established the outline of operation of the shell model, it is easy to apply it to the determination of the excited states of nuclei. In the case of ^{41}Ca (figure 5.10), the ground state is $\frac{7}{2}^-$, since the extra neutron occupies the orbital $f_{\frac{7}{2}}$. The first excited level corresponds to a jump of that neutron to $p_{\frac{3}{2}}$, generating the state $\frac{3}{2}^-$. The second excited

Table 5.3 Determination of the spin of odd-odd nuclei. The 7th column lists the possible values of j predicted by the Nordheim rules and the last column lists the experimental values with their respective parities.

Nucleus	Proton Z	State	Neutron N	State	$\mathcal{N}$	j_{pred}	j_{exp}
^{14}N	7	$p_{\frac{1}{2}}$	7	$p_{\frac{1}{2}}$	-1	0 or 1	1^+
^{42}K	19	$d_{\frac{3}{2}}$	23	$f_{\frac{7}{2}}$	0	2	2^-
^{80}Br	35	$p_{\frac{3}{2}}$	45	$p_{\frac{1}{2}}$	0	1	1^+
^{208}Tl	81	$s_{\frac{1}{2}}$	127	$g_{\frac{9}{2}}$	1	4 or 5	5^+

state, $\frac{3}{2}^+$, is obtained by the passage of a neutron from $1d_{\frac{3}{2}}$ to $1f_{\frac{7}{2}}$, leaving a hole in $1d_{\frac{3}{2}}$. It should be noticed, however, that in a very few cases the single particle model gets to a reasonable prediction of spins and energy of excited states. For such a purpose it is necessary to have a more sophisticated version of the shell model, to be described in the following section.

As a last topic in this section we will examine the situation of one odd-odd nucleus. In this case, two nucleons, a proton and a neutron, are unpaired. If $\mathbf{j}_p$ and $\mathbf{j}_n$ are the respective angular momenta of the nucleons, the angular momentum of the nucleus j can have values from $|j_p - j_n|$ to $j_p + j_n$. L. W. Nordheim proposed, in 1950 [No50], rules to determine the most probable value of the spin of an odd-odd nucleus. Defining the *Nordheim number*

$$\mathcal{N} = j_p - l_p + j_n - l_n, \tag{5.73}$$

those rules establish that

a) If $\mathcal{N} = 0, j = |j_p - j_n|$; this is called the *strong rule*.
b) If $\mathcal{N} = \pm 1, j = j_p + j_n$ or $j = |j_p - j_n|$, the *weak rule*.

The Nordheim rules reveal a tendency (not widespread, because there are exceptions) to an alignment of the intrinsic spins, as in the state $j = 1$ of the deuteron. Table 5.3 shows some practical examples.

When we have an incomplete subshell, the states that it can form with k nucleons is degenerate. The presence of residual forces among those nucleons separates the states in energy; that is, the degeneracy is removed. The angular momentum of each state is one of the possible values that result from adding k angular momenta $\mathbf{j}$. The ground state will be the lowest energy of the group, and its angular momentum will not necessarily be equal to j^2.

---

[2] It can be shown that the forces between nucleons inside closed shells and between the valence nucleons and the closed shells do not modify the ordering of the energy levels.

Table 5.4 Possible values of the projection of the
total angular momentum when two identical
nucleons occupy the level $j = \frac{5}{2}$.

m_1	m_2	m	m_1	m_2	m
$\frac{5}{2}$	$\frac{3}{2}$	4	$\frac{3}{2}$	$-\frac{5}{2}$	-1
$\frac{5}{2}$	$\frac{1}{2}$	3	$\frac{1}{2}$	$-\frac{1}{2}$	0
$\frac{5}{2}$	$-\frac{1}{2}$	2	$\frac{1}{2}$	$-\frac{3}{2}$	-1
$\frac{5}{2}$	$-\frac{3}{2}$	1	$\frac{1}{2}$	$-\frac{5}{2}$	-2
$\frac{5}{2}$	$-\frac{5}{2}$	0	$-\frac{1}{2}$	$-\frac{3}{2}$	-2
$\frac{3}{2}$	$\frac{1}{2}$	2	$-\frac{1}{2}$	$-\frac{5}{2}$	-3
$\frac{3}{2}$	$-\frac{1}{2}$	1	$-\frac{3}{2}$	$-\frac{5}{2}$	-4
$\frac{3}{2}$	$-\frac{3}{2}$	0			

Before we examine the possible values of the angular momentum, it is convenient to discuss the composition of the valence level. When we filled out the levels of the shell model we placed $2j + 1$ protons in a level n, l, j, and also $2j + 1$ neutrons, since the Pauli principle does not restrict the presence of different nucleons in the same quantum state. We can, in this way, think of a "proton" well and a "neutron" well, filled in independent ways. When we have a heavy nucleus, the Coulomb force implies that the proton well is shallower than the neutron well. We have a situation similar to the one shown in figure 5.8b for the case of the Fermi gas. Thus, the shell where we find the last protons and the shell where we find the last neutrons, both close to the Fermi level, can be different, corresponding to quite different wavefunctions and of small space overlap. It is expected that a residual proton-neutron interaction is not important in this case. For light nuclei, on the other hand, the Coulomb effect is small and the inclusion of particle-particle interactions should treat protons and neutrons within a single context.

We will analyze the first case initially and imagine that k "last" nucleons, say, protons, reside in a subshell defined by n, l, j. The configuration for the proton well has the form

$$(n_1 \, l_1 \, j_1)^{2j_1+1} \, (n_2 \, l_2 \, j_2)^{2j_2+2} \ldots (n \, l \, j)^k.$$

If we do not take into consideration the interaction forces, the $2j + 1$ states that compose the last level, where we find k valence protons, are degenerate. The presence of the interaction removes the degeneracy; thus, for example, if we have two protons ($k = 2$) in a level $j = \frac{5}{2}$, we can follow the picture shown in table 5.4. The first two columns list the possible values of m_j for the two particles allowed by the Pauli principle, and the last column lists the value $m_j = m_j(1) + m_j(2)$. We know that the possible values of the resulting angular momentum are located in the range $|j_1 - j_2| < j < j_1 + j_2$, that is, j could, in this example,

Table 5.5 Possible values for the total angular momentum of k identical nucleons placed in a sub-shell of angular moment j. The numbers in parentheses indicate the number of times the value repeats.

j	k	
$\frac{1}{2}$	1	$\frac{1}{2}$
$\frac{3}{2}$	1	$\frac{3}{2}$
	2	0,2
$\frac{5}{2}$	1	$\frac{5}{2}$
	2	0,2,4
	3	$\frac{3}{2},\frac{5}{2},\frac{9}{2}$
$\frac{7}{2}$	1	$\frac{7}{2}$
	2	0,2,4,6
	3	$\frac{1}{2},\frac{3}{2},\frac{5}{2},\frac{7}{2},\frac{9}{2},\frac{11}{2},\frac{15}{2}$
	4	0,2(2),4(2),5,6,8
$\frac{9}{2}$	1	$\frac{9}{2}$
	2	0,2,4,6,8
	3	$\frac{3}{2},\frac{5}{2},\frac{7}{2},\frac{9}{2}(2),\frac{11}{2},\frac{15}{2},\frac{17}{2},\frac{21}{2}$
	4	0(2),2(2),3,4(3),5,6(3),7,8(2),9,10,12
	5	$\frac{1}{2},\frac{3}{2},\frac{5}{2}(2),\frac{7}{2}(2),\frac{9}{2}(3),\frac{11}{2}(2),\frac{13}{2}(2),\frac{15}{2}(2),\frac{17}{2}(2),\frac{19}{2},\frac{21}{2},\frac{25}{2}$

take the values 0, 1, 2, 3, 4, or 5. If, however, we accept the rule that each value of m_j in table 5.4 is the projection of a single value of j, we will see that only the momenta $j = 0$, 2, and 4 can exist.

When we have more than two particles in a level, the situation can become much more complex. The practical method of finding the possible values of the resulting angular momentum shown in table 5.4 can always be used, but it is quite a difficult matter when we have many nucleons. Table 5.5 shows the possibilities for several configurations, where we see new restrictions imposed by the antisymmetrization, or, equivalently, by the Pauli principle, for given values of the total angular momentum J.

If we have k nucleons in an orbital around a light core of $Z = N$, we can form $k + 1$ nuclei by varying the neutron and proton numbers in the level. The value of T for each nucleus is located inside the limits

$$|T_z| \leq T \leq \frac{k}{2}. \tag{5.74}$$

Let us take as an example the magic core $Z = N = 4$ above which two nucleons are placed. We can form the nuclei $^{10}_{4}$Be, $^{10}_{5}$B, and $^{10}_{6}$C. As $k = 2$, T can assume the values 0 and 1. The component T_z for each nucleus has the values -1, 0, and $+1$, respectively. Figure 5.11 exhibits the level diagrams of each of these nuclei. We can see that states of T equal to 1 form a triplet with a representative in each of the nuclei and states of $T = 0$

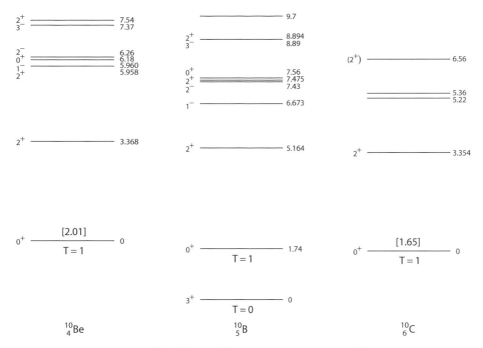

Figure 5.11 States in $T = 1$ of $^{10}_{4}$Be, $^{10}_{5}$B, and $^{10}_{6}$C, above the ground state of $^{10}_{5}$B with $T = 0$. The effect of the Coulomb energy 5.3 is cancelled by adding its values to the binding energy of each isobar. The energy values are given in MeV [Wa93].

exist only in ^{10}B, of component $T_z = 0$. The states that compose each triplet have the same approximate energy in the three nuclei; they are referred to as *isobaric analog states*. As the effect of the Coulomb energy was subtracted, this result is another indication that the nuclear forces are independent of charge. In other words, if the p-p, n-n, and p-n nuclear forces are identical, systems of A nucleons should have the same properties and, in particular, if we discount the Coulomb force, the same energy levels. The existence of states of $T = 0$ for ^{10}B that do not have partners in ^{10}Be and ^{10}C can be understood by the Pauli principle: ^{10}B has five n-p pairs while ^{10}Be and ^{10}C have four n-p pairs plus a pair of identical nucleons. Thus, the last pair of ^{10}B can produce states not allowed for the last pair of ^{10}Be and ^{10}C.

It is now useful to recall the study we did with the deuteron in chapter 2, and to observe that the systems p-p, n-n, and the excited state of the deuteron form a triplet of states similar to $T = 1$ above the ground state of the deuteron with $T = 0$. From this point of view it is directly justified that the p-p and n-n systems do not exist, since the excited state of the deuteron is not bound.

A nucleus characterized by Z and N is also characterized by the isospin component $T_z = (Z - N)/2$. The isospin value T of a given state of this nucleus cannot be smaller than $|T_z|$, since T_z is the projection of $\mathbf{T}$. Observing the example of figure 5.11, we see that the ground states have all the smallest possible values of T for each case, that is, $T = |T_z|$. One can verify that the rule works well for all light nuclei.

5.5 Residual Interaction

With the possible values established for the angular momenta of the levels that are split by the residual interaction, the question that immediately arises is the location in energy of those new levels. The hypothesis of the shell model is, as we have seen, to suppose a small effect of the residual interaction and, in that way, to assume H_{res} in (5.36) as a small perturbation. The energy correction to E (eq. 5.41) is given in first order perturbation theory by

$$\Delta E = \int \Psi^* H_{res} \Psi \, d^3 r, \tag{5.75}$$

that is, ΔE is obtained by the expectation value of the residual interaction, calculated with the nonperturbed wavefunctions.

Let us look for example at the case of two particles in excess of a closed core. We can for this case write the residual Hamiltonian (5.37) as

$$H_{res} = \sum_{i=3}^{A} \sum_{ij=3}^{A} V_{i,j}(\mathbf{r}_i, \mathbf{r}_j) - \sum_{i=3}^{A} U(r_i) + \sum_{i=3}^{A} V_{1,i} - U(r_1) + \sum_{i=3}^{A} V_{2,i} - U(r_2) + V_{1,2}. \tag{5.76}$$

The first two parts refer to the closed core and, if we stipulate that it is inert, they can be ignored. For the four following parts we can suppose to a first approximation that the idea of the model is valid, that is, the interaction of particles 1 and 2 with the core is given by an average potential U. Thus, these parts cancel out and we remain with

$$H_{res} = V_{1,2} \tag{5.77}$$

for use in (5.75).

The interaction potential $V_{1,2}$ of valence nucleons would be, in principle, the nucleon-nucleon potential we studied earlier. That potential is not, however, of immediate application when the particles are bound to the nucleus because the Pauli principle modifies in a drastic way the effect of the interaction for bound particles. To find a nucleon-nucleon potential that works well inside the nucleus, two approaches are usual. In the first, one tries to obtain a potential for bound nucleons from the nucleon potential outside the nucleus. This microscopic treatment is very complicated, and the practical results are not very satisfactory. In the second, a phenomenological treatment, a parametrized form is proposed for the interaction and the parameters are determined by a comparison with experimental data. Other conditions of importance for the choice of an effective potential are the simplicity of its use and the elements of the physics of the problem that it incorporates. An example is the *surface delta interaction* (SDI),

$$V(1, 2) = A\delta(\mathbf{r}_1 - \mathbf{r}_2)\delta(|\mathbf{r}_1| - R_0), \tag{5.78}$$

which tries to incorporate the facts that: 1) the interaction is of short range and 2) inside the nucleus the nucleons are practically free due to the action of the Pauli principle, so the interaction should be located at the nuclear surface of radius R_0. The presence of the delta function makes the calculation of the matrix elements (5.75) relatively simple.

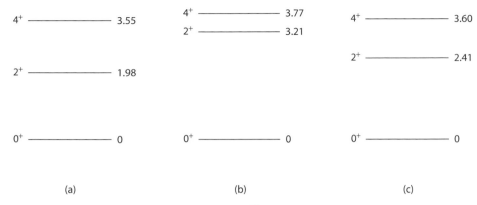

Figure 5.12 Ground state and first excited states of $^{18}_{8}$O: (a) experimental values of the energy (MeV); (b) calculated with the configuration $1d_{\frac{5}{2}}$; (c) calculated with a mixture of the configurations $1d_{\frac{5}{2}}$, $2s_{\frac{1}{2}}$, and $1d_{\frac{3}{2}}$.

Let us take as a practical example $^{18}_{8}$O. For the shell model this nucleus is constituted of a doubly closed core of 8 protons and 8 neutrons plus two valence neutrons in the level $1d_{\frac{5}{2}}$. In agreement with table 5.5, the neutron interaction will split the level $j = \frac{5}{2}$ into three states, of $j = 0, 2$, and 4, all with isotopic spin $T = 1$ (two neutrons) and positive parity ($l = 2$). The corresponding levels appear in the experimental spectrum of $^{18}_{8}$O (figure 5.12a).

Brussaard and Glaudemans [BG77] calculated the matrix elements for this nucleus using a modified surface delta interaction (MSDI),

$$V(1, 2) = A\delta(\mathbf{r}_1 - \mathbf{r}_2)\delta(|\mathbf{r}1| - R_0) + B[\mathbf{t}^{(1)} \cdot \mathbf{t}^{(2)}] + C, \tag{5.79}$$

which is obtained from SDI by adding a constant part and another with the Heisenberg operator (3.16), substantially increasing the agreement with experience. The parameters A, B, and C are given by using (5.79) to determine a great number of binding and excitation energies in a certain region of masses, assuming that these parameters are constant in that region. The values of the matrix elements determined for the three levels

$$\left\langle \left(d_{\frac{5}{2}}\right)^2 |V_{12}| \left(d_{\frac{5}{2}}\right)^2 \right\rangle_{J=0, T=1} = -2.78 \text{ MeV},$$

$$\left\langle \left(d_{\frac{5}{2}}\right)^2 |V_{12}| \left(d_{\frac{5}{2}}\right)^2 \right\rangle_{J=2, T=1} = 0.43 \text{ MeV},$$

$$\left\langle \left(d_{\frac{5}{2}}\right)^2 |V_{12}| \left(d_{\frac{5}{2}}\right)^2 \right\rangle_{J=4, T=1} = 0.99 \text{ MeV}, \tag{5.80}$$

establish the values of the excitation energy of the states 2^+, $E_{2^+} = 0.43 - (-2.78) = 3.21$ MeV, and 4^+, $E_{4^+} = 0.99 - (-2.78) = 3.77$ MeV, which appear in figure 5.12b. The value of the binding energy of ^{18}O can be obtained through the experimental value of the binding energy of ^{16}O:

$$B(^{18}O) = B(^{16}O) + 2S_n(d_{\frac{5}{2}}) - \left\langle (d_{\frac{5}{2}})^2 |V_{12}|(d_{\frac{5}{2}})^2 \right\rangle_{J=0}, \tag{5.81}$$

where S_n is the separation energy of a neutron of the level $d_{\frac{5}{2}}$ that can, in turn, be determined by (4.10),

$$S_n = B(^{17}O) - B(^{16}O), \tag{5.82}$$

from the experimental values $B(^{17}O) = 131.77$ MeV and $B(^{16}O) = 127.62$ MeV. With these values the result of 5.81 is $B(^{18}O) = 138.7$ MeV, in good agreement with the experimental value $B(^{18}O)_{exp} = 139.8$ MeV.

The above results were obtained supposing the two neutrons permanently in $1d_{\frac{5}{2}}$, that is, the *configuration space* was limited to that configuration. If we compare figures 5.12 (a) and (b) we will see that the excitation energy is poorly determined, not leading to a group of parameters that reproduce the relative distance of the levels 2^+ and 4^+. This is a direct consequence of the limitation of our configuration space and, in fact, there is no reason not to admit that the two neutrons can spend part of their time in other configurations close in energy. When we take into account a group of possible configurations for the valence nucleons, we are making a *configuration mixing* and, with that resource, the results are substantially better. Part (c) of figure 5.12 exhibits the same excited states of ^{18}O but this time using a mixture of the configurations $1d_{\frac{5}{2}}$, $2s_{\frac{1}{2}}$ and $1d_{\frac{3}{2}}$. The agreement with experiment is clearly better.

This approach has a price, however. As we increase the number of configurations, especially if there are more than two valence nucleons, we have to solve the problem of the diagonalization of matrices of dimensions that can reach thousands. This requires sufficiently fast computers, and progress of shell model calculations has, in fact, been conditioned to progress in computer science.

5.6 Nuclear Vibrations

We shall now study collective excitation modes that affect the nucleus as a whole and not just a few nucleons. The model in this section assumes that the nuclear surface can accomplish oscillations around an equilibrium form in the same way as happens with a liquid drop. The starting point is to imagine that a point on the surface of the nucleus is now defined by its radial coordinate $R(\theta, \phi, t)$, a function of the polar and azimuthal angles and of time. Let us consider, for simplicity, oscillations around a spherical form that do not alter the volume and the nuclear density. To describe the form of the nucleus at each instant it will be very convenient to use the property that a function of two variables can be expanded in an infinite series of spherical harmonics and write:

$$R(\theta, \phi, t) = R_0 \left[1 + \sum_{\lambda=0}^{\infty} \sum_{\mu=-\lambda}^{\mu=+\lambda} \alpha_{\lambda\mu}(t) Y_{\lambda}^{\mu}(\theta, \phi) \right], \tag{5.83}$$

where the dependence on time is transferred to the coefficients of the expansion.

The application of (5.83) to our problem immediately imposes, conditions on the possible values of λ. Thus, a first examination of this equation shows that the term $\lambda = 0$ just corresponds to a change in the radius of a spherical form, and that is against our hypothesis of volume conservation. This term must, therefore, be removed from the sum. In the same way, the terms with $\lambda = 1$ are not relevant to the description of our model because they just correspond to the center of mass motion of the nucleus. In fact, let X, Y, Z be the coordinates of the center of mass, which we will assume to be fixed at the origin. Thus ($dv = d^3r$),

$$X = \int x \, dv = 0, \quad Y = \int y \, dv = 0, \quad Z = \int z \, dv = 0. \tag{5.84}$$

But

$$\begin{aligned} X = \int x dv &= \int r \sin\theta \, \cos\phi \, r^2 dr d\Omega \\ &= \int \sin\theta \, \cos\phi \left(\int_0^R r^3 \, dr \right) d\Omega \\ &= \frac{1}{4} \int R^4 \sin\theta \, \cos\phi \, d\Omega. \end{aligned} \tag{5.85}$$

In a similar way,

$$Y = \frac{1}{4} \int R^4 \sin\theta \, \sin\phi \, d\Omega \quad \text{and} \quad Z = \frac{1}{4} \int R^4 \cos\theta \, d\Omega. \tag{5.86}$$

If we admit now that the oscillations are small compared with the radius, we can write

$$\begin{aligned} R &= R_0[1 + \epsilon(\theta, \phi)], \\ R^4 &= R_0^4(1 + \epsilon)^4 \cong R_0^4(1 + 4\epsilon), \end{aligned} \tag{5.87}$$

Thus,

$$\begin{aligned} X = 0 &\to \int (1 + 4\epsilon)(Y_1^1 + Y_1^{1*}) \, , d\Omega = 0, \\ Y = 0 &\to \int (1 + 4\epsilon)(Y_1^1 - Y_1^{1*}) \, , d\Omega = 0, \\ Z = 0 &\to \int (1 + 4\epsilon)Y_1^0 \, d\Omega = 0. \end{aligned} \tag{5.88}$$

Expanding the first equation gives

$$\begin{aligned} \int (1 + 4\epsilon)(Y_1^1 + Y_1^{1*}) \, d\Omega &= \int \left(1 + 4 \sum_{\lambda\mu} \alpha_{\lambda\mu} Y_\lambda^\mu \right)(Y_1^1 + Y_1^{1*}) \, d\Omega \\ &= \int (Y_1^1 + Y_1^{1*}) \, d\Omega + 4 \sum_{\lambda\mu} \alpha_{\lambda\mu} \left[\int Y_\lambda^\mu (-Y_1^{-1*}) \, d\Omega + \int Y_\lambda^\mu Y_1^{1*} \, d\Omega \right] \\ &= 4(\alpha_{11} - \alpha_{1-1}) = 0. \end{aligned} \tag{5.89}$$

Thus, $\alpha_{11} = \alpha_{1-1}$. The condition $Y = 0$ implies $\alpha_{11} = -\alpha_{1-1}$; thus, $\alpha_{11} = \alpha_{1-1} = 0$. The condition $Z = 0$ yields $\alpha_{10} = 0$. Therefore, fixing the center of mass at the origin gives $\alpha_{11} = \alpha_{10} = \alpha_{1-1} = 0$, eliminating the terms $\lambda = 1$ from our sum.

Therefore, the vibration modes that are of interest to us begin with $\lambda = 2$; they are the quadrupole oscillations, where the deformations take the nucleus to a form similar to an ellipsoid. For $\lambda = 3$ we have octupole oscillations, for $\lambda = 4$ hexadecapole oscillations, and so on. Figure 5.13 shows the main vibrational modes.

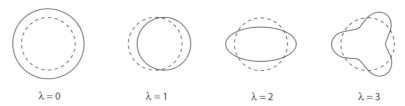

$\lambda = 0$ $\qquad\qquad$ $\lambda = 1$ $\qquad\qquad$ $\lambda = 2$ $\qquad\qquad$ $\lambda = 3$

Figure 5.13 The first vibration modes of the nuclear surface, showing the form of the nucleus for each mode (solid line) in comparison to the original spherical nucleus. The dotted line represents the original "fixed volume" sphere.

The energy problem of an oscillating liquid drop was solved by Lord Rayleigh in 1877 [Ra879]. He considered an incompressible and irrotational fluid, and arrived at the expression

$$T = \frac{1}{2} \sum_{\lambda\mu} B_\lambda \mid \dot{\alpha}_{\lambda\mu}(t) \mid^2 \tag{5.90}$$

for the kinetic energy, where

$$B_\lambda = \frac{\rho R_0^5}{\lambda}, \tag{5.91}$$

ρ being the density and R_0 the radius of the drop. For the potential energy the result was a value

$$V = \frac{1}{2} \sum_{\lambda\mu} C_\lambda \mid \alpha_{\lambda\mu} \mid^2, \tag{5.92}$$

with

$$C_\lambda = S R_0^2 (\lambda - 1)(\lambda + 2) - \frac{3}{2\pi} \frac{Z^2 e^2}{R_0} \frac{\lambda - 1}{2\lambda + 1}, \tag{5.93}$$

where one subtracts the Coulomb energy of the protons reduced by the distortion from the first part of Rayleigh's calculations. The factor S is the surface tension, which can be calculated by the mass formula

$$4\pi R_0^2 S = a_S A^{2/3} \text{ MeV}. \tag{5.94}$$

To adapt the above calculation to our problem, it is necessary to write the Hamiltonian starting from the classical expressions (5.90) and (5.92). If we define the coefficients $\alpha_{\lambda\mu}$ as generalized coordinates, the canonical conjugated momentum of $\alpha_{\lambda\mu}$ is

$$P_{\lambda\mu} = \frac{\partial \mathcal{L}}{\partial \dot{\alpha}_{\lambda\mu}} = B_\lambda \dot{\alpha}_{\lambda\mu}, \tag{5.95}$$

where $\mathcal{L} = T - V$ is the Lagrangian. Thus, the Hamiltonian is given by

$$H = T + V = \sum_{\lambda\mu} \left(\frac{1}{2B_\lambda} P_{\lambda\mu}^2 + \frac{1}{2} C_\lambda \alpha_{\lambda\mu}^2 \right), \tag{5.96}$$

which is a sum of harmonic oscillator Hamiltonians

$$H_{osc} = \frac{p^2}{2m} + \frac{1}{2} k q^2, \tag{5.97}$$

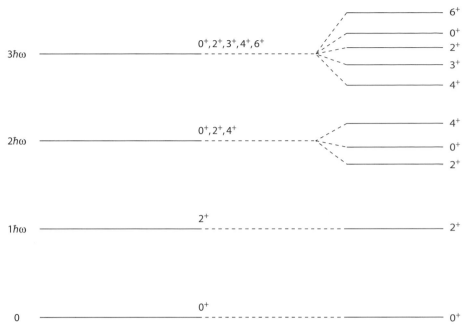

Figure 5.14 Three quadrupole phonons.

whose frequency ω is given by $\sqrt{k/m}$. Thus, the vibrations we are considering can be understood as a sum of harmonic oscillators of frequency

$$\omega_\lambda = \sqrt{\frac{C_\lambda}{B_\lambda}},$$ (5.98)

and our quantization of energy scheme is composed of levels built from these oscillators. When the nucleus passes from a level to a higher or a lower one it is said to have absorbed or emitted a *phonon* of vibration. Thus, quadrupole, octupole, etc., vibration phonons contribute to building the vibration spectrum of the nuclei starting from the ground state; see figure 5.14. This spectrum is referred to as the *vibrational band*.

There is a practical way of establishing which spins result from the coupling of a given phonon number. To see, for example, why in the coupling of two phonons (with $\lambda = 2$) the spins 1 and 3 are absent, consider the scheme of table 5.6. In principle, the total spins could be $I = 0$, 1, 2, 3, and 4. But this scheme all shows the possible combinations of the quantum numbers μ_1 and μ_2 of the two phonons whose sum is μ. Take the spin $I = \lambda = 4$; its possible projections are $\mu = +4, +3 \ldots, -4$. For spin $I = 3$, the maximum projection would be +3. But there is no other value $\mu = 3$ besides the one attributed to spin 4; thus it is not possible for spin 3 to exist. That is, we only obtain three groups of values of μ, which correspond to spins 0, 2, and 4. In the same way $I = 1$ is prohibited. A similar method can be used for the combination of three phonons to justify the absence of $I = 1$ and $I = 5$. It should be kept in mind in this picture that the phonons, having integer spins, do not obey the Pauli principle and, therefore there is no barrier to two phonons having the same quantum numbers.

Table 5.6 The several possible configurations for the states of two phonons, with $\lambda = 2$.

$\mu_{1,2}$															
+2	XX	X	X	X	X										
+1		X				XX	X	X	X						
0			X				X			XX	X	X			
−1				X				X			X		XX	X	
−2					X				X			X		X	XX
μ	+4	+3	+2	+1	0	+2	+1	0	−1	0	−1	−2	−2	−3	−4

A more formal justification for the results of the previous paragraph can be obtained directly from the wavefunction for the coupling of the states that describe the two phonons. If we use the appropriate Clebsch-Gordan coefficients we will see, for example, that the coupling of two phonons of $\lambda = 2$ results in symmetrical functions for spins 0, 2, and 4 and antisymmetric functions for spins 1 and 3. The last ones are inadmissible for the description of a composed system of bosons.

The vibrational model predicts, in the ideal case, a ratio equal to 2 between the energies of the second and first excited states. Figure 5.15 shows the experimental values of this ratio for even-even nuclei, as a function of the neutron number N. There is an oscillation around the value 2 for $40 < N < 80$, but in other regions, around 100 and 140, the energies show a ratio close to 3.3. These last regions are characteristic of large deformations of the nucleus, and the low energy levels are rotational states, which we shall study in the next section.

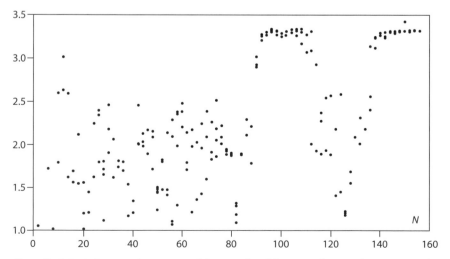

Figure 5.15 Ratio between the energies of the second and first excited states of even-even nuclei as a function of the neutron number N. The sampling is composed of all the even-even nuclei plus the ones of long lifetime for the isotopes with $Z > 82$.

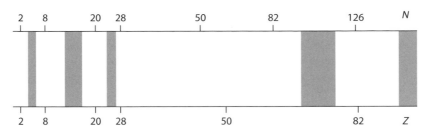

Figure 5.16 Forms of the nucleus in the ground state. The areas in white correspond to spherical nuclei and the dotted areas to deformed nuclei.

To conclude, it is worth mentioning that odd nuclei also admit vibrational bands. If the nucleus is composed of a spherical core plus an extra nucleon, states can be formed by the coupling of the individual orbit j with the vibrational states of the core. Thus, for example, these can be created in ^{63}Cu, whose ground state is $\frac{3}{2}^-$, a quadruplet of states $\frac{1}{2}^-, \frac{3}{2}^-, \frac{5}{2}^-$, and $\frac{7}{2}^-$, resulting from the coupling of $j = \frac{3}{2}^-$ with $\lambda = 2$.

5.7 Nuclear Deformation

In certain regions of the periodic table the nuclei in their ground estate are deformed (figure 5.16). Experimental evidence of this fact is the high values of the quadrupole moment in the region of $150 \lesssim A \lesssim 180$ and $A \gtrsim 240$ (figure 4.5).

The reason that certain nuclei have permanent deformations is the way of filling up their orbits starting from closed shells. Let us imagine, for example, that the first available state above a closed shell is a state with $l = 2$. The angular part of the wavefunction for the three possible situations is shown in figure 5.17. The first nucleon pair will give preference to $m_l = +2$ and $m_l = -2$, because in that case the overlap of their wavefunctions is larger and leads to a larger binding energy, since the force between them is attractive (pairing force). Let us now assume that we have one level g ($l = 4$) just above level d. Since the figure for Y_4^4 is spatially similar to the figure for Y_2^2, the next nucleon will give preference to staying at g with $m = |4|$ (because the overlap of its wavefunction with those of the previous nucleon is larger) over d with $m = \pm 1$ or $m - 0$. The grouping of these nucleons

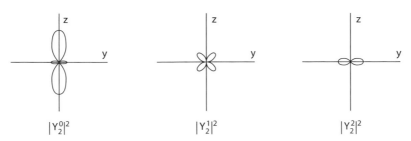

Figure 5.17 Visualization in the y, z-plane of the space distribution of the functions $\left| Y_2^m \right|^2$.

will create a space asymmetry in the wavefunction of the nucleus, and that characterizes a deformation.

Although a more complete understanding of that phenomenon requires sophisticated models and calculations, the qualitative argument above exhibits in a simple way that the deformation is due to a preference for a form that makes the binding energy for the nucleus larger.

5.8 The Nilsson Model

The fact that some nuclei are deformed in the ground state suggests the use of shell models for which the central potential is no longer spherically symmetric. Such a potential was developed by S. G. Nilsson in 1955 [Ni55], proposing employment of the single particle Hamiltonian

$$H = H_0 + Cl \cdot s + Dl^2, \tag{5.99}$$

where H_0 is an anisotropic harmonic oscillator Hamiltonian

$$H_0 = -\frac{\hbar^2}{2m} \nabla^2 + \frac{1}{2} m(\omega_x^2 x^2 + \omega_y^2 y^2 + \omega_z^2 z^2), \tag{5.100}$$

$l \cdot s$ is the usual spin-orbit coupling term of the spherical shell model, and l^2 is a term thought to simulate a flattening of the oscillator potential and make it closer to a real potential. C and D are constants to be determined by adjustment to the experimental results.

If we are just interested in the particular case of axial symmetry, we can write

$$\omega_x^2 = \omega_y^2 = \omega_0^2 \left(1 + \frac{2}{3}\delta\right),$$
$$\omega_z^2 = \omega_0^2 \left(1 - \frac{4}{3}\delta\right), \tag{5.101}$$

where the frequencies are made to be a function of two parameters, ω_0 and δ; these can be related through the condition imposed by the conservation of the nuclear volume (assuming that it can be determined by an equipotential surface). The result of this condition takes us to the relation $\omega_x \omega_y \omega_z = \text{const} = \overline{\omega}_0^3$, where $\overline{\omega}_0$ is the frequency for a zero deformation. Thus,

$$\omega_0(\delta) = \overline{\omega}_0 \left(1 - \frac{4}{3}\delta^2 - \frac{16}{27}\delta^3\right)^{-1/6} \tag{5.102}$$

is a relationship that leaves us with a single parameter, called the *deformation parameter*, δ. Thus, the resulting energy by diagonalization of the Hamiltonian are functions of δ. The procedure employed for the calculation of the energy will not be presented here. The result

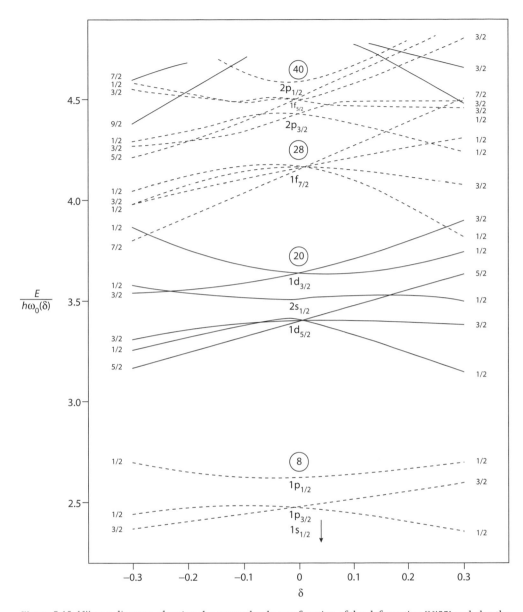

Figure 5.18 Nilsson diagram, showing the energy levels as a function of the deformation [Ni55] and also the number of nucleons at shell closure for $\delta = 0$.

is shown in figure 5.18, where we plot the energy levels as a function of the deformation δ. A straight vertical line drawn on this graph indicates the energy levels allowed for a given value of the parameter δ. Positive values of δ correspond to a prolate nucleus and negative values to an oblate nucleus. It is also important to mention that, for a deformed potential, j and l are no longer good quantum numbers, classically corresponding to the fact that the angular momentum is no longer a constant of motion for a non-spherically symmetric potential.

Table 5.7 Magnetic moments determined by the Nilsson model. The Schmidt values were calculated according to section 4.6, for $j = I \mp \frac{1}{2}$.

Nucleus	δ	μ_{th}	μ_{exp}	Schmidt lines	
^{23}Na	0.5	2.4	2.22	−0.4	3.8
^{173}Yb	0.28	−0.8	−0.65	−1.9	1.2
^{187}Re	0.19	3.7	3.2	0.8	4.8

a) The model determines correctly the spin of all light nuclei and of almost all heavy ones.

b) The determination of magnetic moments of many nuclei is now much more accurate. Table 5.7 lists the experimental and theoretical values for three nuclei, together with the values given by the corresponding Schmidt lines.

c) Electric quadrupole moments for the region $150 < A < 180$ are now calculated in a correct way. Figure 5.19 exhibits values of the deformation parameter δ obtained from the quadrupole moments for nuclei of that region.

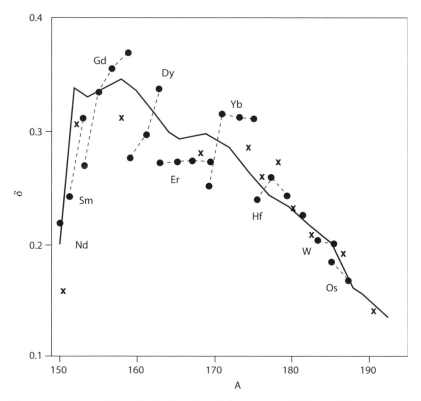

Figure 5.19 Values of the calculated nuclear deformation (solid line) and the measured ones, in the actinide region [MN55].

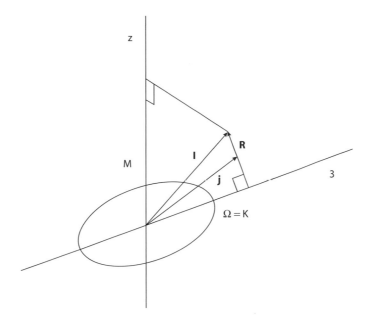

Figure 5.20 Coupling of angular momenta in a deformed nucleus with symmetry axis in the direction 3.

5.9 The Rotational Model

The existence of nuclear deformation opens the possibility of nuclear rotations as a new way for the nucleus to absorb energy. The reason is that it does not make sense to think about rotations of a spherically symmetrical quantum system. In fact, one has observed energy bands in deformed nuclei with level spacing characterized unequivocally as a collective nuclear rotation.

To study these rotational bands, let us look at figure 5.20. The vector **R** represents the angular momentum of rotation, which due to the axial symmetry, is perpendicular to the symmetry axis 3 (we denote by 1, 2, and 3 the system of axes fixed to the nucleus). The vector **j** represents the intrinsic angular momentum, and its projection Ω on 3 coincides with the projection K of the total angular moment $\mathbf{I} = \mathbf{R} + \mathbf{j}$.

Classically, the kinetic rotation energy is given by

$$T_{\text{rot}} = \frac{R^2}{2\tau} = \frac{(\mathbf{I} - \mathbf{j})^2}{2\tau}, \tag{5.103}$$

where τ is the moment of inertia for a rotation around an axis perpendicular to axis 3. Writing 5.103 in terms of angular operators,

$$T_{\text{rot}} = \frac{\hbar^2}{2\tau}(I^2 + j^2 - 2I_3j_3) + T_{\text{coup}}, \tag{5.104}$$

where T_{coup} includes small terms related to the coupling of the intrinsic and rotational parts by the Coriolis force; these are generally negligible at slow rotations.

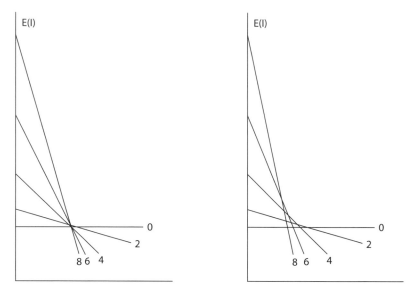

Figure 5.21 Lipkin focal diagrams. The band that corresponds to the diagram (a) has strict agreement with (5.106). In the diagram (b) the agreement is only approximate.

The rotation eigenfunctions, labeled by $| IMK \rangle$, obey the eigenvalue equations

$$I^2 \, | \, IMK \rangle = I(I + 1) \, | \, IMK \rangle,$$

$$I_z \, | \, IMK \rangle = M \, | \, IMK \rangle,$$

$$I_3 \, | \, IMK \rangle = K \, | \, IMK \rangle. \tag{5.105}$$

From the relationships (5.104) and (5.105) we can verify that the energy of a rotational band, that is, the expected values of the operator T_{rot}, has the form

$$E(I) = AI(I + 1) + B, \tag{5.106}$$

where $A = \hbar^2/2\tau$ and B is a constant that involves the projection K and quantities related to the intrinsic part of the system. The energies of the band are obtained by the values of $E(I)$ corresponding to the integer numbers $I \geq K$.

It is possible to see if a sequence of levels obeys 5.106 using *Lipkin focal diagrams* (figure 5.21): the values of $E(I)$ are placed along a vertical axis and a straight line of inclination $I(I + 1)$ to the horizontal axis is traced from those points. If (5.106) is obeyed, the straight lines cross in a single focal point (a); if the agreement is only approximate there is the formation of a caustic curve, (b).

Let us examine now some well-known rotational bands. We will treat three cases separately, since the formation of the levels is very different for each of them.

a) Even-even nuclei. The intrinsic part does not contribute to the angular momentum and, in this way, $K = 0$. We have a band beginning in $I = 0^+$ and with excited states with $l = 2^+, 4^+, 6^+$, etc. The absence of odd terms is due to the form of the wavefunction, and this must incorporate the symmetry with regard to the x, y-plane. If we want the level of

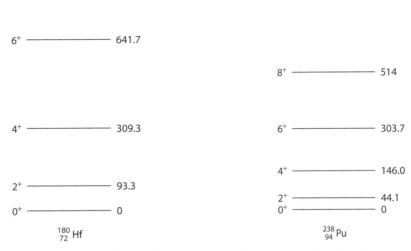

Figure 5.22 Rotational bands of even-even nuclei. The numbers to the right of the levels are the energy values in keV.

zero energy to correspond to the state 0^+, the band will have the simple form

$$E(I) = \frac{\hbar^2}{2\tau} I(I + 1) \qquad I = 0, 2, 4, \ldots \tag{5.107}$$

Figure 5.22 gives two examples, one in the rare earth and the other in the actinide region.

The rotational bands are of easy experimental identification. If, for example, the level 8^+ of one of the bands of figure 5.22 is excited, it will decay to the level 6^+ and so on, forming the *cascade* $8^+ \rightarrow 6^+ \rightarrow 4^+ \rightarrow 2^+ \rightarrow 0^+$. If the intensity is enough, the rotational band is easily reconstructed from the γ-ray spectrum.

b) Odd nuclei with $K \neq \frac{1}{2}$. In this case the band is mounted on the ground state $I = K$, with excited states $I = K + 1, K + 2, K + 3$, etc., with energy levels given by

$$E(I) = \frac{\hbar^2}{2\tau} [I(I + 1) - K(K + 1)], \qquad I = K, K + 1, K + 2, \ldots \tag{5.108}$$

An example is seen in figure 5.23a.

A given nucleus can admit other rotational bands beyond the one associated with the ground state. In an odd nucleus, for example, an excitation of the intrinsic part creates a new state (generally with a new value of K) starting from which a new rotational band can be created.

c) Odd nuclei with $K = \frac{1}{2}$. The coupling term that was neglected in (5.104) has the property of mixing bands in which the values of K differ by 1 unit. As K and $-K$ enter simultaneously in the wavefunction, the band of $K = \frac{1}{2}$ is perturbed to first order, resulting

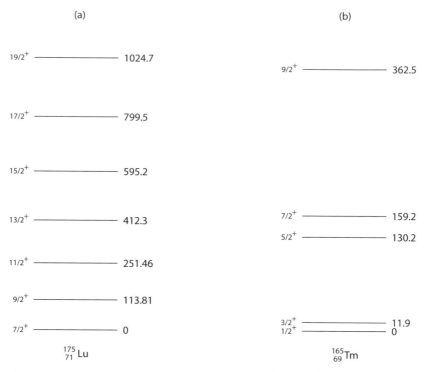

Figure 5.23 Rotational bands for odd nuclei with: a) $K \neq \frac{1}{2}$, b) $K = \frac{1}{2}$. The numbers to the right are the values of the energy in keV.

in a modification of the energy values, which are now calculated by the expression

$$E_{K=\frac{1}{2}}(I) = \frac{\hbar^2}{2\tau}[I(I+1) - a(-1)^{I+\frac{1}{2}}(I+\frac{1}{2})], \qquad I = \frac{1}{2}, \frac{3}{2}, \frac{5}{2}, \dots \qquad (5.109)$$

with a, the *decoupling parameter*, only depending on the intrinsic part. It can be determined, together with τ, if we know the energy of two excited states. Let us observe that, for this band, $I = K = \frac{1}{2}$ is not necessarily the level of lower energy. An example of a band with $K = \frac{1}{2}$ is seen in figure 5.23b. In this example the ordering of the spins is regular but the values of the energy are not proportional to $I(I+1)$.

Expressions (5.107), (5.108), and (5.109) work correctly just for the first levels of a band. As I grows, the predicted results are placed farther and farther below than the experimental values. An improvement can be made if we write the energy dependence in I as the series

$$E(I) = AI(I+1) + B[I(I+1)]^2 + \cdots, \qquad (5.110)$$

where we have used in (5.106) just the first part of (5.110).

Expression (5.110) has a justification. Equations (5.107), (5.108), and (5.109) were obtained from the idea that the nucleus is a rigid body during rotation and that, as a consequence, it has a constant moment of inertia. Observation of the first rotational levels of Hf and Pu, in figure 5.22, shows that this is just approximate. As a result of the deformation of the nucleus with rotation, there is a small increase of the moment of inertia when I increases, with a consequent decrease of the energy $\hbar^2/2\tau$. Using (5.106) we see that this energy falls from 15.55 keV to 15.07 keV when we go from level 2^+ to level 8^+ of Hf, and, from 7.35 keV to 7.14 keV of Pu.

The variation of the moment of inertia τ with rotation can be expressed by writing the rotation energy as

$$E = \frac{1}{2}\tau(\omega)\omega^2, \tag{5.111}$$

with an explicit dependence of τ on the angular frequency of rotation ω. The behavior of the function $\tau(\omega)$ was studied by S. M. Harris [Ha65], extending a model created by D. R. Inglis [In54] to treat the rotation problem microscopically. This model has the same conception as the shell model, where a nucleon moves in the potential created by the other nucleons. The additional condition is that the deformed potential rotates around a fixed axis. Harris chose the series

$$\tau(\omega) = \tau_0 + \alpha\omega^2 + \beta\omega^4 + \cdots \tag{5.112}$$

to describe the dependence of the moment of inertia with the frequency ω. When the rotation is understood as a perturbation, a calculation of the perturbation to second order allows us to determine τ_0 (rigid rotor) in terms of the wavefunctions of the undisturbed nucleus (without rotation). The inclusion of terms of fourth order allows the determination of α.

In terms of practical applications, the angular frequency ω must be extracted from experimental quantities. From the classical formula

$$\omega = \frac{dE}{dI} \tag{5.113}$$

we can write the quantum relationship

$$\omega = \frac{\Delta E}{\hbar \Delta \sqrt{I(I+1)}}. \tag{5.114}$$

For the rotational band of an even-even nucleus we have

$$\hbar\omega = \frac{E_I - E_{I-2}}{\sqrt{I(I+1)} - \sqrt{(I-1)(I-2)}}, \tag{5.115}$$

which, for the highest levels, is reduced to a good approximation to

$$\hbar\omega = \frac{E_I - E_{I-2}}{2}. \tag{5.116}$$

For the first energy levels, the first two terms of the series (5.112) are enough and a graph of τ against ω^2 is a straight line. For growing values of ω, other terms can be important and the curve deviates smoothly upwards. However, we can come to a point where the curve radically assumes a different behavior. This was observed for the first time by A. Johnson, H. Ryde, and S. A. Hjorth [Jo72] when they studied the rotational band of ^{162}Er, induced by the reaction $(\alpha,3n)$ in ^{161}Dy. Figure 5.24 shows the spectrum of the obtained transition energy. The levels have a regular behavior up to $I \cong 10$. From this point on the transition energy decreases in growth and even decreases in value when I passes from 14 to 16. In a graph of τ against ω^2 this effect is shown in the form of a backward shift in the curve (backbending effect) (figure 5.25).

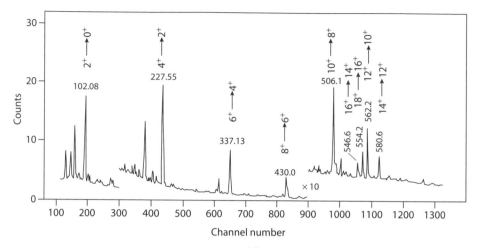

Figure 5.24 Energy spectrum of the γ-transitions of ^{162}Er. Each peak characterizes a transition inside of rotational band. The number of the channels in the horizontal axis is proportional to the energy of the emitted γ-ray [Jo72].

The independent particle model in a rotational potential cannot explain this phenomenon, related as it is to the residual interaction. The rotation, through the Coriolis force, competes with the pairing force, and it can even break the coupling of zero angular momentum of a pair of nucleons, which starts to align its angular momentum with that of the rotating nucleus (rotational alignment). The result is a global increase in the spin and moment of inertia values of the nucleus, the latter approaching the rigid body value.

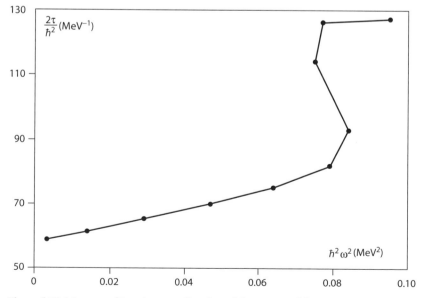

Figure 5.25 Moment of inertia τ as a function of the square of the angular frequency ω, calculated from the rotational band of ^{162}Er.

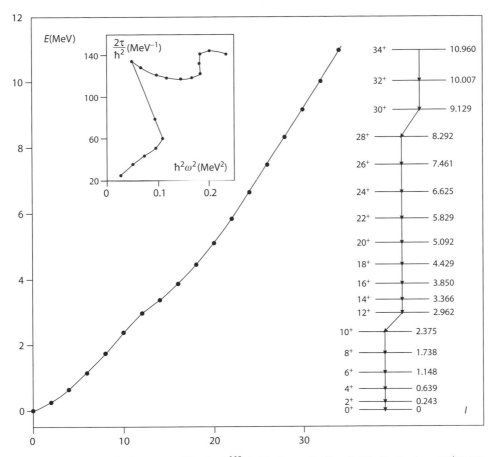

Figure 5.26 *Yrast* line for the rotational bands of $^{160}_{70}$Yb. The first excited band of the line begins at 12^+ (2.962 MeV), and the second at 30^+ (9.129 MeV). Several bands of negative parity, not shown in the figure, begin between 1.5 and 2.8 MeV.

This increase can be substantial in the region of rare earths where valence neutrons are located in the $i\frac{13}{2}$ orbit.

The change in the internal structure gives opportunity to the appearance of a new rotational band associated to the new parameters of the nucleus. The process can repeat, and what one observes is a group of bands, each associated to an internal structure of the nucleus. They are called *excited bands* to distinguish them from the *ground state band* created on the ground state of the nucleus. The state of lower energy for a given angular momentum is called the *yrast state* (a Swedish word that means *that which rotates more*). It is a state of pure rotation, without intrinsic excitation. In a graph of the energy E against the spin I, the sequence of all the *yrast* states is called the *yrast line* (see figure 5.26). The rotating nucleus is generally formed by a nuclear reaction in a state of high excitation, well above the *yrast* line. It decays by neutron emission and gamma radiation until reaching the *yrast* line. From that point on it decays to the ground state, passing by all states on the line.

The region of abrupt changes in the nuclear structure are also often visible in a graph of E against I. Figure 5.26 illustrates the case of $^{160}_{70}$Yb. The ground state band has the lowest energy up to $I = 10$. Starting from there the *yrast* line follows the excited band that begins in 12^+. At 30^+ we have a new turn, with the beginning of a new band. The first turn is visible as a change in the inclination of the *yrast* line, but it appears in a much clearer way in the plot τ against ω^2 inserted in the figure. The second turn is visible only in this last graph.

An even-even nucleus at the ground state has all the nucleon pairs coupled with angular momentum zero and, in this situation, it is equivalent to a *superfluid*. The pairing broken by the centrifugal force corresponds to a transition of the superfluid phase to that of a normal fluid. In this aspect there is an analogy with a superconductor, where a magnetic field provokes the change to normal conduction phase (Meissner effect).

For very high rotation speeds the new forces that govern the nuclear balance can make the surface of the nucleus assume forms very different from the ground state form. In particular, rotational bands where the nucleus has a very large deformation have been discovered since 1985. Today we know more than 160 *superdeformed bands* distributed in four regions, of $A = 80$, 130, 150, and 190. In these bands the nucleus has a prolonged ellipsoidal form with a ratio between the axes of 2:1, 1.5:1, 2:1, and 1.7:1, respectively. Only a few superdeformed bands can have a well established connection with the *yrast* line.

5.10 Microscopic Theories

Microscopic theories for the nuclei start with a fundamental NN interaction and build the properties of the nuclei in a self-consistent way. There are numerous theories which have been developed for different purposes. We will describe only two of these theories whose goal is to describe the static properties of the nuclei, for example, their density profiles, binding energies, etc.

5.10.1 Hartree-Fock theory

Let us consider a system of particles with a central (mean-field) potential, U_0, and a two-body (particle-particle) potential, v, for instance, the atomic electron system

$$H = \sum_i^A \left[T_i + U_0(\mathbf{r}_i) \right] + \frac{1}{2} \sum_{ij} v(\mathbf{r}_i, \mathbf{r}_j), \tag{5.117}$$

where the factor $\frac{1}{2}$ prevents double-counting the two-body interaction energy (see figure 5.27a). The interaction v in (5.117) is the residual interaction of (5.37).

Neglecting $v(r_i, r_j)$, we have

$$\left\{ -\frac{\hbar^2}{2m} \nabla^2 + U_0(\mathbf{r}) \right\} \psi_i(\mathbf{r}) = \epsilon_i \, \psi_i(\mathbf{r}). \tag{5.118}$$

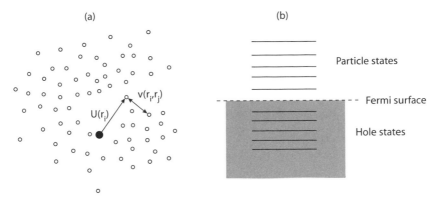

Figure 5.27 (a) Particles, i and j, interacting with a mean field $U_0(r)$ and among themselves with an interaction $v(\mathbf{r}_i, \mathbf{r}_j)$. (b) Labeling of states used in Hartree-Fock calculations.

If antisymmetrization is neglected, the total wavefunction is given by (5.40) and the average interaction felt by particle i due to all other particles is

$$U_1(\mathbf{r}_i) = \sum_{j \neq i}^{A} \int d^3 r_j |\psi_j(\mathbf{r}_j)|^2 v(\mathbf{r}_i, \mathbf{r}_j). \tag{5.119}$$

Now we add this potential to (5.118) and obtain a new wave equation:

$$\left\{ -\frac{\hbar}{2m} \nabla^2 + U_0(\mathbf{r}) + U_1(\mathbf{r}) \right\} \psi_i(\mathbf{r}) = \epsilon_i \psi_i(\mathbf{r}). \tag{5.120}$$

The *Hartree method* is now clear: one begins with the shell model Hamiltonian (5.118) and constructs the total wavefunction (5.40). From this wavefunction one calculates the average two-body interaction (5.119). One then solves the new wave equation (5.120). This sequence (5.40) $\rightarrow$ (5.119) $\rightarrow$ (5.120) is repeated (a process called iteration) until a stable solution is found. In general, only a few iterations are needed if $|U_0(r)| \gg |U_1(r)|$.

The *Hartree-Fock method* is easily derived from a variational method. The total wavefunction is now antisymmetrized as

$$\phi(1, 2, \ldots, A) = \mathcal{A} \, \psi_1(1) \psi_2(2) \ldots \psi_A(A), \tag{5.121}$$

where $\mathcal{A}\psi_1(1)\psi_2(2) \ldots \psi_A(A)$ is the Slater determinant of 5.42.

For a small variation $\delta\phi$, the energy expectation should be stationary. Thus,

$$\delta \langle \phi|H|\phi \rangle = \langle \delta\phi|H|\phi \rangle = 0. \tag{5.122}$$

The variation of $\delta\phi$ should preserve normalization of the wavefunctions:

$$\int |\psi_i(\mathbf{r})|^2 d^3 r = 1. \tag{5.123}$$

Now we solve the many-body Hamiltonian with two-body forces only, that is, with the Hamiltonian of (5.35). The expectation value with the wavefunction (5.121) is

$$\langle\phi|H|\phi\rangle = -\frac{\hbar^2}{2m}\sum_i^A \int \psi_i^*(\mathbf{r})\nabla^2\psi_i(\mathbf{r})d^3r$$

$$+\frac{1}{2}\sum_{ij}^A \iint \psi_i^*(\mathbf{r})\psi_j^*(\mathbf{r}')v(\mathbf{r},\mathbf{r}')\psi_i(\mathbf{r})\psi_j(\mathbf{r}')d^3r d^3r'$$

$$-\frac{1}{2}\sum_{ij}^A \iint \psi_i^*(\mathbf{r})\psi_j^*(\mathbf{r}')v(\mathbf{r},\mathbf{r}')\psi_i(\mathbf{r}')\psi_j(\mathbf{r})d^3r d^3r'. \tag{5.124}$$

The last term is a consequence of antisymmetrization. Applying the variation on $\psi_i(r)$ we obtain

$$-\frac{\hbar^2}{2m}\nabla^2\psi_i(\mathbf{r}) + \sum_j^A \int d^3r' \psi_j^*(\mathbf{r}')v(\mathbf{r},\mathbf{r}')\psi_j(\mathbf{r}')\psi_i(\mathbf{r})$$

$$-\sum_j^A \int d^3r' \psi_j^*(\mathbf{r}')v(\mathbf{r},\mathbf{r}')\psi_j(\mathbf{r})\psi_i(\mathbf{r}') = \epsilon_i\psi_i(\mathbf{r}) \tag{5.125}$$

where ϵ_i is seen as a Lagrange multiplier that enforces the constraint (5.123). It has the significance of a single particle energy.

We can rewrite (5.125) as

$$-\frac{\hbar^2}{2m}\nabla^2\psi_i(\mathbf{r}) + \int d^3r' U(\mathbf{r},\mathbf{r}')\psi_i(\mathbf{r}') = \epsilon_i\psi_i(\mathbf{r}), \tag{5.126}$$

where $U(r,r')$ is the self-consistent field

$$U(\mathbf{r},\mathbf{r}') = \delta(\mathbf{r}-\mathbf{r}')\sum_j^A \int d^3r' v(\mathbf{r},\mathbf{r}')\psi_j(\mathbf{r}')\psi_j^*(\mathbf{r}') - \sum_j^A v(\mathbf{r},\mathbf{r}')\psi_j(\mathbf{r})\psi_j^*(\mathbf{r}'). \tag{5.127}$$

The first term is the direct term (*Hartree field*). The second is called the exchange interaction and is nonlocal. This nonlocality is closely related to the range of the two-body interaction. If we use a δ-force, then the Fock (or exchange) term is also local.

In constructing the HF (Hartree-Fock) determinant one naturally selects the A lowest energy wavefunctions. Thus the HF state corresponds to a Fermi sea of particles with a sharp Fermi surface.

As in the Hartree method, one guesses an initial $U^{(0)}$, solves (5.126), finds ψ_i, calculates (5.127), finds new $\psi_i's$, and so on, until the desired accuracy is achieved.

5.10.2 The Skyrme interaction

The most popular effective interaction used in Hartree-Fock calculations is the *Skyrme interaction*. It is based on the fact that a finite range interaction can be simulated by a momentum dependence. This can be shown by transforming an interaction $V(r)$, where

$r = r_1 - r_2$, into momentum space,

$$\langle \mathbf{p} | v | \mathbf{p}' \rangle = \frac{1}{(2\pi\hbar)^3} \int e^{-i(\mathbf{p}-\mathbf{p}')\cdot\mathbf{r}/\hbar} v(\mathbf{r}) d^3r. \tag{5.128}$$

This integral gives a constant if $v(r) = \delta(r)$. A finite range $v(r)$ will represent a p-dependence in momentum space.

The simplest form that one may find for $\langle p | v | p' \rangle$ which is rotationally invariant is

$$(2\pi\hbar)^3 \langle \mathbf{p} | v | \mathbf{p}' \rangle = v_0 + v_1 \mathbf{p}'^2 + v_1 \mathbf{p}^2 + v_2 \mathbf{p} \cdot \mathbf{p}', \tag{5.129}$$

which in coordinate space is related to the momentum-dependent operator

$$v(\mathbf{r}) = v_0 \delta(\mathbf{r}) + v_1 \left[\hat{\mathbf{p}}^2 \delta(\mathbf{r}) + \delta(\mathbf{r}) \hat{\mathbf{p}}^2 \right] + v_2 \, \hat{\mathbf{p}} \cdot \delta(\mathbf{r}) \hat{\mathbf{p}}. \tag{5.130}$$

The Skyrme interaction is based on this property and is an effective interaction with a three-body term [Sk57, Sk59]

$$v = \sum_{i<j} v(i,j) + \sum_{i<j<k} v(i,j,k). \tag{5.131}$$

For $v(i,j)$ one uses the form (5.130) with

$$v(1,2) = t_0 \, (1 + x_0 \, P^\sigma) \delta(\mathbf{r}_1 - \mathbf{r}_2) + \frac{1}{2} \, t_1 \left[\delta(\mathbf{r}_1 - \mathbf{r}_2) \hat{\mathbf{k}}^2 + \hat{\mathbf{k}}^2 \, \delta(\mathbf{r}_1 - \mathbf{r}_2) \right]$$
$$+ t_2 \, \hat{\mathbf{k}} \, \delta(\mathbf{r} - \mathbf{r}_2) \hat{\mathbf{k}} + i \, W_0 \left[\sigma^{(1)} + \sigma^{(2)} \right] \left[\hat{\mathbf{k}} \times \delta(\mathbf{r}_1 - \mathbf{r}_2) \hat{\mathbf{k}} \right], \tag{5.132}$$

where

$$\hat{\mathbf{k}} = \frac{1}{\hbar} \hat{\mathbf{p}} = \frac{1}{2i} (\nabla_1 - \nabla_2) \quad \text{and} \quad P^\sigma = \frac{1}{2} \left[1 + \sigma^{(1)} \cdot \sigma^{(2)} \right]. \tag{5.133}$$

with σ the Dirac matrices that act on the spin part of the wavefunctions. The three-body part of the Skyrme interaction is also taken as a zero range (δ-function) force,

$$v(1,2,3) = t_3 \delta(\mathbf{r}_1 - \mathbf{r}_2) \delta(\mathbf{r}_2 - \mathbf{r}_3). \tag{5.134}$$

The constants t_0, t_1, t_2, t_3, x_0, and W_0 are manipulated so as to adjust the experimental binding energies and radii. There are several sets of parameters, often called Skyrme I, II, etc. For example, the Skyrme III interaction uses

$$t_0 = -1128.75 \text{ MeV fm}^3, \qquad t_1 = 395 \text{ MeV fm}^5,$$
$$t_2 = -95 \text{ MeV fm}^5, \qquad t_3 = 1.4 \times 10^4 \text{ MeV fm}^6,$$
$$W_0 = 120 \text{ MeV fm}^5, \qquad x_0 = 0.45. \tag{5.135}$$

The parameter t_0 describes a pure δ-force with a spin-exchange; t_1 and t_2 simulate an effective range, as in (5.130). The fourth term in (5.132) represents a two-body spin-orbit interaction. It can be obtained from a normal spin-orbit term in the short range limit.

To implement numerically a Hartree-Fock calculation with the Skyrme potential a little more algebra is necessary. In a long but straightforward calculation, it can be shown [VB72] that the HF equation in coordinate space becomes

$$\left\{-\nabla \frac{\hbar^2}{2m^*(\mathbf{r})}\nabla + U(\mathbf{r}) + \mathbf{W} \cdot \frac{1}{i}(\nabla \times \boldsymbol{\sigma})\right\}\psi_i(\mathbf{r}) = \epsilon_i\,\psi_i(\mathbf{r}), \qquad (5.136)$$

where

$$m^*(\mathbf{r}) = m\left[1 + \frac{2m}{\hbar^2}\frac{1}{16}(3t_1 + 5t_2)\rho\right]^{-1}, \qquad (5.137)$$

$$U(\mathbf{r}) = \frac{3}{4}\,t_0\,\rho + \frac{3}{16}\,t_3\rho^2 + \frac{1}{16}(3t_1 + 5t_2)\tau$$
$$+ \frac{1}{32}(5t_2 - 9t_1)\nabla^2\rho - \frac{3}{4}\,W_0\,\nabla\cdot\mathbf{J}, \qquad (5.138)$$

and

$$\mathbf{W}(\mathbf{r}) = \frac{3}{4}\,W_0\,\nabla\rho. \qquad (5.139)$$

In the equations above,

$$\rho(\mathbf{r}) = \sum_{i,s,t}|\psi_i(\mathbf{r},s,t)|^2 \qquad (5.140)$$

is the nucleon density,

$$\tau(\mathbf{r}) = \sum_{i,s,t}|\nabla\,\psi_i(\mathbf{r},s,t)|^2 \qquad (5.141)$$

is the kinetic energy density, and

$$\mathbf{J}(\mathbf{r}) = -i\sum_{j,t,s,s'}\psi_j^*(\mathbf{r},s,t)\left[\nabla\psi_j(\mathbf{r},s',t)\times\boldsymbol{\sigma}_{ss'}\right] \qquad (5.142)$$

is the "spin-orbit density." s and t denote the spin and isospin quantum numbers, respectively.

We observe that $U(r)$ is local and (5.136) is a pure differential equation. The nonlocality appears only in the r-dependence of $m^*(r)$, the effective nucleon mass. For spherical symmetry (5.136) reduces to a one-dimensional differential equation of second order in the radial coordinate r. Then the spin-orbit term in (5.138) becomes

$$\frac{3}{2}\,W_0\left(\frac{1}{r}\frac{\partial}{\partial r}\rho\right)\mathbf{l}\cdot\mathbf{s}, \qquad (5.143)$$

well known in the shell model.

5.10.3 Relativistic mean field theory

The *relativistic mean field* theory for the nuclear dynamics is based on a Lagrangian density (see Appendix D) that ascribes to each nucleon a Dirac field (or spinor) ψ_i which interacts with meson fields, that is, the nucleon-nucleon interaction is assumed to arise from the exchange of mesons.

The most common Lagrangian used is given by (see [SW86])

$$\mathcal{L} = \overline{\psi}\left(i\gamma_\mu\partial_\mu - m\right)\psi + \frac{1}{2}\partial_\mu\sigma\,\partial_\mu\sigma - U(\sigma) - \frac{1}{4}\,\Omega_{\mu\nu}\,\Omega_{\mu\nu}$$
$$+ \frac{1}{2}\,m_\omega^2\,\omega_\mu\,\omega_\mu - \frac{1}{4}\,\mathbf{R}_{\mu\nu}\,\mathbf{R}_{\mu\nu} + \frac{1}{2}\,m_\rho^2\,\boldsymbol{\rho}_\mu\boldsymbol{\rho}_\mu - \frac{1}{4}\,F_{\mu\nu}F_{\mu\nu}$$
$$- g_\sigma\,\overline{\psi}\sigma\psi - g_\omega\,\overline{\psi}\gamma_\mu\omega_\mu\psi - g_\rho\,\overline{\psi}\gamma_\mu\boldsymbol{\rho}_\mu\tau\psi - e\,\overline{\psi}\gamma_\mu A_\mu\psi, \tag{5.144}$$

where

$$\overline{\psi} = \psi^\dagger\gamma_0; \tag{5.145}$$

σ is the isoscalar-scalar $(T=0,\ S=0)$ σ-meson field

$$\Omega_{\mu\nu} = \partial_\mu\omega_\nu - \partial_\nu\omega_\mu, \tag{5.146}$$

in terms of the isoscalar-vector $(T=0,\ S=1)$ ω-meson field, ω_μ;

$$\mathbf{R}_{\mu\nu} = \partial_\mu\,\boldsymbol{\rho}_\nu - \partial_\nu\,\boldsymbol{\rho}_\mu - g_\rho\left(\boldsymbol{\rho}_\mu \times \boldsymbol{\rho}_\nu\right) \tag{5.147}$$

in terms of the isovector-vector $(T=1,\ S=1)$ ρ-meson field, $\boldsymbol{\rho}_\mu$;

$$F_{\mu\nu} = \frac{\partial A_\mu}{\partial x_\nu} - \frac{\partial A_\nu}{\partial x_\mu} \tag{5.148}$$

is the familiar expression for the electromagnetic fields tensor; m is the nucleon mass; m_σ, m_ω, and m_ρ are the masses of the σ-meson, ω-meson, and ρ-meson, respectively; and the corresponding coupling constants are g_ρ, g_ρ, g_ω.

The σ-meson moves in a potential with self-interacting nonlinear cubic (σ^3) and quartic (σ^4) terms with strength parameters g_2 and g_3, respectively:

$$U(\sigma) = \frac{1}{2}\,m_\sigma^2\sigma^2 + \frac{g_2}{3}\sigma^3 + \frac{g_3}{4}\sigma^4. \tag{5.149}$$

The Euler-Lagrange equations (see Appendix D) lead to the Dirac equation for the nucleons

$$\left\{-i\boldsymbol{\alpha}\cdot\boldsymbol{\nabla} + V(\mathbf{r}) + \beta\left(m + S(\mathbf{r})\right)\right\}\psi_i = \epsilon_i\psi_i, \tag{5.150}$$

where V is a repulsive vector potential

$$V(\mathbf{r}) = g_\omega\omega_0(\mathbf{r}) + g_\rho\,\tau_3\rho_0(\mathbf{r}) + e\,\frac{1+\tau_3}{2}A_0(\mathbf{r}). \tag{5.151}$$

τ_3 is the third component of the isospin operator, defined here[3] as $t_3\chi_p = -\frac{1}{2}\,\chi_p$, $t_3\,\chi_n = \frac{1}{2}\chi_n$, where χ_p, and χ_n are the proton and neutron isospin wavefunctions, respectively. S is the attractive scalar potential

$$S(\mathbf{r}) = g_\sigma\sigma(\mathbf{r}) \tag{5.152}$$

contributing to the effective Dirac mass

$$m^*(\mathbf{r}) = m + S(\mathbf{r}). \tag{5.153}$$

---

[3] Note that, for convenience, this definition is opposite to what we have used before.

The equations for the mesonic fields are also obtained from the Euler-Lagrange equations, leading to the Klein-Gordon equations with source terms involving the baryon densities:

$$
\begin{aligned}
\{-\Delta + m_\sigma^2\}\, \sigma(\mathbf{r}) &= -g_\sigma\, \rho_s(\mathbf{r}) - g_2 \sigma^2(\mathbf{r}) - g_3 \sigma^3(\mathbf{r}), \\
\{-\Delta + m_\omega^2\}\, \omega_0(\mathbf{r}) &= g_\omega\, \rho_v(\mathbf{r}), \\
\{-\Delta + m_\rho^2\}\, \rho_0(\mathbf{r}) &= g_\rho\, \rho_3(\mathbf{r}), \\
-\Delta\, A_0(\mathbf{r}) &= e\, \rho_c(\mathbf{r}).
\end{aligned}
\tag{5.154}
$$

The corresponding source terms are

$$
\rho_s = \sum_{i=1}^{A} \overline{\psi}_i\, \psi_i\,, \qquad \rho_v = \sum_{i=1}^{A} \psi_i^\dagger \psi_i,
$$

$$
\rho_3 = \sum_{p=1}^{Z} \psi_p^\dagger \psi_p - \sum_{n=1}^{N} \psi_n^\dagger \psi_n\,, \qquad \rho_c = \sum_{p=1}^{Z} \psi_p^\dagger \psi_p.
\tag{5.155}
$$

These sets of equations, known as RMF (relativistic mean field) equations, are solved self-consistently by iteration, as in the usual H-F procedure.

A typical set of parameters are

$$
\begin{aligned}
m_\sigma &= 504.89, & g_\sigma &= 9.111, \\
m_\omega &= 780, & g_2 &= -2.304 \text{ fm}^{-1}, \\
m_\rho &= 763, & g_3 &= -13.783, \\
& & g_\omega &= 11.493, \\
& & g_\rho &= 5.507,
\end{aligned}
\tag{5.156}
$$

where the masses are in MeV.

The Lagrangian density (5.144) is based on the hypothesis that the one-pion-exchange potential contribution to the bulk properties of nuclear matter largely averages to zero [SW86].

5.11 Exercises

1. Using a table of masses, determine the percent error in the calculation of the masses of ^{4}He, ^{120}Sn, and ^{208}Pb from (5.8).

2. The nucleus $^{132}_{50}$Sn is not stable, in spite of possessing magic numbers of protons and neutrons. Verify if (5.8) can give an explanation for that fact.

3. Show that the average kinetic energy of the nucleons in the interior of a nucleus given by the Fermi gas model is about 23 MeV.

4. Show that the total kinetic energy of a nucleus with $N = Z$ in the Fermi gas model is given by

$$E_T = 2C_3 A^{-2/3} \left(\frac{A}{2} \right)^{5/3},$$

and find the value of C_3. Repeat the calculation for $N \neq Z$ and show that, in this case,

$$E'_T = C_3 A^{-2/3} (N^{5/3} + Z^{5/3}).$$

5. The group of stable odd-odd nuclei consists of very light nuclei. Is it possible to find a justification for this?

6. The table below shows nuclei with their respective experimental values of spin and parity of the ground state. Compare with the predictions of the extreme shell model for these nuclei and try to justify the discrepancies.

^{7}Be	^{17}F	^{61}Cu	^{91}Zr	^{93}Ni	^{123}Sb	^{159}Tb	^{183}Ta	^{199}Tl	^{209}Pb
$\frac{3}{2}^-$	$\frac{5}{2}^+$	$\frac{3}{2}^-$	$\frac{5}{2}^+$	$\frac{9}{2}^+$	$\frac{7}{2}^+$	$\frac{3}{2}^+$	$\frac{7}{2}^+$	$\frac{1}{2}^+$	$\frac{1}{2}^-$

7. The table below exhibits the orbits attributed to the extra proton and neutron for a series of odd-odd nuclei. a) Try to justify these properties using figure 5.9. b) Determine the spins and parities of these nuclei with help of the Nordheim rule and compare with the experimental values, also shown in the table.

Nucleus	p	n	Spin^P	Nucleus	p	n	Spin^P
^{16}N	$p_{1/2}$	$d_{5/2}$	2^-	^{70}Ga	$p_{3/2}$	$p_{1/2}$	1^+
^{34}Cl	$d_{3/2}$	$d_{3/2}$	0^+	^{90}Y	$p_{1/2}$	$5/2$	2^-
^{38}Cl	$d_{3/2}$	$f_{7/2}$	2^-	^{92}Nb	$g_{9/2}$	$5/2$	7^+
^{41}Sc	$f_{7/2}$	$f_{7/2}$	0^+	^{206}Tl	$s_{1/2}$	$p_{1/2}$	0^-
^{62}Cu	$p_{3/2}$	$f_{5/2}$	1^+	^{202}Bi	$h_{9/2}$	$f_{7/2}$	5^+

8. The spin and parity of ^{9}Be and ^{9}B are $\frac{3}{2}^-$ for both nuclei. Assuming that these values are given by the last nucleon, justify the observed value, 3^+, of ^{10}B. What other combinations of spin-parity can appear? Verify in a nuclear chart the presence of excited states of ^{10}B that could correspond to those combinations.

9. ^{13}C and ^{13}N both have a ground state $\frac{1}{2}^-$ and three excited states below 4 MeV, of spin-parity $\frac{1}{2}^+$, $\frac{3}{2}^-$, and $\frac{5}{2}^+$. The other states are located above 6 MeV. Interpret these four states using the shell model.

10. The numbers 28 and 40 are sometimes treated as *semi-magic*. Would it be possible, examining figure 5.9, to find a justification for that attribute?

11. The ground state of ^{137}Ba possesses spin-parity $\frac{3}{2}^+$. The first two excited states possess spin-parity $\frac{1}{2}^+$ and $\frac{11}{2}^-$. According to the shell model, which levels would be expected for these excited states?

12. Assuming that an extra proton moves on the surface of a spherical core, calculate the electric quadrupole moments of ^{17}O, ^{175}Lu, and ^{209}Bi. Compare with the respective experimental values, given in barns, -0.026, $+7.0$, and -0.37. Try to justify some eventual discrepancy.

13. The ground state of $^{123}_{51}$Sb is a spin-parity state $\frac{7}{2}^+$ with magnetic moment 2.547 μ_N. a) Determine the most probable value of the orbital momentum l of the last nucleon. b) Check the value found with that indicated in figure 5.9. Is there some special reason to expect a good prediction for this nucleus?

14. Using an identical construction to the one of table 5.4, show that the possible values of j resulting from the coupling of three nucleons in a $j = \frac{5}{2}$ level are $\frac{3}{2}$, $\frac{5}{2}$, and $\frac{9}{2}$.

15. Find the volume of the nucleus whose surface is described by (5.83), with $\lambda = 2$ and $\mu = 0$.

16. For an axially symmetric nucleus with density given by

$$\rho(r) = \begin{cases} \rho_0 & \text{for} & r \leq R_0(1 + \beta Y_{20}(\theta, \phi)), \\ 0 & \text{otherwise,} \end{cases}$$

show that the intrinsic (charge) quadrupole moment up to second order in the deformation parameter β is given by

$$Q_0 = \frac{3}{\sqrt{5\pi}} Z R_0^2 \beta (1 + 0.36\beta)$$

and the moment of inertia about the z-axis is given by

$$I = \frac{2}{5} M R_0^2 (1 + 0.31\beta)$$

to first order in β. Here M is the mass of the nucleus, R_0 is the radius of a sphere having the same volume, and ρ_0 may be found from normalization.

17. Building a table similar to table 5.6, show that the allowed final states resulting from the coupling of three quadrupole phonons are 0^+, 2^+, 3^+, 4^+, and 6^+.

18. Find the possible states (spin and parity) resulting from the coupling of a quadrupole phonon ($\lambda = 2$) with an octupole phonon ($\lambda = 3$). Justify.

19. The second state 2^+ of a vibrational band has an energy of 3.92 keV and the first state 4^+ has an energy of 3.55 keV. Estimate the energy of the first 2^+ state and the first 6^+ state.

20. The first excited state of most of the magic nuclei has spin-parity 3^- and is interpreted as a one-phonon vibrational state. a) What multipole order corresponds to such a vibration type? b) Which spins and parities are expected for the resulting states of the coupling of two of these phonons? c) One of those nuclei is $^{208}_{82}$Pb. Its first excited state is found at 2615 keV. Around what energy should we seek the states of item b?

21. Examining the position of the nuclei $^{122}_{52}$Te and $^{160}_{66}$Dy in figure 5.16, estimate the energy of states 4^+ and 6^+ of each nucleus, knowing that the energy of state 2^+ is 563.6 keV for the first and 86.8 keV for the last.

22. The deformed nucleus $^{164}_{68}$Er has its first excited state, 2^+, at 91.4 keV. a) Knowing that this is the first excited state of a rotational band, use (5.107) to determine its moment of inertia. b) What are the spin and parity values of the four next states of the band? c) Calculate the energy of these states and compare with the respective experimental values (in keV): 208.1, 315.0, 410.2, 493.5. d) Calculate again the moment of inertia of the nucleus for each of these experimental values and interpret the results.

23. In a deformed odd nucleus, the first three levels are to 0, 40, and 96 keV. Assuming that they form a rotational band, find in what energy range we should look for the level of spin $\frac{11}{2}$ of that band.

24. The ground state of an odd nucleus has spin $\frac{3}{2}$. This, and three more excited states, 12, 32, and 60 keV, are part of a rotational band. a) What parameters define that band? b) What are the spins and parities of these levels?

25. The *yrast* line of the deformed nucleus $^{158}_{68}$Er is formed by the states listed in the table below. Plot the graphs corresponding to figures 5.25 and 5.26 and try to identify the present rotational bands.

Spin	Energy (keV)	Spin	Energy (keV)	Spin	Energy (keV)
0^+	0	14^+	3193	28^+	8143
2^+	192	16^+	3666	30^+	9020
4^+	527	18^+	4233	32^+	9927
6^+	970	20^+	4892	34^+	10887
8^+	1494	22^+	5632	36^+	11906
10^+	2074	24^+	6438	38^+	12967
12^+	2683	26^+	7284		

6 | Radioactivity

6.1 Introduction

The stable isotopes are located in a narrow band of the nuclear chart called the β-stability line, alongside of which nuclei unstable by β^+ or β^- emission are located. For $A > 150$ the emission of an α-particle is energetically favorable, and in this region one finds several α-emitters. Heavy nuclei also release energy if divided in two nearly equal parts and can, for this reason, fission spontaneously. A *radioactive substance*, which contains some unstable isotope, is in permanent transformation by the action of one or more of these processes. The physics of each of them will be studied later. In this chapter we are only interested in the statistical aspect of the problem: we want to find the evolution in time of the number of atoms, and of the activity of radioactive substances, in a way independent of the process in action.

Thus let us call N the number of atoms of an unstable isotope at the instant t. If we admit that there is a fixed probability λdt of occurrence of a certain process in a nucleus in the time interval dt, the change dN in the number of atoms during the interval dt can be written as

$$dN(t) = -\lambda N\, dt. \tag{6.1}$$

λ is called the *disintegration or decay constant* of that isotope for the process in question. The rate at which an amount of a substance sample disintegrates is measured by the *activity* $A(t)$ of that amount. From (6.1) we obtain

$$A(t) = -\frac{dN}{dt} = \lambda N. \tag{6.2}$$

The number N of atoms at a given instant t can be easily obtained by the integration of (6.1):

$$N(t) = N_0 e^{-\lambda t}, \tag{6.3}$$

where N_0 is the number of atoms at the initial instant. Using (6.3) we can evaluate the *half-life* of a radioactive sample, defined as the time necessary for the decay of half the atoms of the sample, that is, $N(t_{1/2}) = N_0/2$. We obtain

$$t_{1/2} = \frac{\ln 2}{\lambda} = \frac{0.693}{\lambda}. \tag{6.4}$$

If, instead of half, we use the decay to $1/e$ of the initial value, the time τ necessary for that to occur is defined as the *mean lifetime* of the substance, and it is obvious that

$$\tau = 1/\lambda. \tag{6.5}$$

This latter definition is less employed than the former one.

6.2 Multiple Decays—Decay Chain

If for each atom of a radioactive substance there are several decay branches with probabilities λ_1, λ_2, etc., the number of atoms $N(t)$ at each time will be given by (6.3), with $\lambda = \lambda_1 + \lambda_2 + \cdots$. The activity for each decay mode k will be

$$A_k = -\left(\frac{dN}{dt}\right)_k = \lambda_k N = \lambda_k N_0 e^{-\lambda t}. \tag{6.6}$$

Another interesting case is that of decay chains:

$$N_1 \xrightarrow{\lambda_1} N_2 \xrightarrow{\lambda_2} N_3 \xrightarrow{\lambda_3} \cdots \xrightarrow{\lambda_{k-1}} N_k \tag{6.7}$$

Substance 1, with initial number of atoms $N_1(0)$, decays with an activity $\lambda_1 N_1(t)$ into substance 2, which in turn is radioactive and decays with activity $\lambda_2 N_2(t)$ into substance 3, and so on, until one reaches substance k, which is stable. The differential equation satisfied by the i-th $(i \neq 1)$ member of the series is

$$\frac{dN_i}{dt} = \lambda_{i-1} N_{i-1} - \lambda_i N_i. \tag{6.8}$$

With the assumption that in the initial instant only substance 1 was present, that is, $N_2(0) = N_3(0) = \cdots = 0$, Bateman [Ba10] presented, in 1910, the solution of the system (6.8), showing that the number of atoms of the i-th $(i \neq 1)$ element is given by

$$N_i(t) = N_1(0) \left(h_1 e^{-\lambda_1 t} + h_2 e^{\lambda_2 t} + \cdots + h_i e^{-\lambda_i t} \right), \tag{6.9}$$

where the coefficients h are expressed by

$$h_1 = \frac{\lambda_1 \lambda_2 \lambda_3 \ldots \lambda_{i-1}}{(\lambda_2 - \lambda_1)(\lambda_3 - \lambda_1) \ldots (\lambda_i - \lambda_1)},$$

$$h_2 = \frac{\lambda_1 \lambda_2 \lambda_3 \ldots \lambda_{i-1}}{(\lambda_1 - \lambda_2)(\lambda_3 - \lambda_2) \ldots (\lambda_i - \lambda_2)},$$

$$\ldots\ldots\ldots\ldots\ldots\ldots\ldots\ldots\ldots\ldots\ldots\ldots$$

$$\ldots\ldots\ldots\ldots\ldots\ldots\ldots\ldots\ldots\ldots\ldots\ldots$$

$$h_i = \frac{\lambda_1 \lambda_2 \lambda_3 \ldots \lambda_{i-1}}{(\lambda_1 - \lambda_i)(\lambda_2 - \lambda_i) \ldots (\lambda_{i-1} - \lambda_i)}. \tag{6.10}$$

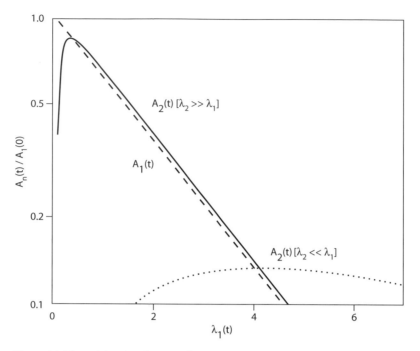

Figure 6.1 The activity given by (6.12) for two extreme cases: $\lambda_2 \gg \lambda_1$ and $\lambda_2 \ll \lambda_1$.

Let us see the application of (6.9) to the first two elements of a chain. In this case the constants for the second element are

$$h_1 = \frac{\lambda_1}{\lambda_2 - \lambda_1} \quad \text{and} \quad h_2 = \frac{\lambda_1}{\lambda_1 - \lambda_2};$$

the number of atoms of substances 1 and 2 will be given by

$$N_1(t) = N_1(0)e^{-\lambda_1 t},$$
$$N_2(t) = N_1(0)\frac{\lambda_1}{\lambda_2 - \lambda_1}\left(e^{-\lambda_1 t} - e^{-\lambda_2 t}\right), \tag{6.11}$$

and the corresponding activities by

$$A_1(t) = \lambda_1 N_1 = \lambda_1 N_1(0)e^{-\lambda_1 t},$$
$$A_2(t) = \lambda_2 N_2 = A_1(t)\frac{\lambda_2}{\lambda_2 - \lambda_1}\left[1 - e^{-(\lambda_2 - \lambda_1)t}\right]. \tag{6.12}$$

Figure 6.1 shows, for two extreme cases, the behavior of the activities of substances 1 and 2 as a function the dimensionless quantity $\lambda_1 t$.

If the third element of the chain is stable, its number of atoms increases according to the expression

$$N_3 = N_1(0)\left(\frac{\lambda_2}{\lambda_1 - \lambda_2}e^{-\lambda_1 t} + \frac{\lambda_1}{\lambda_2 - \lambda_1}e^{-\lambda_2 t} + 1\right). \tag{6.13}$$

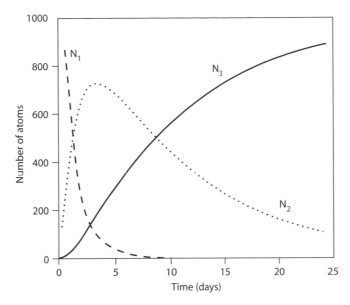

Figure 6.2 Time evolution of the number of atoms of three substances of a decay chain. The first substance has 1000 atoms in the initial instant and the other two, zero. The third element of the chain is stable.

If, for example, the first substance has a half-life of 1 day and the second of 1 week, (6.11) and (6.13) give, the plot of figure 6.2.

6.3 Preparation of a Radioactive Sample

The preparation of a sample containing a radioactive substance is usually done by creating atoms of that substance at a fixed rate. An example is the activation by a constant neutron flux from a reactor. The neutrons are captured by the target nuclei, leading to the formation of unstable nuclei. Let P be the rate at which the substance is produced. The number of atoms $N(t)$ of the substance obeys the equation:

$$\frac{dN}{dt} = -\lambda N + P, \tag{6.14}$$

whose solution is

$$N = \frac{P}{\lambda}\left(1 - e^{-\lambda t}\right), \tag{6.15}$$

with the corresponding activity

$$A = \lambda N = P\left(1 - e^{-\lambda t}\right). \tag{6.16}$$

Figure 6.3 shows this activity as a function of time, indicating that the production process begins to be inefficient after one half-life. After several half-lives, to continue with the process is completely useless since the substance decays with the same rate that it is produced. The most efficient way to produce a radioactive substance is to periodically renew the material to be irradiated with a period equal to some few half-lives.

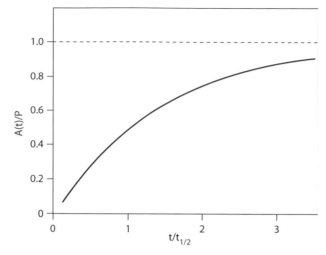

Figure 6.3 Activity of a radioactive substance produced with a constant rate.

6.4 Secular Equilibrium

Let us consider a set of radioactive isotopes that forms the series of decay chains, (6.7). We say that this set is in secular equilibrium when the number of atoms of each species is practically constant, that is,

$$\frac{dN_1}{dt} = \frac{dN_2}{dt} = \cdots \cong 0. \tag{6.17}$$

Looking at (6.8), valid for $i > 1$, we see that this condition implies

$$\lambda_1 N_1 = \lambda_2 N_2 = \lambda_3 N_3 = \cdots . \tag{6.18}$$

The equation of isotope 1 decay, $dN_1/dt = -\lambda_1 N_1$, shows that (6.17) will be nearly satisfied if λ_1 is very small. Thus the secular equilibrium is characterized by a very long half-life isotope decaying successively into isotopes of shorter period. There are several examples of this form of equilibrium occurring in nature, as we shall see in the next section.

6.5 Natural Radioactive Series

If a radioactive isotope has a half-life with order of magnitude equal to or greater than the age of the earth, samples containing this isotope in secular equilibrium with their daughter nuclei can still be found. That is the case of the isotopes $^{238}_{92}\text{U}$, $^{235}_{92}\text{U}$, and $^{232}_{90}\text{Th}$, which form three independent radioactive series, as can be seen in tables 6.1, 6.2, 6.3.

The independence of the series, that is, the fact that they do not have any element in common, is due to the fact that α-disintegration changes the mass number A by 4 units and β-disintegration does not change A. Thus, with n an integer, the mass number of all

Table 6.1 The ^{235}U natural series.

Nuclide	Decay mode	Half-life	Nuclide	Decay mode	Half-life
$^{238}_{92}$U	α	4.5×10^9 y	$^{218}_{85}$At	α	1.73 s
$^{234}_{90}$Th	β^-	24.1 d	$^{214}_{83}$Bi	α,β^-	19.7 min
$^{234}_{91}$Pa	β^-	1.8 min	$^{214}_{84}$Po	α	1.64×10^{-4} s
$^{234}_{91}$Pa	β^-	6.7 h	$^{210}_{81}$Tl	β^-	1.32 min
$^{234}_{92}$U	α	2.5×10^5 y	$^{210}_{82}$Pb	β	22 y
$^{230}_{90}$Th	α	8.0×10^4 y	$^{210}_{83}$Bi	β^-	5 d
$^{226}_{88}$Ra	α	1620 y	$^{210}_{84}$Po	α	138.3 d
$^{222}_{86}$Rn	α	3.82 d	$^{206}_{81}$Tl	β^-	4.2 min
$^{218}_{84}$Po	α,β^-	3.1 min	$^{206}_{82}$Pb	stable	
$^{214}_{82}$Pb	β^-	26.8 min			

isotopes of the first series will be of the form $4n + 2$, and we say for this reason that ^{238}U and its daughter nuclei form the $4n + 2$ series. Likewise, ^{235}U heads the $4n + 3$ series and ^{232}Th the $4n$ series. The series with $4n + 1$ does not have any isotope with half-life long enough to have survived since the formation of the earth, and so it cannot be found in nature. But an artificially produced isotope like $^{237}_{93}$Np, of half-life 2.14×10^6 years, can give place to that series.

Table 6.2 The ^{238}U natural series.

Nuclide	Decay mode	Half-life	Nuclide	Decay mode	Half-life
$^{235}_{92}$U	α	7.1×10^8 y	$^{215}_{84}$Po	α,β^-	1.83×10^{-3} s
$^{231}_{90}$Th	β^-	24.6 h	$^{211}_{82}$Pb	β^-	36.1 min
$^{231}_{91}$Pa	α	3.43×10^4 y	$^{215}_{85}$At	α	10^{-4} s
$^{227}_{89}$Ac	α,β^-	22 y	$^{211}_{83}$Bi	α,β^-	2.16 min
$^{227}_{90}$Th	α	18.6 d	$^{211}_{84}$Po	α	0.52 s
$^{223}_{87}$Fr	β^-	21 min	$^{207}_{81}$Tl	β^-	4.79 min
$^{223}_{88}$Ra	α	11.2 d	$^{207}_{82}$Pb	stable	
$^{219}_{86}$Rn	α	3.92			

Table 6.3 The ²³²Th natural series.

Nuclide	Decay mode	Half-life	Nuclide	Decay mode	Half-life
$^{232}_{90}$Th	α	1.39×10^{10} y	$^{212}_{82}$Pb	β^-	10.6 h
$^{228}_{88}$Ra	β^-	6.7 y	$^{216}_{85}$At	α	3×10^{-4} s
$^{228}_{89}$Ac	β^-	6.13 h	$^{212}_{83}$Bi	α,β^-	47 min
$^{228}_{90}$Th	α	1.9 y	$^{212}_{84}$Po	α	3×10^{-7} s
$^{224}_{88}$Ra	α	3.64 d	$^{208}_{81}$Tl	β^-	2.1 min
$^{220}_{86}$Rn	α	54.5 s	$^{208}_{82}$Pb	stable	
$^{216}_{84}$Po	α,β^-	0.16 s			

6.6 Radiation Units

To establish a set of units of radioactivity, we can look at the intensity of the emission sources or at the effect produced by them, especially in the human body. A quantity that takes into account only the first property is the activity, defined by (6.2), which refers to the number of disintegrations of the sample, no matter what the energy or the type of emitted radiation. The units employed for the activity are

a) The *Curie* (Ci), with 1 Ci = 3.7×10^{10} disintegrations/s.
b) The *Becquerel* (Bq), with 1 Bq = 1 disintegration/s = 0.27×10^{-10}Ci.
c) The *Rutherford* (Rd), with 1 Rd = 10^6 disintegrations/s.

When we also want to compute the effects of radioactivity, other quantities are necessary. A characteristic effect of radiation is ionization, that is, knocking electrons out from atoms, producing free electrons and positive ions. A non-SI quantity that expresses the ionization produced by gamma and X radiation in air is *exposure*. Its unit (now replaced by the name "air kerma"[1]) in the International System (SI) is the Coulomb/kg of air, but a traditional unit still in use is the *Roentgen* (R), defining 1 R as the amount of radiation that produces the charge of 1 esu in 1 cm³ of air at standard temperature and pressure (STP). At 1 atmosphere and 20° Celsius the two are related in the following way:

$$1 \text{ Roentgen} \equiv 1R = 1 \text{ esu}/1 \text{ cm}^3 \text{ of air} = 2.58 \times 10^{-4} \text{ C/kg of air}. \tag{6.19}$$

We can also define a quantity related to the amount of energy produced by the passage of radiation, or particles, in a given material. Such quantity is the *absorbed dose* and its traditional unit is the *rad* (rad = "<u>r</u>adiation <u>a</u>bsorbed <u>d</u>ose"),

$$1 \text{ rad} = 10^{-2} \text{ J/kg} = 100 \text{ ergs/g of the material}. \tag{6.20}$$

---

[1] kerma = <u>k</u>inetic <u>e</u>nergy <u>r</u>eleased to charged particles per unit <u>m</u>ass of the medium.

Table 6.4 Values of the radiation weighting factor, w_R.

Radiation		w_R
X-and X-rays, all energies		1
Electrons and muons, all energies		1
Neutrons	<10 keV	5
	10 keV–100 keV	10
	100 keV–2 MeV	20
	2 MeV–20 MeV	10
	>20 MeV	5
Protons	>2 MeV	5
α-particles, fission fragments, and heavy nuclei		20

although, as for the other quantities, one advises the use of the SI, whose respective unit is the *Gray* (Gy),

$$1 \text{ Gy} = 1 \text{ J/kg} \text{ of the material} = 100 \text{ rad.} \tag{6.21}$$

For a given material the exposure and the absorbed dose are related. So, for instance, for soft living tissue, 1 R $\cong$ 1 rad; for air, 1 R $\cong$ 0.9 rad.

Biological damages caused by radiation depends not only on the energy deposited but also on the nature of the ionizing particle, in particular, the ionization density it lays down. The latter is measured in terms of the linear energy transfer (LET) typically quoted in units of keV μm^{-1}. To have a measure of biological damages that is free from this dependence, one has created a quantity, the *equivalent dose*, obtained from the absorbed dose by multiplying it by a *radiation weighting factor, w_R*, that depends on the nature and energy of the particle or radiation. The units employed are the *rem* (roentgen equivalent man),

$$1 \text{ rem} = w_R \times (\text{dose in rads}), \tag{6.22}$$

and, in SI, the *Sievert* (Sv),

$$1 \text{ Sv} = w_R \times (\text{dose in Gy}) = 100 \text{ rem.} \tag{6.23}$$

For the radiation weighting factors, w_R, values from table 6.4 can be used.

6.7 Radioactive Dating

For the determination of geological ages we use radioactive isotopes whose long half-lives have allowed them to survive since the earth's formation, and which are found today in sufficient amounts for analysis.

Let us examine the problem quantitatively. Let P_0 be the amount of a radioactive isotope at time t_0. This isotope decays, with a disintegration rate λ, into another isotope, whose number of atoms at time t_0 is D_0. If we admit that there is no gain or loss of parent and daughter nuclei in a sample, except for the decay of the parent into its progeny, we can say that

$$P + D = P_0 + D_0, \tag{6.24}$$

where P and D are the respective number of atoms present at time t and

$$P = P_0 e^{-\lambda(t-t_0)}. \tag{6.25}$$

From (6.24) and (6.25) we obtain

$$e^{-\lambda(t-t_0)} = \frac{P}{P_0} = \frac{1 + D_0/P_0}{1 + D/P}. \tag{6.26}$$

Relation (6.26) allows us to know the interval $\Delta t = t - t_0$ if we know the quantities λ and D/P, which can be measured, and D_0/P_0. For the latter one needs a hypothesis, or model, for the ratio between the isotopes at the initial instant. A procedure that makes this hypothesis unnecessary can be developed in the situation in which the sample produces another isotope of the progeny nuclei, with number of atoms D', that is both stable (nonradioactive) and not fed by a long half-life decay. In this case, (6.24) can be rewritten in the form

$$\frac{P + D}{D'} = \frac{P_0 + D_0}{D'_0}, \tag{6.27}$$

since, with our assumptions, $D' = D'_0$. This results in

$$\frac{D}{D'} = \frac{P}{D'} \left[e^{\lambda(t-t_0)} - 1 \right] + \frac{D_0}{D'_0}. \tag{6.28}$$

Let us assume that we have a rock in which geological evidence indicate that it has crystallized from a mixture within a short period. The different minerals present in the rock have different ratios D/D' and P/D', but we can suppose that the isotopic ratio D_0/D'_0 is the same for all of them. If this is true, then relation (6.28) between D/D' and P/D' is linear, and the slope of the straight line constructed from the different rock minerals gives us the age of the rock, $\Delta t = t - t_0$.

The isotopes to be chosen for the application of (6.28) depend on the presumable age of the rock to be analyzed and on its composition. The isotopes employed in geochronology are listed in table 6.5.

For organic dating, and specifically when an organism dies, the equilibrium ends and ^{14}C begins to decay without feedback. Comparing the ratio $^{14}C/^{12}C$ in an animal or vegetal fossil with the equilibrium value in the atmosphere, we can from (6.3) determine the time elapsed after the death of the organism [Lib55]. Of course, the method is only applicable if the ^{14}C activity is not very low; it is impractical for times greater than approximately 10 half-lives. But, with recent techniques, using accelerators like mass spectrographs, counting ^{14}C atoms directly, it is possible to exceed this limit. Since it is a method to determine

Table 6.5 Methods employed in geochronology. The quantities refer to the variables of equation 6.28. ^{40}K has two modes of decay but only the mode shown in table is utilized.

Parent nucleus	Progeny nucleus	Half-life	Stable nucleus
(P)	(D)	$(10^9$ years$)$	(D')
^{238}U	^{206}Pb	4.47	^{204}Pb
^{235}U	^{207}Pb	0.70	^{204}Pb
^{232}Th	^{208}Pb	14.0	^{204}Pb
^{87}Rb	^{87}Sr	48.8	^{86}Sr
^{40}K	^{40}Ar	11.9	^{36}Ar
^{147}Sm	^{143}Nd	106.0	^{144}Nd

relatively recent ages, its results can be compared, in several cases, with historical registers or with counting tree rings. These comparisons usually show very good agreement.

In recent times, the equilibrium between the carbon isotopes in the atmosphere has been upset by the indirect action of humanity. The burning of coal and petroleum has increased the proportion of stable carbon isotopes in the atmosphere, since the fossil fuels are very old to contain ^{14}C. On the other hand, tests with nuclear weapons have increased the amount of ^{14}C in the atmosphere in an uncontrolled way. These facts create obvious difficulties for the application of the method in the future.

6.8 Properties of Unstable States—Level Width

In this and the next section we shall examine some properties of unstable nuclear states, namely, the states responsible for the activity of the radioactive substances.

A quantum system, described by a wavefunction that is a Hamiltonian eigenfunction, is in a well-defined energy state and, if it does not suffer external influences, it will remain indefinitely in that state. But this ideal situation does not prevail in excited nuclei, or in the ground state of unstable nuclei. Interactions of several types can add a perturbation to the Hamiltonian and the pure energy eigenstates no longer exist. In this situation a transition to a lower energy level of the same or of another nucleus can occur.

An unstable state normally lives a long time compared to the fastest nuclear processes, for example, the time spent by a particle with velocity near that of light in crossing a nuclear diameter. In this way we can admit that a nuclear state is approximately stationary, and to write for its wavefunction

$$\Psi(\mathbf{r}, t) = \psi(\mathbf{r})e^{-iWt/\hbar}. \tag{6.29}$$

$|\Psi(\mathbf{r}, t)|^2 dV$ is the probability of finding the nucleus in the volume dV, and, if the state described by Ψ decays with a decay constant λ, it is reasonable to write

$$|\Psi(\mathbf{r}, t)|^2 = |\Psi(\mathbf{r}, 0)|^2 e^{-\lambda t}. \tag{6.30}$$

To obey (6.29) and (6.30) simultaneously, W must be a complex quantity with imaginary part $-\lambda \hbar/2$. We can write:

$$W = E_0 - \frac{\hbar \lambda}{2} i, \tag{6.31}$$

which shows that the wavefunction (6.29) does not represent a well-defined stationary state, since the exponential contains a real part $-\lambda t/2$. However, we can write the exponent of eq. 6.29 as a superposition of values corresponding to well-defined energies E (for $t \geq 0$):

$$e^{-(iE_0/\hbar + \lambda/2)t} = \int_{-\infty}^{+\infty} A(E) e^{-iEt/\hbar} \, dE. \tag{6.32}$$

Functions connected by a Fourier transform relate to each other as

$$f(t) = \frac{1}{\sqrt{2\pi}} \int_{-\infty}^{+\infty} g(\omega) e^{-i\omega t} \, d\omega,$$

$$g(\omega) = \frac{1}{\sqrt{2\pi}} \int_{-\infty}^{+\infty} f(t) e^{i\omega t} \, dt; \tag{6.33}$$

this allows us to establish the form of the amplitude $A(E)$:

$$A(E) = \frac{1}{2\pi \hbar} \int_0^{+\infty} e^{[i(E-E_0)/\hbar - \lambda/2]t} \, dt, \tag{6.34}$$

where the lower limit indicates that the stationary state was created at the time $t = 0$. The integral (6.34) has an easy solution, giving

$$A(E) = \frac{1}{\hbar \lambda/2 - 2\pi i(E - E_0)}.$$

The probability of finding a value between E and $E + dE$ in an energy measurement is given by the product

$$A^*(E) A(E) = \frac{1}{\hbar^2 \lambda^2/4 + 4\pi^2 (E - E_0)^2}, \tag{6.35}$$

and this function of energy has the form of a Lorentzian, with the aspect shown in figure (6.4). Its width at half-maximum is $\Gamma = \hbar \lambda = \hbar/\tau$. The relationship

$$\tau \Gamma = \hbar \tag{6.36}$$

between the half-life and the width of a state is directly connected to the uncertainty principle and shows that the longer a state survives the greater is the precision with which its energy can be determined. In particular, only to stable states can one attribute a single value for the energy.

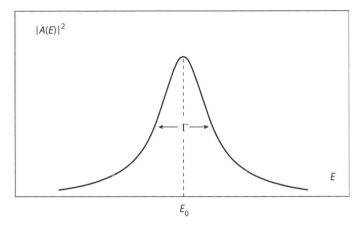

Figure 6.4 Form of the distribution (6.35).

6.9 Transition Probability—Golden Rule

The decay constant λ was presented at the beginning of the chapter as the probability per unit time of occurrence of a transition between quantum states, and its values were supposed known from experimental data. In this section we shall show that a formula to evaluate the decay constant can be obtained from the postulates of perturbation theory.

The previous section has described an unstable state by the addition of a perturbation to an stationary state. Formally we can write

$$H = H_0 + V, \tag{6.37}$$

where H is the Hamiltonian of the unstable state, composed of the nonperturbed Hamiltonian H_0 and a small perturbation V. The Hamiltonian H_0 satisfies an eigenvalue equation

$$H_0 \psi_n = E_n \psi_n, \tag{6.38}$$

whose eigenfunctions form a complete basis in which the total wavefunction Ψ, obeying

$$H\Psi = i\hbar \frac{\partial \Psi}{\partial t}, \tag{6.39}$$

can be expanded:

$$\Psi = \sum_n a_n(t) \psi_n e^{-iE_n t/\hbar}. \tag{6.40}$$

Using (6.37) and (6.40) in (6.39), and with the aid of (6.38), we obtain

$$i\hbar \sum_n \dot{a}_n \psi_n e^{-iE_n t/\hbar} = \sum_n V a_n \psi_n e^{-iE_n t/\hbar}, \tag{6.41}$$

with $\dot{a}_n \equiv \partial a_n(t)/\partial t$. Using the orthogonalization properties of the ψ_n, let us multiply (6.41) to the left by ψ_k^* and integrate it in the coordinate space. From this results

$$\dot{a}_k = -\frac{i}{\hbar} \sum_n a_n V_{kn} e^{i\frac{E_k-E_n}{\hbar}t},$$

(6.42)

where we introduce the matrix element

$$V_{kn} = \int \psi_k^* V \psi_n \, d\tau.$$

(6.43)

Let us make the following assumptions about the perturbation V. It begins to act at time $t = 0$, when the unperturbed system is described by an eigenstate ψ_m. It stays at a very low value, and after a short time interval, becomes zero at $t = T$. These assumptions allow us to say that the conditions

$$\begin{cases} a_m = 1, \\ a_n = 0 \text{ if } n \neq m, \end{cases}$$

(6.44)

are rigorously verified for $t < 0$ and also work approximately for $t > 0$. Thus, (6.42) has only one term, and the value of the amplitude is obtained from

$$a_k = -\frac{i}{\hbar} \int_0^T V_{km} e^{i\frac{E_k-E_m}{\hbar}t} \, dt,$$

(6.45)

whose value must be necessarily small by the assumption that followed (6.44). The above approach is also known as first order perturbation theory. The integral of (6.45) gives

$$a_k = \frac{V_{km}\left(1 - e^{i\frac{E_k-E_m}{\hbar}T}\right)}{E_k - E_m}.$$

(6.46)

We need now to interpret the meaning of the amplitude a_k. The quantity $a_k^* a_k$ measures the probability of finding the system in the state k. This characterizes a transition occurring from the initial state m to the state k, and the value of $a_k^* a_k$ divided by the interval T should be a measure of the decay constant λ_k relative to the state k. The total decay constant is obtained by the sum over all states:

$$\lambda = \sum_{k \neq m} \lambda_k = \frac{\sum |a_k|^2}{T}.$$

(6.47)

Let us now suppose that there are a large number of available states k. We can, in this case, replace the summation in (6.47) by an integral. Defining $\rho(E)$ as the density of available states around the energy E_k, we write

$$\lambda = \frac{1}{T} \int_{-\infty}^{+\infty} |a_k|^2 \rho(E_k) \, dE_k = \frac{4}{T} \int_{-\infty}^{+\infty} |V_{km}|^2 \frac{\sin^2\left[\left(\dfrac{E_k - E_m}{2\hbar}\right)T\right]}{(E_k - E_m)^2} \rho(E_k) \, dE_k.$$

(6.48)

The function $\sin^2 x/x^2$ only has significant amplitude near the origin. In the case of (6.48), if we suppose that V_{km} and ρ do not vary strongly in a small interval of the energy

E_k around E_m, both these quantities can be taken outside the integral, and we obtain the final expression

$$\lambda = \frac{2\pi}{\hbar}|V_{km}|^2 \rho(E_k) \tag{6.49}$$

that we have been looking for. Equation (6.49) is known as *golden rule no. 2* (also known as the *Fermi golden rule*), and allows us to determine the decay constant if we know the wavefunctions of the initial and final states. The result (6.49) can also be used to obtain, to first order, the cross sections of processes induced by particle interaction through a potential.

6.10 Exercises

1. The activity of a given material decreases by a factor of 8 in a time interval of 30 days. What is its half-life, mean lifetime, and decay constant? If the sample initially had 10^{20} atoms, how many disintegrations have occurred in its second month of life?

2. The theories of grand unification predict that the proton is not a stable particle, although it has a long half-life. For a half-life of 10^{33} years, how many proton decays can we expect in one year in a mass of 10^3 tons of water?

3. A radioactive element decays to a stable nucleus. A counting of decays in a sample of this element was done during intervals of 1 minute in each hour. The values found were 93, 60, 49, 41, 27, 28, 20, 18, 11, ... Plot a "semilog" graph of count versus time and obtain from it an estimate of the source half-life. Recalling that the error in N counts is equal to $\sqrt{N}$, do the data seem reasonable?

4. Natural uranium is a mixture of 99.3% ^{238}U and 0.7% ^{235}U. The half-life for α-emission by the first nucleus is 4.5×10^9 years and of the second one is 7×10^8 years. a) How long ago were the amounts of the two isotopes in a sample the same? b) If a sample initially had 10 g of natural uranium, what is the mass of He gas produced since that time by the two isotopes?

5. A nucleus with decay constant λ exists at time $t = 0$. What is the probability that it disintegrates between t and $t + \Delta t$?

6. $^{252}_{98}$Cf has a half-life of 2.64 years. It disintegrates by α-emission in 96.9% of the events and by spontaneous fission in 3.1%. a) What is its mean lifetime? b) What is the ratio between the number of α-particles and number of fissions produced? c) What half-life would this isotope have if it did not have spontaneous fission?

7. Show from (6.12) that: a) The maximum of the activity $A_2(t)$ from substance 2 occurs at the crossing of the curves of $A_1(t)$ and $A_2(t)$. b) When $\lambda_1 \cong \lambda_2$ this crossing occurs at

$t = \tau$, where τ is the mean lifetime common to both substances. c) When $\lambda_1 > \lambda_2$, the ratio $A_2(t)/A_1(t)$ increases without limit. d) When $\lambda_1 < \lambda_2$, the ratio $A_2(t)/A_1(t)$ increases toward the limit $\tau_1/(\tau_1 - \tau_2)$. Next to this limit the ratio between the activities is almost constant (>1), establishing what one calls *transient equilibrium*, from which the secular equilibrium ($\lambda_1 \ll \lambda_2$) is a special case.

8. Explain how (6.16) can be understood as a limiting case of (6.12).

9. The secular equilibrium can also be defined through the condition

$$\frac{d}{dt}\left(\frac{N_2}{N_1}\right) = \frac{d}{dt}\left(\frac{N_3}{N_2}\right) = \frac{d}{dt}\left(\frac{N_4}{N_3}\right) = \cdots = 0.$$

Supposing that $\lambda_1 \ll \lambda_2, \lambda_3, \lambda_4, \ldots$ show explicitly that the relations (6.18) are reproduced. What happens with the final state from the decay chain?

10. Construct what would be the series $4n + 1$ of $^{237}_{93}$Np in a form similar to that presented in tables 6.1, 6.2, 6.3.

11. Using tables 6.1 and 6.2, find: a) for 1 g of natural uranium the α and β activities in Ci; b) the mass of natural uranium that would have a global activity of 10^{-2} Rd.

12. Find the equivalent dose (in mSv) that human tissue receives when located in a place where X-radiation produces an ionization of 2.6×10^{-7} C/kg in air.

13. A wood relic contains 1 g of carbon with an activity of 4×10^{-12} Ci. If in living trees the ratio ^{14}C/^{12}C is 1.3×10^{-12}, what is the age of the relic? The ^{14}C half-life is 5730 years.

7 | Alpha-Decay

7.1 Introduction

The emission of an α-particle is a possible nuclear disintegration process in situations in which (5.12) is satisfied. In contrast with the restricted existence of emitters of light fragments, α-emitter nuclei are largely due to the large binding energy of the α-particle. In turn, the α-emitting process is energetically advantageous in practically all nuclei with $A \gtrsim 150$. Figure 7.1, based on the balance of masses, exhibits the energy available by emission of several nuclei for ^{239}Pu. We see that α-emission is the only energetically possible process.

Very rarely one detects emission of heavier fragments, with $A > 4$. Examples are the emission of ^{14}C for several isotopes of Ra, and ^{24}Ne for Th and U isotopes. The probability of emission of these fragments is several orders of magnitude smaller than the probability of α-decay, and this justifies that these processes were discovered only recently [Pr89].

The important experimental factors in any decay process are the kinetic energy of the emitted particle and the half-life of the process. Figure 7.2 shows that these two quantities are correlated in α-disintegration. In particular, the *Geiger-Nuttall rule* [GN11] establishes an inverse relationship between the half-life and the available total energy for the process.

A theory for the α-emission should first explain why it happens and then establish the reason for the enormous range (about twenty orders of magnitude) of possible values for the half-life and the relationship of this with the energy of the emitted particle, given by the Geiger-Nuttall rule.

7.2 Theory of α-Decay

In this section we shall show that a theory based on concepts of elementary quantum mechanics is enough to explain the main properties of α-decay. This theory was developed in 1928 by Gamow [Ga28] and, independently, by Condon and Gurney [CG28]. It has as its starting point the hypothesis that the α-particle is pre-formed inside the nucleus and that

		n ^{238}Pu −5.69	^{239}Pu
	^{3}H ^{236}Np −9.79	d ^{237}Np −9.42	p ^{238}Np −6.16
^{6}He ^{233}U −5.93	^{5}He ^{234}U −0.95	^{4}He ^{235}U 5.24	^{3}He ^{236}U −8.79	
^{7}Li ^{232}Pa −2.26	^{6}Li ^{233}Pa −2.99	^{5}Li ^{234}Pa −3.45		

Figure 7.1 Available energy for the emission of particles by ^{239}Pu. The figure shows the emitted particle, the residual nucleus and the energy (in MeV) released in the process. Notice that only the α-emission (^{4}He) has a positive energy.

it can leave it crossing the barrier formed by the Coulomb and nuclear potential. In fact, many nuclei present a peculiar structure, as if they were formed by clusters of α-particles. For example, many of the properties of ^{12}C and ^{16}O can be explained supposing that they are clusters of 3 and 4 α-particles, respectively.

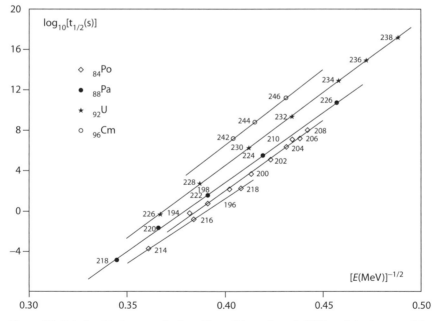

Figure 7.2 Relationship between the logarithm of the α-decay half-life and the inverse of the square root of the energy of disintegration for even-even nuclei. The Geiger-Nuttall Rule establishes that the isotopes of a given element are placed along a straight line. Odd-even and odd-odd nuclei also obey the rule but the linear relationship is less clear. The isotopes of polonium are located along two straight lines, as a result of the shell effect in passage through the magic number of neutrons $N = 126$.

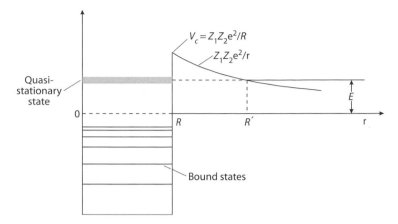

Figure 7.3 Potential barrier for an α-particle inside the nucleus.

A sophistication in the theory is the introduction of the *pre-formation factor* [Pe50], which is, for the present case, the probability that two neutrons and two protons establish the conditions to be emitted by the nucleus in the form of an α-particle. There is no satisfactory way to calculate the pre-formation factor, and we shall limit ourselves to reproduction of the original theory. We shall see that, in spite of that simplification, it leads to results that explain qualitatively the behavior of the half-lives for a great number of α-emitters.

The interaction potential between the α-particle and the rest of the nucleus can be described approximately by figure 7.3. Inside the nucleus the potential is attractive and outside the α-particle ($Z_1 = Z_a$) feels the Coulomb repulsion of the residual nucleus of charge Z_2. The potential of figure 7.3 is the combination of these two potentials. In an unstable nucleus the α-particle occupies initially a quasi-stationary state, with energy $E > 0$. In each collision with the barrier there is a nonzero probability that, due to the tunnel effect, it crosses the barrier and becomes free with kinetic energy E_a, where $E - E_a$ is the recoil energy of the residual nucleus. This last value is, in general, much smaller than the energy of the α-particle. The decay probability of an unstable nucleus for α-emission per unit of time is given by

$$P_\alpha = \lambda \cong \frac{v}{R} T, \tag{7.1}$$

indicating that the disintegration constant is the product of the probability of crossing the barrier (given by the transmission coefficient T) times the number of attempts that the particle makes to cross it (given by the number of collisions with the surface per unit of time, which is approximately equal v/R, where v is the speed of the particle inside the nucleus and R the radius of the nucleus).

To calculate the transmission coefficient, we first solve the ideal problem of the rectangular barrier shown in figure 7.4. Let us treat the simplest case $l = 0$. The radial part of the Schrödinger equation

$$\frac{d^2u}{dr^2} + \frac{2m}{\hbar^2}\left[E - V(r)\right] = 0 \tag{7.2}$$

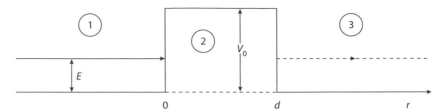

Figure 7.4 Potential barrier in the radial coordinate.

has as solution in region 1

$$u_1(r) = \alpha_1 e^{ik_1 r} + \beta_1 e^{-ik_1 r}, \tag{7.3}$$

where

$$k_1 = \frac{\sqrt{2mE}}{\hbar}, \tag{7.4}$$

with m representing the reduced mass of the particle-nucleus system. In region 2 we have the solution

$$u_2(r) = \alpha_2 e^{+k_2 r} + \beta_2 e^{-k_2 r}, \tag{7.5}$$

where

$$k_2 = \frac{\sqrt{2m(V_0 - E)}}{\hbar}. \tag{7.6}$$

Finally, in region 3 we have the solution

$$u_3 = \alpha_3 e^{ik_1 r}. \tag{7.7}$$

Using the conditions for the continuity of the wavefunction and of its derivative,

$$u_1(0) = u_2(0), \quad u_1'(0) = u_2'(0),$$
$$u_2(d) = u_3(d), \quad u_2'(d) = u_3'(d), \tag{7.8}$$

we get

$$2\frac{\alpha_1}{\alpha_3} = \left(1 + \frac{iq}{2}\right) e^{(ik_1 + k_2)d} + \left(1 - \frac{iq}{2}\right) e^{(ik_1 - k_2)d}, \tag{7.9}$$

where

$$q = \frac{k_2}{k_1} - \frac{k_1}{k_2}. \tag{7.10}$$

This implies

$$\left|\frac{\alpha_1}{\alpha_3}\right|^2 = \left(\frac{\alpha_1}{\alpha_3}\right)^* \left(\frac{\alpha_1}{\alpha_3}\right) = 1 + \frac{1}{4}\left[2 + \left(\frac{k_2}{k_1}\right)^2 + \left(\frac{k_1}{k_2}\right)^2\right] \sinh^2(k_2 d), \tag{7.11}$$

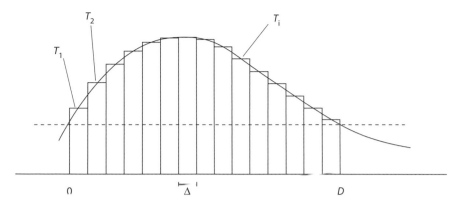

Figure 7.5 Barrier of an arbitrary shape subdivided in rectangular barriers.

where we have used

$$\sinh^2 x = \frac{1}{4}(e^{2x} + e^{-2x}) - \frac{1}{2}. \tag{7.12}$$

Since $(k_2/k_1)^2 = (V_0 - E)/E$, the transmission coefficient will be given by

$$T = \left|\frac{\alpha_3}{\alpha_1}\right|^2 = \left[1 + \frac{V_0^2}{V_0^2 - (2E - V_0)^2} \sinh^2 (k_2 d)\right]^{-1}. \tag{7.13}$$

When the "wavelength" $\lambda_2 = 1/k_2$ is much smaller than the width of the barrier, that is, when

$$k_2 d \gg 1, \quad \text{for} \quad d \gg \frac{1}{k_2} = \lambda_2, \tag{7.14}$$

we have

$$\sinh^2 (k_2 d) \cong \frac{1}{4} e^{2k_2 d} \tag{7.15}$$

or

$$T \cong 4 \frac{V_0^2 - (2E - V_0)^2}{V_0^2} e^{-2k_2 d}. \tag{7.16}$$

The numerical factor will not be of interest to us in the future generalization of the problem, and thus we write for now

$$T \propto e^{-\frac{2}{\hbar} \sqrt{2m(V_0 - E)} d}, \tag{7.17}$$

indicating that the transmission coefficient falls exponentially with the width of the barrier.

A barrier of an arbitrary shape can be subdivided into rectangular barriers of small width, as shown in figure 7.5. The total transmission coefficient will be given by

$$T \cong \text{constant} \cdot T_1 \cdot T_2 \cdot T_3 \cdots$$

$$\cong \text{constant} \cdot \prod_i e^{-\frac{2}{\hbar} \sqrt{2m(V_i - E)} \Delta}$$

and, in the limit $\Delta \to 0$,

$$T \cong \text{constant} \cdot e^{-\frac{2}{\hbar} \int_0^D \sqrt{2m[V(r)-E]}\,dr}. \tag{7.18}$$

A more careful calculation using the WKB theory [PC61] shows that the constant in (7.18) is of the order of 1.

When $l \neq 0$,

$$T \propto e^{-2\int k(r)\,dr}, \tag{7.19}$$

where

$$k^2(r) = \frac{2m}{\hbar^2}\left[V(r) + \frac{l(l+1)}{2mr^2}\hbar^2 - E \right], \tag{7.20}$$

differing from (7.18) only by the addition of the centrifugal potential to $V(r)$.

For practical applications we can suppose, for simplification, that the barrier has the form of a Coulomb potential in the range R and R' (figure 7.3). In the case of the orbital momentum $l = 0$, we have

$$T_a \propto e^{-2G}, \tag{7.21}$$

where we have introduced the *Gamow factor*

$$G = \frac{\sqrt{2m}}{\hbar} \int_R^{R'} \sqrt{\frac{Z_1 Z_2 e^2}{r} - E}\,dr. \tag{7.22}$$

Integrating (7.22) we obtain

$$G = \frac{1}{\hbar}\sqrt{\frac{2m}{E}}Z_1 Z_2 e^2 \gamma(x), \tag{7.23}$$

where

$$\gamma(x) = \arccos\sqrt{x} - \sqrt{x(1-x)}, \tag{7.24}$$

and $x = R/R' = E/V_c$, with

$$V_c = \frac{Z_1 Z_2 e^2}{R} = \frac{Z_1 Z_2 e^2}{r_0\left(A_1^{1/3} + A_2^{1/3}\right)}. \tag{7.25}$$

For the calculation of (7.1) we obtain the value of v observing that

$$\frac{Z_1 Z_2 e^2}{R'} = E = \frac{1}{2}mv^2, \tag{7.26}$$

resulting finally in the disintegration constant

$$\lambda \cong \frac{1}{R}\sqrt{\frac{2E}{m}}\exp\left[-\frac{2}{\hbar}\sqrt{\frac{2m}{E}}Z_1 Z_2 e^2 \gamma\left(\frac{R}{R'}\right) \right]. \tag{7.27}$$

From (7.27) we can understand the Geiger-Nuttall rule (figure 7.2): the logarithm of the disintegration constant λ varies with the inverse of the square root of the energy E,

Table 7.1 Half-lives calculated for two values of r_0 and a comparison with experimental values.

Z	A	E(MeV)	$t_{1/2}$[exp(s)]	$t_{1/2}$[calc(s)] $r_0 = 1.2$ fm	$t_{1/2}$[calc(s)] $r_0 = 1.3$ fm
84	215	7.53	0.18	0.45×10^{-5}	0.18×10^{-6}
84	218	6.11	182.0	1.79	0.072
89	211	7.63	0.25	0.3×10^{-3}	10^{-5}
89	218	9.38	0.3×10^{-6}	0.9×10^{-9}	0.3×10^{-10}
94	238	5.59	2.8×10^9	3.9×10^7	1.2×10^6
94	239	5.25	7.6×10^{11}	5.5×10^9	1.7×10^8

neglecting slow variations due to log E and to the presence of the function $\gamma(R/R')$, which is also a function of E, also varying slowly.

Table 7.1 exhibits the application of (7.27) for some isotopes of heavy elements. The calculation was carried out for two values of the constant r_0 $(R = r_0 A^{1/3})$. We see that the half-lives can vary by a factor of 30 when we vary the radius of the nuclei by 10%. This extreme sensitivity, resulting from the exponential character of (7.27), can have an important connection to the fact that the model was developed for spherical nuclei and a good part of the α-emitters are located in regions where the nuclei are deformed in the fundamental state. In this sense, the discrepancies with the observed experimental values in table 7.1 should contain a contribution that originates from the shape of the initial nucleus. Another important factor not considered in (7.27) is the angular momentum carried by the α-particle, imposing selection rules for the population of excited states of the daughter nucleus. This aspect will be examined in the following section.

7.3 Angular Momentum and Parity in α-Decay

Let I_i be the angular momentum of an α-emitter and I_f the angular momentum of the final state (not necessarily the ground state) of the daughter nucleus. The orbital angular momentum l carried by the α-particle is limited to

$$|I_i - I_f| \leq l \leq I_i + I_f. \tag{7.28}$$

The initial parity, Π_i, and the final one, Π_f, are related to the parity $(-1)^l$, and associated to the wavefunction of the α-particle, by

$$\Pi_i = (-1)^l \, \Pi_f. \tag{7.29}$$

Table 7.2 Energy E, angular momentum I, and parity π of the states of ^{238}Pu and the intensity i of the α-emission of ^{242}Cm feeding them. Parentheses mark doubtful information.

E(keV)	I, π	$i\%$	E(keV)	I, π	$i\%$
0	0^+	74.0	941.5	0^+	0.000052
44.08	2^+	25.0	962.8	1^-	0.000001
145.96	4^+	0.035	968.1	(2^-)	—
303.6	6^+	0.0031	983.0	2^+	0.0000016
513	8^+	0.00002	985.5	2^-	—
605.18	1^-	0.00024	1028.5	2^+	0.0000034
661.4	3^-	0.000012	1069.95	3^+	—
763.2	(5^-)	2×10^{-7}	1078	12^+	—
772	10^+	—	1125.8	4^+	3.4×10^{-7}

The need to obey (7.28) and (7.29) imposes restrictions on the possible states of a nucleus fed by an α-decay. Let us examine, as an example, table 7.2, where we list the levels of $^{238}_{94}$Pu and the intensity of the α-decay of $^{242}_{96}$Cm, which feeds each one of them. To feed an excited state means here to emit an α-particle and to leave the residual nucleus in that excited state. The energy of the corresponding α-particle will be the difference between the maximum energy (for the ground state of ^{238}Pu) and the energy of the corresponding excited state. We see that the excited states of ^{238}Pu are not all populated with the same probability. The intensity of population of each level decreases for levels of higher energy. This is due partly to the centrifugal barrier $l(l+1)\hbar^2/2mr^2$ contained in (7.20), not computed in the special case $l=0$ of (7.27). The centrifugal barrier is effectively equivalent to an increase in the height and thickness of the barrier to be surmounted by the α-particle, causing a decrease in the probability of feeding the levels as the value of the orbital angular momentum l increases. Consider, for example, the levels 2^+, 4^+, 6^+, 8^+, 10^+, and 12^+ of ^{238}Pu (table 7.2), which constitute a rotational band; the values of l are the same, as the particular values of the angular momenta of each state (due to 7.28), and thus the reason for a decrease in probability with increase of the spin of the level is understood. The emission intensity for levels 10^+ and 12^+ is even below the detection limit.

Another important feature is the nature of the initial and final levels in the decay. Significant overlap of the wavefunctions of those states contributes, in agreement with (6.49), in making the passage of the system from one state to another easier. This is not the case, for example, for the group of levels above the rotational band of the fundamental state of ^{238}Pu. Being constituted of vibrational states and of independent particles, it

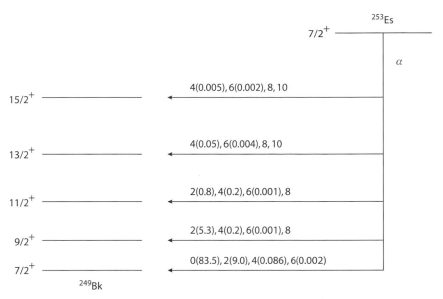

Figure 7.6 Intensities (%) for the several components of angular moment in the α-decay of ^{253}Es to ^{249}Be. For $l \geq 8$ the intensities are very small and not available to measurement [So70].

is described by wavefunctions that are not similar to the wavefunctions of ^{242}Cm. The intensities for those states, even for those with low I_f, are very small.

A final remark on table 7.2 is that the two states 2^- and the state 3^+ are not populated. In this case it is easy to see that a decay to those states is prohibited by (7.28) and (7.29). One should have in mind, however, that decays with very weak intensities and that violate (7.29) can happen by the presence of an additional part in the wavefunction due to the weak interaction (about 10^7 times smaller than the strong interaction), which does not conserve parity.

When the initial and final states are different from zero, the value l of the orbital momentum of the α-particle is not unique, and we have a series of possible values given by (7.28). Figure 7.6 shows the decay of einsteinium-253 to berkelium-249. At the side of each level we show the values of l which are possible and the respective intensities. It is clear that these values cannot be determined by the energy of the corresponding α-particle, because they are the same. Separation of the branches is made, in this case, by analysis of the angular distribution of the α-particle emitted when the emitter is polarized (with the spins of all the nuclei parallel) by application of an electric or magnetic field at low temperatures. Since the angular part of the wavefunction is given by a spherical harmonic Y_l^m, different values of l will give different probabilities of detecting the α-particle as a function of the emission angle.

Measurement of the angular distribution also reflects another aspect, not taken into consideration in (7.27): when the nucleus is deformed, the Coulomb barrier is smaller in the region of larger curvature (peak). This implies that the intensity can be several times greater for emission in angles where the barrier is reduced. This aspect can be of great importance, since the larger part of the α-emitter is constituted of deformed heavy nuclei.

7.4 Exercises

1. Determine the speed and the linear momentum of an α-particle of energy 5 MeV. Is the use of relativistic formulas important in this case?

2. Using a table of masses, show that the energy released in the α-emission by ^{239}Pu is 5.24 MeV. The kinetic energy of the measured α-particle is 5.16 MeV. Verify if one can attribute the difference to the energy of recoil of the ^{235}U.

3. a) Using the value for the depth of the nuclear well shown in figure 5.8, calculate the kinetic energy and the linear momentum of the α-particle inside ^{235}U. b) What is the wavelength of this particle inside the nucleus? Compare with the diameter of the nucleus and interpret. c) What is the wavelength of the particle after being emitted?

4. A well-known application of the uncertainty principle shows that an electron cannot be confined to the interior of a nucleus. Make the same application to the case of an α-particle inside a heavy nucleus and compare the two situations.

5. Use expression (7.25) to calculate the height of the Coulomb barrier for α-emission for the nuclei ^{40}Ca, ^{112}Sn and ^{232}Th.

6. a) Show that the emission of ^{14}C by ^{222}Ra is energetically possible. b) Calculate the half-life of the process supposing that it happens through a mechanism similar to α-emission. c) Compare with the experimental half-life $t_{1/2} = 3200$ years and interpret an eventual discrepancy. d) Verify if the proton and neutron number of the residual nucleus could have some influence in the process.

7. A nucleus X emits α-particles of energies (in MeV): 5.42; 5.34; 5.21; 5.17; and 5.14. Each α-emission can be accompanied by one or more γ-rays of energies (in MeV): 0.20593; 0.0744; 0.166407; 0.21598; 0.13161; and 0.08437. Make a diagram representing the levels of the daughter nuclei, the γ-decay between them, and the respective energy of the α-particle that feeds them.

8. The α-decay from the fundamental state ($\frac{7}{2}^+$) of ^{253}Es leads to a sequence of states of even parity of ^{249}Bk, as shown in figure 7.6. But this decay also feeds a band of negative parity states with $I = \frac{3}{2}, \frac{5}{2}, \frac{7}{2}, \frac{9}{2}, \frac{11}{2}, \frac{13}{2}$. Find the allowed values of I for each state of this band.

8 | Beta-Decay

8.1 Introduction

The most common form of radioactive disintegration is β-decay, detected in isotopes of practically all elements, with the exception, up to now, of the very heavy ones at the extreme of the chart of nuclides. It consists in the emission of an electron and an antineutrino (β^--decay) or in the emission of a positron and a neutrino (β^+-decay), keeping the nucleus, in both cases, with the same number of nucleons according to the equations

$$^A_Z X_N \rightarrow ^{A}_{Z+1} Y_{N-1} + e^- + \bar{\nu} \tag{8.1}$$

and

$$^A_Z X_N \rightarrow ^{A}_{Z-1} Y_{N+1} + e^+ + \nu. \tag{8.2}$$

The mechanisms of α- and β-emission differ in an essential aspect: whereas the nucleons that form the α-particle already reside in the nuclear matter, the electron (or positron) is forbidden to exist inside the nucleus and is created at the instant of the emission by means of the so-called *weak interaction*, one of the four fundamental forces of particle interaction, along with the nuclear, electromagnetic, and gravitational ones. It is responsible for the transmutation of a neutron into a proton with the emission of an electron and an antineutrino, described by (8.1), or the change of a proton into a neutron with the emission of a positron and a neutrino, described by (8.2).

In the same way that the pions are the mediators of the nuclear force, a triplet of particles, W^+, W^-, and Z^0, are responsible for the weak interaction, acting as "quanta" of that force. The existence of these particles was predicted in 1967 by S. Weinberg [We67] and, independently, by A. Salam and S. Glashow [Sa80], and detected in 1983 at the European Nuclear Research Center (CERN) using high energy proton beam collisions. The *intermediate vector bosons*, as they are also called, are very heavy, with masses

$$m_{W^\pm} c^2 = (80.8 \pm 2.7) \text{ GeV}, \tag{8.3}$$

$$m_{Z^0} c^2 = (92.9 \pm 1.6) \text{ GeV}. \tag{8.4}$$

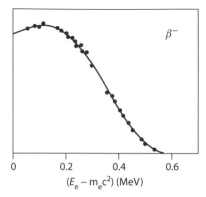 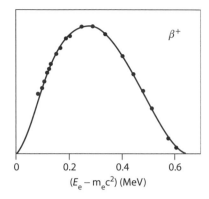

Figure 8.1 Energy distribution of the electron and positron in ^{64}Cu decay. The low energy part of the electron spectrum is enhanced due to the deceleration caused by the nuclear attraction. For the positron one has the opposite effect.

We have seen earlier how a nonvanishing value of the meson masses leads to a finite range of the nuclear force, as a consequence of the uncertainty principle. Owing to the large masses involved, the same argument indicates for the present case that the weak interaction is a force of very short range ($\sim 10^{-3}$ fm). However, we shall see that a basic theory for β-decay can be developed without detailed knowledge of the form of the interaction.

8.2 Energy Released in β-Decay

A disintegration process that leads from a state of energy ϵ_2 of a nucleus to a state of energy ϵ_1 of the daughter nucleus releases a well-defined amount of energy $\epsilon_0 = \epsilon_2 - \epsilon_1$. If only one particle is emitted, as in α-disintegration, it carries this value in the form of kinetic energy (neglecting the recoil of the residual nucleus). When more than one particle is emitted, as in β-decay, there is a continuous spectrum for the kinetic energy of the particles. Figure 8.1 shows typical examples of decay curves of β^- and β^+.

It was based on the existence of a continuum electron energy spectrum that W. Pauli [Pa34] proposed, in 1931, the presence of a second particle in β-decay, later named a *neutrino* by E. Fermi [Fe34], who developed the first theory for the process. Since the energy levels of a nucleus are discrete, the energies of the electron and of the recoiling nucleus in the center of mass system should also be discrete, by conservation of energy and momentum, unless a third particle (the neutrino) is present to share the momentum and energy of the decay. Neutrinos were not known at that time, since they are particles without charge and magnetic moment, which interact only through the weak force and are, for this reason, able to pass through any common experimental apparatus without being detected. They were detected for the first time in 1953, when Reines and Cowan [RC53], using a 1 m^3 liquid scintillator rich in protons, observed the products of the "inverse β-decay"

$$\bar{\nu} + \mathrm{p} \rightarrow \mathrm{n} + e^+, \tag{8.5}$$

where the antineutrinos necessary for the reaction were supplied by the β-decay of the fission products of a nuclear reactor.

Neutrinos are fermions of charge zero and spin $\frac{1}{2}$. The neutrino mass can be obtained, in principle, from the mass balance of a β-disintegration,

$$Q = m_{Z,A}c^2 - m_{Z\pm1,A}c^2 - m_e c^2 - m_\nu c^2, \tag{8.6}$$

where the value of Q is measured. The experimental results show that this mass has a very small or null value. The distinction between these two possibilities has important theoretical implications, as we shall see shortly.

Let us go back to (8.6), assuming now $m_\nu = 0$. This equation can be more appropriately written in terms of the atomic masses $M_{Z,A}$ (the ones effectively measured):

$$M_{Z,A}c^2 = m_{Z,A}c^2 + Zm_e c^2 - \sum_1^Z B_i, \tag{8.7}$$

where B_i is the i-th electron binding energy. In this way, one rewrites (8.6) as

$$Q = M_{Z,A}c^2 - Zm_e c^2 + \sum_1^Z B_i - \left[M_{Z\pm1,A}c^2 - (Z \pm 1)m_e c^2 + \sum_1^{Z\pm1} B_i \right] - m_e c^2. \tag{8.8}$$

Neglecting the binding energies of the last nucleons, we can equalize both sums and write (8.8) for both forms of disintegration,

$$Q_{\beta^-} = (M_{Z,A} - M_{Z+1,A})c^2,$$
$$Q_{\beta^+} = (M_{Z,A} - M_{Z-1,A} - 2m_e)c^2, \tag{8.9}$$

showing that β^--decay occurs whenever the atomic mass of the parent nucleus is larger than that of the daughter nucleus. For β^+-decay there is an additional term of two electronic masses in the computation of Q. In both cases the value of Q is shared in the form of kinetic energy of the electron (positron), of the antineutrino (neutrino), and of a very small term due to the energy of the residual recoiling nucleus.

8.3 Fermi Theory

A simple theory for β-decay was suggested by [Fe34] in 1934. Although this theory is incomplete (it does not permit parity violation, for instance), it is able to describe the spectra of figure 8.1, and gives a qualitative understanding of the values of the decay half-lives. Fermi an analogy of β-decay with the emission of electromagnetic radiation by the nucleus, induced by the time-dependent interaction between the system that irradiates and the electromagnetic field. In the case of β-decay the weak force is the agent responsible for the decay. It can be understood as a perturbation; that is, it is small compared with the forces responsible for maintaining the initial and final quasi-stationary states. Expression (6.49) for the disintegration constant,

$$\lambda = \frac{2\pi}{\hbar} |\mathcal{M}_{if}|^2 \frac{dN}{dE_T}, \tag{8.10}$$

with

$$M_{if} = \int \Psi_f^* \mathcal{V} \Psi_i \, d^3r, \tag{8.11}$$

can be applied, with $\mathcal{V}$ being the operator associated to the weak force, dN/dE_T the density of available final states with disintegration energy E_T, Ψ_i the wavefunction of the parent nucleus, and Ψ_f the wavefunction of the final system, composed by the residual nucleus, the electron (or positron), and the antineutrino (or neutrino):

$$\Psi_f = \Psi_R \Psi_e \Psi_\nu. \tag{8.12}$$

Let us initially examine the integral in (8.11). The wavefunctions of the product in (8.12) must be normalized; if Ψ_e and Ψ_ν are the free particle wavefunctions—the correction due to the influence of the Coulomb field of the nucleus will be commented upon later—the normalization can be done in a cubic box of side a. Since it is irrelevant whether to work with traveling or standing waves, an expression of the plane wave type for Ψ_e and Ψ_ν can be employed. In this case the normalization gives $A = 1/\sqrt{V}$, where $V = a^3$ is the box volume. Thus we can write for the lepton wavefunctions,

$$\Psi_e = \frac{1}{\sqrt{V}} e^{i\mathbf{p}_e \cdot \mathbf{r}/\hbar},$$

$$\Psi_\nu = \frac{1}{\sqrt{V}} e^{i\mathbf{p}_\nu \cdot \mathbf{r}/\hbar}, \tag{8.13}$$

where $\mathbf{p}_e$ and $\mathbf{p}_\nu$ are the electron and neutrino moments. The product that appears in (8.12) can be written in the form of a power series

$$\Psi_e \Psi_\nu = \frac{1}{V} \left[1 + \frac{i(\mathbf{p}_e + \mathbf{p}_\nu) \cdot \mathbf{r}}{\hbar} + \cdots \right]. \tag{8.14}$$

If we take into account that the wavelengths associated with the leptons are very large compared to the nuclear dimensions (a 1 MeV electron, for instance, has $\lambda = 897$ fm), we shall see that in the proximity of the nucleus the first term in (8.14) is largely dominant; Ψ_e and Ψ_ν can be considered constant and their product equal to $1/V$. A partial wave expansion of the plane wave shows that this first term is part of the $l = 0$ component of the expansion, that is, the lepton orbital momentum vanishes and the transitions with $l = 0$ are said to be *allowed transitions*. In some circumstances, however, the matrix element M_{if} of (8.11) vanishes when (8.14) is reduced to just its first term. In this case the remaining terms must be taken into account in the evaluation of M_{if}. Their contribution is, as we have seen, small, and transitions where this happens are said to be *forbidden transitions*, although they are not really forbidden but merely less probable than the allowed ones. If only the first term of (8.14) vanishes, then in (8.11) we have a first forbidden transition, if the first two vanish, a second forbidden transition, and so on. These high order terms correspond to the sum of the lepton orbital momenta equal to 1, 2, etc. The higher the order of forbiddenness, the lower will be the decay constant λ by a factor $pr/\hbar \cong 10^{-4}$. One example is the decay of $^{115}_{49}\text{In} \rightarrow {}^{115}_{50}\text{Sn}$, where the first nonzero term is the fifth one: the decay constant is very small and the half-life of this process is about 10^{14} years.

For the solution of the integral (8.11) it is necessary to know the form $\mathcal{V}$ of the weak inter-action. Fermi did not take into account in his theory the spins of the particles involved in the process (the nucleons, the electron, and the neutrino all have spin $\frac{1}{2}\hbar$). In this case, the matrix element constructed from the interaction v has a more or less simple nonrelativistic expression:

$$\mathcal{M}_{if}^{F} = g_F M_{if}^{F},$$ (8.15)

with

$$\left| M_{if}^{F} \right|^{2} = \sum_{m_f} \left| \int \Psi_f^{*} \left(\sum_k t_{\pm}^{k} \right) \Psi_i \, d^3 r \right|^{2},$$ (8.16)

where (8.15) makes explicit the factor g_F, the coupling constant for Fermi transitions, being a measure of the weak interaction intensity for this case. In this sense it has a role equivalent to the charge in electromagnetic interaction. The matrix element M_{if}^{F} is now dimensionless.

The first sum of (8.16) should be taken over the values of the magnetic quantum number m of the final nucleus and the second over all the nucleons of the initial nucleus, where the operators $t_+ = t_x + i t_y$ and $t_- = t_x - i t_y$ are constructed from isospin operators defined in section 1.6. t_+ transforms a neutron into a proton and should be used in β^--decay. t_- has the opposite effect and is used in β^+-decay. In simple calculations using the shell model, the sum over k reduces to a few (or even 1) valence nucleons. In some of these cases the matrix element M_{if} can be easily obtained, and this allows determination, as we shall see later, of the value of the constant g_F, whose value is around 10^{-4} MeV fm^3.

Carrying on the analysis of (8.10), we can infer how to write the density dN/dE_T. For that, let us initially evaluate λ for a given total relativistic energy E_e of the electron. Thus, dN now represents the possible number of states for the neutrino energy in the interval between E_ν and $E_\nu + dE_\nu$. Here $E_T = E_e + E_\nu$, and with fixed E_e, $dE_T = dE_\nu$. It is then possible to apply the calculation for the number of possible energy states in a Fermi gas contained in a volume V. Thus, using formula (5.19), we have

$$dn(k) = \frac{1}{2} \frac{k^2 \, dk}{\pi^2} V,$$ (8.17)

and recalling that for the neutrino $k = p/\hbar = E/\hbar c$, we immediately arrive at

$$\frac{dN}{dE_T} = \frac{dN_\nu}{dE_\nu} = \frac{V}{2\pi^2 (\hbar c)^3} (E_T - E_e)^2.$$ (8.18)

Before establishing the final form of (8.10), it is convenient to add a correction factor. This is normally introduced to take into account the nuclear Coulomb field effects over the electron wavefunction, which in reality could not be represented by a plane wave. This factor, which depends on the atomic number Z and on the final electron energy E_e, is referred to as the *Fermi function* $F(Z, E_e)$. We have seen that the electron wavefunction $\Psi_e(Z, \mathbf{r})$ is essentially constant inside the nucleus. It can be replaced by its value at the center

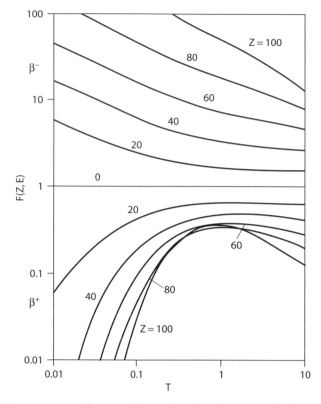

Figure 8.2 Fermi function $F(Z, E)$ plotted as a function of the kinetic energy $T = E - mc^2$ of the electron or positron.

$\Psi_e(Z, 0)$, and the Fermi function is the factor that corrects the probability to find the electron inside the nucleus, that is, $|\Psi(Z, 0)|^2 = F(Z, E_e)|\Psi(0, 0)|^2$, where $\Psi(0, 0)$ is the electron wavefunction without the Coulomb interaction. The Fermi function has a nonrelativistic expression given by

$$F(Z, E_e) = \frac{2\pi\eta}{1 - e^{-2\pi\eta}}, \tag{8.19}$$

where $\eta = \pm Ze^2/\hbar v_e$, and the plus (minus) sign is valid for the electron (positron). v_e is the electron velocity corresponding to E_e. The relativistic calculation of $F(Z, E_e)$ is complicated, and we shall limit ourselves to reproducing the results in figure 8.2.

If we now collect all the factors and omit the index from the electron energy, (8.10) takes the form

$$\lambda(E) = \frac{F(Z, E)}{V\pi\hbar^4 c^3}|\mathcal{M}_{if}|^2(E_T - E)^2, \tag{8.20}$$

where it is explicit that $\lambda(E)$ refers to just one energy E of the emitted electron.

From (8.20) we can get the probability per unit time of the emission of an electron with energy between E and $E + dE$; it is enough to multiply the rate $\lambda(E)$ by the number of

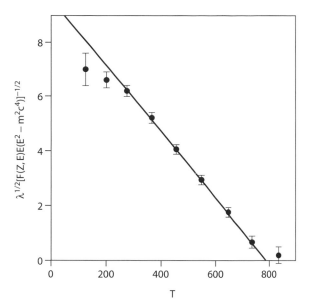

Figure 8.3 Kurie plot for neutron disintegration [Ro51]. Units are arbitrary.

possible states for the electron in that energy interval. Using again (5.19), on this occasion for the electron energy, one finds that

$$\lambda(E)dN = \frac{F(Z, E)|\mathcal{M}_{if}|^2}{2\pi^3 \hbar^7 c^6} E(E^2 - m^2c^4)^{1/2}(E_T - E)^2 dE, \tag{8.21}$$

where m is the electron mass. Equation (8.21) is the expression that should describe the behavior of the curves of figure 8.1. We see that the general features are reproduced, with a maximum in E between zero and the disintegration energy E_T; a more detailed calculation needs knowledge of the matrix element $\mathcal{M}_{if}$. Simple examples will be discussed later. When $v_e \to 0$, $F(Z, E_e) \to 2\pi\eta$ for electrons. The factor $1/_e$ makes $\lambda(E)$ nonzero at the origin, where $E_e \to m_e c^2 + m_e v_e^2/2$; the decay rate is not small for small electron energies, since the Coulomb field for the electrons is attractive. For low energy positrons, $F(Z, F_e) \to 2\pi\eta \, e^{-2\pi\eta}$. The Coulomb field is repulsive for the positrons, and we can recognize the exponential as a "tunneling" factor through the Coulomb barrier, which tends to suppress the emission of low energy positrons.

A better comparison between theory and experiment is done through *Kurie plots*. In these plots, the vertical axis shows the quantity

$$[\lambda(E)/F(Z, E)E(E^2 - m^2c^4)^{1/2}]^{1/2}$$

and the horizontal axis the electron kinetic energy. According to (8.21), an allowed decay would be represented by a straight line, as shown in figure 8.3 for free neutron disintegration.

It is also possible, in many cases, to linearize Kurie plots for forbidden decays. In these cases an appropriate correction factor is added to the vertical scale of figure 8.3; a given degree of forbiddenness is attributed to the spectrum when the plot is linearized by the addition of the factor corresponding to that order. Some first order decays, for instance, have their Kurie plots linearized by multiplying the denominator of the square root by $(p_e^2 + p_\nu^2)$, with $\mathbf{p}_e$ and $\mathbf{p}_\nu$, as in (8.13), being the electron and the neutrino momenta. This value is justified if we verify that the second term of the expansion (8.14) is of the form $p_e + p_\nu$, contributing $p_e^2 + p_\nu^2$ to $|\mathcal{M}_{if}|^2$. The contribution of the cross term $\mathbf{p}_e \cdot \mathbf{p}_\nu$ vanishes when one takes an average over the angles between electrons and neutrinos.

The Kurie plot is also useful for other purposes. If we have an allowed transition, the Kurie plot is, as we have seen, a straight line. However, in the derivation of (8.18) a zero mass was used for the neutrino. If the neutrino mass is not zero, the Kurie plot should deviate from a straight line. This is one of the methods used to verify if the neutrino has mass [Be72, Lu80]. The results are not conclusive, but an upper limit of 18 eV/c^2 for the neutrino mass has been established [Ku86].

8.4 The Decay Constant—The Log *ft* Value

Let us now derive the value of the decay constant λ in (8.21). The integral of (8.21) is

$$\lambda = \frac{m^5 g^2 c^4 |M_{if}|^2}{2\pi^3 \hbar^7} f(Z, E_T),$$ (8.22)

where the function

$$f(Z, E_T) = \frac{1}{m^5 c^{10}} \int_0^{E_T} F(Z, E) E (E^2 - m^2 c^4)^{1/2} (E_T - E)^2 \, dE,$$ (8.23)

known as the *Fermi integral*, is dimensionless and usually presented in curves that are a function of the atomic number Z and of the electron maximum energy E_T. A set of such curves is shown in figure 8.4.

Equation (8.22) allows us to examine the influence of the matrix element M_{if} in the evaluation of the decay constant λ. Using relation (6.4) between λ and the half-life $t_{1/2}$, (8.22) can be rewritten as

$$ft_{1/2} = \frac{1.386\pi^3 \hbar^7}{g^2 m^5 c^4 |M_{if}|^2}.$$ (8.24)

We see that the product $ft_{1/2}$ or, simply, ft, depends only on M_{if}: the larger the matrix element value the more probable is the occurrence of the transition.

In some special cases the matrix element M_{if} is easily evaluated. This occurs, for example, in a β^+ transition from ^{14}O to ^{14}N. It is a $0^+ \to 0^+$ transition with $M_{if} = \sqrt{2}$, where the measured half-life gives for ft a value near 8 hours. This allows us to use (8.24) for determining the coupling constant, resulting in the value $g_F \cong 10^{-4}$ MeV fm^3, mentioned

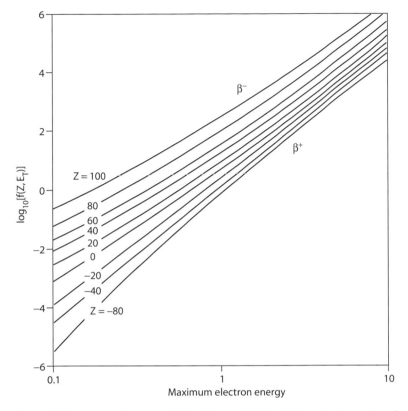

Figure 8.4 Representative curves of Fermi integral (8.23). Z is the atomic number of the product nucleus and is set positive for β^+-decay and negative for β^--decay [FT50]. Units are arbitrary.

in the last section. Transitions in which the value of M_{if} is near unity produce the lowest values of ft and are called *superallowed*.

Forbidden transitions can have ft values several orders of magnitude larger than the allowed ones. This is due to the natural difficulty in creating an electron-neutrino pair with $l > 0$. This can be shown by a simple classical calculation: assume a β-decay with $Q = 1$ MeV. With the assumption that the electron is emitted with the total energy, it would have near the nuclear surface a maximum angular momentum $m_e v R \cong 0.05\hbar$, for the case of a heavy nucleus with $R = 7.4$ fm. This shows how the $l = 1$ value is improbable, and this is even more true for larger l values. From this fact results a very large range of ft values, and it is common to use $\log_{10} ft$, with t given in seconds, as a measure of the decay probability of a given state by β-emission. An experimental distribution of $\log ft$ values is shown in figure 8.5.

The use of $\log ft$ of allows, in a crude way, separation of the several transitions by their degree of forbiddenness, since each degree has a certain range of $\log ft$ values. However, this separation is not perfect since, as we see in figure 8.5, the bands intercept themselves. Another remarkable fact is the paucity of data for transitions with $l \geq 2$. For $l = 4$, only four cases are known, with $\log ft$ in the range 22.5–24.5.

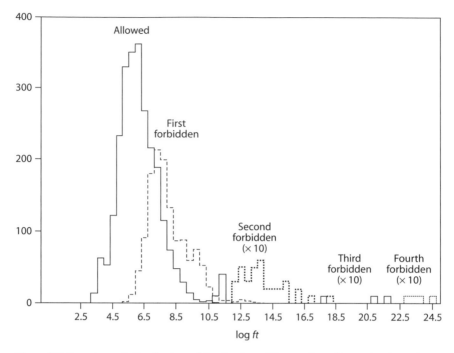

Figure 8.5 Experimental distribution of log ft values. The number of cases in the ordinate includes the electron capture process.

8.5 Gamow-Teller Transitions

When a β-decay leads an initial nucleus of spin I_i to a final nucleus of spin I_f, with angular momentum and parity conservation,[1] this contribution can be very small. We need, in association with this, to create some selection rules for I_i, I_f, and the parities of the initial and final states. To establish these rules it is necessary initially to distinguish between two possible situations. In the first one, the electron and the neutrino have opposite spins and do not contribute to the angular momentum balance. Such transitions are called *Fermi transitions* and for these it is easy to see that

$$I_i = I_f + l \quad \text{(Fermi)}, \tag{8.25}$$

where l is the orbital angular momentum carried by the electron and the neutrino. In turn, for *Gamow-Teller transitions*, the lepton spins are parallel and contribute one unit to the momenta:

$$I_i = I_f + l + 1 \quad \text{(Gamow-Teller)}. \tag{8.26}$$

The Gamow-Teller transitions are not accounted for by Fermi theory, which, as we have seen, does not take into account the spins of the particles. We can show that, with the introduction of spin, the matrix element of (8.16) is modified by the additional presence

---

[1] The nuclear wavefunction actually has a contribution of the weak force that does not conserve parity.

Table 8.1 Selection rules for angular momentum and parity in β-decay.

Transition	$\Delta I = I_i - I_f$	Parity change
Allowed	$0, \pm 1$	No
First forbidden	$0, \pm 1, \pm 2$	Yes
Second forbidden	$\pm 2, \pm 3$	No
....	....	....
nth forbidden	$\pm n, \pm(n+1)$	$(-1)^n[1 = \text{yes}, -1 = \text{no}]$

of the three components of the Pauli spin operator

$$|M_{GT}|^2 = \sum_{m_f} \sum_x \left| \int \Psi_f^* \left(\sum_k t_{\pm}^k \sigma_x^k \right) \Psi_i \, d^3 r \right|^2, \qquad (8.27)$$

where $\sum_x$ represents the sum over the Pauli matrices σ_x, σ_y, and σ_z. The index k means again that the operators t and σ act on the nucleon k of the initial nucleus, whose wavefunction is Ψ_i. If Fermi and Gamow-Teller transitions are both possible, one has to write (8.15) in a complete form: $|\mathcal{M}_{if}|^2 = g_F^2 |M_F|^2 + g_{GT}^2 |M_{GT}|^2$. We have already seen that g_F is of the order of 10^{-4} MeV fm^3. One can show through several examples that the coupling constant g_{GT} for a Gamow-Teller transition has a value a little larger than g_F.

From (8.25) and (8.26) we can establish the spin variation of the emitting nucleus for the several values of l. Thus, in allowed decays ($l = 0$) of the Fermi type, $\Delta I = |I_i - I_f| = 0$. In Gamow-Teller transitions, ΔI, by angular momentum composition, one can assume the values 0 and 1, except if $I_i = 0$ and $I_f = 0$. For this case only Fermi transitions are possible. One example is ^{14}O$\rightarrow^{14}$N* decay, where the oxygen 0^+ level decays to a 0^+ excited state of nitrogen. Examples of allowed decay in which only Gamow-Teller transitions are possible involve nuclear spin variation, as in the case ^{60}Co$\rightarrow^{60}$Ni, where the Co 5^+ initial state decays to a Ni 4^+ state.

In the general case, both (8.25) and (8.26) can be satisfied and the decay proceeds by a mixture of Fermi and Gamow-Teller transitions. A typical example is the free neutron decay, where a state $\frac{1}{2}^+$ decays to another state $\frac{1}{2}^+$ of the proton: 18% of the transitions are Fermi and 82% Gamow-Teller.

Parity conservation has also very clear effects: with the parity of the orbital part given by $(-1)^l$, the allowed transitions and the even degree forbidden ones do not imply parity change, whereas for the odd degree forbidden ones the parity of the daughter nucleus is opposed to that of the initial nucleus. Table 8.1 summarizes the selection rules for β-decay.

Since the transitions of lower l are the most important, useful information would also be to show the lowest possible values of l for each variation ΔI between initial and final states. Table 8.2 shows these values for Fermi and Gamow-Teller transitions.

Table 8.2 Lowest possible value of l for each variation $|\Delta I|$ of nuclear spin and $\Delta \pi$, the parity. The asterisk indicates when the variation is not possible, if $I_i = 0$ or $I_f = 0$. l values higher than 4 have never been observed.

$|\Delta I|$	0		1		2		3		4		5	
$\Delta \pi$	yes	no	yes	no	yes	no	yes	no	yes	no	yes	no
Fermi	1*	0	1	2*	3*	2	3	4*	5*	4	5	6*
G-T	1	0*	1	0	1	2	3	2	3	4	5	4

8.6 Selection Rules

The existence of selection rules allows us, in several cases, to detect disintegrations with a high order forbiddenness owing to the absence of lower order disintegrations. One example is the second forbidden decay $3^+ \to 0^+$ from ^{22}Na to ^{22}Ne, impossible with $l < 2$. This example is not essentially different from the case of ^{115}In mentioned after (8.14). In fact, a transition that violates the selection rules of table 8.1 has its corresponding matrix element of (8.16) equal to zero. One of the simplest examples is to imagine an allowed Fermi transition with parity change. From table 8.1 this transition is not possible. In fact, examining the matrix element of (8.16) we see that it vanishes since it is the integration of the product of two different parity functions.

Besides the spin selection rules, β-decay also has selection rules for the isospin. It is easy to see from (8.16), for example, that states with different values of the isospin T cannot be connected by a Fermi transition, since the operator $\sum t_\pm = T_\pm$ (see (4.65)) changes the z-component of T by 1 unit, and nuclei with $T_i \neq T_f$ will have $M_{if} = 0$. If the transition is $0^+ \to 0^+$, a case in which $M_{GT} = 0$, the associated half-life must be very high. In fact, about 20 transitions $0^+ \to 0^+ (\Delta T \neq 0)$ are listed in [Si98], where the $\log ft$ value varies from 6.5 to 10.5, characterizing a forbiddenness dictated by the isospin selection rule.

8.7 Parity Nonconservation in β-Decay

Symmetries perform an essential role in physics. Fundamental principles such as energy and momentum conservation are connected to space and time symmetries. The homogeneity of space leads to linear momentum conservation, and the nonexistence of favored directions (space isotropy) results in the conservation of the angular momentum of an isolated system. Energy conservation is a consequence of time homogeneity, i.e., the fact that there is no special moment to set the origin of time. The above examples are not unique, and the German mathematician Emmy Noether established that to each symmetry in nature is associated a conservation principle.

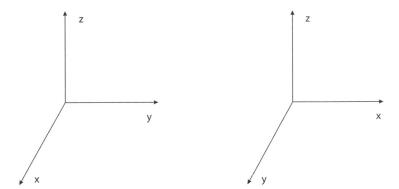

Figure 8.6 Direct (left) and inverse systems employed in the description of a physical situation. Note that one system cannot be obtained from the other by a spatial rotation.

The description of a phenomenon through a system of Cartesian axes should not depend, in principle, on axis orientation. In particular, no property should depend on the fact that the system is direct or inverse (see figure 8.6). To this space symmetry is associated a conservation principle, that of the parity of the wavefunction that describes the system. The parity conservation was established earlier and applied to reactions involving baryons and mesons.

There is another way in which the above described symmetry is present: if some physical event is described by spatial coordinates $\mathbf{r}(t)$, the coordinates $-\mathbf{r}(t)$ should describe an event that obeys the same physical laws. Let us take as an example the particular case of reflection in a plane mirror, which corresponds to the change of sign of only one coordinate. What parity conservation imposes is that the world viewed in the mirror cannot be distinguished from the real world by analysis of the physical laws of each world. In other words, the world seen through the mirror should obey the same physical laws as the real world.

This logical scheme seemed to be of universal validity, but by the middle of the 1950s surprising news appeared. The starting point was the behavior of two particles, known at that time as θ and τ. These particles had the same mass and the same spin zero but different intrinsic parities, since they decayed by, among others, the modes

$$\theta \rightarrow \pi^+ + \pi^0, \tag{8.28}$$

$$\tau \rightarrow \pi^+ + \pi^+ + \pi^-, \tag{8.29}$$

implying the attribute of positive parity for θ and negative parity for τ, since the pions have negative parity and the eventual orbital angular momenta have zero overall contribution to the parity.

The existence of particles with identical characteristics but differing only in parity was something new and intriguing and was called the "$\theta - \tau$ puzzle." This puzzle was solved by two Chinese physicists, T. D. Lee and C. N. Yang [LY56], who in 1956 assumed that θ and τ were the same particle, known today as the K-meson. To explain how the K-meson can decay by the two modes, (8.28) and (8.29), Lee and Yang proposed a revolutionary

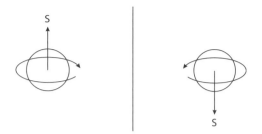

Figure 8.7 A nucleus and its mirror image have opposite spins.

idea, namely that in the weak interaction, which governs decay, parity is not necessarily conserved, as would happen in strong, electromagnetic, and gravitational interactions. Thus, β-decay would be included in reactions that do not conserve parity, and Lee and Yang idealized some experiments where this could be verified.

One of these experiments was carried out by C. S. Wu and collaborators [Wu57] in 1957. ^{60}Co nuclei had their spins (total angular momentum) aligned by the action of a magnetic field acting on a sample kept at a very low temperature, the latter being necessary to avoid the destructive action of thermal motion on the alignment.

^{60}Co is a β-emitting nucleus. What does one expect from the angular distribution of the emitted electrons? In particular, is there any preference for electron emission in the nuclear spin direction or in the opposite direction? Let us examine figure 8.7. The nuclear spin is represented at the left, together with the movement of a particle that symbolically creates that spin. The image of the nucleus is seen in the mirror at the right, and it is easy to verify that the movement of the particle is inverted and the spin of the image must have a direction contrary to the original one. In this context, what would happen if the electrons in a β-decay had a large preference to be emitted, say, in the direction of the spin over the opposite direction? The image we would have in the mirror would be exactly the opposite one, the electrons being emitted contrary to the spin direction. If parity conservation were mandatory for the process, we could not have one physics for the real world and another for the mirrored one. Thus, detection of this asymmetry in the β-emission would point out that parity is not conserved in this process.

The results obtained by Wu and her collaborators have shown that this asymmetry effectively exists, the electrons from the ^{60}Co β-decay being emitted preferentially in the direction opposite to the spin. Figure 8.8 shows the form of the angular distribution, obtained in an improved experiment made 25 years after the original one.

Other experiments of different character (Lederman and collaborators [Ga57], for instance, have analyzed the pion and the subsequent muon decay) have confirmed parity nonconservation in the weak interaction, according to the Lee and Yang predictions.

Observation of the effect of this particular aspect of interaction in the properties of the emerging particles can be done by means of a quantity called *helicity*. Helicity

$$h = \frac{\mathbf{s} \cdot \mathbf{p}}{|\mathbf{s} \cdot \mathbf{p}|} \tag{8.30}$$

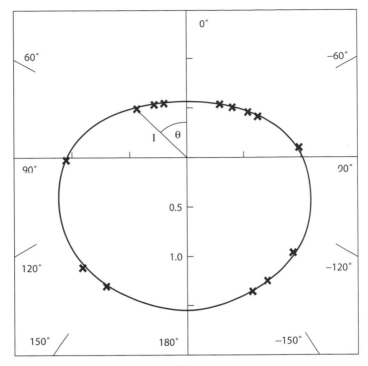

Figure 8.8 Angular distribution of ^{60}Co β-decay. The angle 0° points to the direction of polarization, and the radius I is proportional to the number of electrons emitted in that direction. The crosses mark the experimental values obtained from [Ch80].

is defined as the spin projection over the particle linear momentum **p**. In principle its value depends on the observer, since if a referential sees the vector **p** in a given direction, another referential can see it in the opposite direction (it is enough that this referential has a velocity greater than and in the same direction as the particle). However, this does not happen with particles that travel with the speed of light, as, for instance, photons and, supposedly, neutrinos. For them, helicity is a well-defined quantity independent of the observer. Here resides a surprising fact: while photons can have helicity +1 or −1 (light right or left polarized), neutrinos always have $h = -1$ and antineutrinos $h = 1$.[2] For an inverse system, $h = +1$ for the neutrinos and $h = -1$ for antineutrinos. A neutrino in front of a mirror reveals an image that does not correspond to an existing physical situation, namely, a neutrino with helicity +1 (figure 8.9). A similar situation but with exchanged numbers occurs with the antineutrino.

In β-decay the electrons are also polarized, but the nonzero mass implies that polarization is not pure. Different from the case of neutrinos, helicity is not a good quantum number, but the global result is that the fraction of electrons emitted with positive helicity minus the fraction emitted with negative helicity is equal to $-v/c$, where v is the electron velocity. For positrons this value is $+v/c$.

---

[2] When observed in a direct system of axes.

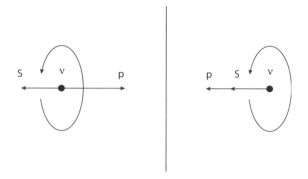

Figure 8.9 A neutrino (to the left) with helicity −1, has as its image a neutrino with helicity +1 (to the right), a particle that does not exist in nature.

The connection between neutrino helicity and the nonconservation of parity in ^{60}Co β-decay can be explained in the following way: this transition leads a 5^+ state of cobalt to a 4^+ state of ^{60}Ni and is of the Gamow-Teller type. Table 8.1 shows that the transition must proceed with $l = 0$. The nuclei are polarized, the greater part of them having maximum spin projection, $M = +5$. To conserve angular momentum the lepton spins must be aligned with the nuclear spin and, for the emitted antineutrino, this means that the nuclear spin is the direction of emission. It happens that for this type of transition there is an angular correlation between the directions of the antineutrino and electron emission, that is, there is a function that relates the emerging angles of the electron and the neutrino, and this function indicates that the two particles leave the nucleus preferably in opposite directions. This leads to the predominant emission of the electron opposite to the spin.

When one analyzes the behavior of other symmetries in the experiments described above, one sees that it is not only the parity that presents unexpected results. The *charge conjugation* operation—the operation of exchange of particles with their respective antiparticles and vice versa—behaves in the same way. This operation, normally designated by C, when applied to some process, generally leads to another process possible in nature. However, this is not the case, for the phenomenon discussed above. Let us take the example of the neutrino. The operation C, of charge conjugation, transforms the neutrino of helicity −1 into an antineutrino with the same helicity, a particle that does not exist. In this sense, both charge conjugation C and parity change P produce results of the same type.

Figure 8.10 illustrates the situation. Isolated application of the operations C and P to the neutrino does not produce real particles. The application of CP, on the other hand, restores the real world. One can show that this result is not restricted to this example but is also valid for the other cases that we have previously analyzed. Thus, one is bound to believe that all physical laws are invariant with regard to a CP operation, since phenomena described by the strong, electromagnetic, and gravitational forces are known to be invariant by C and P operations separately.

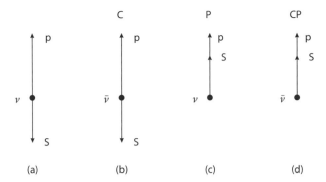

Figure 8.10 Diagram (a) shows a neutrino with its spin projection opposed to the movement direction ($h = -1$). In (b), after the application of the C operator, this results in an nonexistent particle, a antineutrino with $h = -1$. Similarly, there is no particle of the character shown in (c), namely a neutrino with $h = 1$, after the application of the P operator in (a). However, the successive application of both operators, CP, leads to the particle of diagram (d), an antineutrino with correct helicity $h = 1$.

Belief in the universal invariance of all the processes under CP operation had to be revised when, in 1964, Cronin and collaborators [Ch64], analyzing the decay of the mesons K^0 and $\bar{K}^0$, detected a small (0.3%) contribution of events that do not obey that expectation. The CP-violation of the K-meson, as it is commonly referred to, presents new and difficult problems that at the present time have yet to find satisfactory resolution up to the present.

8.7.1 Double β-Decay

A special case of β-decay can occur when a neighbor isobar is not accessible to the decay but the subsequent one is. In this case the transition can be completed with the simultaneous emission of two electrons.

Let us look at the example of figure 8.11. The $A = 130$ isobar chain shows an energy distribution typical of even A (see figure 5.2b). The $^{130}_{52}$Te nucleus cannot decay by β^- to $^{130}_{53}$I, situated at a higher energy, but can reach $^{130}_{54}$Xe by the emission of two electrons.

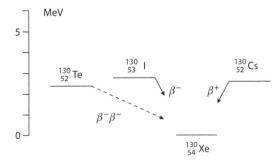

Figure 8.11 Ground state energy diagrams of the $A = 130$ isobars.

Table 8.3 Half-lives for double β-decay [MV94, Ta96, Ar96, Si97]. In the values with double error indication, the first one refers to the true error, the second to the systematic error. In the third column, d and g refer to direct and geochemical methods, respectively [MV94].

Isotope	$T_{1/2}$ (years)	Method
^{76}Ge	$(14.2 \pm 0.3 \pm 1.3) \times 10^{20}$	d
^{82}Se	$(1.08^{+0.26}_{-0.06}) \times 10^{20}$	d
^{96}Zr	$(3.9 \pm 0.9) \times 10^{19}$	g
^{100}Mo	$(6.82^{+0.38}_{-0.53} \pm 0.68) \times 10^{18}$	d
^{116}Cd	$(3.75 \pm 0.35 \pm 0.21) \times 10^{19}$	d
^{128}Te	$(2.2 \pm 0.3) \times 10^{24}$	g
^{130}Te	$(7.9 \pm 1.0) \times 10^{20}$	g
^{150}Ne	$(6.75^{+0.37}_{-0.42} \pm 0.68) \times 10^{18}$	d
^{238}U	$(2.0 \pm 0.6) \times 10^{21}$	g

Possibilities like this occur in several other isobar chains; the phenomenon is easy to detect since the half-lives for the simultaneous emission of two electrons are very long, near or more than 10^{19} years, demanding very specialized experimental techniques.

Two methods are employed: direct count of events and analysis of old rocks that contain the double β-emitter material and should also have correlated quantities of the product nucleus. Table 8.3 shows a few results that have been confirmed up to now. The strong motivation to detect the phenomenon is due, mainly, to the possibility of detecting the existence of double β-decay without the presence of neutrinos. This possibility is connected to the question about the neutrino mass, namely whether it is zero. If the neutrino mass is not zero, helicity is not a good quantum number and the neutrino and antineutrino carry a small component of the "wrong" helicity. Specifically, the neutrino has a character of antineutrino and vice versa, since helicity is the only quantity that distinguishes the two particles. If this is the case, the antineutrino resulting from the decay

$$n \rightarrow p + e^- + \bar{\nu} \tag{8.31}$$

could induce the reaction

$$\nu + n \rightarrow p + e^-, \tag{8.32}$$

whose global result is the double β-decay

$$2n \rightarrow 2p + 2e^- \tag{8.33}$$

without neutrino emission. The occurrence of (8.33) by the two distinct steps (8.31) and (8.32) is extremely improbable and in some cases energetically impossible (see the earlier example of $^{130}_{52}$Te). For these cases we can admit the virtual occurrence of the process, performed in a time interval compatible with the uncertainty principle $\Delta E \Delta T \sim \hbar$. The neutrino would exist for a very short time Δt, characteristic of the weak interaction, and its energy could reach several tens of MeV. This would open a large volume in phase space, making the process in (8.33) several orders of magnitude more probable.

Experimental methods to verify the phenomenon profit from the fact that the electrons are emitted simultaneously and should have total energy equal to the Q value of the reaction. The existence of the double β-decay process without neutrino emission would be revealed by the presence of a well localized peak in the kinetic energy distribution of the emitted electrons. This would not happen, however, for another type of proposed decay: double β-decay without neutrino emission but with the emission of a hypothetical particle, a light neutral boson, the *majoron* (in honor of Ettore Majorana). We know that electrons and neutrinos are leptons (leptonic number $+1$), and positrons and antineutrinos are antileptons (leptonic number -1). This third form of double β-decay restores leptonic number conservation, which is violated in (8.33).

The distinction between the two last processes is only formal. No experiment has shown, up to now, any positive indication of the existence of double β-decay without the presence of neutrinos. One has only established lower limits for the half-lives in some nuclei.

8.8 Electron Capture

The electrons that surround the nucleus are described by wavefunctions that have some spatial extension. In particular, those which are localized in the K shell have a sizable probability of being found inside the nucleus and to give place to a capture reaction

$$^{A}_{Z}X_N + e^- \rightarrow \ ^{A}_{Z-1}Y_{N+1} + \nu, \tag{8.34}$$

competing with β^+-decay, that produces the same final nucleus. The Q value of the reaction

$$Q_\epsilon = (M_{Z,A} - M_{Z-1,A})c^2 - B_K, \tag{8.35}$$

where ϵ represents the capture, takes into consideration the loss of the electron binding energy B_K in the K shell. In units of mc^2, this binding energy has in the nonrelativistic approximation the value

$$B_K = \frac{1}{2}\left(\frac{Ze^2}{\hbar c}\right). \tag{8.36}$$

At first, the process in (8.34) has simpler features than β^+-emission, since there is no emitted positron: the neutrinos are monoenergetic, with an energy very near the value Q_ϵ. On the other hand, the neutrinos are hard to detect and the electron capture process is normally studied by X-ray emission, resulting from the occupation of the vacancy of the captured electron by another electron from outer shells. With regard to the energetics

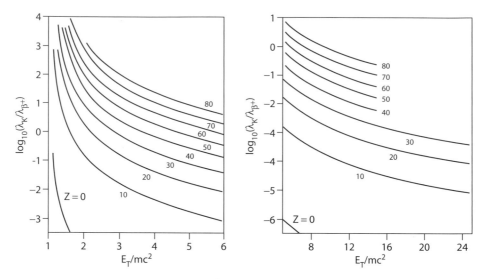

Figure 8.12 Ratio between K-capture and β^+-emission probability as a function of the total energy available [FT50]. The values labeling each curve correspond to the charge of the nucleus.

of the process, the electron capture does not involve energy loss with the creation of a positron and can occur in situations where β^+-emission is energetically forbidden.

The decay constant λ can be evaluated in a simple form. Initially, we have to use the fact that in electron capture there is only the neutrino emission, and (8.18) is the measure of the neutrino final state density

$$\frac{dN_\nu}{dE_T} = \frac{VE_T^2}{2\pi^2(\hbar c)^3}. \tag{8.37}$$

In the present case, we shall use for the electron wavefunction, Ψ_e, (8.12), the value of the electron wavefunction Ψ_K in the K orbit at the origin,

$$\Psi_e = \Psi_K(0). \tag{8.38}$$

Thus we can write (8.10) as

$$\lambda = 2\frac{2\pi}{\hbar}|\mathcal{M}_{if}|^2|\Psi_K(0)|^2\frac{dN_\nu}{dE_T}, \tag{8.39}$$

where $\mathcal{M}_{if}$ is calculated with the nuclear wavefunctions only. The new factor 2 is due to the existence of two electrons in the K orbit. The value of Ψ_K at the origin is given approximately by

$$\Psi_K(0) = \pi^{-1/2}\left(\frac{Zme^2}{\hbar^2}\right)^{3/2}, \tag{8.40}$$

and thus we obtain the value for the decay constant by electron capture in the K orbit as

$$\lambda_K = \frac{2Z^3m^3e^6|\mathcal{M}_{if}|^2E_T^2}{\pi^2\hbar^{10}c^3}. \tag{8.41}$$

The nuclear matrix element $|\mathcal{M}_{if}|$ connects the same initial and final wavefunctions in electron capture and β^+-decay. In this way, it cancels when, using (8.41), one forms the ratio

$$\frac{\lambda_K}{\lambda_{\beta^+}} = \frac{4\pi Z^3 e^6 E_T^2}{m^2 \hbar^3 c^7 f(Z, E_T)} \tag{8.42}$$

between the electron capture and β^+-emission probabilities. In the application of (8.42), note that $E_T = Q_\epsilon + mc^2$. The curves $\lambda_K/\lambda_{\beta^+}$ are shown as a function of E_T in figure (8.12), where more precise relativistic results are used. One feature of these functions is that the process of electron capture increases its relative probability for heavier nuclei. For these, the capture in L orbit can give a significant contribution and reach 15% of the K-capture events.

8.9 Exercises

1. From the intermediate vector boson masses of (8.3) and (8.4), use the uncertainty principle to make an estimate of the range of the weak interaction forces.

2. Since the stable sodium isotope is ^{23}Na, what type of radioactivity would you expect from (a) ^{22}Na and (b) ^{24}Na?

3. ^{150}Eu can decay by β^-, β^+, and ϵ (electron capture). Determine the Q values for the three decay modes.

4. ^{75}Ge decays by β^- to ^{75}As. Using a mass table, determine the maximum energy of the emitted electrons.

5. ^{66}Cu β^--decay occurs, in 80% of the cases, to a ^{66}Zn excited level, emitting electrons of 1.6 MeV maximum energy. Find the excitation energy of that zinc state.

6. Consider the allowed β-decays that have a disintegration energy E_T large compared to $m_e c^2$ (^{8}B decay, for example). In such decays, the Coulomb correction and lepton mass effects are small. Show that, neglecting such effects, a) the average electron energy is $E_T/2$, b) the mean lifetime depends on E_T as E_T^{-5}. Show, on the other hand, that if $E_T \ll m_e c^2$, the average energy is $E_T/3$.

7. Obtain an expression for the electron momentum distribution in β-decay for a neutrino mass $m_\nu \neq 0$.

8. Obtain an expression for the electron energy distribution, assuming that two (zero mass) neutrinos are emitted in the β-decay process.

9. Show that the energy spectrum of the allowed β-decay arrive at the maximum energy with horizontal slope if $m_\nu = 0$ and with vertical slope if $m_\nu \neq 0$.

10. Classify the following decays according to the degree of forbiddenness:

a) $^{15}O(\frac{1}{2}^-) \to {}^{15}N(\frac{1}{2}^-)$ d) $^{99}Tc(\frac{9}{2}^+) \to {}^{99}Ru(\frac{5}{2}^+)$

b) $^{69}Zn(\frac{1}{2}^-) \to {}^{69}Ga(\frac{3}{2}^-)$ e) $^{115}In(\frac{9}{2}^+) \to {}^{115}Sn(\frac{1}{2}^+)$

c) $^{87}Rb(\frac{3}{2}^-) \to {}^{87}Sr(\frac{9}{2}^+)$ f) $^{141}Ce(\frac{7}{2}^-) \to {}^{141}Pr(\frac{5}{2}^+)$

11. a) Depict the Kurie plot for the β^+ and β^- decay of the 1^+ $^{64}_{29}Cu$ ground state, from points extracted from figure 8.1 and using figure 8.2. Say, from the aspect of the figures, if they correspond to allowed or forbidden transitions. b) Obtain from table 8.1 the orbital angular momentum l of the electrons emitted in each case. Are the results coherent with those obtained in (a)? c) How does one obtain, for each case, the linear momentum distribution of the electrons and positrons?

12. The 0^+ ground state of ^{50}Mn decays by β^+ to the ground state of the same spin and parity of ^{50}Cr, with a half-life of 0.286 s. a) Is it a Fermi or Gamow-Teller transition? b) Use a mass table and figure 8.4 to determine the $\log ft$ value for this transition. c) Is the result compatible with the information from table 8.1 and figure 8.5?

13. Show that in the β^+-decay of ^{14}O to ^{14}N, the shell model gives a value for the matrix element of the Fermi transition (8.16) equal to $\sqrt{2}$. Suggestion: treat the two nuclei as a system of two particles.

14. Show that the β-decay between mirror nuclei (the number of protons of one is the number of neutrons of the other and vice versa) of the same J has the transition matrix elements given by

$M_{ij}^2 = 1$ Fermi

$M_{ij}^2 = (j + 1)/j$ Gamow-Teller $j = l + \frac{1}{2}$

$M_{ij}^2 = j/(j + 1)$ Gamow-Teller $j = l - \frac{1}{2}$

Suggestion: Observe that in $\beta^\pm$-decay, $\Psi_i = T_\pm \Psi_f$, and use the commutation rules between isospin operators, identical to those of spin operators.

15. Neutron decay has a measured $\log ft$ value equal to 3.0. Use the result of the previous exercise to show that the ratio g_{GT}/g_F is about 1.2.

16. Prove the statement made after (8.29).

17. Using the preponderant signs of the helicities of the emitted particles, verify in each case if the tendency is the electron (positron) emission in the same or in the opposite direction to the antineutrino (neutrino): a) β^-, $0^+ \to 0^+$; b) β^+, $0^+ \to 0^+$; c) β^-, $1^+ \to 0^+$; d) β^+, $1^+ \to 0^+$.

18. In the electron capture

$$^7_4\text{Be (atom)} \to {}^7_3\text{Li (atom)} + \nu,$$

with the beryllium initially at rest, the recoil energy of the lithium atoms (mass 6536 MeV/c^2) was measured as 55.9 ± 1.0 eV [Da52]. The mass difference between the two atoms is 0.862 MeV/c^2. Show that the result of this experiment limits the neutrino mass to a value less than 160 keV/c^2.

19. Using (8.42) and figure 8.4, find the proportion of electron captures with respect to the β^+-decay of $^{125}_{55}\text{Cs}$ that emits positrons of maximum energy 2.05 MeV. Compare the obtained value with that extracted from figure 8.12 and also with the real value, presented, for instance, in the Table of Isotopes of Lederer and Shirley [LS78].

20. An electron is in the ground state of the tritium atom. The tritium nucleus ^{3}H (one proton + two neutrons) suddenly undergoes β-decay into the helium nucleus ^{3}He (two protons and one neutron); the new electron and antineutrino created in the β-decay carry away the electric charge and energy. Find the probability for the original atomic electron to remain in the ground state of the helium atom.

21. In the 1956 experiment of Wu the β-decay of polarized Co nuclei into Ni nuclei and electron and electron antineutrinos,

$$^{60}_{27}\text{Co}_{33} \to {}^{60}_{28}\text{Ni}_{32} + e^- + \bar{\nu}_e, \tag{8.43}$$

was observed. The angular distribution of the electrons turned out to be

$$N(\theta) \propto 1 + \alpha \cos\theta, \tag{8.44}$$

where the asymmetry coefficient is $\alpha \approx -v/c$ (v is the electron velocity) and θ is the angle between the electron momentum **p** and the spin **J** of the ^{60}Co nuclei. The initial and final nuclei have spins $5\hbar$ and $4\hbar$, respectively. Discuss what symmetries (if any) are violated in the process and explain why the electrons prefer to be emitted in the direction opposite to the polarization direction, $\theta \to \pi$.

9 | Gamma-Decay

9.1 Introduction

The quantum system of A nucleons that form the nucleus has, above its state of lowest energy (ground state), a large number of possible excited states that can be accessed if enough energy is given to the system. The transitions among these states, either through excitation or through de-excitation, are accomplished mainly through γ-radiation, which embraces a high energy region of the electromagnetic spectrum. This region is located basically between 0.1 MeV and 10 MeV, being a γ-ray of 1 MeV of order 3×10^5 times more energetic than violet light.

Besides the energy released, another characteristic parameter is the half-life of the decaying state. Half-lives for γ-emission can vary from 10^{-17}s to 100 years; even the lower limit is still large compared to the shortest nuclear times, for example, the time for a particle with velocity close to the light to cross a nuclear diameter ($\cong 10^{-22}$s). Thus, the states that decay by γ-radiation can be considered quasi-stationary. This is reflected in the level widths calculated from (6.49): even a half-life of 10^{-17}s corresponds to a width of about 70 eV, small compared to the energy spacing of the levels.

9.2 Quantization of Electromagnetic Fields

9.2.1 Fields and gauge invariance

The electromagnetic field is described by the 4-vector of electromagnetic potentials $(\phi, \mathbf{A})$ that determines the electric, $\mathbf{E}$, and magnetic, $\mathbf{B}$, fields,

$$\mathbf{E} = -\nabla\phi - \frac{1}{c}\frac{\partial \mathbf{A}}{\partial t}, \tag{9.1}$$

$$\mathbf{B} = \nabla \times \mathbf{A}. \tag{9.2}$$

The electromagnetic fields carry energy

$$E = \frac{1}{8\pi} \int d^3r \, (\mathbf{E}^2 + \mathbf{B}^2) \tag{9.3}$$

and momentum

$$\mathbf{P} = \frac{1}{c^2} \int d^3r \, \mathbf{S} \equiv \frac{1}{4\pi c} \int d^3r \, [\mathbf{E} \times \mathbf{B}], \tag{9.4}$$

where the *Poynting vector* $\mathbf{S}$ gives the momentum density (flux of electromagnetic energy). Here we use the Gaussian system of units.

The important difference between electric and magnetic fields is in their behavior under discrete space-time transformations, inversion $\mathcal{P}$ and time reversal $\mathcal{T}$ (see Appendix C). The electric vector $\mathbf{E}$ is a *polar* vector, while the magnetic vector $\mathbf{B}$ is an *axial* vector (pseudovector). They behave differently under time reversal as well: electric fields are generated by charges and do not change sign, while magnetic fields are generated by currents and change sign along with them. The Maxwell equations are invariant under both $\mathcal{P}$- and $\mathcal{T}$-transformations. As seen from (9.1) and (9.2), the scalar potential ϕ is a $\mathcal{T}$-even scalar and $\mathbf{A}$ is a polar $\mathcal{T}$-odd vector (it is also generated by currents).

Electromagnetic potentials are not defined uniquely by electromagnetic fields. Keeping intact the vector character of $\mathbf{A}(t, \mathbf{r})$, one can make a *gauge transformation* adding a gradient of an arbitrary single-valued scalar function of time and space coordinates,

$$\mathbf{A}(t, \mathbf{r}) \Rightarrow \mathbf{A}'(t, \mathbf{r}) = \mathbf{A}(t, \mathbf{r}) + \nabla f(t, \mathbf{r}). \tag{9.5}$$

It does not change magnetic field since (curl $\nabla \times \nabla \equiv$ curl ∇) $\equiv 0$. If the scalar potential is simultaneously transformed according to

$$\phi(t, \mathbf{r}) \Rightarrow \phi'(t, \mathbf{r}) = \phi(t, \mathbf{r}) - \frac{1}{c} \frac{\partial f(t, \mathbf{r})}{\partial t}, \tag{9.6}$$

with the same gauge function $f(t, \mathbf{r})$ as in (9.5), the electric field (9.1) also does not change.

We can use the gauge freedom to simplify the consideration of specific problems. For our purposes the *radiation gauge* seems to be convenient. First, we can always get rid of the scalar potential by taking $f(t, \mathbf{r}) = c \int^t dt' \phi(t', \mathbf{r})$ so that $\phi'(t, \mathbf{r}) = 0$. Then, since we are interested in describing the radiation field in free space, where curl $\nabla \cdot \mathbf{E} \equiv$ div $\mathbf{E} = 0$, we can choose the vector potential $\mathbf{A}$ subject to the additional condition

$$\nabla \cdot \mathbf{A} \equiv \text{div} \, \mathbf{A} = 0. \tag{9.7}$$

As a result, only two components of $\mathbf{A}$ are physically independent.

Below we will see that this is equivalent to the statement that electromagnetic waves in free space are *transversely polarized*. The polarization degrees of freedom play the role of the photon spin. The transverse character of the electromagnetic field is related to the fact that the quanta, *photons*, which appear after the field is quantized have zero mass. For massless particles there is no rest frame. Their free motion wavefunction with wave vector $\mathbf{k}$ is axially symmetric with respect to the $\mathbf{k}$-axis, and, for nonzero spin, can be characterized by helicity, the spin projection onto the direction of motion. Helicity is Lorentz invariant for a massless particle. The two possible values of photon helicity, ± 1, correspond to right (left) circular polarization. Their combinations describe two possible *linear* polarizations in the plane perpendicular to the axis of motion. The longitudinal polarization along the wave vector, $\mathbf{A} \| \mathbf{k}$, which would have zero helicity (similar to the motion of a particle along the axis that does not create angular momentum along the same direction), is not allowed for the real massless photon.

9.2.2 Normal modes

Consider a free electromagnetic field in a large normalization volume V. The vector potential satisfies the *wave equation*

$$\left(\frac{1}{c^2} \frac{\partial^2}{\partial t^2} - \nabla^2 \right) \mathbf{A} = 0. \tag{9.8}$$

The plane wave with the wave vector $\mathbf{k}$,

$$\mathbf{A}_\mathbf{k}(t, \mathbf{r}) = \mathbf{A}_\mathbf{k} e^{i(\mathbf{k} \cdot \mathbf{r}) - i\omega t}, \tag{9.9}$$

is a solution of the wave equation (a normal mode in the normalization box) if the wave frequency ω satisfies the dispersion law

$$\omega = \omega_\mathbf{k} \equiv c|\mathbf{k}| \equiv ck. \tag{9.10}$$

The components of the wave vectors are quantized, $k_i = (2\pi/L)n_i$, where L is a size of the volume $V = L^3$ and n_i are integers (positive or negative).

The supplementary gauge condition (9.7) requires that the amplitude vector $\mathbf{A}_\mathbf{k}$ be orthogonal to the wave vector,

$$\mathbf{k} \cdot \mathbf{A}_\mathbf{k} = 0, \tag{9.11}$$

which is just the *transversality* condition. To describe the wave polarization we introduce two unit vectors $\mathbf{e}_{\mathbf{k}\lambda}$ labeled by an additional subscript λ taking, for given $\mathbf{k}$, one of two values. The polarization vectors are perpendicular to the wave vector,

$$\mathbf{e}_{\mathbf{k}\lambda} \cdot \mathbf{k} = 0. \tag{9.12}$$

They are also taken to be mutually orthogonal. In general, they are complex and their scalar products are defined with complex conjugation, as in the case of complex wavefunctions,

$$\mathbf{e}_{\mathbf{k}\lambda}^* \cdot \mathbf{e}_{\mathbf{k}\lambda'} = \delta_{\lambda\lambda'}. \tag{9.13}$$

Two standard choices are *linear* polarizations described by real vectors $\mathbf{e}_{\mathbf{k}1}, \mathbf{e}_{\mathbf{k}2}$ along $x \equiv 1$ and $y \equiv 2$ axes in the plane perpendicular to the wave vector direction $z \equiv 3$, or *circular* polarizations with the complex polarization vectors

$$\mathbf{e}_{\mathbf{k}\pm} = \mp \frac{1}{\sqrt{2}} (\mathbf{e}_{\mathbf{k}1} \pm i\mathbf{e}_{\mathbf{k}2}). \tag{9.14}$$

In both cases three vectors, $\mathbf{e}_{\mathbf{k}\lambda}$ and $\mathbf{k}/k$, form a right-handed triplet. The frequencies (9.10) do not depend on polarization.

Due to the linearity of the equations, one can use the superposition principle to form linear combinations of specific solutions. The solution (9.9) with fixed **k** can be represented by a sum of two polarizations with some scalar amplitudes,

$$\mathbf{A_k}(t, \mathbf{r}) = \sum_{\lambda} A_{\mathbf{k}\lambda} \mathbf{e}_{\mathbf{k}\lambda} e^{i(\mathbf{k}\cdot\mathbf{r})-i\omega_k t}. \tag{9.15}$$

Finally, since the vector potential is real, a physical solution should contain a complex conjugate (c.c.) term along with (9.15). Thus, the general solution of the wave equation is an arbitrary superposition of partial plane wave solutions,

$$\mathbf{A}(t, \mathbf{r}) = \sum_{\mathbf{k}\lambda}(A_{\mathbf{k}\lambda} \mathbf{e}_{\mathbf{k}\lambda} e^{i(\mathbf{k}\cdot\mathbf{r})-i\omega_k t} + A_{\mathbf{k}\lambda}^* \mathbf{e}_{\mathbf{k}\lambda}^* e^{-i(\mathbf{k}\cdot\mathbf{r})+i\omega_k t}). \tag{9.16}$$

The potential (9.16) generates electric and magnetic field according to (9.1) (with $\phi = 0$) and (9.2),

$$\mathbf{E}(t, \mathbf{r}) = \frac{i}{c} \sum_{\mathbf{k}\lambda} \omega_k \left(A_{\mathbf{k}\lambda} \mathbf{e}_{\mathbf{k}\lambda} e^{i(\mathbf{k}\cdot\mathbf{r})-i\omega_k t} - (\text{c.c.})\right), \tag{9.17}$$

$$\mathbf{B}(t, \mathbf{r}) = i \sum_{\mathbf{k}\lambda}\left[\mathbf{k} \times \left(A_{\mathbf{k}\lambda} \mathbf{e}_{\mathbf{k}\lambda} e^{i(\mathbf{k}\cdot\mathbf{r})-i\omega_k t} - (\text{c.c.})\right)\right]. \tag{9.18}$$

9.2.3 Photons

The field can be quantized by a direct analogy with the harmonic oscillator, or nuclear shape vibrations. The time dependence of the coordinate operators for the normal vibrational modes with the quantum numbers $(\lambda\mu)$ of angular momentum and its projection is

$$\alpha_{\lambda\mu}(t) = \text{const}\left\{a_{\lambda\mu} e^{-i\omega_\lambda t} + (-)^\mu a_{\lambda-\mu}^\dagger e^{i\omega_\lambda t}\right\}. \tag{9.19}$$

Here the constant prefactor depends on the normalization; the argument referring to the initial time $t = 0$ of the creation, $a^\dagger$, and annihilation, a, operators in curly brackets is omitted. The expansion (9.16) has the same structure. It is a superposition of independent normal modes with quantum numbers of momentum and polarization instead of the angular momentum quantum numbers in (9.19). The amplitudes in the general solution presented as a sum of normal modes are annihilation and creation operators for quanta with quantum numbers of a given normal mode. The annihilating term carries time dependence $e^{-i\omega t}$ and the creation term carries time dependence $e^{i\omega t}$, where in both cases the energy of a quantum $\hbar\omega$ is positive. All quantum numbers in the creation part are time-reversal with respect to the annihilation part. This is a general recipe for quantizing a system of independent oscillators, and the procedure is applicable to the electromagnetic field as well.

Thus, we postulate that in quantum theory the amplitudes $A_{\mathbf{k}\lambda}$ and $A_{\mathbf{k}\lambda}^*$ are replaced by operators proportional to annihilation and creation operators,

$$A_{\mathbf{k}\lambda} \Rightarrow \hat{A}_{\mathbf{k}\lambda}, \quad A_{\mathbf{k}\lambda}^* \Rightarrow \hat{A}_{\mathbf{k}\lambda}^\dagger. \tag{9.20}$$

Their commutation relations and the corresponding type of statistics are to be established with the help of physical arguments.

After the substitution (9.20), the electric and magnetic fields (9.17), (9.18) also become operators. The quantized energy (9.3) gives the Hamiltonian $\hat{H}$. Using the operator definitions of the fields, we get in (9.3) four terms for electric energy and a similar four terms for magnetic energy with the double summation $\sum_{\mathbf{k}\lambda\mathbf{k}'\lambda'}$. These terms have different operator structures. To calculate them, we first perform spatial integration using the orthogonality of plane waves,

$$\int d^3r e^{i(\mathbf{k}\pm\mathbf{k}')\cdot\mathbf{r}} = V\delta_{\mathbf{k},\mp\mathbf{k}'}. \tag{9.21}$$

In both cases ($\pm$) the frequencies are equal, $\omega_\mathbf{k} = \omega_{\mathbf{k}'}$ so the terms such as $\hat{A}\hat{A}^\dagger$ and $\hat{A}^\dagger\hat{A}$ turn out to be time-independent. The terms with the operators $\hat{A}\hat{A}$ or $\hat{A}^\dagger\hat{A}^\dagger$ depend on time $\propto \exp(\mp 2i\omega_\mathbf{k}t)$, which would contradict energy conservation. However, such terms vanish due to the cancellation of electric and magnetic contributions. Indeed, electric terms have the polarization-dependent amplitudes

$$-\frac{\omega_\mathbf{k}^2}{c^2}(\mathbf{e}_{\mathbf{k}\lambda} \cdot \mathbf{e}_{-\mathbf{k}\lambda'}) = -k^2(\mathbf{e}_{\mathbf{k}\lambda} \cdot \mathbf{e}_{-\mathbf{k}\lambda'}). \tag{9.22}$$

For analogous magnetic terms we obtain with simple vector algebra

$$-[\mathbf{k} \times \mathbf{e}_{\mathbf{k}\lambda}] \cdot [(-\mathbf{k}) \times \mathbf{e}_{-\mathbf{k}\lambda'}] = k^2(\mathbf{e}_{\mathbf{k}\lambda} \cdot \mathbf{e}_{-\mathbf{k}\lambda'}) - (\mathbf{k} \cdot \mathbf{e}_{-\mathbf{k}\lambda'})(\mathbf{k} \cdot \mathbf{e}_{\mathbf{k}\lambda}). \tag{9.23}$$

The last item in (9.23) vanishes because of the field transversality, and the first item cancels the electric contribution (9.22). Hence, the time-dependent terms disappear and the Hamiltonian is a constant of motion as it should be. In the remaining terms the polarization vectors form the scalar product (9.13) and the orthogonality of different polarizations selects $\lambda' = \lambda$. The magnetic and electric contributions are added to

$$\hat{H} = \frac{V}{4\pi c^2} \sum_{\mathbf{k}\lambda} \omega_\mathbf{k}^2(\hat{A}_{\mathbf{k}\lambda}^\dagger \hat{A}_{\mathbf{k}\lambda} + \hat{A}_{\mathbf{k}\lambda}\hat{A}_{\mathbf{k}\lambda}^\dagger). \tag{9.24}$$

We introduce normalized dimensionless operators $a_{\mathbf{k}\lambda}$ and $a_{\mathbf{k}\lambda}^\dagger$ according to

$$\hat{A}_{\mathbf{k}\lambda} = \sqrt{\frac{2\pi\hbar c^2}{\omega_\mathbf{k} V}} a_{\mathbf{k}\lambda}, \quad \hat{A}_{\mathbf{k}\lambda}^\dagger = \sqrt{\frac{2\pi\hbar c^2}{\omega_\mathbf{k} V}} a_{\mathbf{k}\lambda}^\dagger \tag{9.25}$$

to bring the Hamiltonian to the sum of independent photon energies $\hbar\omega_\mathbf{k}$,

$$\hat{H} = \frac{1}{2} \sum_{\mathbf{k}\lambda} \hbar\omega_\mathbf{k}(a_{\mathbf{k}\lambda}^\dagger a_{\mathbf{k}\lambda} + a_{\mathbf{k}\lambda}a_{\mathbf{k}\lambda}^\dagger). \tag{9.26}$$

At this point we have to decide about the type of *statistics*. The decision is easy because the Fermi choice with anticommutators would lead to the Hamiltonian containing no operators at all (just an infinite constant sum over all modes). Therefore we have to select

Bose statistics, postulating the commutation relations

$$[a_{k\lambda}, a_{k'\lambda'}] = [a_{k\lambda}^\dagger, a_{k'\lambda'}^\dagger] = 0, \quad [a_{k\lambda}, a_{k'\lambda'}^\dagger] = \delta_{kk'}\delta_{\lambda\lambda'}. \tag{9.27}$$

The photon number operator is defined for each mode $(k\lambda)$ as usual,

$$\hat{n}_{k\lambda} = a_{k\lambda}^\dagger a_{k\lambda}, \tag{9.28}$$

and the Hamiltonian takes a standard form for the gas of noninteracting Bose-quanta, *photons*,

$$\hat{H} = \sum_{k\lambda} \hbar\omega_k \left(\hat{n}_{k\lambda} + \frac{1}{2} \right). \tag{9.29}$$

Here, in distinction to the oscillator or to nuclear vibrational modes, we quantize a *field*, referring to a system with an *infinite* number of degrees of freedom. Indeed, to specify the initial value of the vector potential $\mathbf{A}(\mathbf{r})$, we should define its magnitude and direction for all points $\mathbf{r}$ of continuous space. Correspondingly we have an infinite number of normal modes that give rise to photons with different quantum numbers $(k\lambda)$. The zero point vibrational energy, $\hbar\omega/2$ for each mode in (9.29), gives an infinite sum. But this does not bring any physical difficulties because this energy is a nonoperator constant that can be identified with the vacuum energy. All physical states with nonzero occupation numbers have positive energy relative to the vacuum state, which is defined in a standard way as a state $|0\rangle$ such that

$$a_{k\lambda}|0\rangle = 0 \tag{9.30}$$

for all modes $(k\lambda)$. The basis states in Fock space $|\{n_{k\lambda}\}\rangle$ are constructed by the action of the creation operators. They are labeled by integer photon numbers $n_{k\lambda}$ as eigenvalues of the operators (9.28).

Similar to the derivation of the Hamiltonian (9.29), we can construct the operator of the total momentum of the field (9.4). Starting with the expressions for electric (9.17) and magnetic (9.18) fields in terms of the quantized operators (9.25), integrating over the volume, and using the transversality property, we get rid of the time-dependent $(\sim e^{\pm 2i\omega t})$ terms that would contain the operator products aa and $a^\dagger a^\dagger$. The remaining terms are diagonal in the number of photons and do not depend on time, as it should be for the total momentum,

$$\hat{\mathbf{P}} = \frac{1}{2} \sum_{k\lambda} \hbar\mathbf{k}(a_{k\lambda}^\dagger a_{k\lambda} + a_{k\lambda} a_{k\lambda}^\dagger). \tag{9.31}$$

Now we can again use the Bose commutators. In contrast to (9.29), the vacuum momentum $(\frac{1}{2}) \sum_{k\lambda} \hbar\mathbf{k}$ has to be considered equal to zero because of cancellation of terms with opposite direction to $\mathbf{k}$ (isotropy of the vacuum state). Finally,

$$\hat{\mathbf{P}} = \sum_{k\lambda} \hbar\mathbf{k}\hat{n}_{k\lambda}. \tag{9.32}$$

This is the obvious result confirming that each photon $(k\lambda)$ carries momentum $\mathbf{p} = \hbar\mathbf{k}$. The dispersion law (9.10) shows that the photons are massless.

The polarization vectors play the same role as spin wave functions for a particle of nonzero spin. They are transformed under rotations as vectors. This supplements the transformation of the momentum in the same way as the transformation of spin variables of a particle goes together with the transformation of its coordinates corresponding to spin and orbital parts of the angular momentum. A normal vector would have three possible states, or three spherical components with the projection of spin momentum 0 and ± 1. We know, however, that, with respect to the natural axis of $\mathbf{k}$ the "spin wave function of the photon," that is, the vector $\mathbf{e}_{\mathbf{k}\lambda}$, has only two allowed components. They correspond to helicity ± 1. This is important for the selection rules at work in the processes of emission and absorption of photons.

The final quantized form of the vector potential is

$$\hat{\mathbf{A}}(t, \mathbf{r}) = \sum_{\mathbf{k}\lambda} \sqrt{\frac{2\pi\hbar c^2}{\omega_{\mathbf{k}} V}} \left(a_{\mathbf{k}\lambda} \mathbf{e}_{\mathbf{k}\lambda} e^{i(\mathbf{k}\cdot\mathbf{r}) - i\omega_{\mathbf{k}} t} + a_{\mathbf{k}\lambda}^{\dagger} \mathbf{e}_{\mathbf{k}\lambda}^{*} e^{-i(\mathbf{k}\cdot\mathbf{r}) + i\omega_{\mathbf{k}} t} \right). \tag{9.33}$$

The field operators are easily obtained from (9.17) and (9.18). All these quantities are linear in creation and annihilation operators. Therefore they have selection rules $\Delta n = \pm 1$, and their matrix elements between the states with definite photon numbers are given by

$$a_{\mathbf{k}\lambda} |n_{\mathbf{k}\lambda}\rangle = \sqrt{n_{\mathbf{k}\lambda}} |n_{\mathbf{k}\lambda} - 1\rangle, \quad a_{\mathbf{k}\lambda}^{\dagger} |n_{\mathbf{k}\lambda}\rangle = \sqrt{n_{\mathbf{k}\lambda} + 1} |n_{\mathbf{k}\lambda} + 1\rangle. \tag{9.34}$$

As for a simple harmonic oscillator, we have the photon absorption, $\sim a$, and spontaneous and stimulated emission, $\sim a^{\dagger}$.

We have quantized the electromagnetic field in plane waves. As we know from the general secondary quantization formalism, it is possible to make a unitary transformation to an arbitrary basis adjusted to any field geometry like spherical, cylindrical waves or waves in a cavity of arbitrary shape (in a cavity we have real boundary conditions on the actual surface instead of our fictitious normalization volume). The procedure is the same: find a complete set of solutions (vector normal modes) for the wave equation in any coordinates, write a general solution as a superposition of all normal modes, and declare the coefficients of the superposition Bose operators. Similar to (9.25), they should be normalized to define the photon energy in accordance with the spectrum of the normal modes in a given geometry. Each mode can be populated by an arbitrary number of Bose quanta.

This recipe works for any quantized field satisfying its own wave equation. The new elements as compared to the photon field might be nonzero mass, complex wavefunctions, in contrast to the real vector potential (9.16), and Fermi statistics for half-integer spins which again comes from the physical requirements to the quantized hamiltonian.

9.3 Interaction of Radiation with Matter

Here we consider a nonrelativistic system of charged particles that interact with electromagnetic fields. The interaction makes the system capable of absorbing and emitting photons. To describe the interaction we turn it on in the so-called *minimal* way. Let

$H_0(\mathbf{r}_a, \mathbf{p}_a)$ be the original hamiltonian of the many-body system with no field. The interaction is described by a substitution $\mathbf{p}_a \Rightarrow \mathbf{p}_a - (e_a/c)\mathbf{A}(\mathbf{r}_a)$ for all particles so that the new hamiltonian includes the terms depending both on field and matter variables,

$$H_0 \Rightarrow H' = H_0\left(\mathbf{r}_a, \mathbf{p}_a - \frac{e_a}{c}\mathbf{A}(\mathbf{r}_a)\right) + \sum_a e_a \phi(\mathbf{r}_a). \tag{9.35}$$

We also added potential energy of charged particles, the last term in (9.35).

This procedure is obviously consistent with relativity since the 4-vector of energy momentum $(E, c\mathbf{p})$ is transformed by adding another 4-vector, that of electromagnetic potential $(\phi, \mathbf{A})$. It is easy to check, for example, that, for a particle in a uniform static magnetic field $\mathbf{B}$ defined by the vector potential $\mathbf{A} = \frac{1}{2}[\mathbf{B} \times \mathbf{r}]$, one gets from (9.35) correct classical or quantum operator equations of motion with the Lorentz force $e(\mathbf{E} + (1/c)[\mathbf{v} \times \mathbf{B}])$. The minimal way of coupling particles to the electromagnetic field is also consistent with gauge invariance. Indeed, the fictitious gauge field f, (9.5), is equivalent to a redefinition of the particle momentum operator, $\mathbf{p} \Rightarrow \mathbf{p}' = \mathbf{p} - (e/c)\nabla \times \nabla f$. However, the new operator $\mathbf{p}'$ has the same commutation relations, and, hence, the same dynamics, as the old operator $\mathbf{p}$. The commutator of two components $[p_i', p_j']$ still vanishes because it reduces to the difference of second derivatives $\partial_i \partial_j$ of a single-valued function f, taken in a different order. The real physical field $\mathbf{A}$ cannot be eliminated by such a redefinition because the analogous commutator would not now vanish, being proportional to the magnetic field,

$$[p_i - (e/c)A_i, p_j - (e/c)A_j] = \frac{ie\hbar}{c}\left(\frac{\partial A_j}{\partial x_i} - \frac{\partial A_i}{\partial x_j}\right) = \frac{ie\hbar}{c}\epsilon_{ijl}\mathcal{B}_l. \tag{9.36}$$

This description is applicable to the electromagnetic field of any origin including the radiation field of the system itself. The general idea of gauge fields, which is placed nowadays as part of the foundation of quantum field theory for the description of fundamental particles and their interactions, is an extension of the minimal principle for the fields A_μ that have discrete intrinsic degrees of freedom and therefore do not commute among themselves.

In the radiation problem, we can omit the scalar potential and write the full Hamiltonian of the system including the interparticle interaction U and the interaction with the radiation field as

$$H = \sum_a \frac{1}{2m_a}\left(\mathbf{p} - \frac{e_a}{c}\hat{\mathbf{A}}(\mathbf{r}_a)\right)^2 + U(\mathbf{r}_a), \tag{9.37}$$

where $\hat{\mathbf{A}}(\mathbf{r}_a)$ is the quantized vector potential (9.33) taken at the point $\mathbf{r}_a$ of the particle a. Here we have assumed that the interaction between particles depends on their coordinates only. In the presence of velocity-dependent interactions, as, for example, spin-orbit forces, the minimal substitution is to be made in the interaction term as well.

The electromagnetic interaction is relatively weak compared to the strong forces. The relevant parameter is the fine structure constant $\alpha = e^2/\hbar c = 1/137$. As a rule, this interaction can be taken into account in the lowest nonvanishing order of perturbation theory.

Looking for the transition probabilities of radiative processes, we keep first the main term of (9.37), linear in the vector potential

$$H^{(1)} = -\sum_a \frac{e_a}{2m_a c} \left(\mathbf{p}_a \cdot \hat{\mathbf{A}}(\mathbf{r}_a) + \hat{\mathbf{A}}(\mathbf{r}_a) \cdot \mathbf{p}_a \right). \tag{9.38}$$

This Hamiltonian is *linear* in creation and annihilation operators describing the *single photon* emission and absorption processes. The first order Hamiltonian for a quantum system in the classical magnetic field has the same form. Using again $\mathbf{A} = \frac{1}{2}[\mathbf{B} \times \mathbf{r}]$, it can be reduced to $-\sum_a g_a^{(l)} (\mathbf{l}_a \cdot \mathbf{B})$, which describes the paramagnetic alignment of orbits by the magnetic field (Zeeman effect). Here $\mathbf{l}$ is the orbital momentum (in units of $\hbar$), and $g_a^{(l)} = e_a \hbar / 2m_a c$ is the orbital gyromagnetic ratio (a magneton) for the particle a. The interaction of particle spins with the electromagnetic field in nonrelativistic theory has to be added separately in the form of spin magnetism,

$$H'_s = -\sum_a g_a^{(s)} \left(\mathbf{s}_a \cdot \mathbf{B}(\mathbf{r}_a) \right), \tag{9.39}$$

similar to orbital magnetism. The spin gyromagnetic ratios $g_a^{(s)}$ are empirical parameters which, in principle, could be predicted by relativistic theory of particle structure.

Now we have to consider the unified system "particles + radiation field." Due to the interaction of radiation with matter, stationary states of the "bare" system are not stationary any more. The particles can absorb and emit photons going to other states. The same is valid for the states of the radiation field with a certain number of photons. This number is also changed by emission and absorption. The former stationary states acquire a finite lifetime τ. This leads to energy uncertainty. The energy of a state is defined only within its width $\Gamma \sim \hbar/\tau$; see below. The energy spectrum can still be considered quasidiscrete if the widths are small compared with level spacings and the levels do not overlap. Strictly speaking, only a ground state of a stable system is stationary.

If the whole system were kept inside a real physical box for a long time, it would make sense to speak about genuine stationary states. They would be superpositions of the bare states "a system with no photons," "a system in another state + one photon," and so forth. Roughly speaking, emitted photons being reflected from the boundaries have a probability to be reabsorbed, reemitted again, and so on. Thus, over a long time one can prepare stationary superpositions, and the total energy of the whole system can be well defined. However, we have only an auxiliary normalization volume, which can be considered arbitrarily large, so the emission process is irreversible and the probability of equilibration via reabsorption is zero. In such a situation the language of quasistationary states with finite lifetime is physically more appropriate. Later we will return to this important physical question in more detail.

We have not yet fixed the gauge in (9.38). The radiation gauge (9.7) is very convenient because under this condition the two terms in (9.38) are equal to each other (their commutator equals div $\mathbf{A}$ and vanishes under this choice). Then

$$H^{(1)} = -\sum_a \frac{e_a}{m_a c} \left(\mathbf{p} \cdot \hat{\mathbf{A}}(\mathbf{r}_a) \right), \tag{9.40}$$

and we can use the quantized potential (9.33) derived in the same gauge. This form is a nonrelativistic version of the relativistic Hamiltonian corresponding to the interaction of the electromagnetic field with the current $\mathbf{j}$,

$$H' = -\frac{1}{c} \int d^3 r \mathbf{j}(\mathbf{r}) \cdot \hat{\mathbf{A}}(\mathbf{r}). \tag{9.41}$$

For the orbital current, the standard symmetrized quantum operator form

$$\mathbf{j}(\mathbf{r}) = \sum_a \frac{1}{2m_a} \left\{ \mathbf{p}_a \delta(\mathbf{r} - \mathbf{r}_a) + \delta(\mathbf{r} - \mathbf{r}_a) \mathbf{p}_a \right\} \tag{9.42}$$

can be used in our expressions (9.38), (9.40).

Equation (9.40) describes the radiation processes induced by the convection current of charged particles. The total current can also include spin and magnetization effects as well as the charge transfer by the virtual particles as charged mesons mediating nuclear forces. All components of the current can radiate according to the general interaction Hamiltonian (9.41). They all come together in relativistic theory; in our consideration of this we will introduce the various terms separately, see below.

9.3.1 Radiation probability

Let us consider a system in an excited initial state $|i\rangle$. We always have in mind the states that would be stationary without the interaction with the radiation field. In this approximation the initial (final) state has certain energy $E_i(E_f)$. To find the probability of radiation with the transition of the system to a final state $|f\rangle$, we use the golden rule derived for weak perturbations. This rule gives the probability of radiation per second (transition rate, w_{fi}); multiplied by the photon energy $\hbar\omega$, it determines the intensity of radiation.

Let us calculate the probability of emitting the photon with quantum numbers $(\mathbf{k}\lambda)$. The joint system of particles and radiation field undergoes the transition

$$|i; n_{\mathbf{k}\lambda}\rangle \rightarrow |f; n_{\mathbf{k}\lambda} + 1\rangle. \tag{9.43}$$

Written explicitly, the golden rule with the perturbation (9.40) leads to the transition rate

$$w_{fi} = \frac{2\pi}{\hbar} \left| \langle f; n_{\mathbf{k}\lambda} + 1 | \sum_a \frac{e_a}{m_a c} (\mathbf{p}_a \cdot \hat{\mathbf{A}}(\mathbf{r}_a)) |i; n_{\mathbf{k}\lambda}\rangle \right|^2 \delta(E_i - \hbar\omega_{\mathbf{k}} - E_f). \tag{9.44}$$

Here the vector potential (9.33) is taken without the time-dependent exponent because the unperturbed time dependence is already fully accounted for in the derivation of the golden rule (it resulted in the energy conservation δ-function).

In the emission matrix element the only contribution comes from the creation operator $a^\dagger_{\mathbf{k}\lambda}$ in (9.33). The matrix element of this operator is, as in (9.34), equal to $\sqrt{n_{\mathbf{k}\lambda} + 1}$, revealing stimulated and spontaneous radiation. Therefore we find

$$w_{fi} = \frac{4\pi^2}{V\omega_{\mathbf{k}}}(n_{\mathbf{k}\lambda} + 1)\left|\left\langle f\left|\sum_a \frac{e_a}{m_a}(\mathbf{p}_a \cdot \mathbf{e}_{\mathbf{k}\lambda}^*)e^{-i(\mathbf{k}\cdot\mathbf{r}_a)}\right|i\right\rangle\right|^2 \delta(E_i - \hbar\omega_{\mathbf{k}} - E_f). \tag{9.45}$$

The emitted photon can have any direction and we have to count available final states in the continuum. The density of final states for a photon ($\omega = \omega_{\mathbf{k}}$) = ck is

$$d\rho_f\, d(\hbar\omega) = \frac{V d^3 k}{(2\pi)^3} = \frac{V k^2}{(2\pi)^3}\, do\, dk = \frac{V\omega^2}{(2\pi c)^3 \hbar}\, do\, d(\hbar\omega). \tag{9.46}$$

Using the δ-function in (9.45) to integrate over photon energy, we come to the differential emission rate for the fixed photon direction in the solid angle do and for the fixed polarization λ,

$$\frac{dw_{fi}}{do} = \frac{\omega_{\mathbf{k}}}{2\pi c^3 \hbar}(n_{\mathbf{k}\lambda} + 1)\left|\left\langle f\left|\sum_a \frac{e_a}{m_a}(\mathbf{p}_a \cdot \mathbf{e}_{\mathbf{k}\lambda}^*)e^{-i(\mathbf{k}\cdot\mathbf{r}_a)}\right|i\right\rangle\right|^2. \tag{9.47}$$

The observable transition rate (9.47) does not depend on the auxiliary volume V. The radiation probability is actually a quantity of the *form-factor* type for the current component in the plane perpendicular to the radiation direction. The exponents in (9.47) give the relative phases for the radiation by different constituents, and their interfering contributions are coherently superimposed.

The states of radiating systems have their own quantum numbers. For a finite system, the initial and final states can be characterized by the total angular momentum and its projection, J_i, M_i and J_f, M_f, respectively. Some of the transitions can be forbidden due to the selection rules related to the properties of the electromagnetic field. For example, all single-photon transitions with $J_f = J_i = 0$ (the so-called *0-0 transitions*) are strictly forbidden by the angular momentum conservation. Indeed, the emitted photon carries helicity ± 1 (it can also be in a superposition of these spiral states). This means that the projection of the angular momentum of the system onto the direction of radiation $\mathbf{k}$ is changed by ± 1, which is impossible for the 0-0 transition.

9.3.2 Long wavelength approximation

An important practical case is connected to the long wavelength radiation: $\lambda \sim 1/k \gg R$ where R is a size of the system. In nuclei $R \approx 1.2\, A^{1/3}$ fm, so the condition

$$kR \ll 1 \tag{9.48}$$

is equivalent to $\hbar\omega \ll 165 A^{-1/3}$ MeV, which is usually fulfilled. This gives us a convenient tool of regular *multipole expansions* using the smallness parameter (9.48).

Electric dipole radiation

In the limit (9.48) one can replace the exponent in the matrix element (9.47) over intrinsic variables by unity since all particles are confined to the volume of nuclear size $\sim R$. The rate of spontaneous radiation in a given direction within a solid angle do is given then by

$$dw_{fi} = \frac{\omega_k}{2\pi c^3 \hbar} \left| \left\langle f \left| \sum_a \frac{e_a}{m_a} (\mathbf{p}_a \cdot \mathbf{e}_{k\lambda}^*) \right| i \right\rangle \right|^2 do. \tag{9.49}$$

In the long wavelength limit the phase differences are lost and particles radiate coherently via the total current $\sum_a (e_a/m_a)\mathbf{p}_a = \sum_a e_a\mathbf{v}_a$. Recall that in the Hamiltonian (9.37) and in the definition of the current (9.42) we neglect the possible effects of velocity-dependent and exchange forces. In this approximation the velocity operator for a particle a in the unperturbed system is simply

$$\mathbf{v}_a = \dot{\mathbf{r}}_a = \frac{1}{i\hbar}[\mathbf{r}_a, H_0] = \frac{1}{i\hbar}\left[\mathbf{r}_a, \frac{\mathbf{p}_a^2}{2m_a}\right] = \frac{\mathbf{p}_a}{m_a}. \tag{9.50}$$

Taking the matrix element $\langle f| \ldots |i\rangle$ of the operator relation (9.50) between the stationary states of the unperturbed Hamiltonian H_0 with energies E_f and $E_i = E_f + \hbar\omega_k$, we link the matrix elements of the particle momentum $\mathbf{p}_a$ to those of the coordinate $\mathbf{r}_a$,

$$(\mathbf{p}_a)_{fi} = \frac{m_a}{i\hbar}(E_i - E_f)(\mathbf{r}_a)_{fi} = -im_a\omega_k(\mathbf{r}_a)_{fi}. \tag{9.51}$$

Thus, the matrix element in (9.49) becomes that of the dipole operator

$$\mathbf{d} = \sum_a e_a \mathbf{r}_a, \tag{9.52}$$

where the particle positions $\mathbf{r}_a$ are given with respect to the center of mass, and we obtain the rate of **electric dipole** (E1) radiation as

$$dw_{fi} = \frac{\omega_k^3}{2\pi c^3 \hbar} |(\mathbf{e}_{k\lambda}^* \cdot \mathbf{d}_{fi})|^2 do, \tag{9.53}$$

or its intensity

$$dI_{k\lambda} = \hbar\omega_k \, dw_{fi} = \frac{\omega_k^4}{2\pi c^3} |(\mathbf{e}_{k\lambda}^* \cdot \mathbf{d}_{fi})|^2 do. \tag{9.54}$$

This result is similar to the one known from classical radiation theory. It could be derived if, instead of the current form of the Hamiltonian (9.40), one were to use from the very beginning the dipole interaction with the electric field,

$$H_{\text{dip}} = -\mathbf{E} \cdot \mathbf{d}, \tag{9.55}$$

and calculate the transition rate with the operator $\mathbf{E}$ given by the quantization of (9.17). Our derivation shows which approximations are necessary along this road to guarantee that the results from the two forms are identical.

Note that in a natural frame with the z-axis along the emitted photon, only transverse components d_x and d_y can radiate. According to the general properties of vector operators, these terms have the selection rules $\Delta M = \pm 1$ with respect to the total angular momentum projection onto the $\mathbf{k}$-axis. The missing projection is carried away as the photon helicity.

Now we can sum over the two polarizations possible for a given wave vector $\mathbf{k}$. The matrix element $\mathbf{d}_{fi}$ is some vector for a fixed pair of states f, i. Let θ be a polar angle of this vector in a frame with $\mathbf{k}$ along the z-axis. We can choose real vectors of linear polarization $\mathbf{e}_{\mathbf{k}1}$ and $\mathbf{e}_{\mathbf{k}2}$ in such a way that the vector $\mathbf{d}_{fi}$ is in the plane formed by the vectors $\mathbf{k}$ and $\mathbf{e}_{\mathbf{k}1}$. Then $\mathbf{e}_{\mathbf{k}2} \cdot \mathbf{d}_{fi} = 0$, and the vector $\mathbf{d}_{fi}$ has components along $\mathbf{k}$, equal to $|\mathbf{d}_{fi}| \cos \theta$, and along $\mathbf{e}_{\mathbf{k}1}$, equal to $|\mathbf{d}_{fi}| \sin \theta$. In this way we obtain the transition rate and the angular distribution of the radiation with respect to the direction of $\mathbf{d}_{fi}$,

$$dw_{fi} = dw_{\mathbf{k}1} = \frac{\omega_k^3}{2\pi c^3 \hbar} |\mathbf{d}_{fi}|^2 \sin^2 \theta \, do. \tag{9.56}$$

The total (angle-integrated) rate of radiation from a given transition is the inverse lifetime

$$\tau_{fi}^{-1} = w_{fi} = \frac{\omega^3}{2\pi c^3 \hbar} |\mathbf{d}_{fi}|^2 \int do \sin^2 \theta = \frac{4\omega^3}{3c^3 \hbar} |\mathbf{d}_{fi}|^2. \tag{9.57}$$

All characteristics (angular distribution, polarization, and total intensity) derived from (9.54) coincide with those for a classical oscillator with the eigenfrequency ω and time-averaged mean square dipole moment

$$\overline{|\mathbf{d}(t)|^2} = 2|\mathbf{d}_{fi}|^2. \tag{9.58}$$

This correspondence gives the necessary connection between classical and quantum mechanics. Periodic motion of a classical dipole can be represented by the Fourier series that contains the main frequency ω and its overtones (*harmonics*) $\omega_n = n\omega$,

$$\mathbf{d}(t) = \sum_{n=-\infty}^{\infty} \mathbf{d}_n e^{-i\omega_n t}. \tag{9.59}$$

Here the zeroth harmonic $n = 0$ is absent and complex amplitudes of a real function are related via the condition $\mathbf{d}_n = \mathbf{d}_{-n}^*$. This series can be transformed as

$$\mathbf{d}(t) = \sum_{n=1}^{\infty} \left(\mathbf{d}_n e^{-i\omega_n t} + \mathbf{d}_n^* e^{i\omega_n t} \right) \tag{9.60}$$

$$= 2 \sum_{n=1}^{\infty} \left(\cos \omega_n t \operatorname{Re} \mathbf{d}_n + \sin \omega_n t \operatorname{Im} \mathbf{d}_n \right). \tag{9.61}$$

Averaging $|\mathbf{d}(t)|^2$ over the period removes the cross terms $\sim \overline{\cos \omega_n t \sin \omega_n t}$ and gives the mean square

$$\overline{|\mathbf{d}(t)|^2} = 4 \sum_{n=1}^{\infty} \left[(\operatorname{Re} \mathbf{d}_n)^2 \overline{\cos^2 \omega_n t} + (\operatorname{Im} \mathbf{d}_n)^2 \overline{\sin^2 \omega_n t} \right] \tag{9.62}$$

$$= 2 \sum_{n=1}^{\infty} \left[(\operatorname{Re} \mathbf{d}_n)^2 + (\operatorname{Im} \mathbf{d}_n)^2 \right] = 2 \sum_{n=1}^{\infty} |\mathbf{d}_n|^2. \tag{9.63}$$

Only a single harmonic of the sum (9.63) contributes to a given transition. Comparing with (9.58) we conclude that in the classical limit quantum matrix elements go into Fourier components of the classical periodic time dependence (9.59). The result (9.58) is necessary for the correct limiting transition.

In (9.56) we found the angular distribution of the radiation relative to the "direction of oscillations" in the quantum antenna. One can also find the angular distribution of radiation in the laboratory frame with an arbitrary z-axis taken as the quantization axis of the angular momentum. In general one has to calculate the Poynting vector $\mathbf{S}$ (see (9.4)), of the emitted wave. This is a straightforward but cumbersome procedure. We just illustrate the results by a simple example.

Let us fix the initial projection M_i, sum the radiation probability over two polarizations, and consider different possibilities for the final projection M_f. As seen from (9.53) and (9.56),

$$\sum_\lambda |(\mathbf{e}_{\mathbf{k}\lambda}^* \cdot \mathbf{d}_{fi})|^2 = |[\mathbf{n} \times \mathbf{d}_{fi}]|^2, \tag{9.64}$$

where $\mathbf{n} = \mathbf{k}/k$ is the direction of radiation. We want to find the distribution of probabilities for the angles of the vector $\mathbf{n}(\theta_k, \varphi_k)$ in the laboratory frame.

It is convenient to use for this purpose spherical components V_μ, $\mu = 0, \pm 1$, defined for any vector $\mathbf{V}$ in a standard way,

$$V_0 = V_z, \quad V_{\pm 1} = \mp \frac{1}{\sqrt{2}}(V_x \pm i V_y). \tag{9.65}$$

The spherical component $(d_{fi})_\mu$ of the vector $\mathbf{d}_{fi}$ generates the transition with $\Delta M = \mu$. For $\Delta M = 0$, only $(d_{fi})_0 = d_{fi}^z$ is effective. The corresponding coefficient in (9.64) is

$$\Delta M = 0: \quad n_x^2 + n_y^2 = \sin^2 \theta. \tag{9.66}$$

The transverse components are

$$d_{fi}^x = \frac{1}{\sqrt{2}}\left[(d_{fi})_{-1} - (d_{fi})_1\right], \quad d_{fi}^y = \frac{i}{\sqrt{2}}\left[(d_{fi})_{-1} + (d_{fi})_1\right]. \tag{9.67}$$

In the process with $\Delta M = 1$, we need only the component $(d_{fi})_1$. Calculating terms in the vector product (9.64) that contain the matrix element $|(d_{fi})_1|^2$, we obtain the angular distribution for this case (the result is the same for $\Delta M = -1$),

$$\Delta M = \pm 1: \quad \frac{1}{2}(2n_z^2 + n_x^2 + n_y^2) = \frac{1}{2}(1 + \cos^2 \theta). \tag{9.68}$$

In all cases the angular distribution has no azimuthal asymmetry; it can appear only if the photon polarization is registered.

The observed angular distribution depends on the initial population of different sub-states M_i. For example, consider the electric dipole radiation for the transition between states with the spin and parity quantum numbers $1^- \to 0^+$. In this case all Clebsch-Gordan coefficients, which, according to the Wigner-Eckart theorem, determine the dependence of matrix elements on magnetic quantum numbers, are equal to unity. If $g(M_i)$ are relative occupancies for $M_i = 0, \pm 1$, the angular distribution takes the form

$$F(\theta) = g(0) \sin^2 \theta + \frac{g(1) + g(-1)}{2}(1 + \cos^2 \theta). \tag{9.69}$$

For the *unpolarized* initial state, $g(0) = g(1) = g(-1) = \frac{1}{3}$, and the angular distribution (9.69) becomes *isotropic*.

The angular distribution for the dipole radiation summed over polarizations in general case has a simple form (9.69)

$$F_1(\theta) \propto 1 + \beta_1 \cos^2\theta, \tag{9.70}$$

with the asymmetry coefficient β_1 depending on the initial nuclear polarization (to generate the asymmetry, the initial state of $J = 1$ has to be *aligned*, $g(0) \neq [g(1) + g(-1)]/2$, but not necessarily *polarized* when $g(1) \neq g(-1)$). For any multipolarity L of the transition, the angular distribution can be presented as a sum over the even powers, $l = 0, 2, \ldots, 2L$, of $\cos\theta$,

$$F_L(\theta) = \sum_{l=0}^{L} \beta_l \cos^{2l}\theta. \tag{9.71}$$

The presence of odd powers of $\cos\theta$ would imply parity nonconservation (interference of operators with even and odd values of L in the same transition).

A dipole transition can connect the states $|f\rangle$ and $|i\rangle$ only if the dipole *selection rules* are fulfilled. For multipolarity equal to 1 we have the restriction $\Delta J = 0, \pm 1$. Apart from this, the parity selection rule establishes that the initial and final states should have opposite parity. If the selection rules for the E1-transitions are not satisfied, the emission is forbidden in this, lowest in the parameter kR, approximation.

Electric quadrupole radiation
When dipole radiation is not allowed, the photon emission can still occur due to the next terms of the expansion of the exponent in (9.47). They contain the extra factor $-i(\mathbf{k} \cdot \mathbf{r})$ in the amplitude and therefore lead to a suppression of the emission probability by $\sim(kR)^2$.

Let us consider a typical next order term of the amplitude (9.47). It can be rewritten as

$$-i\frac{e}{m}(\mathbf{e}^* \cdot \mathbf{p})(\mathbf{k} \cdot \mathbf{r}) \equiv -i\frac{e}{2m}(F_+ + F_-), \tag{9.72}$$

where another item is added and subtracted to get

$$F_\pm = (\mathbf{e}^* \cdot \mathbf{p})(\mathbf{k} \cdot \mathbf{r}) \pm (\mathbf{e}^* \cdot \mathbf{r})(\mathbf{k} \cdot \mathbf{p}). \tag{9.73}$$

Here the longitudinal components of $\mathbf{r}$ come along with the transverse components of $\mathbf{p}$ or vice versa, so that the operators $\mathbf{p}$ and $\mathbf{r}$ effectively commute. By virtue of the equation of motion (9.50) (again, if there are no velocity-dependent forces) we obtain

$$F_+ = m(\mathbf{e}^* \cdot \dot{\mathbf{r}})(\mathbf{k} \cdot \mathbf{r}) + (\mathbf{e}^* \cdot \mathbf{r})m(\mathbf{k} \cdot \dot{\mathbf{r}}) = m\frac{d}{dt}((\mathbf{e}^* \cdot \mathbf{r})(\mathbf{k} \cdot \mathbf{r})). \tag{9.74}$$

Since $(\mathbf{e}^* \cdot \mathbf{k}) = 0$, we can reduce (9.74) to the time derivative of the single-particle electric quadrupole tensor

$$q_{ij} = e(3x_i x_j - \delta_{ij} r^2). \tag{9.75}$$

Indeed,

$$F_+ = me_i^* k_j \frac{d}{dt}(r_i r_j) \tag{9.76}$$

$$= me_i^* k_j \frac{d}{dt}\left(r_i r_j - \frac{1}{3}\delta_{ij}r^2\right) = me_i^* k_j \frac{d}{dt}\left(\frac{1}{3e}q_{ij}\right). \tag{9.77}$$

The corresponding matrix element in the radiation probability (9.47) becomes

$$-\frac{i}{2}\left\langle f \left| \sum_a \frac{e_a}{m_a} F_+(a) \right| i \right\rangle = -\frac{i}{6}e_i^* k_j \langle f | \dot{Q}_{ij} | i \rangle, \tag{9.78}$$

where the total quadrupole moment $Q_{ij} = \sum_a q_{ij}(a)$ has appeared. Using, as in (9.51), the equation of motion and introducing the transition frequency,

$$i\hbar\dot{Q} = [Q, H_0] \rightsquigarrow \dot{Q}_{fi} = -i\omega_k Q_{fi}, \tag{9.79}$$

we come to the spontaneous emission rate (9.47)

$$\frac{dw_{fi}}{do} = \frac{\omega_k}{2\pi c^3 \hbar}\left| -\frac{1}{6}\omega_k(e_{k\lambda}^*)_i k_j (Q_{ij})_{fi}\right|^2 = \frac{\omega_k^3}{72\pi c^3 \hbar}\left|(e_{k\lambda}^*)_i k_j (Q_{ij})_{fi}\right|^2. \tag{9.80}$$

Thus, this term describes the *electric quadrupole* (E2) radiation. Again it is possible to show that the result is in precise correspondence with the classical quadrupole radiation.

Similar to the dipole case, only the components Q_{xz} and Q_{yz} of the quadrupole tensor in the frame with the polar axis along **k** radiate, being again associated with helicity ± 1 of the emitted photon. As for the selection rules related to angular momentum and parity, we know that for the quadrupole tensor $\Delta J = 0, \pm 1, \pm 2$ and parity does not change. Therefore E1 and E2 transitions cannot happen between the same pair of states if parity is conserved.

The transition rate for E2 radiation increases as ω^5, while it is proportional to ω^3 in the E1 case. In both cases, the initial state, having a choice between several final states allowed for the photon emission, selects the state at a larger energy distance if the corresponding matrix elements are comparable. If final states of different parity are available, in the long wavelength limit the dipole radiation is preferable. Very rough estimates with the matrix element evaluated as $(eR)^2$ and $k^2 e^2 R^4$ for the dipole and quadrupole cases, respectively, and all numerical factors neglected, give for the transition rate

$$w_{\text{dip}} \sim \alpha(kR)^2\omega, \quad w_{\text{quad}} \sim \alpha(kR)^4\omega. \tag{9.81}$$

The fine structure constant $\alpha = e^2/\hbar c$ enters as a signature of the relative weakness of electromagnetic interactions.

The rates (9.81) have the dimension of inverse time. The corresponding lifetimes are $\tau = 1/w$. Due to the small factors α and $(kR)^2$, the lifetimes are very long compared to the period $1/\omega$ of the emitted photon. The radiation widths $\Gamma_\gamma = \hbar/\tau$ are small, and the states that decay only by photon emission can usually be considered nearly stationary.

The estimates (9.81) are based on the fact that the transition operators (dipole or quadrupole moment) are one-body quantities. They are the sums of the terms changing only one single-particle orbit, and the estimates (9.81) are made for a *single-particle* transition. The velocity-dependent and exchange interactions bring in two-body currents and make the results more complicated. If the wave functions reflect coherent many-body motion, the particles can make their transitions synchronously, which will cause a strong enhancement of the transition probability. In other cases, correlations between particles can quench the transition strength. The extreme example is the *isoscalar dipole* transition. The corresponding moment $\sum_p \mathbf{r}_p + \sum_n \mathbf{r}_n$ is proportional to the vector of the center of mass which cannot lead to any internal excitation. More examples will be given later.

Magnetic dipole radiation

We have not yet taken into account the antisymmetric term F_- in the second order matrix element, see (9.72). Using vector algebra as in (9.23) we obtain

$$F_- = [\mathbf{k} \times \mathbf{e}^*] \cdot [\mathbf{r} \times \mathbf{p}] = [\mathbf{k} \times \mathbf{e}^*] \cdot \hbar \mathbf{l}, \tag{9.82}$$

where the orbital momentum $\mathbf{l}$ is introduced. As seen from (9.18), the vector $-i[\mathbf{k} \times \mathbf{e}^*_{\mathbf{k}\lambda}]$ is just a part of the quantized magnetic field vector associated with the creation of the photon with quantum numbers $(\mathbf{k}\lambda)$.

The matrix element resulting from the term F_- is the same as the one that would follow from the interaction

$$H_{\text{orb}} = -\boldsymbol{\mu}_l \cdot \mathbf{B} \tag{9.83}$$

of the orbital magnetic moment

$$\boldsymbol{\mu}_l = \sum_a \frac{e_a \hbar}{2 m_a c} \mathbf{l}_a \equiv \sum_a g_a^{(l)} \mathbf{l}_a, \tag{9.84}$$

with the magnetic field $\mathbf{B}$ of the emitted photon. Particles with spin $\mathbf{s}_a$ and gyromagnetic ratio $g_a^{(s)}$ add similarly to the interaction of the spin magnetic moment,

$$H_{\text{spin}} = -\boldsymbol{\mu}_s \cdot \mathbf{B}, \quad \boldsymbol{\mu}_s = \sum_a g_a^{(s)} \mathbf{s}_a, \tag{9.85}$$

this not being taken into account in the simplest expression (9.41) for the current interacting with the radiation field. As we discussed (9.39), this contribution should be considered on an equal footing with the orbital one.

Combining the orbital and spin magnetic moments into the total operator $\boldsymbol{\mu} = \boldsymbol{\mu}_l + \boldsymbol{\mu}_s$, we obtain for the rate of the *magnetic dipole* (M1) radiation

$$dw_{fi} = \frac{\omega_{\mathbf{k}}}{2\pi c^3 \hbar} \left| -ic[\mathbf{k} \times \mathbf{e}^*_{\mathbf{k}\lambda}] \cdot \boldsymbol{\mu}_{fi} \right|^2 = \frac{\omega_{\mathbf{k}}^3}{2\pi c^3 \hbar} \left| [\mathbf{n} \times \mathbf{e}^*_{\mathbf{k}\lambda}] \cdot \boldsymbol{\mu}_{fi} \right|^2. \tag{9.86}$$

where $\mathbf{n} = \mathbf{k}/k$. This result is almost identical to that for the electric dipole radiation (9.53). The only difference is that the magnetic vector $\boldsymbol{\mu}_{fi}$ is projected onto the vector corresponding to the polarization which is complementary to that of the emitted photon. Again only $\Delta M = \pm 1$ is possible for the angular momentum projection along $\mathbf{k}$; all properties of the classical magnetic dipole radiation are reproduced. As in the E1 case, the probability grows proportionally to ω^3 with the radiation frequency. Angular distributions (for the same initial population of magnetic substates) are always identical for electric and magnetic radiation of the same multipolarity.

The angular momentum and parity restrictions for M1 radiation follow from the character of the magnetic moment as an axial vector. The selection rules are $\Delta J = 0, \pm 1$; parity does not change. In many cases both M1 and E2 radiations are allowed between the same initial and final states. Then we have their interference for the emission in a given direction. As in the scattering problem, the interference terms vanish in the total intensity integrated over angles.

Making the same rough estimate as in (9.81) we obtain

$$w_{\text{magn}} \sim \alpha (kR)^2 \left(\frac{\hbar/mc}{R} \right)^2 \omega. \tag{9.87}$$

The ratio of the Compton wavelength $\hbar/mc$ of the particles to the size R of the system can be estimated as $\bar{p}/mc$ via the typical momentum $\bar{p} \sim \hbar/R$ from the uncertainty relation. Then the extra factor in (9.87) is simply $(v/c)^2$, which would suppress the M1 probability compared to E2. This estimate fails for the Fermi system, where $R \sim r_0 A^{1/3}$ and the typical momentum is $p_F \sim \hbar/r_0 \gg \bar{p}$. The lowest order magnetic transitions (ML) in general are comparable to the next order electric transitions (EL + 1).

9.4 Quantum and Classical Transition Rates

Although we were able to get through all calculations for E1, E2, and M1 multipoles using only elementary techniques, the task becomes more complicated for higher multipoles. The processes of high multipolarity are much slower due to the increasing power of the small parameter (kR). The probability of the radiation of multipolarity (EL) is proportional to $(kR)^{2L+1}$ as follows from simple single-particle estimates. By similar estimates, the magnetic multipole radiation (ML) is suppressed by a factor $\sim (v/c)^2$ compared to the (EL) radiation. Usually the lowest allowed multipolarity is the most probable, albeit with important exceptions, especially in the case of coherent many-body enhancement. But, due to the selection rules in angular momentum and parity, the low order multipole radiation can be forbidden. The general formalism of multipole expansion is needed.

We limit ourselves to explaining the general framework of this theory. The starting point is the Hamiltonian (9.41) for the current interacting with the radiation field. The physical properties of the system become apparent in specific features of the current. In contrast to this, the electromagnetic field is expressed universally. We use the quantized form (9.33). Here the exponent plays the role of the "orbital" wavefunction of the photon, whereas the polarization vectors $\mathbf{e}_{k\lambda}$ characterize the intrinsic state of the photon, that is, its spin, which is equal to 1 (**e** is a vector with three components, equivalent by rotational properties to a tensor of rank 1), but the transversality of the field places additional restrictions.

Thus, the photon is an object with momentum **k** and spin $\mathbf{e}_{\mathbf{k}\lambda}$. We can expand the plane wave into partial waves with a certain orbital momentum l. Each partial wave can be combined with the spin vector **e** into a state with the total angular momentum of the photon **L**, which can take values $L = l, l \pm 1$. These states have parity $(-)^l$ determined by the transformation of the given partial wave. The component with $L = l$ and parity $(-)^L$, after multiplication in the matrix element by the polar vector of current **j**, which changes sign under spatial inversion, gives the magnetic multipole operator (ML) with multipolarity L and parity $(-)^{L+1}$. The components with $L = l \pm 1$ and parity $(-)^{L+1}$ are combined in the matrix element into the electric multipole operator (EL) with multipolarity L and parity $(-)^L$ in accordance with the standard properties of static electric multipoles.

The final expression for the probability of the radiative transition $i \to f$ integrated over angles of the emitted photon and summed over its polarizations is

$$w_{fi} = \frac{8\pi(L+1)}{L[(2L+1)!!]^2\hbar}k^{2L+1}\sum_{LM}\{|(\mathcal{M}(\text{EL}, M))_{fi}|^2 + |(\mathcal{M}(\text{ML}, M))_{fi}|^2\}, \tag{9.88}$$

where the multipole matrix elements for electric and magnetic transitions are expressed via the current operator $\mathbf{j}(\mathbf{r})$,

$$\mathcal{M}(\text{ML}, M) = \int d^3r(\mathbf{j} \cdot \hat{\mathbf{l}})\Phi_{LM}, \tag{9.89}$$

$$\mathcal{M}(\text{EL}, M) = \int d^3r\left[(\nabla \cdot \mathbf{j})\frac{1}{k}\frac{\partial}{\partial r}r - k(\mathbf{r} \cdot \mathbf{j})\right]\Phi_{LM}, \tag{9.90}$$

$\mathbf{l} = -i[\mathbf{r} \times \nabla]$ is the orbital momentum operator and the function Φ_{LM} arises from the partial wave expansion of the plane wave,

$$\Phi_{LM} = -i\frac{(2L+1)!!}{L+1}\frac{1}{ck^L}j_L(kr)Y_{LM}(\mathbf{n}), \tag{9.91}$$

where $j_L(kr)$ are spherical Bessel functions. Of course, for a given pair of states f, i and a given multipolarity L, only one term, either electric or magnetic, in (9.88) works if parity is conserved.

In the long wavelength approximation we can use the limiting values of spherical Bessel functions $j_L(x) \approx x^L/(2L+1)!!$ at small arguments x. Then the magnetic multipoles (9.89) become

$$\mathcal{M}(\text{ML}, M) = -\frac{i}{c(L+1)}\int d^3r\left(\mathbf{j}(\mathbf{r}) \cdot \hat{\mathbf{l}}\right)r^L Y_{LM}(\mathbf{n}). \tag{9.92}$$

For the orbital current (9.42) this expression coincides with the orbital part of the static magnetic moment. The presence of spin implies magnetization currents. In macroscopic electrodynamics such a current is $c[\nabla \times \mathbf{m}]$, where **m** stands for the magnetization density.

The analogous quantity for a quantum particle is

$$\mathbf{m}(\mathbf{r}) = \sum_a g_a^s \mathbf{s}_a \delta(\mathbf{r} - \mathbf{r}_a). \tag{9.93}$$

The induced spin current is

$$\mathbf{j}_{\text{spin}} = c \sum_a g_a^s [\nabla \times \mathbf{s}_a] \delta(\mathbf{r} - \mathbf{r}_a). \tag{9.94}$$

When substituted into (9.92), this current reproduces the spin part of static magnetic multipoles as in (9.85).

In the long wavelength limit, the second term of the electric multipole (9.90) is smaller by a factor $(kR)^2$ than the first one. In the first term $\nabla \cdot \mathbf{j}$ reduces to the time derivative of the charge density ρ^{ch} with the aid of the continuity equation. This time derivative can be expressed through the operator equation of motion as in (9.79); the difference of energies between the initial and final states gives $\hbar\omega_{\mathbf{k}}$. Performing the expansion of the spherical Bessel functions we come to

$$\mathcal{M}(\text{EL}, M) = \int d^3r \rho^{ch}(\mathbf{r}) r^L Y_{LM}(\mathbf{n}), \tag{9.95}$$

which is a standard definition of the electric multipole.

It is easy to check that our results derived earlier coincide with those given by general theory. For example, for the dipole radiation we get from (9.88)

$$w_{fi} = \frac{16\pi}{9\hbar} k^3 \sum_M |(\mathcal{M}(\text{E1}, M))_{fi}|^2. \tag{9.96}$$

The multipole operator $\mathcal{M}(\text{E1})$ in (9.96) differs from the dipole operator (9.52) by the factor $\sqrt{3/4\pi}$. The sum over projections M in (9.96) is the same as the scalar product of the vector $\mathbf{d}_{fi}$ by itself in (9.57). Then (9.96) coincides with (9.57).

Usually the angular momentum projection M_f of the final state is not measured. Then we have to sum the transition rate over all M_f defining the *reduced transition probability*

$$B(\text{EL}; i \to f) = \sum_{MM_f} |(\mathcal{M}(\text{EL}, M))_{fi}|^2. \tag{9.97}$$

According to the Wigner-Eckart theorem (see Appendix B), the entire dependence of the matrix element on the magnetic quantum numbers is concentrated in the Clebsch-Gordan coefficients of vector addition. Using instead the Wigner 3j-symbols, we have for any tensor operator T_{LM}

$$\langle J_f M_f | T_{LM} | J_i M_i \rangle = (-)^{J_f - M_f} \begin{pmatrix} J_f & L & J_i \\ -M_f & M & M_i \end{pmatrix} (f \| T_L \| i), \tag{9.98}$$

where the reduced matrix element $(f \| T_L \| i)$ does not depend on projections. We can use the orthogonality of the vector coupling coefficients and perform the summation over M

and M_f. The reduced transition probability is then related to the reduced matrix element of the multipole operator,

$$B(EL; i \to f) = \frac{1}{2J_i + 1} |\langle f \| \mathcal{M}(EL) \| i \rangle|^2. \tag{9.99}$$

This quantity is convenient because it does not depend on the initial population of various projections M_i. Note that for the inverse transition induced by the same operator, the *detailed balance* relation is valid,

$$B(EL; f \to i) = \frac{2J_i + 1}{2J_f + 1} B(EL; i \to f). \tag{9.100}$$

The reduced transition probability determines the partial lifetime of a given initial state with respect to a specific radiative decay (all M_f summed up),

$$\tau_{i \to f}^{-1} = \sum_{M_f} w_{fi} = \frac{8\pi}{\hbar} \frac{L+1}{L[(2L+1)!!]^2} k^{2L+1} B(EL; i \to f). \tag{9.101}$$

Here the kinematic factors are singled out. They are associated with the geometry and phase space volume of the emitted photon. Information concerning structure of the radiating system is accumulated in the reduced transition probability. With the substitution EL→ML, the same expressions are valid for magnetic multipoles.

The calculation of the decay rate for γ-emission requires the use of a quantum theory for the radiation. But it is instructive to present the correct quantum results as an extension of calculations based on classical electrodynamics.

Classically, the electromagnetic radiation emitted by a system is the result of the variation in time of the charge density or of the distribution of charge currents in the system. The energy is emitted in two types of multipole radiation: the electric and the magnetic. Each one of them is expressed as a function of the corresponding multipole moments, being the quantities that contain the variables (charge and current) of the system. If the wavelength of the emitted radiation is long in comparison to the dimensions of the system (which is valid for a γ ray of $\lesssim 10$ MeV energy) the power emitted by each multipole is given by ([Ja75, chap. 16]):

$$P_E(lm) = \frac{8\pi (l+1)c}{l[(2l+1)!!]^2} \left(\frac{\omega}{c}\right)^{2l+2} |Q_{lm}|^2 \tag{9.102}$$

for electric multipole radiation and

$$P_M(lm) = \frac{8\pi (l+1)c}{l[(2l+1)!!]^2} \left(\frac{\omega}{c}\right)^{2l+2} |M_{lm}|^2 \tag{9.103}$$

for the corresponding magnetic radiation. Q_{lm} and M_{lm} are the electric and magnetic multipole moments, respectively, calculated by the expressions

$$Q_{lm} = \int r^l Y_l^{m*}(\theta, \phi) \rho(\mathbf{r}) \, dV, \tag{9.104}$$

$$M_{lm} = -\int r^l Y_l^{m*}(\theta, \phi) \frac{\nabla \cdot [\mathbf{r} \times \mathbf{j}(\mathbf{r})]}{c(l+1)} \, dV. \tag{9.105}$$

In formulas (9.102)–(9.105), ρ is the charge density, $\mathbf{j}$ the current density, l the multipolar order ($l = 1$, dipole; $l = 2$, quadrupole, etc.), and m can take the values $-l, -l+1, \ldots, l$ for each value of l. The double factorial $(2l + 1)!!$ is defined as the product $(2l + 1)(2l - 1)$ $(2l - 3) \ldots 1$.

The expressions (9.102) and (9.105) for the multipole moments become particularly simple to calculate when $l = 1$, $m = 0$. In this case, we obtain the values

$$Q_{1,0} = \left(\frac{3}{4\pi}\right)^{1/2} \int z\rho(\mathbf{r})dV \tag{9.106}$$

and

$$M_{1,0} = \left(\frac{3}{4\pi}\right)^{1/2} \frac{1}{2c} \int (xj_y - yj_x)\, dV \tag{9.107}$$

and, if we take into consideration that $\mathbf{j} = \rho\mathbf{v}$, we will see that the magnetic multipole moment is smaller than the electric one by a factor of the order of v/c. As the speed of the charges inside the nucleus is much smaller than the speed of light, the magnetic multipole moments are in general much smaller than the corresponding electric moments of the same order, except when selection rules forbid the existence of decay through the corresponding electric multipole (see next section).

In quantum mechanics, the energy is emitted not continually but in packets of energy $\hbar\omega$. In a quantum calculation the disintegration constant is the same as the number of *quanta* emitted per unit of time when the power is given by the classical expressions (9.102) and (9.103). Thus,

$$\lambda_E(lm) = \frac{P_E(lm)}{\hbar\omega} = \frac{8\pi(l+1)}{\hbar l[(2l+1)!!]^2}\left(\frac{\omega}{c}\right)^{2l+1}|Q_{lm}|^2 \tag{9.108}$$

and

$$\lambda_M(lm) = \frac{P_M(lm)}{\hbar\omega} = \frac{8\pi(l+1)}{\hbar l[(2l+1)!!]^2}\left(\frac{\omega}{c}\right)^{2l+1}|M_{lm}|^2. \tag{9.109}$$

The decaying nucleus should also be treated as a quantum system. In this sense, the expressions (9.104) and (9.105) for the multipole moments Q_{lm} and M_{lm}, which are part of (9.108) and (9.109), continue to have validity, if we use for the charge and current densities the quantum expressions

$$\rho(a, b, \mathbf{r}) = \Psi_b^*(\mathbf{r})\Psi_a(\mathbf{r}), \tag{9.110}$$

$$\mathbf{j}(a, b, \mathbf{r}) = -\frac{ie\hbar}{2m}\left[\Psi_b^*(\nabla\Psi_a) - (\nabla\Psi_b^*)\Psi_a\right], \tag{9.111}$$

where a (b) denotes the initial (final) state described by the wavefunction Ψ_a (Ψ_b). Equations (9.110) and (9.111) refer to a single nucleon with mass m that emits radiation in its passage from state a to state b. Thus, a sum over all the protons should be incorporated in the result, when we do the substitution of (9.110) and (9.111) in (9.104) and (9.105) ([Ja75, chap. 16]):

$$Q_{lm}(a, b) = e \sum_{k=1}^{Z} \int r_k^l Y_l^{m*}(\theta_k, \phi_k) \Psi_b^* \Psi_a \, d\tau, \tag{9.112}$$

$$M_{lm}(a, b) = \frac{e}{2mc(l+1)} \sum_{k=1}^{Z} \int r_k^l Y_l^{m*}(\theta_k, \phi_k) \nabla \cdot [\mathbf{r}_k \times \mathbf{j}_k(a, b, \mathbf{r}_k)] \, d\tau. \tag{9.113}$$

The expression (9.113) can be rewritten in another form, using the operator

$$\mathbf{L} = -i\mathbf{r} \times \nabla \tag{9.114}$$

and noticing, through an integration by parts, that the two portions of (9.111) are identical. In this way, (9.113) is rewritten as

$$M_{lm}(a, b) = \frac{e\hbar}{2mc(l+1)} \sum_{k=1}^{Z} \int r_k^l Y_l^{m*}(\theta_k, \phi_k) \nabla \cdot [\Psi_b^* \mathbf{L}_k \Psi_a] \, d\tau. \tag{9.115}$$

Expressions (9.112) and (9.115) only refer to the contribution of the motion of the protons. The spins of both protons and neutrons also give a contribution, but this will not be discussed here.

We saw that the values of the magnetic multipole moments are small in comparison to the electric moments of same order. We will show now that the transition probability decreases quickly with increasing l, restricting the multipole orders that give a significant contribution. To achieve this goal, it is sufficient to observe that the product $(\omega/c)^l Q_{lm}$ in (9.108) is at most equal to $Ze(\omega R/c)^l$, where R is the radius of the nucleus. For the energy that we consider, $\omega R/c$ is very small, implying that, for the larger powers of l, the disintegration constant is also very small. These facts imply, in principle, that the electric dipole is always the dominant radiation. But the selection rules that we will see next can modify this situation.

9.5 Selection Rules

In a similar way to α- and β-disintegration, the conservation of angular momentum and parity can prohibit certain γ-transitions between two states. The selection rules for the γ radiation are easy to establish if we accept the fact that a *quantum* of radiation carries an angular moment $\mathbf{l}$ of module $\sqrt{l(l+1)}\,\hbar$ and component z equal to $m\hbar$, where l is the multipolar order. Thus, in transition between an initial spin $\mathbf{I}_i$ and a final spin $\mathbf{I}_f$ the conservation of angular momentum imposes $\mathbf{I}_i = \mathbf{I}_f + \mathbf{l}$; in this way, the possible values for the multipole order l should obey

$$|I_i - I_f| \le l \le I_i + I_f, \tag{9.116}$$

where $|\mathbf{I}_i| = \sqrt{I_i(I_i+1)}\,\hbar$, etc. A special case is the transition $0^+ \to 0^+$: as multipole radiation of order zero, these transitions do not exist, and they are effectively impossible

$$\Psi_a = R_a(r)Y_l^m(\theta,\phi)$$

$$l = 0 \qquad \Psi_b = R_b(r)/\sqrt{4\pi}$$

Figure 9.1 De-excitation of a proton to a level with $l = 0$.

through the emission of a γ-ray. But in this case a process of *internal conversion* can happen, where the energy is released by the ejection of an atomic electron. This process will be studied latter.

With regard to the parity, transitions between states of same parity can only be accomplished by electric multipole radiation of even number or by magnetic radiation of odd number. The inverse is valid for transitions where there is parity change. Why this happens can be understood by examining (9.112) and (9.115). The functions that compose the integrand have definite parity and it is necessary that the integrand has even parity; otherwise the contribution in $\mathbf{r}$ cancels with the contribution in $-\mathbf{r}$ and the integral over the whole space vanishes. Let us look at the case of (9.112): r^l is always positive and the spherical harmonic $Y_l^m(\theta, \phi)$ is even if l is even. For (9.112) to be nonzero, Ψ_a should have the same parity of Ψ_b for l even and opposite parity for l odd. This justifies the transition rule for the electric multipole radiation. A similar procedure applied to (9.115) (exercise 1) justifies the selection rules for the magnetic multipole radiation.

Let us take an example. If the initial state is 3^+ and the final state 2^-, the possible values of l will be 1, 2, 3, 4, and 5. But the change of parity restricts the transitions to E1, M2, E3, M4, and E5, where E1 symbolizes an electric dipole transition ($l = 1$) and so on. It is convenient to be reminded that, due to what we have seen previously, the E1 transition will give the most significant contribution.

9.6 Estimate of the Disintegration Constants

The use of (9.108) and (9.109) for the calculation of the transition probabilities in a real nucleus has the difficulty that wavefunctions that appear in (9.112) and (9.115) are not known. But a prediction of their order of magnitude for the several modes can be done. A very simple situation was proposed by Weisskopf [We51]: the proton decaying from an excited state is described by the wavefunction Ψ_a to the final state b of $l = 0$. The restriction $l = 0$ makes the calculation easier but, as a simple estimate, this is ignored in the applications. It is described by a wavefunction Ψ_b in agreement with the scheme shown in figure 9.1.

For Ψ_a and Ψ_b one can use, in principle, wavefunctions based on the shell model. But these do not describe well the excited states of the nucleus. For an approximate calculation it is enough to use

$$R_a(r) = \text{const.} = R_a \qquad (r < R),$$
$$R_a(r) = 0 \qquad (r > R), \tag{9.117}$$

where R is the nuclear radius, and to use the same approach for $R_b(r)$. The normalization yields immediately the values for the constants R_a and R_b:

$$R_a = R_b = \sqrt{\frac{3}{R^3}}. \tag{9.118}$$

In this way, it is not difficult to calculate the electric multipole moment Q_{lm} in (9.112):

$$Q_{lm} = e \int r^l Y_l^{m*}(\theta, \phi) \frac{3}{R^3} \frac{Y_l^m(\theta, \phi)}{\sqrt{4\pi}} r^2 \, dr \, d\Omega, \tag{9.119}$$

which yields

$$Q_{lm} = \frac{3eR^l}{\sqrt{4\pi}(l+3)}. \tag{9.120}$$

Thus, in this approximation, the disintegration constant in (9.108) is

$$\lambda_E = \frac{2(l+1)}{l[(2l+1)!!]^2} \left(\frac{3}{l+3}\right)^2 \frac{e^2}{\hbar} \left(\frac{E_\gamma}{\hbar c}\right)^{2l+1} R^{2l}, \tag{9.121}$$

where we wrote explicitly the disintegration energy $E_\gamma = \hbar\omega$. A similar calculation for the magnetic disintegration constant in (9.109) yields

$$\lambda_M = \frac{20(l+1)}{l[(2l+1)!!]^2} \left(\frac{3}{l+3}\right)^2 \frac{e^2}{\hbar} \left(\frac{E_\gamma}{\hbar c}\right)^{2l+1} R^{2l} \left(\frac{\hbar}{mcR}\right)^2, \tag{9.122}$$

where m is the mass of a nucleon. For a typical nucleus of intermediate mass $A = 120$, it is easy to see that $\lambda_E/\lambda_M \cong 100$, independently of the multipolar order l. Evidently, both constants decrease rapidly when the value of l increases.

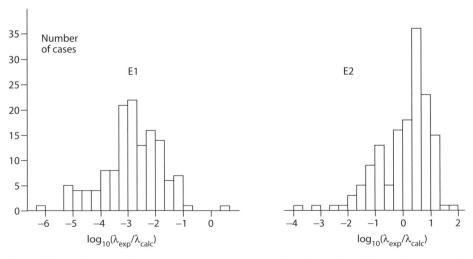

Figure 9.2 Distribution of the ratio between the experimental and theoretical disintegration constants for transitions of E1 and E2 type [Sk66].

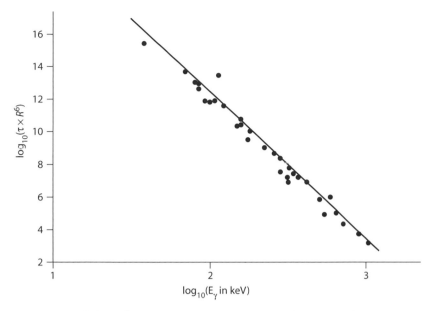

Figure 9.3 τ (half-life) $> R^6$ (R = nuclear radius) as a function of the energy of the γ-ray for a series of transitions of the M4 type. One sees good agreement between the theoretical estimate (straight line) and the experimental data[GS51].

Expressions (9.121) and (9.122) are estimates done for a special case, with the intent of having a first evaluation of the magnitude of the disintegration constant, and disagreements of several orders of magnitude between the result of the calculation and the corresponding experimental values can happen. In particular, experimental disintegration rates smaller than the ones predicted by (9.121) and (9.122), can mean that (9.117) is not very reasonable and that the small overlap of the wavefunctions Ψ_a and Ψ_b decreases the values of λ. Experimental values higher than predicted by (9.121) and (9.122) can mean, on the other hand, that the transition involves the participation of more than one nucleon or even a collective participation of the whole nucleus.

Figures 9.2 and 9.3 illustrate the two situations. In figure 9.2 the experimental values of λ for transitions of E1 multipolarity are orders of magnitude smaller than those calculated from (9.121). The opposite happens with the E2 multipolarity, where in most cases the experimental rate is larger than calculated; this is due to the fact that E2 transitions are common among levels of collective bands, especially rotational bands in deformed nuclei. In figure 9.3, on the other hand, one notices very good agreement between theoretical values and experimental ones for M4. This behavior is typical of transitions of high multipolarity.

9.7 Isomeric States

The transition between two states of very different spins can only happen, in agreement with (9.116), with a high multipolarity. This type of transition has, in agreement with eqs. (9.108) and (9.109), a very long half-life.

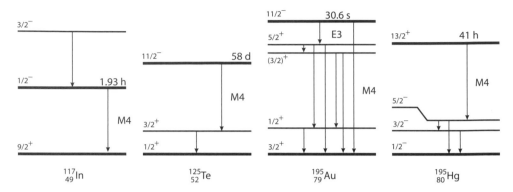

Figure 9.4 Examples of nuclei that contain an isomeric state. Note that a state lowered from another shell, of different parity, can be the initial or final state of an isomeric transition. The energy scales of the four schemes are independent.

Examining the sequence of levels of the shell model (figure 5.9) we see that adjacent levels appear with great spin variations. That is a consequence of the lowering of levels of high spin caused by the spin-orbit interaction. Such levels are $1g\frac{9}{2}$, $1h\frac{11}{2}$ and $1i\frac{13}{2}$, which complete the shells with 50, 82, and 126 protons (or neutrons), respectively. A single particle excitation to one of those levels leaves the nucleus in an excited state of long half-life. These are called *isomeric states*, with an arbitrary value of 10^{-15} s, generally taken as a limit to define a long half-life.

The states belonging to the level $1g\frac{9}{2}$ can be excited states in a series of odd nuclei of Z (or N) below 50, similar to the states $1h\frac{11}{2}$ and $1i\frac{13}{2}$ for nuclei below 82 and 126, respectively. The groups of nuclei that contain one of these isomeric states form the *islands of isomerism* and the level scheme for inhabitants of these islands is shown in figure 9.4.

In these examples the isomeric state decays by an E3 or M4 transition. This is a quite typical behavior: besides the high multipolarity there is always a parity change since the state of high spin is coming from a shell with opposite parity from the states of the shell where it is inserted.

With the existence of islands of isomerism it is easy to understand figure 9.3. Being basically due to isomeric states of these islands, transitions of type M4 have disintegration constants in good agreement with (9.122), deduced from single particle transitions.

9.8 Internal Conversion

The emission of a γ-ray when the nucleus passes from an excited state to a state of lower energy is not the only means of releasing the energy difference between the states. In a process referred to as *internal conversion*, an atomic electron is ejected with an energy equal to that difference, a process that is possible due to the interaction of the electron with the electromagnetic field of the nucleus. It is important to notice that this interaction is direct and does not result from a photoelectric effect caused by a pre-emitted γ-ray. The

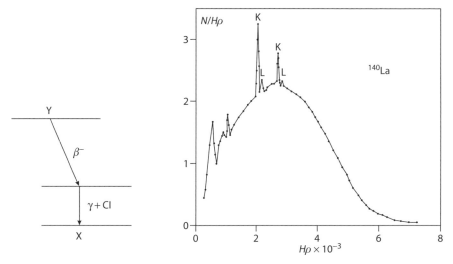

Figure 9.5 The excited state of nucleus X, fed by the β-decay of nucleus Y, can decay by γ-emission and by internal conversion. The spectrum of emitted electrons reveals the presence of a continuous part, characteristic of β-emission, on top of which there are superpositions of internal conversion peaks of the decay from the excited states of X. The example to the right is from the distribution of momenta of the β^--decay of ^{140}La [Be49].

participation of an atomic electron in the deflagration of a nuclear de-excitation process was already discussed in the phenomenon of electron capture.

The kinetic energy of the electron emitted by internal conversion,

$$T = E - B, \tag{9.123}$$

is the result of the difference between the de-excitation energy E and the binding energy of the electron, B. The small energy of nuclear recoil is neglected. As B only assumes discrete values, the spectrum of energy (or momentum distribution) of the conversion electrons of a given nuclear transition presents a series of peaks, characteristic of the atomic levels of a decaying element. A typical example is shown in figure 9.5, where the continuous spectrum of electrons emitted by β-decay also contains peaks due to internal conversion of an excited state of the daughter nucleus.

The binding energy B of the atomic electron depends on the shell or sub-shell where it is. Table 9.1 exhibits the values of B for the K, L, and M shells of lead. Notice the proximity

Table 9.1 Binding energy values of the electrons in the several sub-shells of the K, L, and M shells in lead.

Shell	Binding energy (keV) in each sub-shell				
K	88.0				
L	15.86	15.20	13.04		
M	3.85	3.55	3.07	2.59	2.48

of the energy inside each shell; equipment of smaller resolution cannot distinguish the peaks of the several subshells, showing them as a single peak.

The double option for the decay of an excited state, by γ-emission or by internal conversion, requires a quantity that measures the probability for one or the other decay. To obtain this quantity, let us begin by observing that we can write a total disintegration constant,

$$\lambda_t = \lambda_\gamma + \lambda_e, \tag{9.124}$$

as a sum of the partial constants for each process. Defining

$$\alpha = \frac{\lambda_e}{\lambda_\gamma}, \tag{9.125}$$

we have

$$\lambda_t = \lambda_\gamma(1 + \alpha), \tag{9.126}$$

which in turn can be split into parts for the several shells

$$\lambda_t = \lambda_\gamma(1 + \alpha_K + \alpha_L + \alpha_M + \cdots). \tag{9.127}$$

The ratio α is called the *internal conversion coefficient* and it is, obviously, a measure of the probability that a internal conversion occurs with respect to the emission of a γ-ray.

The theory for the determination of internal conversion coefficients will not be presented here. We will limit ourselves to showing the results of a calculation [BW52, p. 618] that, in spite of being nonrelativistic, is capable of correctly predicting many of the characteristics of this phenomenon.

Let us consider (electric) transitions of multipolarities $L > 0$ that allow comparison with radiative transitions. In contrast to E0 transitions where the electron probes the nuclear volume, here the main events take place outside of the nucleus. The standard multipole expansion of the electrostatic potential in the region outside the nucleus gives

$$H_C = \sum_{LM} \frac{4\pi e}{2L+1} \frac{1}{r^{L+1}} Y_{LM}^*(\mathbf{n}) \mathcal{M}(\mathrm{EL}, M) \tag{9.128}$$

where $\mathbf{n} = \mathbf{r}/r$. The monopole, $L = M = 0$, term of (9.128) creates a normal spherical Coulomb field binding the electrons. The higher terms induce the synchronized transitions in the nucleus and in the atomic shells. For the nuclear transition $i \to f$, the matrix element over nuclear wave functions is the same as that determining the radiation probability (9.97).

Let us estimate the probability of internal conversion with the K-electron (initial energy ϵ_0) emitted into a continuum (a final state c with energy ϵ). Having two electrons in the K-orbit of a heavy atom (this is equivalent to summation over electron spin states) and using the golden rule (6.49), we obtain for the EL transition

$$dw_{fi} = 2\frac{2\pi}{\hbar}|(H_C)_{fi}|^2\delta(\epsilon - \epsilon_0 - E_{if})d\rho_e d\epsilon \tag{9.129}$$

$$= \frac{4\pi}{\hbar}\left(\frac{4\pi e}{2L+1}\right)^2 \left|\sum_M \left\langle c \left| \frac{1}{r^{L+1}} Y_{LM}^* \right| K \right\rangle (\mathcal{M}(\mathrm{EL}, M))_{fi} \right|^2 d\rho_e. \tag{9.130}$$

For the sake of simplicity we neglect the electron interaction with the atom in the final state, taking the state c as a plane wave with momentum $\hbar\mathbf{q}$. The magnitude q is determined by the energy conservation and the corresponding level density in the continuum for the nonrelativistic electron (this approximation works only if $E_{if} = \hbar\omega \ll m_e c^2$) is $d\rho_e = V m_e \hbar q \, d\Omega / (2\pi\hbar)^3$.

The K orbit wavefunction does not depend on angles,

$$\psi_K = \frac{1}{\sqrt{\pi a_0^3}} e^{-r/a_0}, \tag{9.131}$$

where the orbit radius for the nucleus of charge Z is

$$a_0 = \frac{a_B}{Z} = \frac{\hbar^2}{m_e e^2 Z}. \tag{9.132}$$

The angular integral in the electron matrix element in (9.130) picks up the partial wave with quantum numbers (LM) from the expansion of $\exp[-i(\mathbf{q} \cdot \mathbf{r})]$. This reflects the transfer of the angular momentum L from the nucleus to the electron. After the angular integration we obtain

$$\left\langle c \left| \frac{1}{r^{L+1}} Y_{LM}^* \right| K \right\rangle = (-i)^L 4 \sqrt{\frac{\pi}{V a_0^3}} Y_{LM}^*(\mathbf{q}) \int \frac{dr}{r^{L-1}} j_L(qr) e^{-r/a_0}. \tag{9.133}$$

Under our conditions, the electron wavelength $\sim 1/q$ is small compared to the atom radius (9.131), $qa_0 > 1$. Moreover, we will consider $qa_0 > L$ because we are mostly interested in low multipolarities. Then the function $j_L(qr)$ suppresses the integrand before we reach the atomic boundary $r \sim a_0$. Therefore the exponent can be replaced by unity, and the integral is equal to

$$\int_0^\infty \frac{dr}{r^{L-1}} j_L(qr) = q^{L-2} \int_0^\infty \frac{dx}{x^{L-1}} j_L(x) = \frac{q^{L-2}}{(2L-1)!!}. \tag{9.134}$$

Now we substitute (9.133), (9.134), along with the final electron density of states, into (9.130), sum over projections M_f of the nuclear final state, and integrate over the angles of $\mathbf{q}$ the product of spherical functions $\int d\Omega_q Y_{LM}^*(\mathbf{q}) Y_{LM'}(\mathbf{q}) = \delta_{M'M}$. The result is the total probability of the internal conversion for a given nuclear multipole transition,

$$w_{\text{conv}}(\text{EL}; fi) = 128\pi \frac{m_e e^2}{\hbar^3 a_0^3} \frac{q^{2L-3}}{[(2L+1)!!]^2} \sum_{M_f M} |(\mathcal{M}(\text{EL}, M))_{fi}|^2. \tag{9.135}$$

The radiation probability for the same nuclear transition is given by (9.88), where one needs to sum over the projections M_f. It allows one to find the *conversion coefficient* for the K-shell

$$\alpha_K(\text{EL}) = \frac{w_{\text{conv}}(\text{EL}; fi)}{w_{\text{rad}}(\text{EL}; fi)}. \tag{9.136}$$

In the ratio, information on nuclear matrix elements disappears. In our approximation, the electron kinetic energy is large compared to its binding energy, $\hbar\omega \approx \hbar^2 q^2 / 2m$, so that the result can be written as

$$\alpha_K(\text{EL}) = \frac{16L}{L+1} \frac{m_e e^2}{\hbar^2 a_0^3} \frac{q^{2L-3}}{k^{2L+1}} = \frac{L}{L+1} Z^3 \alpha^4 \left(\frac{2m_e c^2}{\hbar\omega} \right)^{L+5/2}. \tag{9.137}$$

On the right-hand side α stands for the fine structure constant. The internal conversion coefficients strongly increase with the nuclear charge and multipolarity of the transition; this shows preference for low frequency transitions.

The result (9.137) is derived under many simplifying assumptions. It is easy to see that they are mutually consistent but restrict the domain of validity of (9.137).

(i) The initial atomic wavefunction was considered in the nonrelativistic approximation, $v_e/c \sim Z\alpha \ll 1$. This does not permit use of the results for large Z.

(ii) The emitted electron is nonrelativistic, $\hbar^2 q^2 < m_e c^2$; this leaves us with low energy nuclear excitations only.

(iii) $\hbar\omega < m_e c^2 \approx 500$ keV. A stronger limitation comes from using the *nonretarded interaction*,

(iv) $(2\pi c/\omega) > a_0$. It leads to $\hbar\omega < 2\pi Z\alpha m_e c^2 \approx Z \times 25$ keV. This practically excludes small Z because of absence of nuclear excitations of such low energy.

(iv) We neglected the electron binding energy and its Coulomb interaction in the final state which is justified for $\hbar\omega > (m_e e^4 Z^2/2\hbar^2) = Z^2 \times 13.6$ eV. For $Z \approx 50$, this limits nuclear excitation energies from below, $\hbar\omega > 35$ keV.

(v) The approximation $qa_0 > 1$ in computing the integral (9.133) is valid if (iv) is fulfilled.

All these approximations can be lifted. The internal conversion coefficients for various atomic shells have been calculated and tabulated using relativistic electron wavefunctions, taking into account the finite nuclear size (recall that the expansion (9.128) can be used only outside the nucleus). The relative magnitudes of the conversion coefficients for different shells turn out to be rather sensitive to the multipolarity of the nuclear transition; their measurement helps establish this multipolarity.

In agreement with these general calculation, the g values of α are obtained from

$$\alpha(EL) \cong \frac{Z^3}{n^3} \left(\frac{L}{L+1}\right) \left(\frac{e^2}{\hbar c}\right)^4 \left(\frac{2m_e c^2}{E}\right)^{L+\frac{5}{2}}, \tag{9.138}$$

$$\alpha(ML) \cong \frac{Z^3}{n^3} \left(\frac{e^2}{\hbar c}\right)^4 \left(\frac{2m_e c^2}{E}\right)^{L+\frac{3}{2}}, \tag{9.139}$$

where EL means electric transition of order L, with similar meaning for ML. Z is the atomic number, n the main quantum number (K shell, $n = 1$; L, $n = 2$, etc.), and E the available energy for the transition that, neglecting the nuclear recoil, is the same as the difference of energy between the initial and final levels.

Several properties revealed by (9.138) and (9.139) can be made explicit. The first is that the internal conversion is a phenomenon more characteristic of heavy nuclei, shown by the dependence in Z^3. The dependence on the main quantum number n shows, in turn, that conversion electrons are ejected preferentially from the internal atomic shells, which is physically expected. Finally, the last factor of (9.138) and (9.139) indicates that the internal conversion is larger for small de-excitation energy and high multipolarities. All these considerations have experimental confirmation, although the expressions (9.138) and (9.139) are approximate and they are not appropriate for precise numerical determinations. Results of elaborate calculations are presented for the coefficient α_K in figure 9.6.

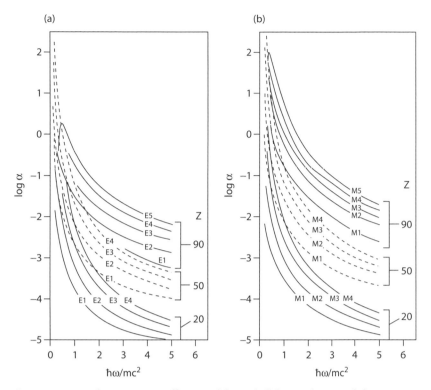

Figure 9.6 Internal conversion coefficients of the K shell for (a) electric and (b) magnetic multipole transitions in the elements Ca, Sn, and Th [Pr62].

These results are sufficiently reliable that one can use them with confidence in the interpretation of experimental data. The character of a transition (E or M) and the value of its multipolarity can be obtained, for example, starting from the knowledge of α_K for that transition. This allows us to determine the relative parity between initial and final states.

A final word should be said with regard to the $0^+ \rightarrow 0^+$ transitions. Since there is no multipole radiation of order zero, only the internal conversion process is capable of promoting this transition. The process necessarily occurs inside the nucleus. As in the case of electronic capture, only the K shell gives a significant contribution.

9.9 Resonant Absorption—The Mössbauer Effect

For long-lived states, the presence of narrow widths makes the resonance curves very sharp and therefore sensitive to small perturbations. The famous Mössbauer effect is based on this sensitivity and the fact that momentum conserved, leads to the recoil of the radiating nucleus. Since the center of mass motion carries kinetic energy, the photon energy is not exactly equal to the level spacing $\Delta \equiv E_i - E_f$ for a given transition. For example, for the emitted photon with wave vector $\mathbf{k}$ and an initial nucleus of mass M at rest, the recoil momentum is $\mathbf{P}_f = -\hbar\mathbf{k}$. Energy conservation determines the recoil energy,

$$E_{rec} = \frac{\hbar^2\omega^2}{2Mc^2} = \Delta - \hbar\omega. \tag{9.140}$$

Thus, the frequency of the emitted photon is in fact less than the energy spacing,

$$\hbar\omega = \Delta - \frac{(\hbar\omega)^2}{2Mc^2} \approx \Delta - \frac{\Delta^2}{2Mc^2}. \tag{9.141}$$

This shift is small because of the heavy nuclear mass, so it is possible, instead of solving the quadratic equation (9.141), to substitute $\hbar\omega \approx \Delta$ in the recoil term. For a nucleus with $A \sim 50$ and excitation energy $\Delta \sim 1$ MeV, the recoil shift is about 10 eV.

The small recoil effect becomes important if one tries to observe the *resonance fluorescence*, the absorption of the photon, emitted in a nuclear transition $i \to f$, by the identical nucleus that would undergo the opposite excitation process to the same initial level, $f \to i$. The absorption implies the momentum transfer from the photon so that, for the absorber nucleus originally at rest, the excitation energy will be again reduced because of recoil, by approximately the same amount as in (9.141). Hence, the fluorescence process includes the detuning, $E_{rec} \simeq \Delta^2/Mc^2$, of the frequencies of the primary and secondary photon. However, this shift is typically much greater than the natural width of the absorption line. For rather fast γ-transitions, let us say $\tau \simeq 10^{-15}$ s, the width $\Gamma < 1$ eV. This means that the recoil shifts the secondary photon away from the absorption width, making the resonance fluorescence impossible. In contrast to this, in atomic radiation with $\Delta \simeq 1$ eV and the same nuclear mass, the recoil shift is very small, $\sim 10^{-8}$ eV, so that the secondary photon is still well within the resonance curve of the absorber.

In order to observe the nuclear resonance fluorescence, one can bring the two nuclei into relative motion in such a way that the Doppler shift of photon energy compensates the recoil shift. The corresponding velocity is determined by $v/c \simeq E_{rec}/\Delta \simeq \Delta/Mc^2$ and varies from mm/s to m/s. In gases the Doppler shift exists naturally because of the thermal velocity distribution which nevertheless is getting narrower as temperature decreases. Mössbauer discovered that the situation is opposite in crystals where the effect of the resonance fluorescence increases at low temperature. (In fact, an effect of the same physical nature was observed much earlier for neutron interaction in crystals.) The nucleus in the crystal lattice is in the bound state of the center-of-mass motion. The radiation of the photon is accompanied by the recoil factor $\exp[-i(\mathbf{k} \cdot \mathbf{R})]$ acting on the center of mass wave function. For a nucleus in free motion this would lead to momentum conservation. For a bound nucleus, the center-of-mass part of the matrix element is to be taken between the states $|s\rangle$ of the lattice. Therefore the total radiation probability includes the factor

$$W_{s's} = \left| \langle s' | e^{-i(\mathbf{k}\cdot\mathbf{R})} | s \rangle \right|^2 \tag{9.142}$$

related to the change of the quantum state of the lattice as a result of the radiative transition of a nucleus at a specific site. The momentum of the emitter nucleus is not conserved; instead, it is passed to the crystallic excitations (sound waves, phonons). The probability

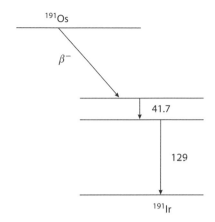

Figure 9.7 Excited states of ^{191}Ir fed by the decay of ^{191}Os. The energies are in keV.

W_{ss} of radiation without changing the crystal state is also present. This implies recoilless radiation, and, with the same probability, recoilless absorption by another bound nucleus. In reality, the probability W_{ss} has to be averaged over the actual thermal state of the lattice. This average quantity $\overline{W}$ is called the Debye-Waller factor. In the harmonic approximation for the vibrations of the cubic lattice, the result can be exactly calculated,

$$\overline{W} \equiv \overline{\langle e^{i(\mathbf{k}\cdot\mathbf{R})} \rangle} = e^{-k^2 \langle R^2 \rangle / 2},$$ (9.143)

where $\langle R^2 \rangle$ is the mean square displacement of the nucleus from the equilibrium position in the lattice; $\overline{W}$ grows at low temperature. The effect is strongly reduced because of the destructive interference when this displacement exceeds the photon wavelength, $k^2 R^2 > 1$.

Strictly speaking, for a real finite crystal, we need to separate the center of mass and the relative coordinates of atoms of the entire crystal. Then momentum conservation will still be fulfilled for the whole piece but the corresponding mass is that of the macroscopic crystal, and the corresponding recoil energy is negligibly small. It is interesting to note that the average excitation energy transferred to the crystal is given by

$$E_{\text{exc}} = \sum_{s'} (E_{s'} - E_s) W_{s's}.$$ (9.144)

Applying this result we see that the average excitation energy of the crystal,

$$E_{\text{exc}} = \frac{\hbar^2 k^2}{2M},$$ (9.145)

is equal to the recoil energy of the radiating nucleus.

The relationship between the width of a level, ΔE, and the half-life $t_{1/2}$ is

$$\Delta E \, t_{1/2} \cong \hbar.$$ (9.146)

The application of (9.146) to the decay of the 129 keV excited state of ^{191}Ir, with half-life 10^{-10}s (figure 9.7) indicates a width $\Delta E \cong 5 \times 10^{-6}$ eV, very small compared with the transition energies involved.

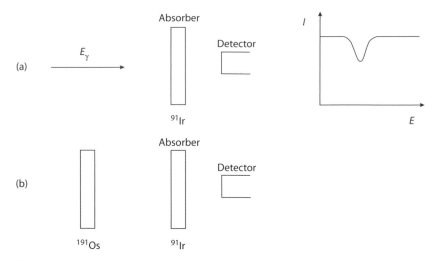

Figure 9.8 "Experiments" for the detection of resonant absorption. In (a), an idealized situation is presented, where a γ-ray beam of variable energy produces the curve shown in the graph. In (b), an attempt at obtaining the minimum value of the curve is shown, using an emitter identical to the absorber.

Let us keep that number and now imagine an ideal experiment shown in figure 9.8. A beam of γ-radiation of arbitrary monoenergetic energy (which one cannot obtain in practice) of around 129 keV hits a sample of ^{191}Ir, which should be transparent for all radiation of energy different from the energy of its excited states. However, at the energy of 129 keV, absorption of part of the beam to excite this state should causes a fall in the measured intensity of the detector, as the figure indicates. This characterizes the phenomenon of *resonant absorption*. In (b) an experimental apparatus is set up with the idea of obtaining the condition of resonant absorption, a sample of ^{191}Os emitting γ-rays of 129 keV, as we saw in figure 9.7.

Let us try to predict the result. If we take into account initially that the emission and absorption of the γ-ray are accompanied by the recoil of the decaying and absorbing nuclei, a first calculation would indicate that one can have resonant absorption because the recoil is of the order of 0.05 eV. The γ-ray arrives at the absorber with 0.05 eV less than the value of the transition energy, when in fact it should arrive with additional 0.05 eV, to compensate for the recoil of the absorber. This establishes a distance of 0.1 eV between narrow peaks of 5×10^{-6}eV, as indicated in figure 9.9a, and no overlap among the peaks would be possible.

There is, however, an additional effect that modifies the situation. The nuclei in the crystal are not at rest but have a vibration characteristic of the temperature of the crystal. This produces, due to the Doppler effect, a broadening in the energy distribution, for the emitter as well as for the absorber; we can see that in part (b) of figure 9.9. At room temperature the width of the peak becomes 0.07 eV, creating an interception area, dashed in the figure, that it is capable of producing events of resonant capture.

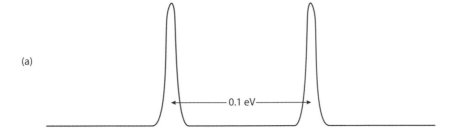

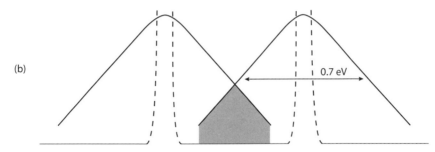

Figure 9.9 (a) Taking into account the recoil but not the Doppler effect, there is no interception between the emission and absorption peaks. (b) With the Doppler effect at room temperature, the broadening of the peaks produces an intersection area.

In 1958, Rudolf Mössbauer [Mo58] sought to perform a resonant capture experiment by cooling the source and the absorber to a temperature of $88°$ K. The expected situation, in this case, would be the disappearance of the phenomenon, since the low temperature causes a narrowing in the distribution of the Doppler effect. The results he observed, however, were an increase in the intensity of the events of resonant capture. The explanation for this behavior requires an analysis of what can happen to an atom of a crystalline lattice when it absorbs energy. If the energy is sufficient, a first possible process is to knock the atom from its position in the lattice. Another possibility is to take it to a state of higher energy. If, however, the available energy and the temperature are low enough, a third possibility is that the energy is absorbed by the whole crystal, or, more specifically, by a great number of atoms ($\sim 10^8$) linked by an acoustic wave that spreads through the crystal. This last possibility is the explanation for the existence of resonant capture under those conditions: the recoil energy of the nuclei due to the emission of a γ-ray is now practically negligible, because the recoil mass is very large.

The events of resonant capture are, in that case, not due to the Doppler effect, which is very small at that temperature, but to the interception of the peaks of energy themselves with their natural widths, as shown in figure 9.9a. In the *Mössbauer effect*, as it is called, the separation of 0.1 eV marked in figure 9.9a disappears, and an apparatus used originally by Mössbauer allows a measurement of the natural width of the peak. This is seen in

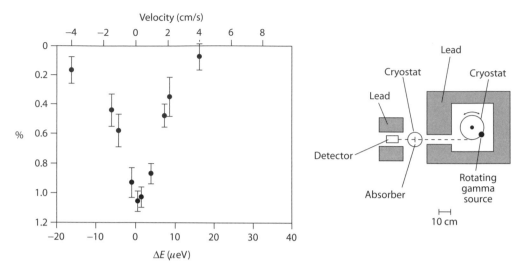

Figure 9.10 Measurement of the natural width of the 129 keV level of ^{191}Ir with the experimental arrangement of Mössbauer. The ordinate indicates the percent reduction of the intensity in the detector.

figure 9.10: a source of ^{191}Os (see figure 9.7) is placed in a circular support that can rotate, with the aim of producing an artificial Doppler effect. The radiation comes from a small opening through the lead protection and reaches a detector placed behind an absorber of ^{191}Ir. Source and absorber are immersed in cryostats that maintain them at low enough temperatures to obtain the desired effects, namely the realization, under special conditions, of the experience described in figure 9.8.

The graph of figure 9.10 exhibit the percent of the intensity lost in the absorber. For zero speed of the source the resonant absorption consumes about 1% of the radiation. When the source moves, the absorption does not fall to zero due to the natural width of the level and, in this way, the experience allows a measurement of this width. We see that, due to the fact that the natural width is very small (5×10^{-6} eV), speeds of a few cm/s are necessary to obtain the data.

Due to the extreme precision with which one can measure differences of energy, the Mössbauer effect has large application in nuclear physics. As an example, one can mention study of the hyperfine structure of nuclear levels, which suffer the same displacement ($\Delta E \cong 10^{-7} - 10^{-6}$ eV) as the atomic levels, but whose ratio $\Delta E/E$, E being a typical nuclear energy, makes it a much more difficult measurement.

Outside the field of nuclear physics, one example of extreme precision making use of resonant absorption was carried out by Pound and Rebka [PR60] in 1959. They measured the energy variation of a photon caused by the gravitational terrestrial field. The measurement was done in a tower 22.5 m high at Harvard University. γ-rays from a source of ^{57}Fe placed in the soil reached the absorber and the detector placed at the top of the building. The energy variation caused by the gravitational field could be detected. The influence of the gravitational field on a photon is a prediction of the theory of general relativity and it had already been confirmed by astronomical means. The results of Pound and Rebka confirmed the theoretical prediction and they were the first ones obtained in terrestrial laboratory conditions.

One phenomenon, related to radiation recoil, and fundamental to an important branch of modern experimental technique, is based on laser cooling and trapping of atoms [MS94]. One of the instructive examples of this physics was observed as a strong reduction of light scattering by atoms forming the Bose condensate in a trap [Sta99]. The atoms form a weakly interacting Bose gas, a system where the low-lying excitations are in fact not the individual freely moving atoms but the *phonons*, sound waves with a linear dispersion $\epsilon(p)$. If the atom recoil velocity after the photon absorption is less than the speed of sound, the phonon cannot be excited, and the absorption is strongly suppressed.

9.10 Exercises

1. Show, through the analysis of the parity of the functions that compose the integrand of eq. 9.115, that magnetic multipole transitions of even order can only happen between states of different parity and vice-versa.

2. Give, for the following γ transitions, all the allowed multipoles and indicate which multipole should be more intense in the emitted radiation.

(a) $\frac{7}{2}^- \rightarrow \frac{3}{2}^-$ (d) $4^+ \rightarrow 2^+$

(b) $\frac{7}{2}^+ \rightarrow \frac{7}{2}^-$ (e) $0^- \rightarrow 0^+$

(c) $0^- \rightarrow 3^-$ (f) $\frac{11}{2}^- \rightarrow \frac{5}{2}^+$

3. $^{167}_{68}$Er has the following sequence of states, beginning with the ground state:

$$\frac{7^+}{2}, \frac{3^+}{2}, \frac{1^-}{2}, \frac{3^-}{2}, \frac{5^-}{2}$$

a) Which of them can be an isomeric state? b) Of the ten possible transitions, which have the smallest chance of happening? c) Make a sketch of the level scheme and of the transitions, indicating the multipolarity of each one.

4. The ground state $\frac{3}{2}^-$ of $^{247}_{97}$Bk decays by α-emission to $^{243}_{95}$Am, which has, above the ground state $\frac{5}{2}^-$, a sequence of excited states $\frac{5}{2}^-, \frac{7}{2}^-, \frac{5}{2}^+, \frac{9}{2}^-, \frac{7}{2}^+, \frac{9}{2}^+, \frac{11}{2}^+, \frac{3}{2}^-, \frac{5}{2}^-$. Establish some plausible ways for decay until one reaches the ground state of the americium and try to infer which is the most probable.

5. Repeat exercise 4 for the β^+-decay of the ground state $\frac{9}{2}^-$ of $^{207}_{83}$Bi. The nucleus $^{207}_{82}$Pb has, above the ground state $\frac{1}{2}^-$, the sequence of excited states $\frac{5}{2}^-, \frac{3}{2}^-, \frac{13}{2}^+, \frac{7}{2}^-$.

6. Give an explanation for the fact that γ-decay among levels of a rotational band have, in many cases, the aspect seen in figure 9.11(a) for odd nuclei and in figure 9.11(b) for even-even nuclei.

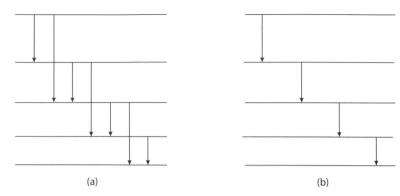

Figure 9.11 Gamma-decay transitions.

7. The nucleus $^{154}_{64}$Ga has, above its ground state 0^+, the excited states listed below (energy in keV).

Spin	Energy	Spin	Energy	Spin	Energy
2^+	123	6^+	718	4^+	1048
4^+	371	2^+	816	3^+	1128
0^+	681	2^+	996	8^+	1145

a) Determine the multipolarities and the energies of the γ-rays emitted when the 1145 keV level, 8^+, decays through the most probable path to the ground state. b) Repeat the same for the 816 keV level, 2^+. c) Repeat the same for the level 0^+, with 681 keV. Is it possible for that level to decay directly to the ground state?

8. The isomeric state $\frac{13}{2}^+$ of ^{199}Hg decays by γ-emission of multipolarity M4 to a state $\frac{5}{2}^+$, 368 keV below it. Determine the value of λ_M, eq. 9.122, and see if the result is coherent with the value of the experimental half-life of 44 minutes of this isomeric level. Do not forget to take into account that this same state of Hg also decays by internal conversion, with a coefficient $\alpha = 4.4$.

9. Using a table of isotopes, e.g., Lederer and Shirley [LS78], try to verify the extension of the islands of isomerism. Try to locate transitions with higher multipolarities than E3 and M4.

10. In a study of conversion electrons emitted in a decay process, the following energies (in keV) of the electrons were measured: 176.8, 310.7, 363.5, 540.2, 592.3, 645.9, 653.7, 966.6, 1019.2, 1206.7. The binding energy of the electrons are (in keV): 63.3 (K), 10.5 (L) and 2.3 (M). What is the smallest number of γ-rays that can produce the groups of observed electrons, and what are their energies?

11. Study of the transition between two levels of ^{108}Pd, separated by 434 keV from each other, shows that in each 1124 γ-decay there is the emission of 10 conversion electrons.

a) Use the formulas (9.125)–(9.139) to obtain the value of α_K. b) With the aid of the graph in figure 9.6, verify if the result is compatible with the information that this transition is a de-excitation of a state belonging to a rotational band of Pd.

12. The transition with 14.4 keV, 98 ns, of ^{57}Fe is much used to produce the Mössbauer effect. For this transition calculate: (a) its natural width; (b) the recoil energy; (c) the Doppler width at room temperature; (d) the Doppler width at the temperature of liquid helium. Use for the last two items the formula for Doppler width: $D = 2\sqrt{T_R kT}$, where T_R is the recoil energy, k is the Boltzmann constant, and T the absolute temperature.

13. Express the total momentum P of the electromagnetic field in terms of photon annihilation and creation operators.

14. Show by explicit consideration of eq.(9.47) that $0-0$ transitions are absolutely forbidden.

15. Calculate the reduced probability of the single-particle magnetic dipole transition between the members of the spin-orbit doublet $j = l \pm 1/2$.

16. List allowed multipolarities of electric and magnetic radiation in the single-particle transitions $s_{1/2} \leftrightarrow s_{1/2}, p_{1/2} \leftrightarrow p_{1/2}, h_{11/2} \leftrightarrow d_{5/2}, i_{13/2} \leftrightarrow f_{7/2}, s_{1/2} \leftrightarrow d_{3/2}, i_{13/2} \leftrightarrow f_{5/2}$.
   (a) A nucleus undergoes a cascade of γ-transitions from the initial state 1 with $J_1 = 0$ to the intermediate state 2 with $J_2 = 1$ and then to the final state 3 with $J_3 = 0$,

$$1 \rightarrow \mathbf{k}_1 \rightarrow 2 \rightarrow \mathbf{k}_2 \rightarrow 3. \tag{9.147}$$

What is the probability distribution for the angle θ between the directions of the photons $\mathbf{k}_1$ and $\mathbf{k}_2$?
   (b) Perform the same calculation for the case when in the first step an alpha-particle is emitted instead of the photon (the second step and all quantum numbers are the same as in a).

17. Calculate the reduced E2 transition probabilities for γ-radiation from the triplet $(0^+, 2_2^+, 4^+)$ of the states with two quadrupole phonons to the one-phonon state 2_1^+.

10 | Nuclear Reactions—I

10.1 Introduction

The collision of two nuclei can give place to a nuclear reaction where, similarly to a chemical reaction, the final products can be different from the initial ones. This process happens when a target is bombarded by particles coming from an accelerator or from a radioactive substance. It was in the latter way that Rutherford observed, in 1919, the first nuclear reaction produced in the laboratory,

$$\alpha + {}^{14}_{7}\text{N} \rightarrow {}^{17}_{8}\text{O} + \text{p}, \tag{10.1}$$

using α-particles coming from a ${}^{214}\text{Bi}$ sample and using as the target nitrogen contained in a reservoir.

As in (10.1), other reactions were induced using α-particles, the only projectile available initially. With the development of accelerators around 1930, the possibilities multiplied, as much for the energy as for the mass of the projectile; it is already possible to bombard a target with protons of energy close to 1 TeV, and beams of particles as heavy as uranium are available for study of reactions with heavy ions.

We can sometimes have more than two final products in a reaction, as in the examples

$$\text{p} + {}^{14}\text{N} \rightarrow {}^{7}\text{Be} + 2\alpha,$$
$$\text{p} + {}^{23}\text{Na} \rightarrow {}^{22}\text{Ne} + \text{p} + \text{n}, \tag{10.2}$$

or just one, as in the capture reaction

$$\text{p} + {}^{27}\text{Al} \rightarrow {}^{28}\text{Si}^{*}, \tag{10.3}$$

where the asterisk indicates an excited state, which usually decays emitting γ-radiation. Under special circumstances, more than two reactants are possible. Thus, for example, the reaction

$$\alpha + \alpha + \alpha \rightarrow {}^{12}\text{C} \tag{10.4}$$

can take place in the overheated plasma of the stellar interior.

Also, the final products can be identical to the initial ones. This case characterizes a process of nuclear scattering, which can be elastic, as in the simple example

$$p + {}^{16}O \rightarrow p + {}^{16}O, \tag{10.5}$$

where there is only transfer of kinetic energy between projectile and target, or inelastic, as in

$$n + {}^{16}O \rightarrow n + {}^{16}O^*, \tag{10.6}$$

where part of the kinetic energy of the system is used in the excitation of ^{16}O.

Naturally, nuclear reactions are not limited to nuclei. They can involve any type of particle, and also radiation. Thus, the reactions

$$\gamma + {}^{63}Cu \rightarrow {}^{62}Ni + p,$$
$$\gamma + {}^{233}U \rightarrow {}^{90}Rb + {}^{141}Cs + 2n \tag{10.7}$$

are examples of nuclear processes induced by gamma-radiation. In the first a γ-ray knocks a proton off ^{63}Cu and in the second it induces the process of nuclear fission in ^{233}U, with the production of two fragments and the emission of two neutrons.

It is essential to highlight that, unlike a chemical reaction, the products resulting from a nuclear reaction are not determined univocally: starting from two or more reactants there can be dozens of possibilities of composition of the final products with an unlimited number of available quantum states. As an example, the collision of a deuteron with a nucleus of ^{238}U can give rise, among others, to the following reactions:

$$d + {}^{238}U \rightarrow {}^{240}Np^* + \gamma,$$
$$\rightarrow {}^{239}Np + n,$$
$$\rightarrow {}^{239}U + p,$$
$$\rightarrow {}^{237}U + t. \tag{10.8}$$

In the first of these the deuteron is absorbed by the uranium target, forming an excited nucleus of ^{240}Np that de-excites by emitting a γ-ray. The two subsequent reactions are examples of *stripping reactions*, where a nucleon is transferred from the projectile to the target. The final reaction exemplifies the inverse process: the deuteron captures a neutron from the target and emerges from the reaction as a ^3H (triton). This is referred to as a *pick-up reaction*. Another possibility would be, in the first reaction, that the nucleus ^{240}Np fissions instead of emitting a γ-ray, contributing dozens of possible final products from the reaction.

Each of the branches of the reaction which can occur, with well-defined quantum states of the participants, is referred to as a *channel*. In (10.8), for the *entrance channel* $d + {}^{238}U$, there are four possible *exit channels*. Notice that a different exit channel would be reached if some of the final products (other than $^{240}Np^*$) were in an excited state. The probability of a nuclear reaction taking place through a certain exit channel depends on the energy of the incident particle and is measured by the *cross section* for that channel.[1] The theory

---

[1] For a definition of cross section, see section 2.7.

of nuclear reactions, besides elucidating the mechanisms that determine the occurrence of the different processes, must also evaluate the cross sections corresponding to all exit channels.

Besides the initial and final constituents, it is also important to discuss the processes through which a nuclear reaction can take place. In this latter aspect, there are two important mechanisms of opposite action. In *direct reactions* the projectile and the target have an interaction of short duration, with possible exchange of energy or particles between them. In the other mechanism there is a fusion of the projectile with the target, the available energy being distributed to all the nucleons, forming a highly excited *compound nucleus*. Decay of the compound nucleus leads to the final products of the reaction. These mechanisms will be detailed in subsequent sections.

10.2 Conservation Laws

Several conservation laws contribute to restricting the possible processes that take place when a target is bombarded with a given projectile, some of them having already been commented upon earlier.

1) *Baryonic number.* There is no experimental evidence of processes in which nucleons are created or destroyed without the creation or destruction of corresponding antinucleons. This fact was discussed in chapter 1 when defining the baryonic number B. The application of this principle to low energy reactions is still more restrictive. Below the threshold for production of mesons ($\sim$140 MeV), no process related to the nuclear forces is capable of transforming a proton into a neutron and vice versa, and processes governed by the weak force (responsible for β-emission of nuclei) are very slow in relation to the times involved in nuclear reactions ($\sim 10^{-22}$ to 10^{-16}s). In this way, we can speak separately of proton and neutron conservation, which should show up with the same amounts on both sides of a nuclear reaction.

2) *Charge.* This is a general conservation principle in physics, valid in any circumstances. In purely nuclear reactions it is computed by obtaining the sum of the atomic numbers, which should be identical on both sides of the reaction.

3) *Energy and linear momentum.* These are two of the most applied principles in the study of the kinematics of reactions. Through them, angles and velocities are related to the initial parameters of the problem.

4) *Total angular momentum.* This is always a constant of motion. In the reaction

$$^{10}\text{B} + {}^{4}\text{He} \rightarrow {}^{1}\text{H} + {}^{13}\text{C}, \tag{10.9}$$

for example, ^{10}B has $I = 3$ in the ground state, whereas the α-particle has zero angular momentum. If it is captured in an s-wave ($l_i = 0$), the intermediate compound nucleus is in a state with $I_c = 3$. Both final products have angular momenta equal to $\frac{1}{2}$; thus their sum is 0 or 1. Therefore the relative angular momentum of the final products will be $l_f = 2, 3$, or 4.

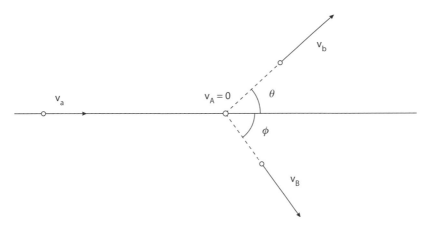

Figure 10.1 Nuclear reaction $a + A \rightarrow b + B$, seen in the laboratory system.

5) *Parity*. This is always conserved in reactions governed by the nuclear interaction. In the previous example, ^{10}B, ^{4}He, and the proton have even parities, while ^{13}C has odd parity. Therefore, if $l_i = 0$, we necessarily have $l_f = 3$. Thus, the orbital momentum of the final products of (10.8) is determined by the joint conservation of total angular momentum and of parity.

6) *Isospin*. This is an approximate conservation law that is applied to light nuclei, where the effect of the Coulomb force is small. A nuclear reaction involving these nuclei not only conserves the z-component of the isospin (a consequence of charge and baryonic number conservation) but also the total isospin **T**. Reactions that populate excited states not conserving the value of **T** are strongly inhibited. An often cited example is the reaction $d + {}^{16}O \rightarrow \alpha + {}^{14}N$, where the excited state 0^+, with 2.31 MeV, of ^{14}N is about a hundred times less populated than the 1^+ ground state. Conservation of energy, angular momentum, and parity does not impose any prohibition for that channel, whose low occurrence can only be justified by isospin conservation: the ground states of the four participant nuclei in the reaction all have **T** = 0 and the state 1^+ of ^{14}N has **T** = 1.

10.3 Kinematics of Nuclear Reactions

We shall study the kinematics of a typical reaction, where the projectile a and the target A give rise to two products, b and B, respectively. This can also be expressed in the notation that we have thus far used,

$$a + A \rightarrow b + B, \tag{10.10}$$

or even in a more compact notation,

$$A(a, b)B. \tag{10.11}$$

Figure 10.1 exhibits the parameters related to the reaction of (10.10).

In the most common situation a and b are light nuclei and A and B heavy ones; the nucleus b has its emitting angle θ and its energy registered in the laboratory system and the recoil nucleus B has short range and cannot leave the target. Thus, we can for convenience eliminate the parameters of B in the system of equations that describe the conservation of energy and momentum:

$$E_a + Q = E_b + E_B,$$
$$\sqrt{2m_a E_a} = \sqrt{2m_b E_b} \cos\theta + \sqrt{2m_B E_B} \cos\phi,$$
$$\sqrt{2m_b E_b} \sin\theta = \sqrt{2m_B E_B} \sin\phi, \tag{10.12}$$

where the *Q-value of the reaction* measures the energy gained (or lost) due to the difference between the initial and final masses:

$$Q = (m_a + m_A - m_b - m_B)c^2. \tag{10.13}$$

Eliminating E_B and ϕ from (10.12) we can relate Q to the parameters of collision that interest us:

$$Q = E_b\left(1 + \frac{m_b}{m_B}\right) - E_a\left(1 - \frac{m_a}{m_B}\right) - \frac{2}{m_B}\sqrt{m_a m_b E_a E_b}\cos\theta. \tag{10.14}$$

A useful relation for the analysis of a nuclear reaction is obtained by observing that (10.14) is an equation of the second degree in $\sqrt{E_b}$, whose solution is

$$\sqrt{E_b} = \frac{1}{m_b + m_B}\left\{\sqrt{m_a m_b E_a}\cos\theta \right.$$
$$\left. \pm \sqrt{m_a m_b E_a \cos^2\theta + (m_b + m_B)[E_a(m_B - m_a) + Q m_B]}\right\}. \tag{10.15}$$

If we place in a graph the energy E_b of the emitted particle, observed at an angle θ, as a function of the energy E_a of the incident particle, we obtain a set of curves, one for each value of θ. Figure 10.2 exhibits the curves obtained for reaction $^{12}C + ^{14}N \rightarrow ^{10}B + ^{16}O$, where $Q = -4.4506$ MeV.

Two things are evident when observing figure 10.2. In the first case, since Q is negative for that reaction, an energy threshold exists for the incident particle E_t as a function of the angle θ, below which nuclei b are not observed at that angle. At these energies the discriminant of (10.15) vanishes, and this condition yields

$$E_t = \frac{-Q m_B(m_B + m_b)}{m_a m_b \cos^2\theta + (m_B + m_b)(m_B - m_a)}. \tag{10.16}$$

These thresholds are just due to the nuclear force and in fact, they could be smaller when the Coulomb repulsion is taken into account.

The smallest value of E_t in (10.16),

$$E_t = \frac{-Q(m_B + m_b)}{m_B + m_b - m_a}, \tag{10.17}$$

occurs for $\theta = 0$, and is the absolute threshold of the reaction, that is, the smallest value of the incident energy E_a for which the reaction can occur. If $Q > 0$ (*exoergic*, or *exothermic reactions*) the threshold is negative and the reaction occurs for any incident energy. If $Q < 0$ (*endoergic*, or *endothermic reactions*), an incident energy $E_a = E_t$ begins to produce events for $\theta = 0°$. With the increase of E_a starting from E_t, other angles become accessible.

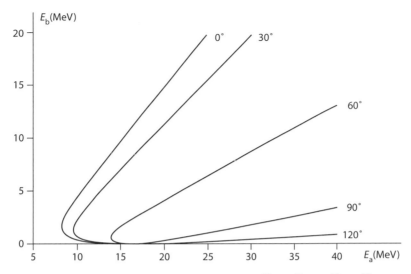

Figure 10.2 Energy E_b of boron nuclei in the reaction $^{12}C + ^{14}N \rightarrow ^{10}B + ^{16}A$, as a function of the energy E_a of the incident ^{12}C nuclei, for several scattering angles.

The second important observation in figure 10.2 is with respect to the unique relation between E_b and E_a. For each energy E_a there exists only one value of E_b for each angle θ, except in the region of energy between 8.26 MeV and 17.82 MeV. The first number corresponds to the energy threshold, (10.17). The second can be determined setting the numerator of (10.15) to zero; this implies

$$E'_a = \frac{-m_B Q}{m_B - m_a}, \qquad (10.18)$$

independent of the value of the angle θ. Thus, the curves for all θ angles cut the horizontal axis at the same point.

The region of double values of E_b only exists for endoergic reactions; when $Q > 0$, the correspondence between E_b and E_a is also unique for all energies and any value of θ. This is a clear consequence of (10.18) and is illustrated in figure 10.3, where the inverse reaction to the one of the figure 10.2 is plotted.

One also sees from (10.16) and (10.18) that the energy region where double values of E_b happen is wider when the participants in the reaction have comparable masses.

It is worthwhile to observe that only values of θ within 0° and 90° admit a region of double values for the energy; for $\theta > 90°$ the first part of (10.15) is negative and it is not possible to have more than one value of E_b.

To end this section we shall try to understand how a reaction is seen for an observer placed in the center of mass system (CMS). An essential point is the value of the available energy in the CMS, because the total momentum in that system is zero before and after the reaction. Thus, the energy threshold of an endoergic reaction is the same as the value of Q, which does not happen if it is computed in the laboratory system (LS) because, in this system, part of the initial energy is used in the conservation of the momentum.

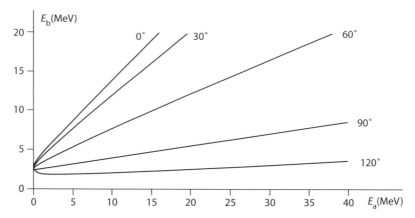

Figure 10.3 Energy of the nucleus ^{12}C emitted in the reaction ^{10}B + ^{16}O → ^{12}C + ^{14}N.

If we assume that b is at rest in the LS, the energy ϵ_i available initially in the CMS is given by the sum of the energy of the two nuclei. With the aid of (2.158) and (2.160), we obtain

$$\epsilon_i = \frac{1}{2}m_a v_a^2 \left(1 - \frac{m_a}{m_a + m_A}\right)^2 + \frac{1}{2}m_A v_a^2 \left(\frac{m_a}{m_a + m_A}\right)^2 = E_a \frac{m_R}{m_a}. \tag{10.19}$$

The final kinetic energy is given, naturally, by

$$\epsilon_f = \epsilon_i + Q, \tag{10.20}$$

because Q measures exactly the energy gained with the rearrangement during the reaction. Using (10.19) in (10.20) gives

$$\epsilon_f = E_a \frac{m_R}{m_a} + Q. \tag{10.21}$$

If E_a is the same as the energy threshold (10.20), the final energy in the CMS is written

$$\epsilon_f = -Q\frac{m_a + m_A - Q/c^2}{m_A - Q/c^2}\frac{m_A}{m_a + m_A} + Q \cong 0, \tag{10.22}$$

where we made use of (10.13). The amount Q/c^2 was neglected in the end result (10.22) since it is small compared with the involved masses. The result shows that, at the threshold, the particles have zero final kinetic energy at the CMS. All the initial kinetic energy was consumed to supply the mass gain, that is, the Q-value of the reaction. The final energy, (10.22), is not strictly zero because (10.12) and the equations derived from it were not written in a fully relativistic form. When this is done [Mi67], (10.17) should be replaced by

$$T_t = \frac{-Q(m_a + m_A + m_b + m_B)}{2m_A}, \tag{10.23}$$

which produces the exact result $\epsilon_f = 0$ when, in the passage from the LS to the CMS, the appropriate relativistic transformations are used.

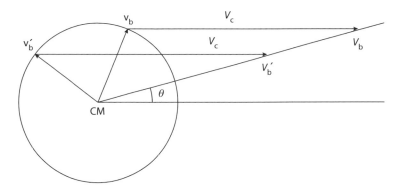

Figure 10.4 Diagram of velocities showing how starting from two different velocities v_b and v_b' in the CMS, one can arrive at the corresponding velocities V_b and V_b' in the LS, causing one to observe particles with different energies at the same angle θ. V_c is the speed of the LS in the CMS.

The analysis of the collision in the CMS is also convenient in seeking to understand physically the possibility of two values of the energy E_b at a given angle for a single incident energy E_a. In the CMS, for a given value of E_a there corresponds a single value of E_b that respects the conservation of zero total momentum and of energy. Particles with energy E_b are detected at all angles and this gives opportunity to the appearance of a double value for the energy in the LS. The diagram of velocities in figure 10.4 illustrates how this happens.

It is also easy to understand from the diagram why the duplicity of energy is limited to angles smaller than 90°. Also, it is not difficult to see that the phenomenon disappears at a certain energy value of the incident particle.

10.4 Scattering and Reaction Cross Sections

When a beam of particles, represented in quantum mechanics by a plane wave, hits a nucleus, we can have, as we saw in previous sections, some processes that are more complex than simple wave scattering, as described in chapter 2 for the nucleon-nucleon case. In order to establish in which way one can calculate the cross section for these processes, we shall reexamine some of the concepts studied earlier.

The incident plane wave was written asymptotically (2.98) as a sum of ingoing and outgoing spherical waves. Expression (2.99) shows that the wavefunction is modified by the presence of a scattering potential $V(r)$, responsible for the appearance of a phase in the outgoing part of the wave. We saw, however, that elastic scattering is just one of the channels for which the reaction can be processed, and we refer to it as the *elastic channel*. The inelastic scattering channel and all the other channels are grouped in the *reaction channel*.

The occurrence of a nuclear reaction proceeding through a given reaction channel leads to a modification of the outgoing part of (2.99), now not only by a phase factor, but also by a factor that changes its magnitude, indicating that there is a loss of particles in the elastic channel. This can be expressed by

$$\Psi \sim \frac{1}{2i} \sum_{l=0}^{\infty} (2l+1) i^l P_l(\cos\theta) \frac{\eta_l e^{i(kr-l\pi/2)} - e^{-i(kr-l\pi/2)}}{kr}, \tag{10.24}$$

where the complex coefficient η_l is the factor mentioned above. To calculate the elastic cross section we should place Ψ in the form of (2.82); using (2.98) and (10.24) we get

$$f(\theta) = \frac{1}{2k} \sum_{l=0}^{\infty} (2l+1) i (1-\eta_l) P_l(\cos\theta), \tag{10.25}$$

which results in the differential scattering cross section

$$\frac{d\sigma_e}{d\Omega} = |f(\theta)|^2 = \frac{1}{4k^2} \left| \sum_{l=0}^{\infty} (2l+1)(1-\eta_l) P_l(\cos\theta) \right|^2 . \tag{10.26}$$

The total scattering cross section is calculated using the orthogonality of the Legendre polynomials, which results in

$$\sigma_e = \pi \bar{\lambda}^2 \sum_{l=0}^{\infty} (2l+1) |1-\eta_l|^2, \tag{10.27}$$

with $\bar{\lambda} = \lambda/2\pi = 1/k$.

To calculate the reaction cross section it is necessary to compute initially the number of particles that disappear from the elastic channel, measured by the flux of the current of probability vector through a spherical surface of large radius centered at the target, calculated with the total wavefunction of (10.24):

$$j_r = -\frac{\hbar}{2im} \int \left(\Psi^* \frac{\partial \Psi}{\partial r} - \Psi \frac{\partial \Psi^*}{\partial r} \right) r^2 \, d\Omega. \tag{10.28}$$

The negative sign indicates that an absorption corresponds to the incoming flux in the sphere. The cross section will be the ratio between j_r and the current of probability for the incident plane wave, $j_i = \hbar k/m$. From this, one finds

$$\sigma_r = \pi \bar{\lambda}^2 \sum_{l=0}^{\infty} (2l+1)(1-|\eta_l|^2). \tag{10.29}$$

From (10.25), (10.27), and (10.29) we can easily see that the total cross section, $\sigma_t = \sigma_e + \sigma_r$, is linked to the scattering amplitude at zero scattering angle through the relationship (2.106), showing that the optical theorem is still valid with the presence of reaction channels different from the elastic channel.

Let us examine (10.27) and (10.29). When $|\eta_l| = 1$ the reaction cross section is zero and we have pure scattering. The contrary, however, cannot happen, as the vanishing of σ_e also implies the vanishing of σ_r. In general there is a region of allowed values of η_l for which the two cross sections can coexist. Such a region is located below the curve shown in figure 10.5.

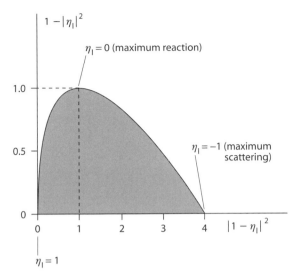

Figure 10.5 Area (below the curve) of coexistence of the scattering and reaction cross sections.

The maximum of σ_r happens for $\eta_l = 0$, corresponding to total absorption. Let us suppose that the absorption potential is limited to the surface of a nucleus with radius $R \gg \bar{\lambda}$, that is, all the particles with impact parameter smaller than the radius R are absorbed. This is equivalent to saying that all particles are absorbed for $l \leq R/\bar{\lambda}$. In this case

$$\sigma_r = \pi \bar{\lambda}^2 \sum_{l=0}^{R/\bar{\lambda}} (2l + 1) = \pi (R + \bar{\lambda})^2. \tag{10.30}$$

This is the value that from an intuitive approach would be adequate for the total cross section, that is, equal to the geometric cross section (the part $\bar{\lambda}$ can be understood as an uncertainty in the position of the incident particle). But we saw above that the presence of scattering is always obligatory. As $\eta_l = 0$, the scattering cross section is identical to the reaction one, producing a total cross section

$$\sigma = \sigma_r + \sigma_e = \pi (R + \bar{\lambda})^2 + \pi (R + \bar{\lambda})^2 = 2\pi (R + \bar{\lambda})^2, \tag{10.31}$$

twice the geometric cross section!

The presence of the scattering part, which apparently has the effect of making (10.31) somewhat strange, can be interpreted as the effect of diffraction of the plane waves at the nuclear surface (figure 10.6). This effect leads to a "shadow" behind the nucleus, decreasing its apparent diameter so that, at a certain distance, the perturbation caused by the presence of the nucleus disappears and the plane wave is reconstructed. In this situation we can say that the part of the beam that is diffracted has to be the same as the part that is absorbed, justifying the equality of σ_r and σ_e. We can also calculate the extension of the shadow formed behind the nucleus using the uncertainty principle: the limitation of the absorption to a circle with radius R creates an uncertainty in the transverse momentum $\Delta p = \hbar/R$,

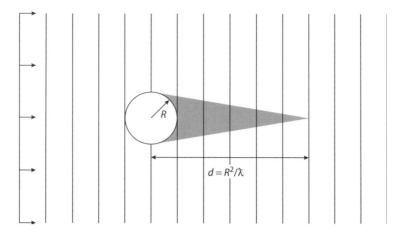

Figure 10.6 Shaded area produced by the total absorption of particles from the incident beam by a nucleus with radius R.

and the distance d shown in figure 10.6 is the result of the ratio $R/d = \Delta p/p$, using $p/\hbar = k = 1/\bar{\lambda}$. This is the same result obtained in the study of Fraunhofer diffraction in the absorption of a light beam by a dark object. The only difference is that, in the latter case the distances d are large for objects of common use in optics, while in the nuclear case d can be of the order of some nuclear diameters if the energy of the incident particles is sufficiently small. The diffraction phenomenon appears clearly in the elastic scattering or inelastic angular distribution (differential cross section as function of the scattering angle): figure 10.7 exhibits angular distributions for the elastic scattering of 30 MeV protons on ^{40}Ca, ^{120}Sn, and ^{208}Pb. The oscillations in the cross sections are characteristic of a Fraunhofer diffraction figure, similar to light scattering by an opaque disk. The angular distances $\Delta\theta$ between the diffraction minima obey in a reasonable way the expression $\Delta\theta = \hbar/pR$ that is a consequence of the considerations above.

We shall express the cross sections now starting from specific conditions of the nuclear surface. Equations (10.27) and (10.29) show that the scattering and reaction cross sections are completely defined with knowledge of the coefficients η_l. Our aim is to relate η_l with the internal properties of the nucleus. This study will be done for neutrons with $l = 0$, and, at least for now, we are going to ignore the spins of the neutron and the target. In this case (2.117)

$$\frac{d^2 u_0}{dr^2} + k^2 u_0 = 0 \quad (r \geq R) \tag{10.32}$$

is valid for the radial wavefunction u_0 at distances r larger than the *channel radius* $R = R_a + R_A$, with R_a and R_A being the radii of the projectile and the target, respectively. The solution of (10.32) is, using (2.94) and (10.24),

$$u_0 = \eta_0 e^{ikr} - e^{-ikr} \quad (r \geq R). \tag{10.33}$$

A radial wavefunction inside the nucleus should connect the external function (10.33) with a continuous function and its derivative at $r = R$. For the application of this principle it

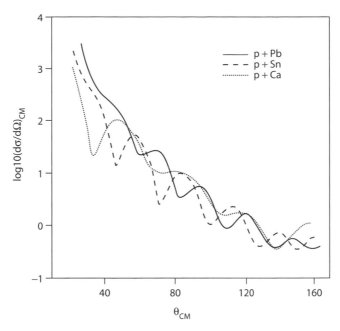

Figure 10.7 Elastic scattering angular distribution of protons of 30 MeV on Ca, Sn, and Pb. The curves are adjusted to the experimental data of Ridley and Turner [RT64].

is essential that we work with the hypothesis that the nucleus has a well-defined surface. The function

$$f_l \equiv R \left[\frac{du_l/dr}{u_l} \right]_{r=R} \tag{10.34}$$

must have identical values if calculated with the internal or the external function, and this condition creates a relationship between f_l and η_l. Thus, knowledge of f_l leads to knowledge of the cross sections. For our example of neutrons with $l = 0$, the application of (10.34) results in

$$f_0 = ikR \frac{\eta_0 + e^{-2ikR}}{\eta_0 - e^{-2ikR}}, \tag{10.35}$$

from which we extract

$$\eta_0 = \frac{f_0 + ikR}{f_0 - ikR} e^{-2ikR}. \tag{10.36}$$

If f_0 is a real number, then $|\eta_0|^2 = 1$. The reaction cross section (10.29) will be zero and we have pure scattering.

Using (10.36), the scattering cross section (10.27) can be written as

$$\sigma_{e,0} = \pi \bar{\lambda}^2 |A_{\text{res}} + A_{\text{pot}}|^2, \tag{10.37}$$

with

$$A_{\text{res}} = \frac{-2ikR}{f_0 - ikR} \tag{10.38}$$

and

$$A_{\text{pot}} = \exp(2ikR) - 1. \tag{10.39}$$

Separation of the cross section into two parts has physical justification: A_{pot} does not contain f_0 and thus does not depend on the conditions inside the nucleus. It represents the situation where the projectile does not penetrate the nucleus, being scattered by its external potential. This is clearly seen in the idealized situation where the nucleus is considered to be an impenetrable hard sphere. In this case the wavefunction is zero inside the nucleus and u_0 vanishes at $r = R$, implying $f_0 \to \infty$ and $A_{\text{res}} \to 0$. In this way, A_{pot} is the only contribution responsible for the scattering.

Applying now (10.36) to the reaction cross section (10.29) and using

$$f_0 = f_R + if_I, \tag{10.40}$$

we have

$$\sigma_{r,0} = \pi \bar{\lambda}^2 \frac{-4kRf_I}{f_R^2 + (f_I - kR)^2}, \tag{10.41}$$

an equation that will be useful when we study the presence of resonances in the excitation functions (cross section as a function of the energy).

10.5 Resonances

To understand why there are resonances, we shall use again the simple model of a single particle subject to a square-well potential. We know that inside the well the Schrödinger equation only admits solution for a discrete group of values of energy, $E_1, E_2, \ldots E_n$. A qualitative way of understanding why this happens is to have in mind that a particle is confined to the interior of the well by the reflections that it suffers at the surface of the well. In these reflections the wave that represents the particle should be in phase before and after the reflection, and this only happens for a finite group of energies. Outside the well the Schrödinger equation does not impose restrictions and the energy can have any value. But we know, from the study of the passage of a beam of particles through a potential step, that the discontinuity of the potential at the step provokes reflection even when the total energy of the particles is larger than the step, a situation where classically there would not be any difficulty for the passage of the particles. This reflection is partial and becomes larger the closer the total energy is to the potential energy. We can say that a particle with energy which is only slightly positive is almost as confined as a particle inside the well. From this fact results the existence of almost bound states of positive energy called *quasi-stationary states* or *resonances*. These resonances appear as prominent peaks

$r = 0$ $r = R$

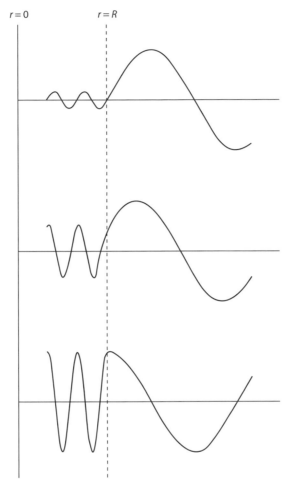

Figure 10.8 Connection of the internal and external wavefunctions of the nucleus, showing that a large amplitude of the internal function is obtained at a resonance.

in the excitation function, a peak at a given energy meaning that energy coincides with a given resonance of the nucleus.

The existence of resonances can also be inferred from the properties of the wavefunction. We shall examine the simplified case where there is only elastic scattering, with the other channels closed. Let us observe figure 10.8. The internal and external wavefunctions are both sine functions, the first with wavenumber $k = \sqrt{2mE}/\hbar$ and the second with $K = \sqrt{2m(E - V_0)}/\hbar$. If E is small and V_0 is about -35 MeV, we have $K \gg k$. The internal and external parts should join at $r = R$ with continuous function and derivatives. As the internal frequency is much larger than the external one, the internal amplitude is quite reduced. Only at the proximity of the situation in which the derivative is zero, there is a perfect matching between both and the internal amplitude is identical to the external one. The energy for which this happens is exactly the energy of resonance.

Resonances appear in the excitation function at relatively low energies, where the number of open channels is not very large and the hypothesis that there is no return to the entrance channel is not valid. To arrive at an expression of the cross section that describes a resonance, we shall rewrite the wavefuntion in the inner region as

$$u_l \sim \exp{(-iKr)} + b \, \exp{(iKr)} \qquad (r < R), \tag{10.42}$$

this time containing a second part, which takes into consideration the part of the wave that returns. This second part allows the existence of resonant scattering, where the incident particle is re-emitted with the same energy as when it entered, after forming the compound nucleus. The complex amplitude b always has a module smaller than 1, since more particles cannot leave from the area $r < R$ than enter.

The use of (10.42) in (10.34) produces, naturally, the same results as the external function (10.33). We shall analyze, in particular, the reaction cross section (10.41). The second parenthesis in the denominator is never zero, since the numerator forces f_l to be always negative. If for a certain energy f_R vanishes, $\sigma_{r,0}$ passes by a maximum in that energy. We can tentatively identify these energies as the energy of the resonances. Let us take the extreme case of a single resonance at the energy E_R, that is, $f_R = 0$ for $E = E_R$. We can expand f_R in a Taylor series in the neighborhood of a resonance,

$$f_R(E) = (E - E_R) \left(\frac{df_R}{dE} \right)_{E=E_R} + \cdots . \tag{10.43}$$

Keeping just the first term of the expansion and applying (10.37) and (10.41), we get

$$\sigma_{e,0} = \pi \bar{\lambda}^2 \left| \exp{(2ikR)} - 1 + \frac{i\Gamma_\alpha}{(E - E_R) + i\Gamma/2} \right|^2, \tag{10.44}$$

$$\sigma_{r,0} = \pi \bar{\lambda}^2 \frac{\Gamma_\alpha(\Gamma - \Gamma_\alpha)}{(E - E_R)^2 + (\Gamma/2)^2}, \tag{10.45}$$

where we define

$$\Gamma_\alpha = -\frac{2kR}{(df_R/dE)_{E=E_R}} \quad \text{and} \quad \Gamma = \frac{2kR - 2f_l}{(df_R/dE)_{E=E_R}}. \tag{10.46}$$

The cross section for formation of a compound nucleus with an energy E close to a resonance should be proportional to the probability of the resonant state to exist at that energy. This probability is given by (6.35), and we get

$$\sigma_{R,0} \propto \frac{1}{(E - E_R)^2 + (\Gamma/2)^2},$$

from which it is deduced that the energy Γ, defined in (10.46), which appears in (10.45), is the total width of the resonance, $\Gamma = \Gamma_\alpha + \Gamma_\beta + \cdots$, that is, the sum of the widths for all the possible processes of decay of the nucleus, starting from the resonant state. The interpretation of Γ_α is obtained by a comparison of (10.45) with (10.44). From the latter, the contribution of the resonant part for the scattering cross section contains the factor $\Gamma_\alpha \Gamma_\alpha$ and (10.45) the factor $\Gamma_\alpha(\Gamma - \Gamma_\alpha)$. It is fair to interpret Γ_α as the entrance channel width, with $\Gamma - \Gamma_\alpha$ the sum of the widths of all the exit channels except α. If we restrict the exit channels to a single channel β or, to put it in another way, we designate β as the group of exit channels except α, (10.45) is rewritten as

$$\sigma_{\alpha,\beta} = \pi \bar{\lambda}^2 \frac{\Gamma_\alpha \Gamma_\beta}{(E - E_R)^2 + (\Gamma/2)^2}, \tag{10.47}$$

which is the usual way of presenting the Breit-Wigner formula that describes the form of the cross section close to a resonance. Let us recall that (10.47) refers to an incident particle of $l = 0$, without charge and without spin. If the spins of the incident and target particles are s_a and s_A, respectively, and the incident beam is described by a single partial wave l_0, the cross section (10.47) should be multiplied by the statistical factor

$$g = \frac{2I + 1}{(2s_a + 1)(2s_A + 1)}, \tag{10.48}$$

where I is the quantum number of the total angular momentum $\mathbf{I} = \mathbf{s_a} + \mathbf{s_A} + \mathbf{l}$ of the compound nucleus. g reduces, naturally, to the value 1 in the case of zero intrinsic and orbital angular momenta.

If the exit channel is the same as the entrance channel α, the cross section should be obtained from (10.44) and its dependence on energy is more complicated because in addition to the resonant scattering there is the potential scattering, and the cross section (10.44) will contain, beyond those two that we already have, an interference term between both. The presence of these three terms results in a peculiar aspect of the scattering cross section, which differs from the simple form (10.47) for the reaction cross section. This is seen in figure 10.9, which shows the form a resonance can take in the scattering cross section.

The region of energy where resonances show up can extend to 10 MeV in light nuclei, but ends well before this in heavy nuclei. Starting from this upper limit the increase in the level density with energy implies that the average distance between the levels is smaller than the width of the levels and individual resonances cannot be observed experimentally. They form a continuum and that region is referred to as the *continuum region*. But, in the continuum region, in spite of there being no characteristic narrow peaks of the individual resonances, the cross section does not vary monotonically; peaks of large width and very wide spacing can be observed. Their presence is mainly due to interference phenomena between the part of the incident beam that passes through the nucleus and the part that passes around it [Mv67].

10.6 Compound Nucleus

The compound nucleus model is a description of atomic nuclei proposed in 1936 by Niels Bohr [Boh36] to explain nuclear reactions as a two-stage process comprising the

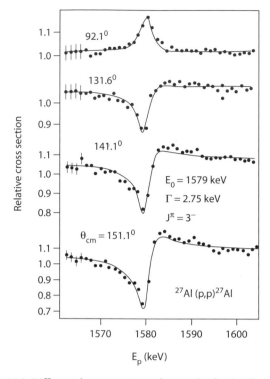

Figure 10.9 Differential cross section at four angles for the elastic scattering of protons off ^{27}Al, in units of the Rutherford cross section, in the neighborhood of the resonance of 1579 keV. For the larger angles we have typical interference figures between the resonant scattering and the potential scattering [Tv72].

formation of a relatively long-lived intermediate nucleus and its subsequent decay. First, a bombarding particle loses all its energy to the target nucleus and becomes an integral part of a new, highly excited, unstable nucleus, called a *compound nucleus*. The formation stage takes a period of time approximately equal to the time interval for the bombarding particle to travel across the diameter of the target nucleus:

$$\Delta t \sim \frac{R}{c} \sim 10^{-21} \text{ s.}$$

Second, after a relatively long period of time (typically 10^{-19}–10^{-15} second) and independent of the properties of the reactants, the compound nucleus disintegrates, usually into an ejected small particle and a product nucleus. For the calculation of properties for the decay of this system by *particle evaporation*, one may thus borrow from the techniques of statistical mechanics. The energy distribution of the evaporated nucleons have sharp resonances, whose width is much smaller than those known from potential scattering. Qualitatively, but physically correctly, this may be understood through the time-energy uncertainty relation: the relaxation to equilibrium and the subsequent "evaporation" of a nucleon take a long time, much longer than the period needed for a nucleon to move

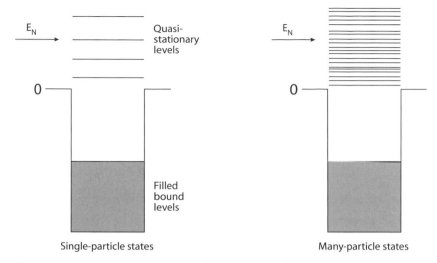

Figure 10.10 Energy levels in a single-particle and compound system.

across the average potential. In other words, the emission of the nucleon with energy equal to that of the incident one is a concentration of the excitation energy on a single particle through a complicated process that needs a long time. Hence the associated energy width will be quite narrow. For very slow neutrons the widths of these resonances are in the range of a few eV!

The basic idea is shown in figure 10.10. First we assume that the incident projectile moves in an attractive real potential so that the quasi-stationary levels are the single-particle levels in the potential with positive energy, and a resonance will occur whenever the incident energy corresponds to the energy of one of these single-particle levels. For a simple square well with depth V_0 and range R, adapted to reproduce the main characteristics of the bound nuclear states, the spacing of these quasi-stationary levels is about 10 MeV. It is evident that this model contains none of the features associated with the fine resonances appearing in the compound nucleus reactions. If the fine resonances are also to be associated with quasi-stationary levels, the levels in question must be closely spaced levels of a many-particle system (right-hand side of figure 10.10). That is, we must abandon the single-particle picture and assume that many nucleons participate in the formation of the compound system.

In a real example, the compound nucleus ^{28}Si is formed by bombarding ^{27}Al with protons. This compound nucleus is excited in a high energy state, and may decay into (a) ^{24}Mg and ^{4}He (an alpha particle), (b) ^{27}Si and a proton, (c) a more stable form of ^{28}Si and a γ-ray, or (d) ^{24}Na plus three protons and one neutron.

The compound-nucleus model is very successful in explaining nuclear reactions induced by relatively low-energy bombarding particles (that is, projectiles with energies below about 50 MeV). In the next several sections we show how microscopic, formal, and empirical concepts are used to explain the rich physics behind the compound nucleus formation and decay.

10.7 Mean Free Path of a Nucleon in Nuclei

Because of the analogies to optical phenomena, it may be no surprise that the potentials used in nuclear scattering are called "optical potentials" and that one speaks of "optical models."

Let us for the moment not worry about how and why such potentials might be justified from a microscopic point of view. Instead, we just take as given a non-Hermitian Hamiltonian. For simplicity let us take a one-dimensional case and let us further assume that V (real part) and W (imaginary part) are constants. Then a solution of the time-dependent Schrödinger equation can be written as

$$\psi(x,t) = \exp\left[(-iV - W)\,t/\hbar\right] \exp\left[-i\frac{p^2}{2m_N}t\right] \psi(x,0). \tag{10.49}$$

For $W > 0$ the amplitude of this wavefunction decays exponentially, as it should for an absorptive process. The stationary solutions obey the equation

$$\left[-\frac{\hbar^2}{2m_N}\frac{\partial^2}{\partial x^2} + V - iW\right]\phi(x) = E\phi(x) \quad (\text{with } E = E^*). \tag{10.50}$$

This equation is satisfied by the ansatz $\phi \sim \exp(iKx) = \exp[(ik - \kappa)x]$. The two quantities k and κ, the real and imaginary parts of K, can be determined by

$$-\frac{\hbar^2}{2m_N}\left(k^2 - \kappa^2\right) = E - V, \quad \text{and} \quad \frac{\hbar^2}{2m_N}k\kappa = W. \tag{10.51}$$

It is seen that κ gets positive for positive W. This means that a wave traveling in the x-direction experiences an attenuation according to which the density behaves like

$$\left|\phi(x)\right|^2 \sim \exp\left[-2\kappa x\right]. \tag{10.52}$$

One may thus define a mean free path

$$\lambda = \frac{1}{2\kappa} = \frac{\hbar^2}{2m_N}\frac{k}{W}. \tag{10.53}$$

This has an analogy with electromagnetic waves traveling in optical media. Such waves get attenuated as soon as the index of refraction n is complex, which in most cases is due to a complex dielectric constant. One may introduce an absorption coefficient α being proportional to the imaginary part of n (see, e.g., [Jac99]). In analogy to (10.52) the connection from this α to our κ would be $\alpha = 2\kappa$. How big the attenuation actually is depends on the frequency of the electromagnetic wave. Likewise, in our case it depends on the energy (or on the real part k of our wavenumber K). Moreover, as we will see soon, W itself will vary with energy.

For interactions of neutrons and nuclei, we expect that the imaginary part increases with energy in the interior of the nucleus. This feature is not difficult to understand on account of our intuitive reasoning regarding the physical nature of absorption. The larger the energy of the incoming nucleon, the easier it will be to excite the nucleus as a whole. The magnitude of this potential, on the other hand, is much more difficult to understand.

Let us look at the mean free part, which for the interior of the nucleus may be estimated on the basis of our considerations above. Using a typical value of $W = 10$ MeV in (10.53) one finds the nucleonic mean free path to be of the order of the size of a heavier nucleus. For the example of a neutron of 40 MeV, one gets $\lambda \simeq 5$ fm $\cong R$. This is an astonishingly large number on account of the large value of the angle integrated cross section σ for the NN scattering. For a relative kinetic energy of the order of 40 MeV one would get a value of about $\sigma \simeq 17$ fm^2. Estimating the mean free path, as one would do if the nucleons behaved like a system of classical particles, one would find

$$\lambda_{\text{class}} \simeq \frac{1}{\rho\sigma} \approx 0.4 \text{ fm} \ll \lambda. \qquad (10.54)$$

In such a case the nucleons would behave like the molecules or atoms of a classical gas or liquid, or like an ensemble of billiard balls. Equation (10.54) assumes the subsequent collisions of the nucleons to be independent of each other, which actually is not the case, as we shall argue below. Within such a picture, (10.54) is understood immediately if one realizes that λ_{class} defines that volume in which on average a moving nucleon meets one scattering partner, for which reason one must have $\rho\sigma\lambda_{\text{class}} \simeq 1$. To get the number shown in (10.54) the density has been taken to be 0.17 nucleons per fm^3.

There are various lessons one may learn from this discussion. First of all, it becomes apparent that the nucleons inside a nucleus do not at all behave like the constituents of a free classical gas. This is not such a big surprise on behalf of our expectation that we are dealing with a quantum many-body system. Therefore the nucleons inside the nucleus must be described by a many-body wavefunction $\Psi(\mathbf{r}_1, \ldots, \mathbf{r}_A)$. The first important feature to be considered then is the *Pauli principle*. This cannot be all one needs to account for the behavior; otherwise, one would be able to strictly justify the picture of independent particles. But for the moment let us just look at the exclusion principle. There are two phenomena which diminish its importance and which are relevant in the present context. First of all, its effects decrease with decreasing density. This fact offers an explanation why, for the lower energies, the $W(r)$ gets much larger in the nuclear surface. There the density is smaller than in the interior, implying that the scattering of the nucleons with each other resembles more the one in free space. Second, the impact of the Pauli principle becomes weaker when the nucleus as a whole is heated up, which may be known from quantum statistical mechanics.

10.8 Empirical Optical Potential

In the previous section we developed a formal theory of the optical potential. In practice, however, the complications involved in inclusing all relevant reaction channels are avoided by the use of empirical optical potentials. In this model the interaction between the nuclei

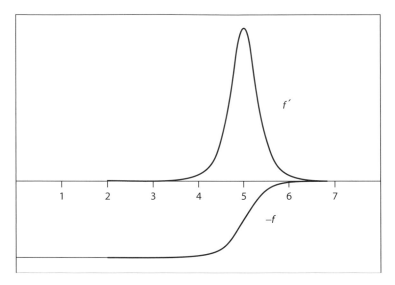

Figure 10.11 The Woods-Saxon form factor f (10.57) and of its derivative f'.

in a reaction is described by a potential $U(r)$, r being the distance between the center of mass of the two nuclei. This idea is similar to the one of the shell model. It replaces the complicated interaction a nucleon has with the rest of the nucleus with a potential that acts on the nucleon. The potential $U(r)$ includes a complex part that, as we shall see, takes into account the absorption effects, that is, the inelastic scattering. The nuclear scattering is treated in similar form as the scattering of light by a translucent glass sphere, and the name of the model derives from this analogy. In the case of the light, the absorption is included by using a complex refraction index.

In its most commonly used form, the optical potential is written as the sum

$$U(r) = U_R(r) + U_I(r) + U_D(r) + U_S(r) + U_C(r), \tag{10.55}$$

which contains parameters that can vary with the energy and the masses of the nuclei and that should be chosen by an adjustment to the experimental data. Obviously, the optical potential $U(r)$ will only make sense if these variations are small for close masses or neighboring energies.

The first part of (10.55),

$$U_R(r) = -Vf(r, R, a), \tag{10.56}$$

is real and represents a nuclear well with depth V, multiplied by a Woods-Saxon form factor

$$f(r, R, a) = \{1 + \exp[(r - R)/a]\}^{-1}, \tag{10.57}$$

where R is the radius of the nucleus and a measures the diffuseness of the potential, that is, the width of the region where the function f is sensibly different from 0 or 1 (see figure 10.11). This produces a well with round borders, closer to reality than a square well. In (10.56) and (10.57), V, R, and a are treated as adjustable parameters.

The absorption effect or, in other words, the disappearance of particles from the elastic channel, is taken into account including the two following imaginary parts:

$$U_I(r) = -iWf(r, R_I, a_I) \tag{10.58}$$

and

$$U_D(r) = 4ia_I W_D \frac{d}{dr} f(r, R_I, a_I). \tag{10.59}$$

An imaginary part produces absorption. It is easy to see this for the scattering problem of the square well: if an imaginary part is added to the well,

$$U(r) = V_0 - iW_0 \quad (r < R)$$
$$= 0 \qquad\qquad (r > R), \tag{10.60}$$

it appears in the value of $K = [2m(E + V_0 + iW_0)]^{1/2}/\hbar$. This will produce a term of decreasing exponential type in the internal wavefunction, as in (10.52). Thus, it corresponds to an absorption of particles from the incident beam.

Expression (10.58) is responsible for the absorption in the whole volume of the nucleus, but (10.59), built from the derivative of the function f, acts specifically in the region close to the nuclear surface, where the form factor f suffers its largest variation (see figure 10.11). These two parts have complementary goals. At low energies there are no available unoccupied states for nucleons inside the nucleus and the interactions are essentially at the surface. In this situation $U_D(r)$ is important and $U_I(r)$ can be ignored. On the other hand, at high energies the incident particle has larger penetration and in this case the function $U_I(r)$ is important.

As with the shell model potential, a *spin-orbit interaction* term is added to the optical potential (see appendix A for discussion of bound states in a real potential). This term, which is the fourth part of (10.55), is usually written in the form

$$U_S(r) = \mathbf{s} \cdot \mathbf{l} \left(\frac{\hbar}{m_\pi c^2}\right)^2 V_s \frac{1}{r} \frac{d}{dr} f(r, R_S, a_S), \tag{10.61}$$

incorporating a normalization factor that contains the mass of the pion m_π. $\mathbf{s}$ is the spin operator and $\mathbf{l}$ the angular orbital momentum operator. As with $U_D(r)$, the part $U_S(r)$ is also only important at the surface of the nucleus since it contains the derivative of the form factor (10.57). The values of V_S, R_S, and a_S must be adjusted by the experiments. The spin orbit interaction leads to asymmetric scattering due to the different signs of the product $\mathbf{s} \cdot \mathbf{l}$ as the projectile passes by one or the other side of the nucleus (see figure 3.1).

The presence of the term (10.61) is necessary to describe the effect of *polarization*. Through experiments of double scattering it can be verified that proton or neutron beams suffer strong polarization at certain angles. This means that the quantity

$$P = \frac{N_u - N_d}{N_u + N_d}, \tag{10.62}$$

where N_u and N_d are the number of nucleons in the beam with spin upward and downward, respectively, has a value significantly different from zero at these angles. With the inclusion

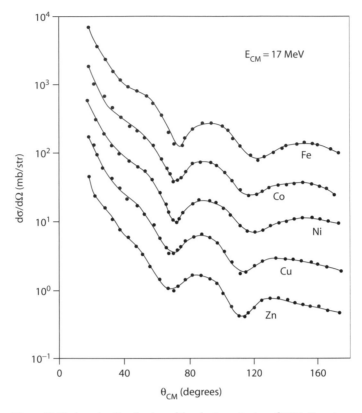

Figure 10.12 Angular distribution of the elastic scattering of 17 MeV protons on nuclei in the region $Z = 26$–30.

of the term (10.61), the optical model is able in many cases to reproduce the experimental values of the polarization (10.62).

Finally, a term corresponding to the Coulomb potential is added to (10.55) whenever the scattering involves charged particles. It has the form

$$U_C(r) = \frac{Z_1 Z_2 e^2}{2R_c} \left(3 - \frac{r^2}{R_c^2} \right) \quad (r \leq R_c)$$

$$= \frac{Z_1 Z_2 e^2}{r} \qquad\qquad (r > R_c), \tag{10.63}$$

where it is assumed that the nucleus is a homogeneously charged sphere of radius equal to the *Coulomb barrier radius* R_c, which defines the region of predominance of each one of the forces—nuclear or Coulomb.

Figures 10.12 and 10.13 exhibit the result of the application of (10.55) to the elastic scattering of 17 MeV protons on several light nuclei. The angular distribution viewed in figure 10.12 is very well reproduced by the model, which also reproduces correctly (figure 10.13) the polarization (10.62) for copper as a function of the scattering angle.

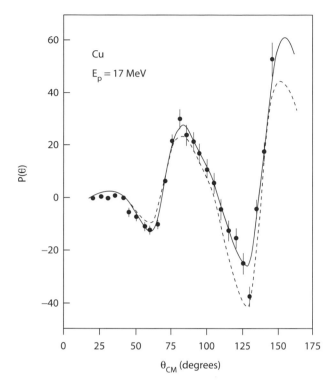

Figure 10.13 The polarization value (10.62) in the scattering of 9.4 MeV protons on copper. The curves are, in both cases, obtained with fits from the optical model [Pe63].

Note that in the case of inclusion of a spin-orbit potential the elastic cross section is calculated as

$$\frac{d\sigma}{d\Omega} = |F(k,\theta)|^2 + |G(k,\theta)|^2,$$
(10.64)

where k and θ are the magnitude of the momentum and the scattering angle, respectively. The so-called spin-no-flip and spin-flip amplitudes F and G can be expanded in partial waves as

$$F(k,\theta) = \sum_{l \geq 0} \left[(l+1) f_{l+}(k) + l f_{l-}(k) \right] P_l(\cos\theta),$$

$$G(k,\theta) = \sum_{l \geq 1} \left[f_{l+}(k) - f_{l-}(k) \right] \sin\theta \, \frac{d}{d\cos\theta} P_l(\cos\theta),$$
(10.65)

where P_l is a Legendre polynomial and the partial wave amplitudes $f_{l\pm}$ are related to the phase shifts by

$$f_{l\pm}(k) = \frac{1}{2ik} \left[\exp\left(2i\delta_{l\pm}\right) - 1 \right].$$
(10.66)

$\delta_{l\pm}$ are the nuclear phase-shifts for $J = \pm\frac{1}{2}$.

The optical model has a limited group of adjustable parameters and is not capable in describing abrupt variations in the cross sections, as it happens for isolated resonances. However, it can provide a good description of the cross sections in the presence of the oscillations of large width in the continuous region, as it treats these as an undulatory phenomenon.

For the scattering of low energy nucleons off heavy nuclei, the surface diffuseness parameters of the optical potentials given above is usually taken as

$$R = r_0 A^{1/3}, \quad \text{with}$$

$$r_0 = 1.25 \text{ fm}, \quad a = 0.65 \text{ fm}, \quad \text{and} \quad a_I = a_S = 0.47 \text{ fm}. \tag{10.67}$$

For nuclei heavier than oxygen ($A \geq 16$), the parameters V, W, U_I, and U_D are practically independent of particle number. However, they depend on the nucleon's energy E. For neutrons with energy $E < 100$ MeV one finds

$$V \simeq -50 \text{ MeV} - 48 \text{ MeV} \left(\frac{N-Z}{A}\right) \hat{t}_z + 0.3 \left(E - U_C(r)\right),$$

$$W(E) \simeq \max \{0.22E - 2 \text{ MeV}, 0\},$$

$$W_D \simeq \max \left\{ 12 \text{ MeV} - 0.25E + 24 \text{ MeV} \left(\frac{N-Z}{A}\right) \hat{t}_z, 0 \right\},$$

$$V_S \simeq 30 \text{ MeV fm}^2. \tag{10.68}$$

A few remarks are in order here these potentials and the values of the parameters. First of all it, must be stressed that we are dealing with a *phenomenological ansatz*. As always such an ansatz is largely governed by the desire for simplicity. Already from this fact it becomes evident that we must not expect the parameters to be known precisely. However, the uncertainties turn out to be not larger than about 10 to 15%, which is quite small, in particular with respect to our present purpose of learning about gross features. Indeed, they allow us to understand many basic physical aspects of nuclear physics.

10.9 Compound Nucleus Formation

It is time to analyze the mechanisms that play a role during a nuclear collision leading to the formation of a compound nucleus. The action of these mechanisms depends on the collision energy in a very pronounced way and results of the study of a given nuclear reaction only have validity for certain range of energy.

We shall assume initially that the incident particle is a neutron of low energy (<50 MeV). When such a neutron enters the field of nuclear forces it can be scattered or begin a series of collisions with the nucleons. The products of these collisions, including the incident particle, will continue along their course, leading to new collisions and new changes of energy. During this process one or more particles can be emitted and they form with the residual nucleus the products of a reaction that is referred to as *pre-equilibrium*. But at low energies the largest probability is the continuation of the process so that the initial energy is distributed to all the nucleons, with no emitted particle. The final nucleus with $A + 1$ nucleons has an excitation energy equal to the kinetic energy of the incident neutron together with the binding energy that the neutron has in the new, highly unstable nucleus. It can, among other processes, emit a neutron with the same or smaller energy as that which was absorbed. The de-excitation processes are not necessarily immediate and the excited nucleus can live a relatively long time. In this situation we say that there is the formation of a *compound nucleus* as intermediary stage of the reaction. In the final stage the compound nucleus can evaporate one or more particles, fission, etc., and so on. In our notation, for the most common process in which two final products are formed (the evaporated particle plus the residual nucleus or two fission fragments, etc.), we write

$$a + A \rightarrow C^* \rightarrow B + b, \tag{10.69}$$

the asterisk indicating that the compound nucleus C is in an excited state.

The compound nucleus lives enough time to "forget" the way it was formed, and the de-excitation to the final products b and B only depends on the energy, angular momentum and parity of the quantum state of the compound nucleus. An interesting experimental verification was accomplished by S. N. Ghoshal in 1950 [Gh50]. He studied two reactions that take a route to the same compound nucleus, ^{64}Zn*, and he measured the cross sections of three different forms of decay, as shown in the scheme

$$
\begin{array}{ccc}
p + {}^{63}\text{Cu} \searrow & \nearrow & {}^{63}\text{Zn} + n \\
& {}^{64}\text{Zn}^* \longrightarrow & {}^{62}\text{Cu} + n + p \\
\alpha + {}^{60}\text{Ni} \nearrow & \searrow & {}^{62}\text{Zn} + 2n
\end{array} \tag{10.70}
$$

The ^{64}Zn can be formed by the two reactions and decay through the three ways indicated in (10.70). If the idea of the compound nucleus is valid and if we choose the energy of the proton and of the incident α-particle to produce the same excitation energy, then the cross section for each one of the three exit channels should be independent of the way the compound nucleus was formed, that is, the properties of the compound nucleus do not have any relationship with the nuclei that formed it. This is confirmed in figure 10.14, where one sees clearly that the cross sections depend practically only on the exit channels. This is called the *independence hypothesis* for the compound nucleus formation.

Another particularity that should happen in reactions in which there is formation of a compound nucleus refers to the angular distribution of the fragments, or evaporated particles: it should be isotropic in the center of mass, and this is verified experimentally. We know, however, that the total angular momentum is conserved and cannot be "forgotten." Reactions with large transfer of angular momentum, as happens when heavy ions are used as projectiles, can show a non-isotropic angular distribution in the center of mass system.

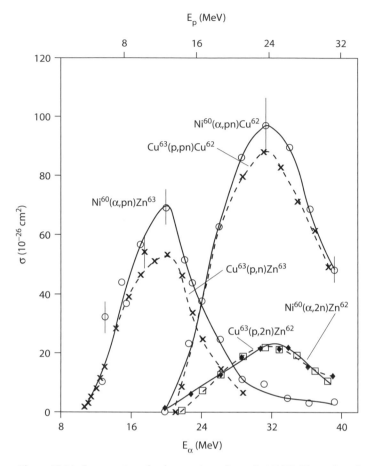

Figure 10.14 Cross sections for the reactions shown in (10.70). The scales of the upper axis (energy of the protons) and lower axis (energy of the α-particle) were adjusted to correspond to the excitation energy of the compound nucleus [Gh50].

The occurrence of a nuclear reaction in two stages allows that the cross section for a reaction $A(a, b)B$ be written as the product,

$$\sigma(a, b) = \sigma_{CN}(a, A)P(b), \tag{10.71}$$

where $\sigma_{CN}(a, A)$ is the cross section of formation of the compound nucleus starting from the projectile a and the target A and $P(b)$ is the probability of the compound nucleus to emit a particle b leaving a residual nucleus B. If not just the particles but the quantum numbers of entrance and exit channels are well specified, that is, if the reaction begins at an entrance channel α and ends at an exit channel β, (10.71) can be written as

$$\sigma(\alpha, \beta) = \sigma_{CN}(\alpha)P(\beta). \tag{10.72}$$

We can associate the probability $P(\beta)$ to the width Γ_β of the channel β and write

$$P(\beta) = \frac{\Gamma(\beta)}{\Gamma}, \tag{10.73}$$

where Γ is the total width, that is, $\tau = \hbar/\Gamma$ is the half-life of disintegration of the compound nucleus. $\Gamma(\beta)$ is the partial width for the decay through channel β, and

$$\Gamma = \sum_\beta \Gamma(\beta).$$

In the competition between the several channels β, the nucleons have clear preference over γ-radiation whenever there is available energy for their emission, and among the nucleons the neutrons have preference as they do not have the Coulomb barrier as an obstacle. Thus, in a reaction where there is no restriction for neutron emission we can say that

$$\Gamma \cong \Gamma_n, \tag{10.74}$$

where Γ_n includes the width for the emission of one or more neutrons.

In some cases it is also useful to define the *reduced width*

$$\gamma_\beta^2 = \frac{\Gamma_\beta}{2P(\beta)},$$

where the penetrability $P(\beta)$ is the imaginary part of the logarithmic derivative of the outgoing wave in the channel β evaluated at the surface $r = R$ and multiplied by the channel radius R. For s-wave neutrons the outgoing wave is just e^{ikr} multiplied by the appropriate S-matrix element, and hence $P = kR$. Thus the reduced width determines the probability that the components specified by β will appear at the surface so that the compound system can decay through this mode.

The study of the function $P(\beta)$ is done in an evaporation model that leads to results in many aspects similar to the evaporation of molecules of a liquid, with the energy of the emitted neutrons approaching the form of a Maxwell distribution

$$I(E) \propto E \exp\left(-\frac{E}{\theta}\right) dE, \tag{10.75}$$

with I measuring the number of neutrons emitted with energy between E and $E + dE$. The quantity θ, with dimension of energy, has the role of a *nuclear temperature*. It is related to the density of levels ω of the daughter nucleus B by

$$\frac{1}{\theta} = \frac{dS}{dE}, \tag{10.76}$$

with

$$S = \ln \omega(E), \tag{10.77}$$

where, to be used in (10.75), dS/dE should be calculated for the daughter nucleus B at the maximum excitation energy that it can have after the emission of a neutron, that is, in the limit of emission of a neutron with zero kinetic energy.

The *level density* $\omega(E)$ is a measure of the number of states of available energy for the decay of the compound nucleus in the interval dE around the energy E. In this sense, the relationship (10.77) is, neglecting the absence of the Boltzmann constant, identical to the thermodynamic relationship between the entropy S and the number of states available for the transformation of a system. The entropy S defined in this way has no dimension, and

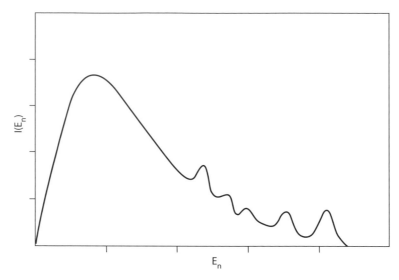

Figure 10.15 Energy spectrum of neutrons evaporated by a compound nucleus.

(10.76) is the well-known relation between the entropy and the temperature. The last one has in the present case the dimension of energy.

The distribution of energy of neutrons emitted by a compound nucleus has the aspect of the curve shown in figure 10.15 (see also figures 10.18 and 10.19). Notice that only the low energy part obeys (10.75). The reason is simple: the emission of a low energy neutron leaves the residual nucleus with a large excitation energy, a situation where the density of levels is very high. That large density of final states available makes the problem tractable with the statistical model that leads to (10.75). The inverse situation is provided by the contributing states of low energy of the residual nucleus. These isolated states appear as prominent peaks in the tail of the distribution. When the emission is of a proton or of another charged particle, the form of figure 10.15 is distorted, the part of low energy of the spectrum being suppressed partially by the Coulomb barrier.

The cross section for the formation of a compound nucleus $\sigma_{CN}(\alpha)$ can be determined in a simple way if some additional hypotheses about the reaction can be done. We shall first suppose that there is only either elastic scattering or formation of compound nucleus. If the projectiles are neutrons, this is true for neutron energies $E_n < 50$ MeV. Then we shall assume that the elastic scattering is purely potential, without resonant elastic scattering, that is, there is no re-emission of neutrons with the same energy as the incident energy. This is equivalent to saying that the probability for the exit channel to be the same as the entrance channel is very low. We shall see that this is not true for certain special values of the energy of the incident particle.

We can still write the wavefunction inside the nucleus as just an incoming wave

$$u_0 \cong \exp(-iKr) \quad (r < R), \tag{10.78}$$

where $K = \sqrt{2m(E - V_0)}/\hbar$ is the wavenumber inside the nucleus, and it is assumed that the neutron with total energy E is subject to a negative potential V_0. The expression

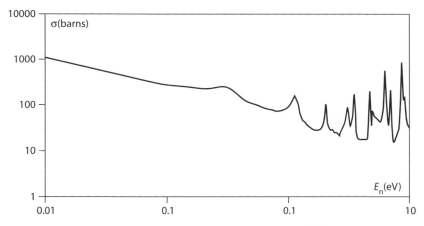

Figure 10.16 Total cross section for neutrons of low energy hitting ^{235}U.

(10.78) is clearly a crude simplification for a situation where the incident neutron interacts in a complicated way with the other nucleons in the nucleus. It allows us, however, to explain the average behavior of the cross sections at low energies. Starting from (10.78), one determines the value of $f_I = R\left[du_0/dr\right]_{r=R}/u_0$:

$$\mathcal{L}^I = -iKR. \tag{10.79}$$

This means that $f_R = 0$ and $f_I = -KR$ in (10.40). Inserting this result in (10.41) we get the expression

$$\sigma_{CN} = \frac{\pi}{k^2}\frac{4kK}{(k+K)^2} \tag{10.80}$$

for the cross section of compound nucleus formation for neutrons with $l = 0$. At low energies, $E \ll |V_0|$, and thus $k \ll K$. Under these conditions, $\sigma_{CN} = 4\pi/kK$. Thus, σ_{CN} varies with $1/k$, that is,

$$\sigma_{CN} \propto \frac{1}{v}, \tag{10.81}$$

where v is the velocity of the incident neutron. This is the well-known $1/v$ *law* that governs the behavior of the capture cross section of low energy neutrons (Wigner's law). Figure 10.16 exhibits the *excitation function* (cross section as function of energy) for the reaction n+^{235}U. The cross section decays with $1/v$ up to 0.3 eV, where a series of resonances start to appear. This abrupt behavior of the cross sections does not belong to the theory of the compound nucleus, and the resonances appear exactly when is not possible to sustain the hypothesis that there is no return to the entrance channel, a hypothesis used in (10.78).

We have already arrived at an expression for the cross section that describes a resonance, (10.44). That formula describes elastic scattering in a resonant situation. In the case of compound nuclei we shall call it the *compound elastic* cross section. A similar expression can be obtained for the reaction, or absorption cross section. Let us briefly review the concepts.

We will study resonance processes involving s-wave neutron scattering and assume that the nucleus has a well-defined surface. The nucleon does not interact with the nucleus at separation distances larger than the *channel radius R*. The logarithmic derivative $\mathcal{L}^I = R\left[du_0/dr\right]_{r=R}/u_0$ for a neutron wavefunction is of the form

$$u_0 = \frac{i}{2k}\left[e^{-ikr} - \eta_0 e^{ikr}\right] \quad (r \geq R), \tag{10.82}$$

so that we have

$$\eta_0 = \frac{\mathcal{L}^I + ikR}{\mathcal{L}^I - ikR}e^{-i2kR}. \tag{10.83}$$

If $\mathcal{L}^I$ is real, $|\eta_0|$ is unity and there is no reaction, but if Im $\mathcal{L}^I < 0$, then $|\eta_0| < 1$.

The scattering amplitudes and the total and reaction widths were obtained earlier. The total width is given by

$$\Gamma = -\frac{2(b + kR)}{a'(E_r)}, \tag{10.84}$$

so that the *reaction* (or absorption) *width* is

$$\Gamma_r = \sum_{\beta \neq \alpha}\Gamma_\beta = \Gamma - \Gamma_\alpha = -\frac{2b}{a'(E_r)}. \tag{10.85}$$

The compound elastic scattering cross section is given in our present notation as

$$\sigma_{ce,\alpha} = \frac{\pi}{k^2}\frac{\Gamma_\alpha^2}{(E - E_r)^2 + \Gamma^2/4}, \tag{10.86}$$

and the absorption cross section is

$$\sigma_{abs} = \frac{\pi}{k^2}\frac{\Gamma_r\Gamma_\alpha}{(E - E_r)^2 + \Gamma^2/4}. \tag{10.87}$$

The cross section for compound nucleus formation is obtained by adding the cross sections for those processes which involve formation of the compound nucleus through channel α, that is,

$$\sigma_{CN} = \sigma_{abs} + \sigma_{ce,\alpha} = \frac{\pi}{k^2}\frac{\Gamma\Gamma_\alpha}{(E - E_r)^2 + \Gamma^2/4}, \tag{10.88}$$

and from (10.72)–(10.73) the cross section for the process $\alpha \longrightarrow \beta$ is

$$\sigma_{abs} = \frac{\pi}{k^2}\frac{\Gamma_\beta\Gamma_\alpha}{(E - E_r)^2 + \Gamma^2/4}, \tag{10.89}$$

as in (10.47).

If the exit channel is the same as the entrance channel α, the cross section should be obtained from (10.44). Its dependence on energy is more complicated because in addition to the resonant scattering there is potential scattering, and the cross section will contain, beyond these two terms, an interference term between both. The presence of these three terms results in a peculiar aspect of the scattering cross section, differing from the simple

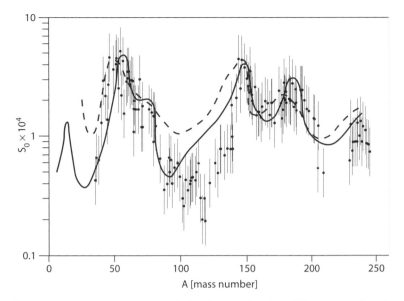

Figure 10.17 s-wave neutron strength function as a function of the mass number A. The solid line is the calculation of [Ch58], while the dashed line is the calculation following [BP62] using deformed optical potentials.

form (10.88) for the compound nucleus cross section. This is seen in figure 10.9, which shows the form that a resonance can take in the scattering cross section.

The region of energy where resonances show up can extend to 10 MeV in light nuclei, but it ends well before this in heavy nuclei. Starting from this upper limit the increase in the density of levels with the energy implies that the average distance between the levels is smaller than the width of the levels and individual resonances cannot be observed experimentally. They form a continuum and that region is referred to as the *continuum region*. But, in the continuum region, in spite of the fact that there is no characteristic narrow peaks of the individual resonances, the cross section does not vary monotonically; peaks of large width and very spaced can be observed, as we show in the example of figure 10.17. Their presence is mainly due to interference phenomena between the part of the incident beam that passes through the nucleus and the part that passes around it [Mv67].

For the decay of the compound nucleus through the channel β we can write (10.80) as

$$\sigma_{CN} = \frac{\pi}{k^2} \frac{4kK}{(k+K)^2} \frac{\Gamma_\beta}{\Gamma}. \tag{10.90}$$

The quantity $4kK/(k+K)^2$ is called the *transmission coefficient* T_0 for s-wave neutrons. If $\langle \Gamma_\alpha \rangle$ is the mean width for resonances due to particles in channel α and D is the mean spacing of levels within an energy interval I, we may write

$$\sigma_{CN}(\alpha) = \frac{1}{I} \int_{E-I/2}^{E+I/2} \frac{\pi}{k^2} \sum_r \frac{\Gamma^r \Gamma_\alpha^r}{(E-E_r)^2 + (\Gamma^r)^2/4} \, dE$$

$$= \frac{\pi}{k^2} \frac{2\pi}{I} \sum_r \Gamma_\alpha^r = \frac{\pi}{k^2} 2\pi \frac{\langle \Gamma_\alpha \rangle}{D}, \tag{10.91}$$

where the energy interval is chosen so that the variation of k^2 can be neglected.

Combining with (10.80) we have

$$\frac{\langle \Gamma_\alpha \rangle}{D} = \frac{1}{2\pi} \frac{4kK}{(k+K)^2} \approx \frac{2k}{\pi K}. \tag{10.92}$$

This quantity is called the s-*wave strength*. To study the A dependence of this quantity, the comparison must be made at a fixed energy, and since the strength function is measured for the different nuclei at different incident energies, one has to transform the values thus measured to those corresponding to a fixed energy conventionally chosen to be $E_0 = 1$ eV. From (10.92) we see that the energy dependence at low energies is

$$\frac{\langle \Gamma_\alpha \rangle}{D} \propto \sqrt{E}.$$

The strength function at the conventional energy $E_0 = 1$ eV is thus related to that measured at an energy E, by the relation

$$\frac{\langle \Gamma_{0,\alpha} \rangle}{D} = \left(\frac{E_0}{E} \right)^{1/2} \frac{\langle \Gamma_\alpha \rangle}{D}.$$

In figure 10.17 we show the neutron s-wave strength function $\langle \Gamma_{0,\alpha} \rangle / D$ as a function of A. The optical model reaction cross section, and thus Γ_0, for a fixed A as a function of E (and also for a fixed E as a function of A) must have a resonant behavior with broad maxima whenever the energy of the neutron in the potential well coincides with that of a single neutron state. This resonant behavior, at an incident energy of $E \approx 1$ eV, is clearly evident in the figure. If one evaluates the energy of the single-particle states in a potential like (10.55) (with the imaginary part set to zero) as a function of A, one finds that the s-states have energy approximately equal to that of the potential well depth, that is, practically equal to that of a neutron energy of 1 eV in the continuum, for $A \approx 13, 58, 160$. Thus the calculation made using a spherical Woods-Saxon potential shows, for $A \geq 20$, two broad maxima centered at $A \approx 58$ and 160. This calculation reproduces only the most prominent features of the experimental data. A more accurate reproduction of the data is given by a calculation with a deformed complex potential for $A \approx 140 - 200$ [BP62].

10.10 Compound Nucleus Decay

At low incident energies the compound nucleus states are excited individually and each produces a resonance in the cross section that may be described by the Breit-Wigner theory (10.88). As the incident energy increases, compound nucleus states of higher energy are excited and these are closer together and of increasing width. Eventually they overlap and it is no longer possible to identify the individual resonances. The cross section then fluctuates.

This fluctuating behavior is due to the interference of the reaction amplitudes corresponding to the excitation of each of the overlapping states, which vanish in the energy average of the cross section since these amplitudes are complex functions with random modulus and phase. The energy average of the cross sections thus shows a weak energy dependence predictable by the theory. To develop such a theory we consider a reaction that proceeds from the initial channel c through the compound nucleus to the final channel c'. If we forget, for the moment, that the compound nucleus may be created in states of different angular momentum J, the hypothesis of the independence of formation and decay of the compound nucleus, according to (10.72 and 10.73) then gives for the cross section

$$\sigma_{cc'} \sim \sigma_{\mathrm{CN}}(c)\frac{\Gamma_{c'}}{\Gamma},\tag{10.93}$$

where $\sigma_{\mathrm{CN}}(c)$ is the cross section for formation of the compound nucleus and $\Gamma_{c'}$ and Γ are, respectively, the energy-averaged width for the decay of the compound nucleus in channel c' and the energy-averaged total width.

We now use the *reciprocity theorem*, derived in the next section, that relates the cross section $\sigma_{cc'}$ to the cross section for the time-reversed process $c' \to c$:

$$g_c k_c^2 \, \sigma_{cc'} = g_{c'} k_{c'}^2 \, \sigma_{c'c},\tag{10.94}$$

where $g_c = 2I_c + 1$ and $g_{c'} = 2I_{c'} + 1$ are the statistical weights of the initial and final channels, I_c and $I_{c'}$ are the spin of the projectile and the ejectile, and k_c and $k_{c'}$ their wavenumbers. This gives

$$g_c k_c^2 \, \sigma_{\mathrm{CN}}(c)\Gamma_{c'} = g_{c'} k_{c'}^2 \, \sigma_{\mathrm{CN}}(c')\Gamma_c\tag{10.95}$$

or, equivalently

$$\frac{\Gamma_c}{g_c k_c^2 \, \sigma_{\mathrm{CN}}(c)} = \frac{\Gamma_{c'}}{g_{c'} k_{c'}^2 \, \sigma_{\mathrm{CN}}(c')}.\tag{10.96}$$

Since the channels c and c' are chosen arbitrarily, this relation holds for all possible channels. So,

$$\Gamma_c \propto g_c k_c^2 \, \sigma_{\mathrm{CN}}(c).\tag{10.97}$$

Since the total width is obtained by summing the Γ_c's over all open channels

$$\Gamma = \sum_c \Gamma_c,\tag{10.98}$$

the cross section (10.93) becomes

$$\sigma_{cc'} = \sigma_{\mathrm{CN}}(c)\frac{g_{c'} k_{c'}^2 \, \sigma_{\mathrm{CN}}(c')}{\sum_c g_c k_c^2 \, \sigma_{\mathrm{CN}}(c)}.\tag{10.99}$$

Ejectiles with energy in the range $E_{c'}$ to $E_{c'} + dE_{c'}$ leave the residual nucleus with energy in the range $U_{c'}$ to $U_{c'} + dU_{c'}$, where

$$U_{c'} = E_{\mathrm{CN}} - B_{c'} - E_{c'},\tag{10.100}$$

and E_{CN} and $B_{c'}$ are respectively the compound nucleus energy and the binding energy of the ejectile in the compound nucleus. Introducing the density of levels of the residual

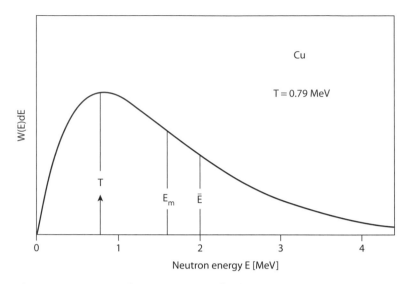

Figure 10.18 Spectrum of neutrons evaporated in the reaction on Cu at 7 MeV incident energy. The residual nucleus temperature T is about 0.79 MeV, $\bar{E} = 2\,T$ is the mean energy of the evaporated neutrons, and $E_m = 1.7\,T$ is the spectrum median energy [Le68].

nucleus $\omega(U_{c'})$, (10.99) becomes

$$\sigma_{cc'}dE_{c'} = \sigma_{CN}(c)\frac{g_{c'}k_{c'}^2\,\sigma_{CN}(c')\omega(U_{c'})dU_{c'}}{\sum_c g_c k_c^2\,\sigma_{CN}(c)\omega(U_c)dU_c}, \tag{10.101}$$

or, since $k^2 = 2\mu E$,

$$\sigma_{cc'}(E_{c'})dE_{c'} = \sigma_{CN}(c)\frac{(2I_{c'} + 1)\mu_{c'}E_{c'}\,\sigma_{CN}(c')\omega(U_{c'})dU_{c'}}{\sum_c \int_0^{E_c^{max}}(2I_c + 1)\mu_c E_c\,\sigma_{CN}(c)\omega(U_c)dU_c}, \tag{10.102}$$

where μ_c is the reduced mass of the ejectile c. This is the *Weisskopf-Ewing formula* for the angle-integrated cross sections.

To a good approximation, the level density $\omega(U) \propto \exp(U/T)$, so the ejectile spectrum given by the Weisskopf-Ewing theory is Maxwellian. It rises rapidly above the threshold energy, attains a maximum and then falls exponentially as shown in figure 10.18 (see also figure 10.15).

Since (10.97) is a proportionality relation it does not allow one to evaluate the absolute values of the decay widths. This may be obtained by use of the *detailed balance principle*, which in addition to the invariance for time reversal leading to the reciprocity theorem (10.94) implies the existence of a long-lived compound nucleus state. This principle states that two systems a and b, with state densities ρ_a and ρ_b, are in statistical equilibrium when the depletion of the states of system a by transitions to b equals their increase by the time-reversed process $b \rightarrow a$.

If $W_{ab} = \Gamma_{ab}/\hbar$ is the decay rate (probability per unit time) for transitions from a to b and $W_{ba} = \Gamma_{ba}/\hbar$ is the decay rate for the inverse process, this equality occurs when

$$\rho_a \Gamma_{ab} = \rho_b \Gamma_{ba}, \qquad (10.103)$$

In the case we are interested in, a is the compound nucleus with energy E_{CN} and, if one neglects the spin dependence, its state density $\rho_{CN}(E_{CN})$ coincides with its density of levels $\omega_{CN}(E_{CN})$ (this is not true for a system with spin J since $(2J+1)$ states correspond to each level). Γ_{ab} is the width $\Gamma_{c'}$ for decay in channel c'. System b is constituted by the ejectile c' with energy from $E_{c'}$ to $E_{c'} + dE_{c'}$ and a residual nucleus with excitation energy from $U_{c'}$ to $U_{c'} + dU_{c'}$. Thus ρ_b is the product of the density of continuum states of c' and the density of levels of the residual nucleus:

$$\rho_b = \rho_{c'}(E_{c'})\omega(U_{c'}). \qquad (10.104)$$

The density of the continuum states of c' is given by the Fermi gas model expression (5.20),

$$\rho_{c'}(E_{c'}) = \frac{\mu_{c'} E_{c'} V}{\pi^2 \hbar^3 v_{c'}} g_{c'} dE_{c'}, \qquad (10.105)$$

where $\mu_{c'}$ and $v_{c'}$ are, respectively, the reduced mass and the velocity of c' and V is the space volume. The decay rate for the inverse process is

$$W_{c'c} = \frac{v_{c'} \sigma_{c'}(E_{c'})}{V}, \qquad (10.106)$$

where $\sigma_{c'}(E_{c'})$ is the cross section for the inverse process (formation of the compound nucleus with energy E_{CN} from channel c') which may be evaluated with the optical potential model.

From relations (10.103)–(10.106) one finally gets

$$\Gamma_{c'} = \frac{1}{\omega_{CN}(E_{CN})} \frac{(2I_{c'}+1)\mu_{c'}E_{c'}}{\pi^2 \hbar^2} \sigma_{c'}(E_{c'})\omega(U_{c'})dE_{c'}. \qquad (10.107)$$

The Weisskopf-Ewing theory provides a simple way of estimating the energy variation, at low incident energies, of the cross sections of all available final channels in a particular reaction; an example of such a calculation is shown in figure 10.19.

The Weisskopf-Ewing theory depends only on the nuclear level density and on the compound nucleus formation cross section, which may be easily obtained from optical model potentials. It is thus simple to use, but it has the disadvantage that it does not explicitly consider the conservation of angular momentum and does not give the angular distribution of the emitted particles. This is provided by the *Hauser-Feshbach theory*. This theory takes into account the formation of the compound nucleus in states of different J and parity π. The compound nucleus states may be of both positive and negative parity. Since parity is conserved, one must take into account that the parity of compound and residual nucleus states may impose restrictions on the values of the emitted particle angular momentum. Thus, positive parity compound nucleus states decay to positive parity states of the residual only by even angular momenta and to negative parity residual nucleus states by odd angular momenta.

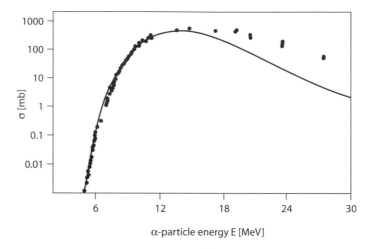

Figure 10.19 Excitation function for the ^{58}Ni(α,p) reaction compared with Weisskopf-Ewing calculations. The disagreement between experimental data and theoretical calculation at the higher energies is due to the emission of *pre-equilibrium* protons.

10.11 Exercises

1. Complete the reactions

$$p + \;\rightarrow\; ^{28}Si + n,$$
$$^{197}Au + {}^{12}C \rightarrow \;+ \gamma,$$
$$^{235}U + n \rightarrow \;^{100}Mo + + 3n.$$

2. What spin and parities can be expected in ^{20}Ne formed from the reaction $\alpha + {}^{16}$O?

3. Obtain the threshold for the production of ^{10}B in the reaction of figure 10.3 for the scattering angles a) $0°$ and b) $60°$. Verify if the results are compatible with the values of the figure.

4. a) Write a relativistic expression that relates the CMS energy with the LS energy for a system of two particles, the second one at rest in the LS. b) Show that if the energy of the incident particle in the LS is the value T_t given by (10.23), the total kinetic energy in the CMS after the collision is zero, showing that T_l is the threshold energy of the reaction.

5. Describe what happens in the diagram of figure 10.4 when the energy of the incident particle is equal: a) to the threshold energy; b) to the energy E'_a [equation 10.18]. Still using the diagram, say why the duplicity of energy does not happen in endoergic reactions.

6. Find the equation of the curve of figure 10.5.

7. Verify how much the curves of figure 10.7 resemble the case of the diffraction of light by an opaque object, calculating the quantities $\Delta\theta = \pi/kR$ that, in a Fraunhofer diffraction, measures, for the first maxima, the angular distance between two sequential maxima.

8. When the reaction $p + {}^{65}Cu \rightarrow {}^{65}Zn + n$ is produced by protons of 15 MeV, a continuous energy distribution is observed for the emerging neutrons, in which a sharp peak at the energy $E_n = 5.57$ MeV is superposed, associated with an excitation energy of 7.3 MeV of ${}^{65}Zn$. What models can be used to justify this result?

9. Show that in the capture of slow neutrons the total cross section at the resonance energy has the value $\sigma = \lambda\sqrt{g\sigma_e/\pi}$, where σ_e is the scattering cross section at the same energy and g the statistical factor of the Breit-Wigner formula.

10. Shows that the loss of flux of a localized beam of particles in an region where a complex potential $U(\mathbf{r})$ exists is given by $\nabla \cdot \mathbf{j}(\mathbf{r}) = \frac{2}{\hbar}|\Psi(\mathbf{r})|^2 \operatorname{Im}[U(\mathbf{r})]$.

11. Build a reaction in which a projectile of $A = 20$ produces a compound nucleus ${}^{258}_{104}Rf^*$. Using the approximate expression for the Coulomb barrier, $B_c = Z_1 Z_2 e^2/1.45(A_1^{1/3} + A_2^{1/3})$, determine the smallest possible excitation energy for the compound nucleus. Repeat the procedure for a projectile with $A = 50$ and compare the results with figure 10.1.

12. Suppose that the meson π^- (spin 0 and negative parity) is captured from the orbit P in a pionic atom, giving rise to the reaction $\pi^- + d \rightarrow 2n$. Show that the two neutrons should be in a singlet state.

13. Verify how much the curves of figure 10.12 resemble the case of the diffraction of light by an opaque object, calculating the quantities $\Delta\theta = \pi/kR$ that, in a Fraunhofer diffraction, measure, for the first maxima, the angular distance between two sequential maxima.

14. In the reactions (10.70), a) which is the excitation energy of ${}^{64}Zn^*$ when it is formed by the collision of protons of 13 MeV with ${}^{63}Cu$? b) What energy should an α-particle have in colliding with ${}^{60}Ni$ to produce the same excitation energy? Compare the result with those of figure 10.14. c) What percent of the kinetic energy of the incident particles contributes to the excitation energy of Zn?

15. When the reaction $p + {}^{65}Cu \rightarrow {}^{65}Zn + n$ is produced by protons of 15 MeV, a continuous energy distribution is observed for the emerging neutrons, in which a sharp peak at the energy $E_n = 5.57$ MeV is superposed, associated to an excitation energy of 7.3 MeV of ${}^{65}Zn$. What models can be used to justify this result?

16. Show that in the capture of slow neutrons the total cross section at the resonance energy has the value $\sigma = \lambda\sqrt{g\sigma_e/\pi}$, where σ_e is the scattering cross section at the same energy and g the statistical factor of the Breit-Wigner formula.

17. ^{109}Ag has a neutron resonance absorption cross section at 5.1 eV with a peak value of 7600 barn and a width of 0.19 eV. Evaluate the expected value of the compound elastic cross section at the resonance.

18. The Breit-Wigner formula for the single-level resonant cross section in nuclear reactions is

$$\sigma_{ab} = \frac{\pi}{k^2} g \frac{\Gamma_a \Gamma_b}{(E - E_0)^2 + \Gamma^2/4},$$

where

$$g = \frac{2J + 1}{(2I_A + 1)(2I_B + 1)},$$

J being the spin of the level, I_A the spin of the incident particle, and I_B the spin of the target.

At neutron energies below 0.5 eV the cross section in ^{235}U ($I_B = 7/2$) is dominated by one resonance ($J = 3$) at a kinetic energy of 0.29 eV with a width of 0.135 eV. There are three channels, which allow the compound state to decay by neutron emission, by photon emission, or by fission. At resonance the contributions to the neutron cross sections are i) elastic scattering (resonant) ($\ll 1$ barn); ii) radiative capture (70 barns); iii) fission (200 barns). a) Calculate the partial widths for the three channels. b) How many fissions per second will there be in a sheet of ^{235}U, of thickness 1 mg/cm^{-2}, traversed normally by a neutron beam of 10^5 per second with a kinetic energy of 0.29 eV? c) How do you expect the neutron partial width to vary with the energy?

19. Evaluate the A dependence of the neutron s-wave strength function for a square-well potential of depth 43 MeV and width 1.3 $A^{1/3}$ fm.

20. From the relations

$$\frac{1}{T} = \frac{dS}{dE} \quad \text{and} \quad \rho(E) = \frac{\exp[S(U)]}{(2\pi T^2 \, dE/dT)^{1/2}},$$

and assuming that $U = aT^n$, show that

$$\rho(E) \propto \frac{E^{-1}}{(1 + 1/n)} \exp\left[\frac{n}{n-1} a^{1/n} E^{1-1/n}\right].$$

(Hint: Assume that E can be represented as a power series in T and $dE/dT \rightarrow 0$, as $T \rightarrow 0$, then if follows that the expansion must start with a term at best quadratic in T.).

21. From the statistical model show that

$$\frac{\Gamma_p}{\Gamma_n} = \frac{\int E_p \, \sigma(E_p) \, w(U_p) \, dE_p}{\int E_n \, \sigma(E_n) \, w(U_n) \, dE_n}.$$

22. Show that in a compound nuclear reaction, the angular distribution of emitted particles with respect to the direction of incidence is symmetric about 90°.

23. There are several methods to associate a temperature with a nucleus in an ideal gas law statistically, such as

$$E = aT^2 \quad \text{and} \quad E = \frac{1}{11}AT^2 - T + \frac{1}{8}A^{2/3}T^{7/3},$$

where A is the mass number of the nucleus [LeC54]. Compare these using $A = 25$.

24. Prove that in the black-nucleus approximation the inverse cross section for emission of a charged particle is given by $\sigma \approx \pi R^2(1 - V_C/E)$, where V_C is the particle Coulomb barrier. Say whether E is the laboratory particle energy or its center of mass energy (the energy of the particle relative to the nucleus). Give an expression for the maximum value of the impact parameter and the compound nucleus angular momentum.

25. Evaluate the average angular momentum carried out by neutrons, protons, and α-particles evaporated by ^{135}Tb at 22 MeV excitation energy. In this and in the following problems, when necessary, assume the nuclear temperatures to be approximately given by $T \approx \sqrt{8E/A}$ MeV and the radii for evaluating the Coulomb barriers by $R \approx 1.5\,(A_R^{1/3} + A_p^{1/3})$, where A_R and A_p are the residual and the emitted particle mass.

26. Evaluate the mean energy of neutrons, protons and α-particles evaporated by ^{55}V, ^{140}Cs and ^{227}Ra at 18 MeV excitation energy.

27. Assuming that emission of neutrons, protons, and α-particles are the dominant decay modes of the compound nuclei of the previous exercise, evaluate the compound nucleus total and partial decay widths.

11 Nuclear Reactions—II

11.1 Direct Reactions

We have already mentioned the existence of reactions that occur within a short duration of the projectile-target interaction. Several of these mechanisms of direct reaction are known. This reaction type becomes more probable as one increases the energy of the incident particle: the wavelength associated with the particle decreases and localized areas of the nucleus can be "probed" by the projectile. In this context, importance in placed on peripheral reactions, where only a few nucleons of the surface participate. These direct reactions happen during a time of the order of 10^{-22}s; reactions in which the formation of a compound nucleus can be up to six orders of magnitude slower. We should notice that one reaction form for a given energy is not necessarily exclusive; the same final products can be obtained partly by events in a direct way and partly through the formation and decay of a compound nucleus.

Besides elastic scattering, which is not of interest in the present study, we can describe two characteristic types of direct reactions. In the first, the incident particle suffers inelastic scattering and the transferred energy is used to excite a collective mode of the nucleus. Rotational and vibrational bands can be studied in this way. The second type involves a modification in the nucleon composition. Examples are the transfer reactions of nucleons, such as *pick-up* and *stripping* reactions. Another important reaction of the second type is the *knock-out* reaction, where the incident particle (generally a nucleon) picks up a particle of the target nucleus and continues on its path, resulting in three reaction products. Reactions with nucleon exchange can also be used to excite collective states. An example is a pick-up reaction where a projectile captures a neutron from a deformed target and the product nucleus is in an excited state belonging to a rotational band.

Some types of direct reaction exhibit peculiar forms of angular distribution that allows us to extract information on the reaction mechanism with the employment of simple models. Typical examples are the stripping reactions (d,n) and (d,p), where the angular

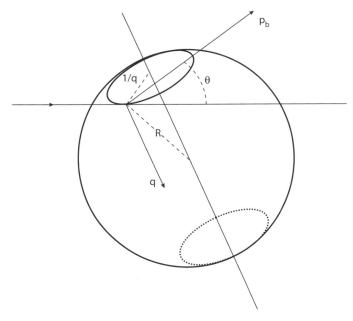

Figure 11.1 Representation of a direct reaction (d,n), where a deuteron with momentum $\mathbf{p}_a$ hits a nucleus, transferring a proton with momentum $\mathbf{q}$ and releasing a neutron with momentum $\mathbf{p}_b$.

distribution of the remaining nucleon presents a forward prominent peak and smaller peaks at larger angles, with the characteristic aspect of a diffraction figure.

With the increase of energy of the incident particle, it would be interesting initially to test the applicability of semiclassical methods in the treatment of a direct reaction. Let us take as an example the reaction (d,n) (figure 11.1): a deuteron of momentum $\mathbf{p}_a$, following a classical straight line trajectory, hits the surface of the nucleus, where it suffers a stripping, losing a proton to the interior of the nucleus and releasing a neutron of momentum $\mathbf{p}_b$, which leaves it forming an angle θ with the incident direction. The proton is incorporated to the nucleus with momentum $\mathbf{q}$, transferring to it an angular momentum $\mathbf{l} = \mathbf{R} \times \mathbf{q}$, that is, approximately $|\mathbf{l}|/\hbar$ units of angular momentum.

Let us concentrate on a certain energy level of the product nucleus produced by a transfer of l units of angular momentum. The energy of the emitted neutron and the vector $\mathbf{q}$ are determined for a given observation angle θ. This reduces the regions of the surface of the nucleus where the reaction takes place to circles of radius l/q, as marked in figure 11.1. The fact that the radii of the circles cannot be larger than the radius of the nucleus imposes a minimum value for θ, a value that increases with l.

In this way we can sketch the prediction of this semiclassical approach for the behavior of the differential cross section. The starting point is to assume that in a direct reaction the particles leave the nucleus with a considerable momentum at forward angles and we should expect a decrease of the cross section at larger angles. With the mentioned restriction for the angular distribution, the valid region for observation is located above

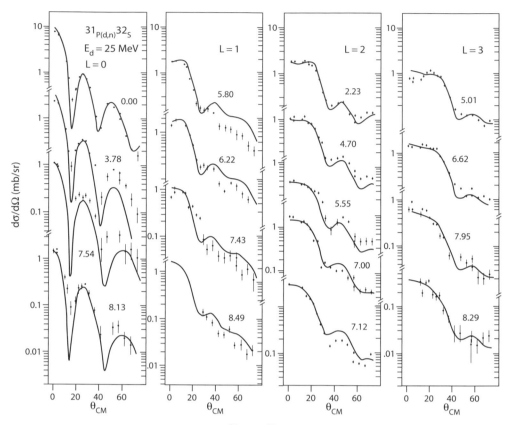

Figure 11.2 Angular distribution of the reaction ^{31}P(d,n)^{32}S, with the transfer of a proton to several states of ^{32}S. The curves are results of DWBA calculations for the indicated l values [Mi87].

the limit angle θ_l. The value of θ_l results from the simple application of conservation of linear momentum:

$$q^2 = p_a^2 + p_b^2 - 2p_a p_b \cos\theta, \tag{11.1}$$

where p_a and p_b are obtained from the initial and final energy of the particles and $q = l/R$. For a typical situation where an incident particle of dozens of MeV excites the first low energy states of the product nucleus, we see from (11.1) that the dependence with θ is essentially proportional to the magnitude of the transferred momentum, q, and in the limit $q = l/R$, the corresponding angle θ_l grows with l. Finally, we should expect that the angular distribution is a function with maxima and minima, resulting from the interference between events that happen in the upper and lower circles represented in figure 11.1.

Figure 11.2 shows experimental results for the reaction ^{31}P(d,n)^{32}S [Mi87]: angular distributions of the detected neutrons corresponding to each energy level of ^{32}S are exhibited for every corresponding angular momentum. We see that the behavior of the cross sections is in agreement with the qualitative predictions: the curves exhibit a first and more important peak at a value of θ that grows with l; starting from this other smaller peaks occur as θ increases. The increase of θ_l with l is an important characteristic that can be used, as we shall see ahead, to identify the value of the transferred momentum in a given angular distribution.

11.1.1 *Theory of direct reactions*

In the next section we will give a detailed account of the theory of direct reactions. But, for the moment, let us see how we can apply a simple quantum treatment to a direct reaction. For this purpose let us first obtain an expression for the differential cross section using the concept of transition probability that was used to establish golden rule 2, expression (6.49).

For the application of (6.49), the entrance channel will be understood as an initial quantum state constituted of particles of mass m_a hitting a target A of mass m_A. The final quantum state is the exit channel, where particles of mass m_b move away from the nucleus B of mass m_B. The essential hypothesis for the application of the golden rule is that the interaction responsible for the direct reaction can be understood as a perturbation among a group of interactions that describe the system as a whole.

If we designate as λ the transition rate of the initial quantum state to the final quantum state, we can write a relationship between λ and the total cross section:

$$\lambda = v_a \sigma = \sigma \frac{k_a \hbar}{m_a}, \tag{11.2}$$

where v_a is now the velocity of the particles of the incident beam. If $d\lambda$ is the part relative to the emission in the solid angle $d\Omega$, then

$$d\sigma = \frac{m_a}{k_a \hbar} d\lambda \tag{11.3}$$

and, using (5.20),

$$d\sigma = \frac{m_a}{k_a \hbar} \frac{2\pi}{h} |V_{fi}|^2 d\rho(E_f), \tag{11.4}$$

where the indices i and f refer to the initial and final stages of the reaction.

A calculation similar to the one in chapter 5 allows us to use the expression (5.20) to obtain the value of ρ:

$$\rho = \frac{dN}{dE} = \frac{pm_b s^3}{2\pi^2 \hbar^3}. \tag{11.5}$$

If we assume an isotropic distribution of momentum for the final states, the fraction $d\rho$ that corresponds to the solid angle $d\Omega$ is

$$d\rho = \frac{pm_b s^3}{2\pi^2 \hbar^3} \frac{d\Omega}{4\pi} = \frac{k_b m_b s^3}{8\pi^3 \hbar^2} d\Omega . \tag{11.6}$$

Using (11.6) in (11.4), we obtain an expression for the differential cross section:

$$\frac{d\sigma}{d\Omega} = \frac{m_a m_b k_b}{(2\pi \hbar^2)^2 k_a} |V_{fi}|^2, \tag{11.7}$$

involving the matrix element

$$V_{fi} = \int \Psi_b^* \Psi_B^* \chi_\beta^{(-)*}(\mathbf{r}_\beta) V \Psi_A \Psi_a \chi_\alpha^{(+)}(\mathbf{r}_\alpha) d\tau. \tag{11.8}$$

Ψ_a, Ψ_b, Ψ_A, Ψ_B are the internal wavefunctions of the nuclei a, b, A, and B. $\chi_\alpha^{(+)}$, $\chi_\beta^{(-)}$ are the wavefunctions of the relative momentum in the entrance channel α and in the exit channel β. The volume s^3 was not written in (11.7) because it can be chosen to be unity, since the wavefunctions that appear in (11.8) are normalized to 1 particle per unit volume. V is the perturbation potential that causes the "transition" from the entrance to the exit channel. It can be understood as an additional interaction to the average behavior of the potential and, in this sense, can be written as the difference between the total potential in the exit channel and the potential of the optical model in that same channel.

The simplest treatment we can do for a direct reaction is to consider the incident beam as a plane wave whose only interaction with the target is through the perturbation potential that causes the reaction. The emerging beam is also treated as a plane wave. The use of plane waves for $\chi_\alpha^{(+)}$ and $\chi_\beta^{(-)}$ in (11.8) is referred to as the *first Born approximation*. With it we can arrive at an approximate expression for the behavior of the differential cross section. Thus, we shall start with the fact that the nuclear forces are of short range, what allows us to restrict the integral (11.8) to regions where $\mathbf{r}_\alpha \cong \mathbf{r}_\beta = \mathbf{r}$. This leads to

$$V_{fi} \cong \int d\mathbf{r} \exp(i\mathbf{q} \cdot \mathbf{r}) \left\{ \int \Psi_b^* \Psi_B^* V \Psi_a \Psi_A \, d\tau' \right\}, \tag{11.9}$$

where $\mathbf{q} = \mathbf{k}_\alpha - \mathbf{k}_\beta$. The global variables in $d\tau$ have been separated into variables $d\mathbf{r}$ and $d\tau'$.

The choice of the interaction V in the formula above depends on the direct reaction process. The physics involved appears more transparent if we discuss the pick-up process. In particular, consider the reaction $A(p,d)B$; then $d = n + p$ and $A = B + n$, and the neutron is transferred. The proton picks up the neutron through their mutual interaction V_{pn}. The transition amplitude $T_{p,d}$ is given by (11.8) with $a \rightarrow p$, $b \rightarrow d$, and $V = V_{pn}$.

Now, we shall use the expansion (2.71) for the plane wave of (11.9):

$$V_{fi} \cong \sum_{l=0}^{\infty} i^l (2l+1) \int j_l(qr) P_l(\cos\theta) F(\mathbf{r}) \, d\mathbf{r}, \tag{11.10}$$

where $F(\mathbf{r}) = \int \Psi_b^* \Psi_B^* V \Psi_a \Psi_A \, d\tau'$ concentrates all the internal properties and can be considered the *form factor* of the reaction. Let us restrict the action of V to the surface of the nucleus, which is reasonable: outside the nucleus the action of V is limited by the short range of the nuclear forces and inside the nucleus there is a strong deviation to the absorption channel. Expression (11.10) then should be rewritten with $r = R$ and we obtain

$$V_{fi} \cong \sum_{l=0}^{\infty} c_l j_l(qR), \tag{11.11}$$

where the coefficients c_l include the constants that multiply the spherical Bessel function in (11.10), besides the integral over the form factor $F(\mathbf{r})$. The calculation of this factor can be laborious even for simpler cases [Bh52], but what we want to retain is the dependence in j_l. The index l can be identified as the angular momentum transferred and, for a reaction that involves a single value of l, we can write for the differential cross section:

$$\frac{d\sigma}{d\Omega} \propto |j_l(qR)|^2, \tag{11.12}$$

where the dependence in θ is contained here again in q;

$$q^2 = p_\alpha^2 + p_\beta^2 - 2p_\alpha p_\beta \cos\theta. \tag{11.13}$$

We have an oscillatory behavior for the angular distribution, the maximum being separated by π from each other in the axis qR. Incidentally, this same result is obtained with the previously developed semiclassical approach (figure 11.1), in the case of small scattering angles and small radii of the circles. It is, in fact, the result corresponding to the interference of two slits separated by $2R$.

The Born approximation with plane waves predicts for certain cases the correct place of the first peaks in the angular distribution but without correctly reproducing the intensities. Considerable progress can be made in the perturbative calculations if in (11.8), instead of plane waves we use *distorted* waves that contain, besides the plane wave, the part dispersed elastically by the optical potential of the target. The Born approximation with distorted waves, or DWBA (*distorted wave Born approximation*), became a popularly employed tool in the analysis of experimental results of direct reactions. With it one can try to extract with a certain reliability the value of the angular momentum l transferred to the nucleus, or removed from it: stripping or pick-up. With this aim it is enough to compare the experimental angular distributions for a given exit energy with the DWBA calculations and to verify if a given value of l reproduces better the experimental data. An example of this is the already mentioned stripping reaction, $^{31}P(d,n)^{32}S$, for deuterons of 25 MeV. For the energy levels shown in figure 11.2 the attribution of the value of l for the level is, in most cases, unequivocal.

An example of the angular distribution in a direct reaction is shown in figure 11.3. The theoretical curve was obtained with a variant of the theory presented in this section.

11.2 Validation of the Shell Model

The angular momentum l transferred in a direct reaction generally modifies the value of the total angular momentum of the nucleus. If J_i is the spin of the target nucleus, the spin J_f of the product nucleus is limited to the values

$$\left| |J_i - l| - \frac{1}{2} \right| \leq J_f \leq J_i + l + \frac{1}{2}, \tag{11.14}$$

and the initial and final parities obey the relationship

$$\pi_i \pi_f = (-1)^l. \tag{11.15}$$

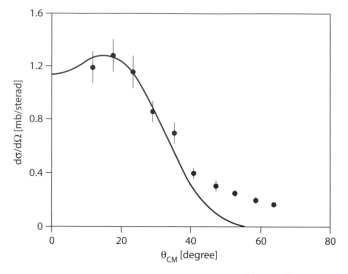

Figure 11.3 Angular distributions of deuterons from $^{16}O(p,d)^{15}O$ ground state. The solid line is a plane wave Butler curve with $l = 1$ and $r_0 = 5.2$ fm. The proton energy was 20 MeV. Data are from [Leg62].

The relationship (11.15) allows, with the knowledge of the target nucleus and of the transferred angular momentum, the determination of the parity of the product state formed, and (11.14) is a guide for the determination of its spin.

Knowledge of the transferred angular momentum value in a direct reaction presents the possibility of testing the predictions of the shell model for the structure of ground states of nuclei. In a direct reaction one assumes that the nucleon is located in an orbit of the nucleus with the same angular momentum as the transferred momentum in the reaction. Table 11.1 exhibits a series of nuclei in the region $28 < A < 43$, where the deformations

Table 11.1 Products formed by stripping (+) or pick-up (−) reactions involving the transfer of a proton (p) or neutron (n). The last column indicates the experimental value of the angular orbital momentum carried by the transferred nucleon.

^{28}Si	+	p	2	^{40}Ca	−	n	3
^{29}P	+	p	0	^{41}Sc	+	p	3
^{31}S	−	n	0	^{42}K	+	n	3
^{35}P	−	p	0	^{42}Ca	+	p	2
^{37}K	+	p	2	^{43}K	−	p	2
^{40}Ar	−	p	2	^{43}Sc	+	p	3

are small and the simple shell model works well. The nuclei are formed starting from pick-up or stripping reactions, and the respective value of l is identified by calculations identical to those leading to the curves of figure 11.2. A comparison with figure 5.9 exhibits that in all cases the value corresponds exactly to the predicted by the shell model. To illustrate, we shall examine the two examples of the first line of table 11.1. In the first, ^{28}Si is formed from a stripping reaction with the transfer of a proton and of two units of angular momentum. If we observe figure 5.9 we shall see that the 14th proton will locate in orbit $1d\frac{5}{2}$, which corresponds to $l = 2$. In the second case, ^{40}Ca is formed by a *pick-up* reaction that picks up the 21st neutron, and figure 5.9 exhibits that it is located in $1f\frac{7}{2}$, justifying the value $l = 3$ found experimentally. The success of a comparison such as this not only supports once again the correctness of the basic assumptions of the shell model but it also gives guarantee to the simple ideas of the model that describes the direct reactions.

If the shell model represented the final theory about the distribution of energy levels in the nucleus, then the single-particle states above Z (or N) should appear as states of energy of the product of a stripping reaction that adds a proton (or neutron) to target of mass A. In fact, the spins of the final states can be obtained from the quantum numbers j and l that identify the single particle states. If, for example, a proton is deposited above the closed shell of $Z = 50$ of an even-even nucleus and we identify the values of l associated to final states as 0, 4, and 5, then we could say that the single-particle states $3s\frac{2}{2}$, $1g\frac{7}{2}$, and $1h\frac{11}{2}$ (see figure 5.9) would appear as $\frac{1}{2}^{+}$, $\frac{7}{2}^{+}$, and $\frac{11}{2}^{+}$ states in the residual nucleus. For $l=2$ an uncertainty would exist between $2d\frac{5}{2}$ and $2d\frac{3}{2}$, which can give rise to states $\frac{5}{2}^{+}$ and $\frac{3}{2}^{+}$. In these cases, the spins of the states should be determined from other processes.

We know, however, that the real situation is more complicated, due to the presence of the residual interactions that give rise to configuration mixing. The consequence is that a single particle state $1g\frac{7}{2}$, for example, will not necessarily generate a single final state $\frac{7}{2}^{+}$. It can mix with other configurations of the same angular momentum and parity (in this process even states of different nature can be involved, as collective states) to generate a series of states $\frac{7}{2}^{+}$ of the final nucleus, each of them spending just part of its time in the configuration $1g\frac{7}{2}$. As result, the cross section for the formation of a state i of the product nucleus is related to the calculated one with DWBA for the formation from a single-particle state by

$$\left(\frac{d\sigma}{d\Omega}\right)_{\text{exp}} = \frac{2J_f + 1}{2J_i + 1} S_{ij} \left(\frac{d\sigma}{d\Omega}\right)_{\text{DWBA}}, \tag{11.16}$$

where the *spectroscopic factor* S_{ij} measures the weight of the configuration j used in the DWBA calculation, in the final state i measured experimentally, with the sum of the contributions limited to

$$\sum_i S_{ij} = n_j. \tag{11.17}$$

The sum (11.17) embraces all the states i of the product nucleus that contains a given configuration j, with total number of nucleons equal to n_j. The statistical weight $(2J_f+1)/(2J_i+1)$

that appears in the DWBA calculation,[1] involving the angular momentum of the target nucleus J_i and final nucleus J_f, is explicitly given in (11.16), because the spectroscopists prefer to work with the product

$$S' = \frac{2J_f + 1}{2J_i + 1}S, \qquad (11.18)$$

denominated *spectroscopic strength*. The advantage of the definition (11.18) is that S' can be determined by (11.16) even when the final angular momentum J_f is not known. Usually, the values of S' enter as parameters for a better adjustment between the calculated curves and the experimental ones.

To illustrate we shall show the results of the reaction $^{36}S(d,p)^{37}S$, where a neutron is located above the closed shell $N = 20$ of $^{36}_{16}S$. Table 11.2 exhibits the energy values of the first levels of ^{37}S formed by the reaction, as well as the value of the transferred angular momentum and of the spectroscopic strength calculated from (11.16). The spectroscopic factor and the spin and parities deduced are also shown.

Let us examine table 11.2 using the information of figure 5.9. In agreement with the figure, the neutron occupies the orbits above the closed shell of 20 neutrons, and values of the transferred angular momentum $l = 1$ and $l = 3$ are expected. Column 2 confirms, with one exception, this expectation. Let us notice that now it is not possible to determine unequivocally the spins of the final states only with the values of l, since two values of j exist for each l. Thus, we can resort to the spectroscopic strength S'. Let us take the ground state, which is $\frac{7}{2}^-$ or $\frac{5}{2}^-$, formed by the transfer of a neutron with $l = 3$ to the orbit $1f\frac{7}{2}$ or $1f\frac{5}{2}$. The value of S' is 7.328 and to this value correspond values 0.916 and 1.221 for S. The last one overruns the sum (11.17) by an amount above the experimental uncertainty, which allows us to affirm that the ground state of ^{37}S is $\frac{7}{2}^-$. It is an almost pure $1f\frac{7}{2}$ state, because S almost exhausts the sum (11.17). In the same way, the state with energy 646 keV can be deduced as being $\frac{3}{2}^-$. These attributions have been confirmed by experiments where the degree of polarization of the emerging proton is measured and that are capable of distinguishing contributions of different values of j for the same l. Similarly one determines the spins of the states with 1992 keV, 2638 keV, and 3262 keV, which could not be obtained only from the value of the spectroscopic factor.

11.3 Photonuclear Reactions

A photonuclear reaction is a reaction resulting from the interaction of the electromagnetic radiation with a nucleus or, more specifically, with the protons of the nucleus. Figure 11.4 shows a typical photonuclear absorption cross section.

The essential difficulty in the study of this reaction type is to obtain a beam of monoenergetic photons with high intensity. Thus, the first experimental works used primarily

---

[1] In a pick-up reaction the statistical weight is equal to 1.

Table 11.2 Energy (E), transferred angular momentum l and spectroscopic strength S' for the first levels of ^{37}S populated by the reaction ^{36}S(d,p) ^{37}S. The corresponding spectroscopic factor S and the spin and parities $J(\pi)$ are shown in the last columns [PS90], [Th84]. Parentheses indicate doubtful values.

E(keV)	l	S'	S	J^π
0	3	7.328	0.916	$\frac{7}{2}^-$
646	1	2.796	0.699	$\frac{3}{2}^-$
1398	2	0.212	0.053	$\frac{3}{2}^+$
1992	1	0.3	0.075	$\frac{3}{2}^-$
2023	(3)	0.08	(0.01; 0.013)	$\left(\frac{7}{2}^-,\frac{5}{2}^-\right)$
2515	(3)	0.272	(0.034; 0.045)	$\left(\frac{7}{2}^-,\frac{5}{2}^-\right)$
2638	1	1.662	0.831	$\frac{1}{2}^-$
3262	1	0.568	0.142	$\frac{3}{2}^-$
3355	1	0.120	(0.03; 0.06)	$\left(\frac{3}{2}^-,\frac{1}{2}^-\right)$
3442	3	0.598	(0.075; 0.1)	$\left(\frac{7}{2}^-,\frac{5}{2}^-\right)$
3493	1	0.340	(0.085; 0.17)	$\left(\frac{3}{2}^-,\frac{1}{2}^-\right)$

bremsstrahlung photons (radiation from decelerated charged particles). This radiation has a continuous distribution of energy, from zero to the energy of the incident particles that produce it. The study of the behavior of a reaction for a given energy E of the photons is obtained by the subtraction of two irradiations, with energy E and $E + \Delta E$.

When the energy of the photons is located a little above the separation energy of a nucleon, the cross section of photoabsorption reveals the presence of characteristic sharp resonances. But, when the incident energy reaches the range of 15–25 MeV, a new behavior appears in the cross section, with the presence of a wide and large peak, called the *giant electric dipole resonance*. The mechanism for this will be discussed in section 11.3.3.

11.3.1 Cross sections

Let us consider the absorption of photons by a nonrelativistic system of particles. A photon of the incident beam is annihilated in the interaction with the system. This is described by the first term of the quantized vector potential (9.33). The observable characteristic of the process is the cross section of photoabsorption which is the ratio of the transition rate to the incident photon flux.

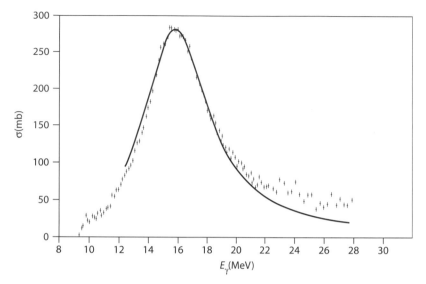

Figure 11.4 Giant resonance in the absorption of photons by ^{120}Sn [Le74].

The process we are interested in can be written as (compare with 9.43)

$$|i; n_{k\lambda}\rangle \rightarrow |f; n_{k\lambda} - 1\rangle, \tag{11.19}$$

with the final state $|f\rangle$ of the system having increased energy $E_f = E_i + \hbar\omega_k$. The matrix element of the photon absorption is equal to $\sqrt{n_{k\lambda}}$. The golden rule (see section 6.9) determines the transition rate

$$
\begin{aligned}
w_{fi} &= \frac{2\pi}{\hbar} \left| \left\langle f; n_{k\lambda} - 1 \left| \sum_a \frac{e_a}{m_a c} (\mathbf{p}_a \cdot \hat{\mathbf{A}}(\mathbf{r}_a)) \right| i; n_{k\lambda} \right\rangle \right|^2 \delta(E_\gamma - E_{fi}) \\
&= \frac{4\pi^2}{V\omega_k} n_{k\lambda} \left| \left\langle f \left| \sum_a \frac{e_a}{m_a} (\mathbf{p}_a \cdot \mathbf{e}_{k\lambda}) e^{i(\mathbf{k}\cdot\mathbf{r}_a)} \right| i \right\rangle \right|^2 \delta(E_\gamma - E_{fi}),
\end{aligned}
\tag{11.20}
$$

where $E_\gamma = \hbar\omega_k$ and $E_{fi} = E_f - E_i$. The incident flux is the photon density multiplied by the speed of light, $(n_{k\lambda}/V)c$. The photoabsorption cross section for the specific transition $i \rightarrow f$ is therefore

$$\sigma_{fi}(E_\gamma) = \frac{4\pi^2\hbar}{E_\gamma c} \left| \sum_a \frac{e_a}{m_a} \langle f|(\mathbf{p}_a \cdot \mathbf{e}_{k\lambda}) e^{i(\mathbf{k}\cdot\mathbf{r}_a)}|i\rangle \right|^2 \delta(E_\gamma - E_{fi}). \tag{11.21}$$

Being an observable, the cross section does not depend on the normalization volume. The interaction of particle spins with the magnetic field of the incident wave can be added as in (9.86).

11.3.2 Sum rules

The δ-function in (11.21) shows an infinitely narrow energy dependence of the resonant absorption. The infinite cross section is a result of our approximation. In reality the excited states have a finite lifetime τ and, consequently, energy uncertainty, or width $\Gamma \sim \hbar/\tau$, because they later re-emit the absorbed photon or get rid of the excitation in another way. Therefore the sharp δ-function in (11.21) or other similar expressions should be replaced by a smooth bell-shape resonance curve of finite height. Here we assume that this *absorption line* is narrow and all matrix elements and kinematic factors can be treated as constant within this line. Then the total transition strength is preserved and one can integrate over photon energy. Using (11.21), we find the integral cross section

$$\int dE_\gamma \, \sigma_{fi}(E_\gamma) = \frac{4\pi^2 \hbar}{E_{fi} c} \left| \sum_a \frac{e_a}{m_a} \langle f | (\mathbf{p}_a \cdot \mathbf{e}_{k\lambda}) e^{i(\mathbf{k}\cdot\mathbf{r}_a)} | i \rangle \right|^2 . \tag{11.22}$$

As in the radiation problem, we can utilize the long wavelength approximation, expanding the exponent in (11.22) and using the equations of motion (9.51). The lowest order term is given by the dipole absorption

$$\int dE_\gamma \, \sigma_{fi} = \frac{4\pi^2 E_{fi}}{\hbar c} |(\mathbf{e}_{k\lambda} \cdot \mathbf{d}_{fi})|^2 . \tag{11.23}$$

In nuclear interactions, the exciting electromagnetic field usually is not monochromatic. It is useful to be able to estimate the total photoabsorption cross section summed over all transitions $|i\rangle \rightarrow |f\rangle$ from the initial, for instance ground, state. Such estimates are given by the *sum rules* (SR) that approximately determine quantities of the following type:

$$\mathcal{S}_i^{(n)}[Q] = \frac{1}{2} \sum_f (E_f - E_i)^n \{ |\langle f|Q|i\rangle|^2 + |\langle f|Q^\dagger|i\rangle|^2 \} . \tag{11.24}$$

Here the transition probabilities for an arbitrary pair of mutually conjugate operators Q and $Q^\dagger$ are weighted with a certain (positive, negative, or zero) power n of the transition energy. For a hermitian operator $Q = Q^\dagger$ the two terms in (11.24) are equal and the factor $\frac{1}{2}$ is cancelled.

The exact result follows immediately for non-energy-weighted SR, $n = 0$, based on the completeness of the set of the states $|f\rangle$,

$$\mathcal{S}_i^{(0)}[Q] = \frac{1}{2} \langle i|Q^\dagger Q + QQ^\dagger|i\rangle . \tag{11.25}$$

Often it turns out to be possible to get a good estimate for the expectation value in the right-hand side of (11.25), or to extract it from data.

For the *energy-weighted* sum rules (EWSR), $\mathcal{S}^{(1)}$, a reasonable estimate can be derived for many operators under certain assumptions about the interactions in the system. First, using again the completeness of the intermediate states $|f\rangle$, we can identically write down $\mathcal{S}^{(1)}$ as an expectation value in the initial state $|i\rangle$ of the double commutator

$$\mathcal{S}_i^{(1)}[Q] = \frac{1}{2} \left\langle i \left| [[Q, H], Q^\dagger] \right| i \right\rangle , \tag{11.26}$$

where H is the total Hamiltonian with energies E_i and E_f as its eigenvalues. Thus, we again need to know the properties of the initial state only. (The same is valid for higher $S^{(n)}$, but the commutator structure becomes more and more complex.) Now we choose the operator Q as a one-body quantity depending on coordinates $\mathbf{r}_a$ of the particles,

$$Q = \sum_a q_a, \quad q_a = q(\mathbf{r}_a). \tag{11.27}$$

Apart from that we assume that the Hamiltonian does not contain momentum-dependent interactions. Then only the kinetic part of the Hamiltonian contributes to $[Q, H]$, and the result can be found explicitly,

$$[Q, H] = \left[\sum_a q_a, \sum_b \frac{\mathbf{p}_b^2}{2m_b} \right] = \sum_a \frac{i\hbar}{2m_a} [(\nabla_a q_a), \mathbf{p}_a]_+, \tag{11.28}$$

where $[\dots, \dots]_+$ denotes the anticommutator. The outer commutator in (11.26) leads now to the simple result

$$S_i^{(1)}[Q] = \sum_a \frac{\hbar^2}{2m_a} \left\langle i \middle| |\nabla_a q_a|^2 \middle| i \right\rangle. \tag{11.29}$$

As an example, we take the charge form factor

$$Q \Rightarrow \rho_{\mathbf{k}} = \sum_a e_a e^{i(\mathbf{k} \cdot \mathbf{r}_a)}. \tag{11.30}$$

The sum rule in this case is universal for any initial state $|i\rangle$,

$$S^{(1)}[\rho_{\mathbf{k}}] = \hbar^2 k^2 \sum_a \frac{e_a^2}{2m_a}. \tag{11.31}$$

Taking $\mathbf{k}$ along an (arbitrary) z-axis and considering the long wavelength limit, $kR \ll 1$, we get from the first nonvanishing term in the expansion of the exponent the EWSR for the dipole operator (9.52)

$$S^{(1)}[d_z] = \sum_a \frac{\hbar^2 e_a^2}{2m_a}. \tag{11.32}$$

This is an extension of the old *Thomas-Reiche-Kuhn* (TRK) dipole SR in atomic physics. For a neutral atom, in the center of mass frame attached to the nucleus of charge Z (here m is the electron mass),

$$S^{(1)}[d_z] = \frac{\hbar^2 e^2}{2m} Z. \tag{11.33}$$

The atomic TRK SR is essentially exact (up to relativistic velocity-dependent corrections). In (11.30) e_a are in fact arbitrary numbers. For intrinsic dipole excitations we have to exclude the center of mass motion. Therefore our z-coordinates should be intrinsic coordinates, $z_a \Rightarrow z_a - R_z$, where $R_z = \sum_a z_a/A$. Hence, the intrinsic dipole moment is

$$d_z = \sum_a e_a(z_a - R_z) = e \sum_p z_p - \frac{Ze}{A}\left(\sum_p z_p + \sum_n z_n\right). \tag{11.34}$$

This operator can be rewritten as

$$d_z = e_p \sum_p z_p + e_n \sum_n z_n, \tag{11.35}$$

where protons and neutrons carry *effective charges*

$$e_p = \frac{N}{A}e, \quad e_n = -\frac{Z}{A}e. \tag{11.36}$$

Now (11.32) gives the dipole EWSR

$$\mathcal{S}_i^{(1)}[d_z] \equiv \sum_f E_{fi}|d_{fi}^z|^2 \tag{11.37}$$

$$= \frac{\hbar^2 e^2}{2M}\left[Z\left(\frac{N}{A}\right)^2 + N\left(\frac{-Z}{A}\right)^2\right] \tag{11.38}$$

$$= \frac{\hbar^2 e^2}{2M}\frac{NZ}{A}. \tag{11.39}$$

The factor (NZ/A) is connected to the reduced mass for relative motion of neutrons against protons as required at the fixed center of mass. This result does not include the dipole strength related to nuclear motion as a whole. According to the classical SR (11.33), this contribution is $(m \to AM, e \to Ze)$

$$\mathcal{S}^{(1)}[\text{c.m.}] = \frac{\hbar^2 (Ze)^2}{2AM}. \tag{11.40}$$

The sum of the global (11.40) and intrinsic (11.39) dipole strength recovers the full TRK SR (11.33),

$$\mathcal{S}_i^{(1)}[\text{tot. dip}] = \frac{\hbar^2 e^2}{2M}\left(\frac{NZ}{A} + \frac{Z^2}{A}\right) = \frac{\hbar^2 e^2}{2M}Z. \tag{11.41}$$

In contrast to the atomic TRK case, the nuclear dipole EWSR (11.39) cannot be exact. Velocity-dependent and exchange forces are certainly present in nuclear interactions. Nevertheless (11.39) gives a surprisingly good estimate of the realistic dipole strength, which is not fully understood. In a similar way one can consider SR for other multipoles, but the results are not universal and in general depend on the initial state. Later we will meet additional examples of the application of the SR.

The EWSR (11.39) is what we need to evaluate the sum of integral dipole cross sections (11.23) over all possible final states $|f\rangle$. Taking in (11.23) the polarization vector along the z-axis, we come to the total dipole photoabsorption cross section

$$\sigma_{tot} = \sum_f \int dE_\gamma \sigma_{fi} = 2\pi^2 \frac{e^2\hbar}{Mc}\frac{ZN}{A}. \tag{11.42}$$

This universal prediction,

$$\sigma_{tot} = 0.06 \frac{ZN}{A} \text{barn} \cdot \text{MeV}, \tag{11.43}$$

on average agrees well with experiments in spite of crudeness of approximations made in the derivation. One should remember that it includes only dipole absorption.

11.3.3 Giant resonances

The first indication of the presence of a giant resonance appeared in W. Bothe and W. Gentner's work [BG37], who used photons of 17.6 MeV from the reaction $^7\text{Li}(p,\gamma)$ in several targets. Baldwin and Kleiber [BK47] confirmed these observations with photons from a betatron. Figure 11.4 exhibits the excitation function of photoabsorption of ^{120}Sn around the electric dipole giant resonance at 15 MeV.

The giant resonance happens in nuclei along the whole periodic table, with the resonance energy decreasing with A without large oscillations (see figure 11.5) starting at $A = 20$. This shows that the giant resonance is a property of the nuclear matter and not a characteristic phenomenon of nuclei of a certain type. The widths of the resonances are almost all in the range between 3.5 MeV and 5 MeV and can reach 7 MeV in a few cases.

The first proposal of explanation for the resonance came from Goldhaber and Teller in 1948 [GT48]. The photon, through the action of its electric field on the protons, takes the nucleus to an excited state where a group of protons oscillates in opposite phase against a group of neutrons. In such an oscillation, these groups interpenetrate, keeping constant the incompressibility of each group separately. A classic calculation using this hypothesis leads to a vibration frequency that varies with the inverse of the square root of the nuclear radius, that is, the resonance energy varies with $A^{-1/6}$.

Later on, H. Steinwedel and J. H. Jensen [SJ50] developed a classic study of the oscillation in another way, already suggested by Goldhaber and Teller, in which incompressibility is abandoned. The nucleons move inside a fixed spherical cavity with the proton and neutron densities being a function of position and time. The nucleons at the surface have fixed position with respect to each other, and the density is written in such a way that, at a given instant, excess of protons on one side of the nucleus coincides with lack of neutrons on that same side, and vice versa. Such a model leads to a variation of the resonance energy with $A^{-1/3}$.

The behavior of the energy of the giant electric dipole resonance E_{GDR} does not agree exactly with either of the two models. But a calculation developed in [My77], which assumes a simultaneous contribution of the two models, obtains an expression for E_{GDR} as a function of the mass number A,

$$E_{GDR}(\text{MeV}) = 112 \times \left[A^{2/3} + (A_0 A)^{1/3}\right]^{-1/2}, \tag{11.44}$$

where $A_0 \cong 274$. With the exception of some very light nuclei, this expression reproduces the behavior of the experimental values very well, as we can see in figure 11.5. An examination of (11.44) shows that the Gamow-Teller mode prevails broadly in light nuclei, while

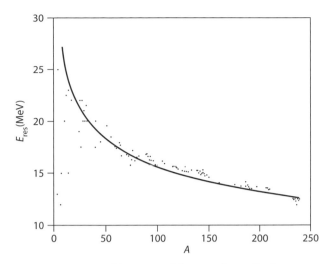

Figure 11.5 Location of the energy of the giant electric dipole resonance given by (11.44) (continuous curve), compared with experimental points [DB88].

the contribution of the Steinwedel-Jensen mode is negligible. The latter mode increases with A, but it only becomes predominant at the end of the periodic table, at $A = A_0$.

The giant electric dipole resonance arises from an excitation that transmits 1 unit of angular momentum to the nucleus ($\Delta l = 1$). If the nucleus is even-even, it is taken to a 1^- state. What one verifies is that the transition also changes the isospin of 1 unit ($\Delta T = 1$) and, due to that, it is also called an *isovector resonance*. For many years this was the only known giant resonance. In the 1970s, giant isoscalar resonances ($\Delta T = 0$) of electric quadrupole ($\Delta l = 2$) [PW71] and electric monopole ($\Delta l = 0$) [Ma75] were observed in reactions with charged particles. The first is similar to the vibrational quadrupole state created by the absorption of a phonon of $\lambda = 2$, since both are, in even-even nuclei, states of 2^+ vibration. But the giant quadrupole resonance has a much larger energy. This resonance energy, in the same way as argued for the dipole, decreases smoothly with A, obeying the approximate formula

$$E_{\text{GQR}}(\text{MeV}) \cong 62A^{-1/3}. \tag{11.45}$$

In the state of giant electric quadrupole resonance the nucleus oscillates between the spherical (supposing that this is the form of the ground state) and ellipsoidal form. If protons and neutrons act in phase we have an isoscalar resonance ($\Delta T = 0$), and if they oscillate in opposite phase the resonance is isovector ($\Delta T = 1$). Figure 11.6 illustrates these two possible vibration modes. The existence of isoscalar electric quadrupole resonances is firmly established with a large number of measured cases. The electric quadrupole isovector resonance has been identified in numerous experimental studies.

The giant monopole resonance is a very special way of nuclear excitation where the nucleus contracts and expands radially, maintaining its original form but changing its volume. It is also called the *breathing mode*. It can also happen in isoscalar and isovector

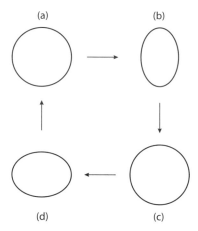

Figure 11.6 Four stages in the vibration cycle of a giant quadrupole resonance. In an isoscalar resonance, protons and neutrons vibrate in phase, while in an isovector resonance, the vibrations occur in opposite phase. For opposite phases, when the protons are at stage (b), the neutrons will be at stage (d), and vice versa.

forms and is an important way to study the compressibility of nuclear matter. Again here, the isoscalar form has a reasonable number of measured cases, the location of the resonance energy being given by the approximate expression

$$E_{GMR}(\text{MeV}) \cong 80A^{-1/3}. \tag{11.46}$$

In 1996 evidence for the existence of a giant isoscalar electric dipole resonance was presented, using the scattering of α-particles of 200 MeV on ^{208}Pb [Da96]. Results for the observation of octupole giant resonances were also published [Wo87].

Besides the electric giant resonances, associated to a variation in the form of the nucleus, magnetic giant resonances exist, involving what one calls *spin vibrations*. In these, nucleons with spin up move out of phase with nucleons with spin down. The number of nucleons involved in the process cannot be very large because it is limited by the Pauli principle.

The magnetic resonances can also separate in isoscalar resonances, where protons and neutrons of same spin vibrate against protons and neutrons of opposite spin, and isovector resonances, where protons with spin up and neutrons with spin down vibrate against their corresponding ones with opposite spins. These last cases, originate from reactions of charge exchange. They are also referred to as giant Gamow-Teller *resonances*. Figure 11.7 exhibits a schematic diagram of the two types of magnetic resonance for dipole vibrations. Magnetic resonances of monopole and quadrupole type have also already been observed [Lin79].

Another important aspect of the study of giant resonances is the possibility that they can be induced in already excited nuclei. This possibility was analyzed theoretically by D. M. Brink and P. Axel [Ax62] for giant resonances excited "on top" of nuclei rotating with high angular momentum, resulting in the suggestion that the frequency and other properties

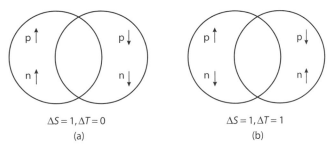

$\Delta S = 1, \Delta T = 0$ (a) $\Delta S = 1, \Delta T = 1$ (b)

Figure 11.7 Magnetic dipole resonances: a) isoscalar; b) isovector.

of giant resonances are not affected by the excitation. A series of experiences in the decade of the 1980s (see [BB86]) gave support to this hypothesis.

A special case happens when the giant resonance is excited on top of another giant resonance. Understanding the excitation of a giant resonance as the result of the absorption of one phonon, we can view these double giant resonances as states of excitation with two vibrational phonons. The double giant dipole resonance was observed for the first time in reactions with double charge exchange induced by pions in ^{32}S [Mo88]. As first shown in [BB86b] a much better possible way to study multiple giant resonances is by means of Coulomb excitation with relativistic heavy projectiles. Later on, this was indeed verified experimentally and several properties of multiple giant resonances have been studied theoretically (for a theoretical review, see [BP99]).

11.4 Coulomb Excitation

Electromagnetic excitation of a nucleus by the field of another nucleus is a powerful tool for studying nuclear structure, especially excited collective states. In a collision of two nuclei with large impact parameter, there is no overlap of nuclear densities and short range nuclear forces are not efficient. To avoid a violent nuclear collision, the impact parameter should be larger than the sum of nuclear radii $R_1 + R_2$.

We discuss shortly the nonrelativistic problem, although currently major experimental efforts are applied to physics of relativistic electromagnetic excitation [BB88]. In contrast to that, in Coulomb excitation we assume that nuclei perturb each other only through long range electromagnetic interactions. Let v be the relative velocity of two nuclei at infinity, which determines the energy of relative motion $E = mv^2/2$, where m is the reduced mass. The strength of the Coulomb interaction can be measured by the parameter

$$\eta = \frac{Z_1 Z_2 e^2}{\hbar v}, \tag{11.47}$$

where Z_1, Z_2 are charges of the nuclei. For $Z_1 Z_2 > 137$, the parameter (11.47) is $\eta = c/v > 1$ and can easily be $\eta \gg 1$. This means that the Coulomb interaction is effectively strong and cannot be accounted for by perturbation theory.

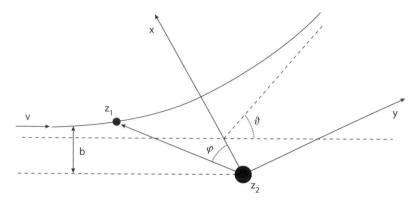

Figure 11.8 Classical Coulomb trajectory for a nucleus-nucleus collision. The path follows a hypothetical trajectory, and b is called the impact parameter, representing, in a head-on collision, the closest distance of approach.

This situation allows for the use of the semiclassical approximation when the Coulomb interaction is taken into account exactly in determining the classical Rutherford trajectory of relative motion $\mathbf{R}(t)$ where $\mathbf{R}$ is the distance between the centers of the colliding nuclei, figure 11.8. The relative energy E is assumed to be large enough to be able to neglect the feedback from the intrinsic excitations to relative motion. Then the trajectory is fixed by energy and impact parameter or deflection angle.

The classical distance of closest approach,

$$R_0 = 2Z_1 Z_2 e^2 / mv^2, \tag{11.48}$$

is larger than $R_1 + R_2$ at relative energy lower than the Coulomb barrier

$$E_B = \frac{Z_1 Z_2 e^2}{R_1 + R_2}. \tag{11.49}$$

The excitation is generated by the time-dependent field, and the probability of the process is determined by the presence in this field of Fourier harmonics with excitation frequencies $\omega(E_f - E_i)/\hbar$. If the motion is too slow, the field acts adiabatically, the intrinsic wave function changes reversibly, and the probability of excitation is low. The corresponding parameter of adiabaticity is the ratio ξ of the time scales for the Coulomb collision, $\sim R_0/v$, and for the nuclear excitation, $\sim 1/\omega$. At $\xi = R_0 \omega v > 1$, the situation is adiabatic and transition probabilities are small.

The interaction Hamiltonian responsible for the excitation processes in the nonrelativistic case can be written as

$$H' = \int d^3 r_1 d^3 r_2 \frac{\rho_1^{ch}(\mathbf{r}_1 - \mathbf{R}_1)\rho_2^{ch}(\mathbf{r}_2 - \mathbf{R}_2)}{|\mathbf{r}_1 - \mathbf{r}_2|} - \frac{Z_1 Z_2 e^2}{R(t)}, \tag{11.50}$$

where charge densities of unperturbed nuclei depend on the distances from the corresponding centers. We subtract the interaction between nuclei as a whole which determines the trajectory (time dependence of $\mathbf{R} = \mathbf{R}_1 - \mathbf{R}_2$) but does not contribute to intrinsic excitations. We introduce the intrinsic coordinates for each nucleus $\mathbf{x}_{1,2} = \mathbf{r}_{1,2} - \mathbf{R}_{1,2}$,

$$H' = \int d^3 x_1 d^3 x_2 \frac{\rho_1^{ch}(\mathbf{x}_1)\rho_2^{ch}(\mathbf{x}_2)}{|\mathbf{R}(t) + \mathbf{x}_1 - \mathbf{x}_2|} - \frac{Z_1 Z_2 e^2}{R(t)}, \tag{11.51}$$

and carry out the multipole expansion for a large distance between the centers, $R \gg x_{1,2}$,

$$H' = \int d^3 x_1 d^3 x_2 \rho_1^{ch}(\mathbf{x}_1)\rho_2^{ch}(\mathbf{x}_2) \sum_{L>0,M} \frac{4\pi}{2L+1} \frac{x_{12}^L}{R^{L+1}(t)} Y_{LM}(\mathbf{x}_{12}) Y_{LM}^*(\mathbf{R}). \tag{11.52}$$

Here $\mathbf{x}_{12} \equiv \mathbf{x}_1 - \mathbf{x}_2$.

This Hamiltonian is rather complicated due to the correlations associated with the mutual excitation of the nuclei. It becomes much simpler if we are interested in the excitation of one of the partners only. Let the "projectile" 2 be not excited and we can neglect effects related to its structure as an extended object. Then $\rho_2^{ch}(\mathbf{x}_2) \approx Z_2 e\delta(\mathbf{x}_2)$, and the hamiltonian is expressed in terms of the electric multipole moments (9.95) of the "target" 1,

$$H' = Z_2 e \sum_{L>0,M} \frac{4\pi}{2L+1} \frac{1}{R^{L+1}(t)} Y_{LM}^*(\mathbf{R}(t)) \mathcal{M}(\mathrm{E}L, M). \tag{11.53}$$

The Hamiltonian (11.53) is time-dependent since the trajectory is considered as a function of time. According to (6.45) the transition amplitude $i \to f$ with excitation by $\hbar\omega = E_f - E_i$ is the Fourier component for the transition frequency of the interaction hamiltonian taken along the unperturbed trajectory,

$$S_{fi} = -\frac{i}{\hbar} \int_{-\infty}^{\infty} dt H_{fi}'(t) e^{i\omega t}. \tag{11.54}$$

For the unpolarized initial nuclei and with no final polarization registered, the transition rate is to be averaged over initial projections and summed over final projections of the target,

$$w_{fi} = \frac{1}{2J_i + 1} \sum_{M_f M_i} |S_{fi}|^2; \tag{11.55}$$

the polarization state of the projectile is assumed to be unchanged.

The trajectory enters the result via the time integral

$$I_{LM}(\omega) = \int_{-\infty}^{\infty} dt \frac{1}{R^{L+1}(t)} Y_{LM}(\mathbf{R}(t)) e^{i\omega t}. \tag{11.56}$$

This Fourier component becomes small, is proportional to $\exp(-\text{const} \cdot \xi)$, if the trajectory changes at too slow a rate compared to the needed transition frequency and the parameter of adiabaticity $\xi > 1$.

The intrinsic matrix elements of multipole moments appear in the transition rate (11.55) in sums over magnetic quantum numbers

$$\Sigma_{LM,L'M'} = \frac{1}{2J_i + 1} \sum_{M_f M_i} (\mathcal{M}(EL, M))_{fi} (\mathcal{M}(EL', M'))_{fi}^*. \tag{11.57}$$

The summation in (11.57) selects $L' = L$, $M' = M$, and the result does not depend on M. Then it is the same as the reduced probability (9.97),

$$\Sigma_{LM,L'M'} = \frac{1}{2J_i + 1} \sum_{M_f M_i} |(\mathcal{M}(EL, M))_{fi}|^2 \delta_{LL'} \delta_{MM'} = \frac{B(EL; i \to f)}{2L + 1} \delta_{LL'} \delta_{MM'}. \tag{11.58}$$

Coulomb excitation is especially useful for studying collective states with enhanced reduced probabilities. The total excitation probability is therefore

$$w_{fi} = \left(\frac{4\pi Z_2 e}{\hbar}\right)^2 \sum_{L>0} \frac{B(EL; i \to f)}{(2L + 1)^3} \sum_M |I_{LM}(\omega_{fi})|^2. \tag{11.59}$$

From the viewpoint of the projectile the process is inelastic scattering. The Coulomb trajectory defines the deflection angle θ and the *Rutherford cross section*, which, in terms of the closest approach distance (11.48), is

$$d\sigma_R = \left(\frac{R_0}{4\sin^2(\theta/2)}\right)^2 d\Omega. \tag{11.60}$$

In our approximation, the trajectory is not influenced by the target excitation so that the inelastic cross section is factorized into the product of the Rutherford cross section (11.60) and the excitation probability (11.59),

$$d\sigma_{fi} = d\sigma_R w_{fi} = \sum_{L>0} d\sigma_{fi}(EL), \tag{11.61}$$

where the cross section for the excitation of multipolarity L is equal to

$$\frac{d\sigma_{fi}(L)}{do} = \left(\frac{\pi Z_2 e R_0}{\hbar \sin^2(\theta/2)}\right)^2 \frac{B(EL; i \to f)}{(2L + 1)^3} \sum_M |I_{LM}(\omega_{fi})|^2. \tag{11.62}$$

We can make a crude estimate of the cross section of Coulomb excitation. The trajectory integral (11.56), after changing the variable to $dR = vdt$, gives the dimensional factor R^{-L}/v. It has to be taken near the closest approach point (11.48) which is the most effective for excitation. The constants from the Rutherford cross section can be combined into the Coulomb parameter (11.47). As a result,

$$\sigma(EL) \simeq \eta^2 \frac{B(EL)}{R_0^{2L-2}} f_L(\xi), \tag{11.63}$$

where the function $f_L(\xi)$ depends smoothly on L but contains the exponential cut-off at large values of the adiabaticity parameter ξ. We remember (9.121), that in photoabsorption each consecutive multipole is suppressed by a factor $(kR)^2$. The situation for exciting higher multipoles is easier in the Coulomb excitation because here

$$\frac{\sigma(\text{EL}+1)}{\sigma(\text{EL})} \sim \frac{B(\text{EL}+1)}{B(\text{EL})R_0^2} \sim \left(\frac{R}{R_0}\right)^2. \tag{11.64}$$

This ratio is significantly larger than $(kR)^2$.

This theory can be extended, considering quantum scattering instead of classical trajectories, using relativistic kinematics, taking into account magnetic multipoles, which become equally important for relativistic velocities, and including higher order processes of sequential excitation of nuclear states. The last generalization is necessary for excitation of rotational bands and overtones of giant resonances (quantum states with several vibrational quanta). The mutual excitation of the projectile and of the target can also be studied. The fundamental review paper [Ald56] contains a great deal of information on the subject and even now does not look obsolete, 40 years after its publication.

11.5 Fission

The discovery of the neutron in 1932 led to intense investigations of nuclear reactions induced by this new projectile. Of particular interest was the possibility of obtaining a transuranic element in neutron capture by uranium, the heaviest known element occurring in nature and with the highest atomic number known at the time. In fact, E. Fermi and his team soon experimentally verified that the element neptunium, $Z = 93$, can be produced in this way, after the capture of a neutron by uranium and the emission of an electron.

But the new discovery revealed a much more complex situation. Besides the presence of neptunium, activities were detected that could not be clearly attributed to any element in the neighborhood of uranium. O. Hahn and F. Strassmann showed, in 1939 [HS39], after careful radiochemical study, that the observed activities were due to several elements with roughly half the mass of uranium. This led L. Meitner and O. R. Frisch [MF39] to finally give the correct interpretation of the phenomenon. They proposed that, in capturing the neutron, the uranium can divide into two fragments of comparable masses. Meitner and Frisch called this process *fission*, a term employed in biology to describe cellular division. These same authors soon recognized, by an analysis of the binding energy of the participants, that the process releases a large amount of energy, close to 200 MeV per fission. Still in 1939, as the result of several studies, it was established that fission does not occur preferentially into two equal fragments but with masses distributed around $A = 95$ and $A = 138$.

Nuclear fission is nowadays a very well established phenomenon. One knows that it can be induced by a vast combination of projectile-target-energies, and that it can also occur spontaneously in some elements. Moreover, the large amount of energy released in the process gave rise to the development of devices of large explosive power, the so-called "atomic bomb," and also of nuclear reactors as a source of energy.

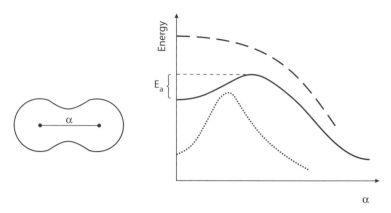

Figure 11.9 Deformation energy of a nucleus as a function of a parameter α that characterizes the distance between the pre-fragments. The solid line refers to a heavy nucleus, the dotted line to a lighter one, and the dashed line to a situation where a fission barrier no longer exists for a very heavy nucleus. E_a is the activation energy.

The first factor to be examined in a possible nuclear disintegration is the energy balance. If we observe figure 5.1, we see that the division of a heavy nucleus into two fragments of comparable masses is a highly exoergic process. Let us examine the typical case, the fission of ^{236}U* into two fragments with mass numbers 95 and 138, with the release of 3 neutrons. We can evaluate for ^{236}U* a total binding energy of about 1770 MeV and a value close to 1976 MeV for the sum of the total binding energies of the daughter nuclei. Since the final products are more bound that the initial one, the resulting binding energy is released in the form of kinetic energy of the fragments and of the neutrons, about 200 MeV being available for this.

The question then arises: if fission is an exoergic process, why does it not occur spontaneously and immediately? The answer is the same as that for α-disintegration of a heavy nucleus: before the process can release energy there is a barrier to overcome. In the case of fission this barrier is manifest through the increase in the potential energy of the nucleus in response to a small deformation of its surface. Figure 11.9 is a schematic representation of what occurs. It shows the deformation energy of a charged drop as a function of a parameter that measures the deformation (a simplistic view of the problem since it is necessary to include more than one parameter in order to better describe the form of the nucleus). The behavior of a nucleus like ^{236}U can be represented by the solid line. A small deformation starting from $\alpha = 0$ increases its energy, and the restoration force tends to bring it back to the initial situation. However, when the deformation is large enough, an increase in deformation decreases the energy, and the curve begins to fall after passing to a maximum. From that point on the process follows spontaneously and the nucleus splits into two fragments. The tail of the curve represents the electrostatic potential energy of the two charged fragments, with a distance α between their centers.

How can a fission process occur? There are two possibilities. The first is to give the nucleus an excitation energy greater than E_a. This allows the onset of oscillations that can lead to the necessary deformation to overcome the barrier. This is the case of $^{236}U^*$, excited by a neutron capture in ^{235}U. The other possibility is spontaneous fission. In this case the barrier is surpassed by a tunnel effect starting from the nucleus ground state.

The above processes have variable importance along the periodic table. For intermediate mass nuclei the activation energy is very large: induced fission only proceeds with high energy projectiles, and spontaneous fission is practically nonexistent. Conversely, very heavy nuclei can fission without any barrier, as shown by the dashed curve of figure 11.9. Elements that are close to this situation have very small spontaneous fission half-lives. This situation can be approximately established by imagining that with a small deformation a spherical nucleus assumes the form of an ellipsoid with major semi-axis $a = R(1 + \epsilon)$, where R is the radius of the original spherical nucleus. If the deformation process proceeds with constant volume, the minor semi-axis has a length $b = R(1 + \epsilon)^{-1/2}$; with this choice, the volume part of the binding energy (5.7) remains constant. One shows, on the other hand [BW39], that the surface and Coulomb energy terms change with deformation according to

$$E_S = E_{S_0}\left(1 + \frac{2}{5}\epsilon^2 + \cdots\right), \quad E_C = E_{C_0}\left(1 - \frac{1}{5}\epsilon^2 + \cdots\right), \tag{11.65}$$

where $E_{S_0} = B_2$ and $E_{C_0} = B_3$ are the respective zero deformation energies given by (5.2) and (5.3). Thus the total energy gained with the deformation is

$$\Delta E = E_S - E_{S_0} + E_C - E_{C_0}$$
$$= -a_S A^{2/3}\left(\frac{2}{5}\epsilon^2 + \cdots\right) - a_C Z^2 A^{-1/3}\left(-\frac{1}{5}\epsilon^2 + \cdots\right). \tag{11.66}$$

The limit being sought occurs when $\Delta E = 0$, that is, when

$$\frac{Z^2}{A} \cong \frac{2a_S}{a_C} \cong 50, \tag{11.67}$$

where in (11.66) we have neglected powers of ϵ higher than the second. The quantity Z^2/A therefore has a relevant role in estimating the probability of a nuclear spontaneous fission and is called the *fissionability parameter*. Figure 11.10 shows the spontaneous fission half-life as a function of this parameter. It is clear that the half-lives decrease toward the limiting value (11.67), but there are no available experimental data close to that value.

11.6 Mass Distribution of Fission Fragments

The way a nucleus divides itself in a fission process is not identical for all events. Thus a nucleus like ^{252}Cf can fission spontaneously through, among others, the following modes:

$$^{252}Cf \rightarrow {}^{112}_{46}Pd + {}^{138}_{52}Te + 2n,$$
$$^{252}Cf \rightarrow {}^{101}_{42}Mo + {}^{148}_{56}Ba + 3n,$$
$$^{252}Cf \rightarrow {}^{105}_{45}Rh + {}^{146}_{53}I + n, \text{ etc.} \tag{11.68}$$

with different probabilities for each mode. In this way we can, with a large number of events, build a curve like that of figure 11.11, where the probability of occurrence of each

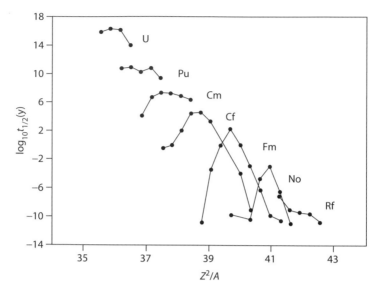

Figure 11.10 Spontaneous fission half-lives of even-even nuclei.

fragment is represented. The curve shows two peaks, one corresponding to lighter fragments centered at $A \cong 110$ and the other to heavier fragments, with a maximum at $A \cong 140$. We see that symmetrical fission corresponds to a minimum of the curve and is about 600 times less probable than the most favorable asymmetric division. The reason for such behavior is the influence of the shell structure of the nucleus. Hence the effect should be less important for large excitation energies, where the high level density has to include an average over a large number of states and not just a few isolated ones. It is in fact experimentally verified that asymmetric fission tends to the symmetric case when the projectile energy increases. This is seen in figure 11.12 for the case of ^{238}U bombarded with protons from 10 MeV to 200 MeV. We see that from 150 MeV on, the minimum in the

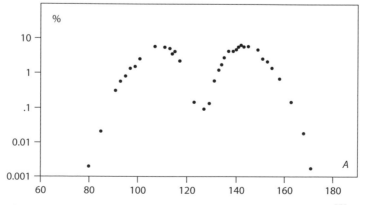

Figure 11.11 Percent yield of fragments in the spontaneous fission of ^{252}Cf [Ne60].

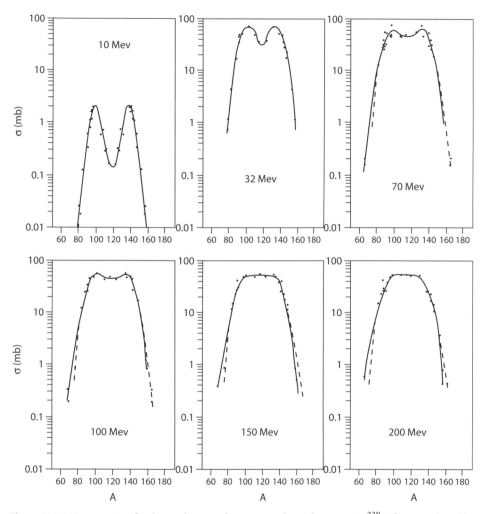

Figure 11.12 Cross section for the production of a mass number A fragment in ^{238}U fission induced by protons of various monoenergetic energies [Hy71].

symmetric fission disappears, giving place to a plateau where the production of $A = 100$ to $A = 130$ fragments are equally probable.

The behavior of ^{238}U is typical of a highly fissionable nuclide, as normally are the ones of very large mass. Lighter nuclei such as radium and actinium have, near the threshold energy for fission, a mass distribution whose characteristic is the presence of three peaks of approximately the same height, indicating that symmetric and very asymmetric fission are the most probable modes. A typical example is seen in figure 11.13, where the yield of each fragment in the fission of ^{226}Ra induced by 11 MeV protons is plotted. For even lighter nuclei, like lead and bismuth, symmetric fission is already dominant from threshold, the mass distribution having a single peak near the mass number $A = 100$.

The shell effect, which leads to asymmetric spontaneous fission of nuclei with $A \gtrsim 256$, can produce the opposite effect when the mass number goes beyond that value. Nuclei like

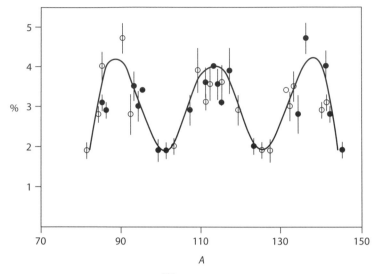

Figure 11.13 Mass distribution in ^{226}Ra fission induced by 11 MeV protons [JF58]. The open circles are the reflection of the experimental points (solid circles) relative to $A = 111$.

^{258}Fm, ^{259}Fm, ^{259}Md, ^{260}Md, ^{258}No, ^{262}No, and ^{262}Rf have a prominent symmetric fission peak in the mass distribution of the fragments [La96]. In these cases, the proximity of the ^{264}Fm nucleus is responsible for the effect, since this nucleus can divide into two double magic fragments ^{132}Sn.

For spontaneous fission, or fission induced by low energy projectiles, division into two fragments is the only experimentally observed mode. Nuclear fission into three fragments of comparable masses (*ternary fission*), has been reported for several nuclei [RH50, Mu67] but later work [St70] has placed serious doubts on those results.

A different situation occurs when the third fragment is an α-particle. α-emission during fission is a relatively common process, occurring in 0.2%–0.4% (depending on the nucleus) of all fissions. This α-particle has, on the average, more energy (peak at 15 MeV) than in ordinary α-emission, and its occurrence is explained by decrease of the barrier due to the formation of a neck between the two fragments.

High energy projectiles can leave the nucleus with a high excitation energy. If this energy is greater than about 5 MeV/nucleon (approximate value of the total binding energy of the nucleus), a process of nuclear rupture into several fragments (*nuclear fragmentation*) can occur. However, the fragmentation involves mechanisms that have no similarity with the ones present in the deformation of a drop in low energy fission.

11.7 Neutrons Emitted in Fission

Fission fragments are necessarily rich in neutrons. This is expected as the fissioning nucleus has a neutron to proton ratio larger than that necessary in balancing the fragments,

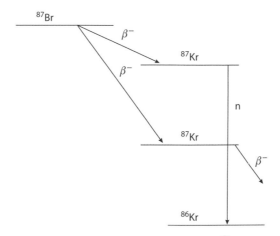

Figure 11.14 Emission of a delayed neutron in ^{87}Br decay.

which have much smaller atomic number. As a result, the fission fragments are β^--radioactive. In a short time after the division the fragments can also go toward the β-stability line emitting a neutron. In contrast to these *prompt neutrons*, emitted by the fragments in a time of at most 10^{-16} s after fission, are the *delayed neutrons*, whose emission proceeds after at least one β^--emission by the fragment. Since decay by neutron emission is a very fast process, the emission time is conditioned to the half-life of the β-decay that precedes it.

Figure 11.14 shows an event of this type: ^{87}Br can decay into ^{87}Kr in two ways, the upper one having enough energy to release one neutron, decaying into the stable isotope ^{86}Kr. The delayed neutrons are only 1% of all the neutrons emitted in fission, but have an important role in the control of a reactor.

The number of neutrons emitted in fission is of utmost importance in *chain reactions*, where neutrons proceeding from one fission are used to induce new fissions. This number can vary from 1 to 8, but this upper limit occurs very rarely, with an average between 2 and 4 for the great majority of events produced by spontaneous and thermal neutron induced fission. This average increases a little with the energy of the incident particle, reaching, for example, 4.75 for ^{238}U fission induced by 15 MeV neutrons.

As important as neutron number is the neutron kinetic energy distribution. In figure 11.15 we see that this distribution is wide, with a maximum around 0.5 MeV. The fitted curve [Wa52]

$$N(E) = \exp(-1.036E) \sinh \sqrt{2.29E} \tag{11.69}$$

well represents the experimental values from 0 to 10 MeV.

11.8 Cross Sections for Fission

The behavior of the fission cross section will be presented for the important examples of ^{235}U and ^{238}U bombarded by neutrons from 0 to 1 MeV. In figure 11.16 we see that

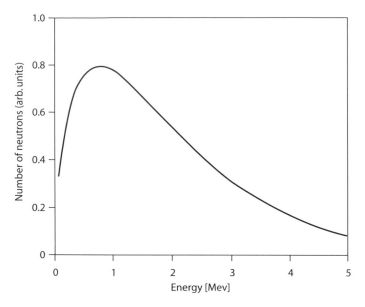

Figure 11.15 Kinetic energy distribution of prompt neutrons.

neutrons with energies close to zero are already able to fission ^{235}U, which indicates that the ^{236}U activation energy is lower than the neutron binding energy. In this way the fission barrier is overcome and the compound nucleus fissions. With increase in energy, the cross section reduces until reaching, at 1 MeV, a value 600 times less than the corre-

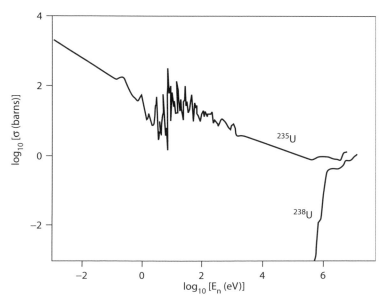

Figure 11.16 Fission cross sections of ^{235}U and ^{238}U bombarded by 0 to 10 MeV neutrons.

sponding thermal neutron value, after crossing a region of resonances, between 0.1 eV and 1 keV.

On the other hand ^{238}U does not fission by slow neutrons, with its fission threshold near 1 MeV. The energy gained by the binding of the incident neutron is not enough, in this case, to reach the summit of the barrier, and it is also necessary for the contribution of the neutron kinetic energy to reach the activation energy.

How can we explain this difference in behavior in nuclei very close to each other? The reason is found in the pairing energy, which contributes with the term (5.5) to the mass formula. This energy implies that an even-even nucleus is, on the average, more bound than its odd neighbors. When ^{235}U absorbs a neutron, the binding energy per nucleon increases by δ in (5.7), being converted into excitation energy; in this way the summit of the fission barrier is more easily reached. The absorption of a neutron by ^{238}U, on the other hand, decreases the binding energy per nucleon by δ, and part of the excitation energy will be used increasing the ^{239}U mass per nucleon. This difference has to be delivered by the neutron incident energy and this explains why there is no fission of ^{238}U by neutrons of energy lower than 1 MeV.

From the above argument it is clear that this effect is not restricted to uranium. However, this element is important because ^{235}U is the only isotope found in nature that is fissionable by slow neutrons, although existing in small proportion (0.7%) compared to ^{238}U (99.3%) in a natural sample of uranium ore.

11.9 Energy Distribution in Fission

The fission of a heavy nucleus releases about 200 MeV. The larger part of this energy ($\cong$ 165 MeV) is spent as kinetic energy of the fragments, the light fragment carrying the larger part. This is due to linear momentum conservation, which implies that the ratio between the energies of the fragments

$$\frac{\frac{1}{2}m_1 v_1^2}{\frac{1}{2}m_2 v_2^2} = \frac{v_1}{v_2} = \frac{m_2}{m_1} \tag{11.70}$$

is equal to the inverse ratio of the masses. Thus the kinetic energy distribution of the fragments has the aspect of figure 11.17, where the 92.5 MeV peak corresponds to the light fragment and the 62.0 MeV peak to the heavy one.

The rest of the available energy is shared by neutrons, γ-particles, and also the electrons of β^--decay. Typical numbers are: the neutrons consume about 5 MeV, the result of an average energy of 2 MeV multiplied by an average number of 2.5 neutrons by fission. The β^--decay releases an approximate energy of 7 MeV, while the prompt γ-rays and the products of the fragment decay carry a total of 15 MeV. We have to also mention that the neutrinos emitted together with the electrons are responsible for 12 MeV of the total energy. However, owing to the weak interaction of these particles, they cross all experimental apparatus without being detected.

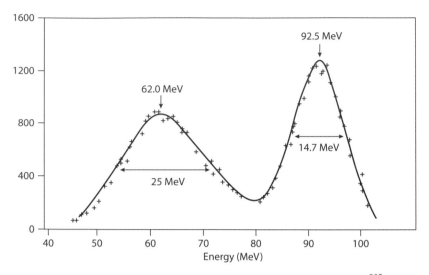

Figure 11.17 Kinetic energy distribution of the fragments in the fission of ^{235}U. The ordinate indicates the number of counts of each experimental point [FR47].

11.10 Isomeric Fission

In 1962, S. M. Polikanov and colleagues published [Po62] results from a spontaneous fission study of very heavy short half-life isotopes. The isotopes were formed by the bombardment of ^{238}U with ions of ^{22}Ne and ^{16}O of several energies around 100 MeV. The motivation for the work was the appearance of a fission half-life of about 14 ms coming from a product of the reaction $^{242}_{94}$Pu + $^{22}_{10}$Ne, studied previously by the same group with the idea of creating the $Z = 104$ element. In the experimental apparatus used, the reaction products, of high linear momentum, were pulled out from the uranium target, reaching a high velocity disk that transported them rapidly to two ionization chambers able to detect fission fragments. The authors could measure fission half-lives of (0.02 ± 0.01) s in the irradiations with neon and (0.013 ± 0.008) s in the irradiations with oxygen. These values were, taking into account the experimental error, identical to the half-life of 14 ms obtained previously, discarding in so doing the element 104 as being responsible for the fission events. In fact, a further work has shown that the detected fragments had their origin in ^{242}Am. The half-life in question was several orders of magnitude lower than that expected for the ground state of possible reaction products, and the conclusion was that the fission occurred from an isomeric excited state. It is hard to understand, however, how this state resists γ-decay for so long a time and how it exhibits such a drastic reduction in the fission half-life compared with that of the ground state. A possible explanation, suggested by the authors, and which would be shown to be correct later, was that the isomery is connected to a strong nuclear deformation.

Complete understanding of the mechanism came in 1967, with work of V. M. Strutinsky [St67] concerning the shell effects in the determination of nuclear masses and deformation

energy. There was the notion that the liquid drop model described well the nuclear properties that depend smoothly on the number of nucleons, as in, for instance, the general trend
of the curve of binding energy versus A (figure 5.1). The shell model, in turn, explained the
nuclear properties that depend strongly on the arrangement of the last nucleons, that is,
the nucleons in the neighborhood of the Fermi surface. It was natural to attempt building
a model that incorporates both effects. Such a model starts with the nucleus total energy

$$E = E_{\text{LDM}} + E_{\text{SM}}, \tag{11.71}$$

where the first term gives the contribution of the liquid drop model and the second one
accounts for the small correction due to shell effects. Strutinsky elaborated an elegant and
efficient method to evaluate this correction, assuming that it should be obtained from the
difference

$$E_{\text{SM}} = U - \tilde{U} \tag{11.72}$$

between the total energy given by the shell model,

$$U = \sum_i E_i, \tag{11.73}$$

where the sum must be extended to all occupied states, and the energy corresponding to
an average situation

$$\tilde{U} = \int_{-\infty}^{E_F} E \tilde{g}(E) \, dE, \tag{11.74}$$

where $\tilde{g}(E)$ is a function that tries to reproduce the average behavior of the level density
$g(E)$. The latter in the shell model is given by

$$g(E) = \sum_i \delta(E - E_i), \tag{11.75}$$

which has as boundary condition the conservation of the number of particles N:

$$\int_{-\infty}^{E_F} \tilde{g}(E) dE = N, \tag{11.76}$$

where E_F is the Fermi energy. Summarizing, (11.71) and (11.72) propose to calculate the
total energy of the nucleus using the shell model energy, but with the step of replacing
the average value of this energy by the energy coming from the liquid drop model.

The Strutinsky method is of a great value in the precise determination of the ground state
masses. More than this, it established that the potential energy as a function of deformation
has more than one minimum. The first minimum corresponds to the ground state and
the second, the reason for the existence of the shape isomeric states.

A qualitative understanding of the presence of the second well can be obtained in the
following way: when we plot the deformation energy of a nucleus in the actinide region
versus a parameter that measures the deformation, what we obtain in principle is a function
described by the curve (a) of figure 11.18. To this simple curve we can add small shell
corrections. Toward this end, we must remember that the distribution of levels filled by the
nucleons change with deformation (Nilsson model—see chapter 5). When the deformation

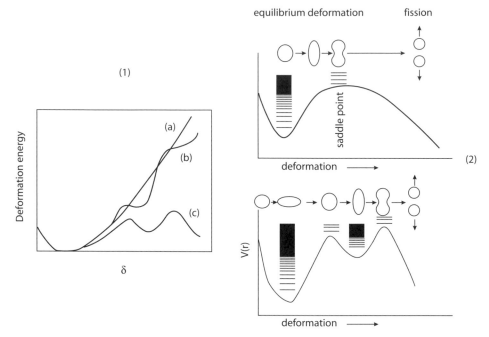

Figure 11.18 (1) Potential energy as a function of deformation: a) for small deformations of an ellipsoidal nucleus; b) with the addition of the shell correction; c) with the existence of a neck. (2) The specific decay rate for fission is determined by the density of states at the fission barrier. The energy released in the descent from saddle point to scission is assumed to appear as kinetic energy of the fragments (adiabatic limit).

increases, the contribution of the shell part changes since the level distribution changes. This contribution is more significant when the deformation reaches such a point that energy levels that were above the last filled level begin to cross it and go below it, creating new options for the nucleons at lower energies. This tends to decrease the energy of the system and decrease the potential energy slope. Increasing the deformation of other level crossings causes new oscillations and the aspect of the function can be that of curve (b) of figure 11.18. To draw the real curve we have yet to consider the appearance of a neck that makes the deformation easier, reducing therefore the deformation energy. In a simple liquid drop treatment the decrease in potential energy from the presence of the neck leads to a single peak, as in figure 11.9. But with the shell correction the final aspect is that of curve (c) of figure 11.18. We see the appearance of a second minimum in the curve, which is the origin of the isomeric fission.

When a nucleus in the actinide region absorbs a nucleon and receives an excitation energy above its fission barrier, the most common result is immediate fission. With a lesser probability the process can end in a *radiative capture* (capture of a nucleon followed by γ-ray emission), leading the nucleus to a state of the first well. But, a third possibility is imprisonment in the region of the second minimum, which occurs in 0.0001% to 0.01% of the cases. The state of lowest energy of the second minimum is an excited isomeric state of the nucleus, which can decay to the first well or fission spontaneously. The fission barrier in this case is much narrower than the barrier the nucleus feels in the first well and this explains the much shorter fission half-life than the half-life of the ground state.

11.11 Exercises

1. It was first pointed out by Bethe and Butler [BB52] that one can verify and determine the shell model single-particle level spectrum through the study of stripping reactions. Using the appropriate selection rules, illustrate this in the case of $^{10}B(d,p)^{11}B$ and compare with the actual experimental data [Aj60].

2. A typical direct reaction is $^{44}Ca(d,p)^{45}Ca$. Assuming the Q-value of the reaction to be 3.30 MeV and the energy of the incident deuteron to be 7.0 MeV, calculate the momentum transfer and cross section using expressions given in this chapter.

3. Build a reaction in which a projectile of $A = 20$ produces a compound nucleus $^{258}_{104}Rf^*$. Using the approximate expression for the Coulomb barrier,

$$B = \frac{Z_1 Z_2 e^2}{1.45(A_1^{1/3} + A_2^{1/3}) \text{ fm}},$$

determine the smallest possible excitation energy for the compound nucleus. Repeat the procedure for a projectile with $A = 50$.

4. With the semi-empirical mass formula (5.7), estimate the binding energy per nucleon for ^{10}B, ^{27}Al, ^{59}Co, and ^{236}U. Compare with the experimental values found in the literature.

5. Using the semi-empirical mass formula, show that the fission of a Z, A nucleus into two equal fragments. a) is energetically possible only for $Z^2/A \geq 18$; b) releases more energy than in different fragments. Do not take into account the pairing energy in these calculations.

6. Show that the energy S_n required to separate a neutron from the nucleus (A,Z) is given approximately by (neglect the pairing term in (5.7)).

$$S_n \approx a_V - \frac{2}{3} a_S A^{-1/3} - a_A \left[1 - \frac{4Z^2}{A(A-1)}\right].$$

Discuss the factors that would lead to corrections to this estimate.

7. Some nuclei with large neutron excess (e.g., $^{31}_{11}Na$) have been produced and have a relatively long lifetime (16 ms). How would such a nucleus decay? Estimate the mass number of the Na nucleus that is just unstable against neutron emission.

8. Evaluate the energy released in each of the processes (11.68) and compare with the value for a symmetric division with the emission of 2 neutrons.

9. Use the chart of nuclides to follow the decay of each of the fragments formed in (11.68) and determine the stable nuclei obtained at the end of the chain.

10. Estimate the maximum relative angular momentum of the two ions when 300 MeV ^{16}O ions bombard a target of ^{107}Ag.

11. In a simple classical model it is assumed that the relative motion of the projectile and target is affected only by the Coulomb force between them if their distance of closest approach d is greater than a critical distance R, while they produce a reaction if $d < R$. By calculating the distance of closest approach as a function of the impact parameter for a Rutherford orbit, show that the total reaction cross section has the form:

$$\sigma_R = \sigma_0(1 - V/E_{cm}) \quad \text{if } E_{cm} > V,$$

where E is the energy in the center of mass coordinate system. Find expressions for the constants σ_0 and V.

12. The table below shows some experimental values of the fusion cross section for ^{32}S on ^{24}Mg as a function of the energy E_{lab} of the ^{32}S nuclei in the laboratory coordinate system. Use these data to test the hypothesis that $\sigma_R = \sigma_{fus}$ at these energies.

E_{lab} [MeV]	70	80	90	110
σ_{fus} [mb]	150	414	670	940

13. Although practically unobserved, one can admit the existence of spontaneous fission processes where the nucleus divides into more than two fragments. Using figure 5.1, find the largest value of n for which the breaking of an $A = 240$ nucleus into n equal fragments is energetically favorable.

14. Thermal neutrons are neutrons in thermal equilibrium with the nuclei of the material that contains them. In this case the velocity of the neutrons follows a Maxwell distribution

$$n(v)dv = 4\pi n \left(\frac{m}{2\pi kT}\right)^{3/2} v^2 \exp\left(-\frac{mv^2}{2kT}\right) dv, \tag{11.77}$$

where $n(v)dv$ is the number of neutrons with velocity between v and $v + dv$, n is the total number of neutrons, m the neutron mass, k the Boltzmann constant, and T the absolute temperature. a) Show that at 20°C the energy corresponding to the most probable velocity of the neutrons is $E_n = 0.025$ eV. b) What is the average energy of these neutrons?

15. Use (11.69) to estimate the average energy of prompt neutrons and the energy for which the probability of neutron emission is a maximum.

16. A 1 MeV neutron incident in a natural uranium sample can fission a ^{235}U or ^{238}U nucleus. Compare the probabilities for the two cases taking into account the proportion of each isotope in the sample and the curves of figure 11.16.

17. A thin film of 0.1 mg of uranium oxide (U_3O_8), made with natural uranium, deposited over a 6 cm x 3 cm mica sheet, is able to register fission fragment tracks coming from the film. The mica is then submitted to a flux of 10^7 thermal neutrons/cm^2/s for 1 hour. Find the number of tracks registered by the mica per cm^2, knowing that the ^{235}U fission cross section for thermal neutrons is 548 b and the proportion of it in natural uranium is 0.7%.

18. Transforming heat into electricity with a conversion efficiency of 5 percent, evaluate the consumption rate of ^{235}U in a nuclear reactor producing electricity with a power of 1000 MAW.

19. a) Using figure 5.1, find the approximate energy released in the spontaneous fission of $^{238}_{92}$U into two fragments of equal masses. b) Which percentage is that energy relative to the $^{238}_{92}$U total binding energy?

20. Using the value 200 MeV as the energy released in each fission, evaluate the energy in joules produced by the fission of 1 g of ^{235}U. How far would 1 g of ^{235}U drive a car which consumes 1 liter of gasoline (density = 0.7 gm/cm^3) for each 10 km? The combustion heat of octane is 5500 kJ/mole, and the combustion engine has an efficiency of 18%.

21. Estimate if fusion of deuterium into helium releases more or less energy per gram of material consumed than fission of uranium.

12 | Nuclear Astrophysics

12.1 Introduction

The hydrogen, deuterium, and most of the helium atoms in the universe are believed to have been created some 20 billion years ago in a primary formation process referred to as the Big Bang, while all other elements have been formed—and are still being formed—in nuclear reactions in the stars. These reaction processes can only be understood in an astrophysical context, as briefly outlined in this chapter, which also describes how nuclear science has provided much understanding about the universe, our solar system and our planets.

The evolution of the universe is the object of study of cosmology and astrophysics; nuclear astrophysics studies the synthesis of heavy nuclei starting from lighter ones in temperature and pressure conditions existing in the stars. Nuclear physics studies the behavior of nuclei under normal conditions or in excited states, as well as the reactions among them. Chemistry studies the structure of the atomic molecules and the reactions among them. Finally, biology studies the formation and development of great molecular agglomerates that compose living beings. In any of these sciences the objective is to understand complex structures starting from simpler structures and from the interactions among them.

Nuclear astrophysics is the field concerning "the synthesis and evolution of atomic nuclei, by thermonuclear reactions, from the Big Bang to the present. What is the origin of the matter of which we are made?" Our high entropy universe, presumably resulting from the Big Bang, contains many more photons than particles of matter with mass, for example, electrons, protons, and neutrons. Because of the high entropy and the consequent low density of matter (on terrestrial or stellar scales) at any given temperature as the universe expanded, there was time to manufacture elements only up to helium, and the major products of cosmic nucleosynthesis remained hydrogen and helium. Stars formed from this primordial matter and used these elements as fuel to generate energy like a giant nuclear reactor. In the process, the stars could shine and manufacture the higher atomic number elements like carbon, oxygen, calcium, and iron of which we and our world are made. The heavy elements are either dredged up from the core to the surface of the star, from which they are dispersed by stellar wind or directly ejected into the interstellar medium when a (massive) star explodes. This stardust is the source of heavy elements for new generations of stars and sunlike systems.

The sun is slowly burning a light element, hydrogen, into a heavier element, helium. It is not exactly the same now as when it first started burning hydrogen in its core and will start to look noticeably different once it exhausts all the core hydrogen. In other words, nuclear reactions in the interiors of stars determine the evolution or the life cycle of the stars, apart from providing them with internal power for heat and light and manufacturing all the heavier elements that the early universe could not.

The emphasis here is on the nuclear reactions in stars and how these are calculated, rather than how stars evolve. The latter usually forms a core area of stellar astrophysics. There is a correspondence between the evolutionary state of a star, its external appearance and internal core conditions, and the nuclear fuel it burns—a sort of a mapping between the astronomer's Hertzsprung-Russell diagram and the nuclear physicist's chart of the nuclides, until nuclear burning takes place on too rapid a timescale.

12.2 Astronomical Observations

12.2.1 The Milky Way

The stars we directly can see all belong to our galaxy, the Milky Way, which is a spiral galaxy, about 30 kpc across and 1 kpc thick. (The kpc, kiloparsec, is the common astronomical unit of distance; 1 parsec is 3.1×10^{16} m, or 3.26 light years, ly.) Thus light travels across our galaxy in about 100,000 years. The Milky Way contains some 200×10^9 stars, and interstellar dust and gas ($\sim$200 pc thick) that spreads out to a diameter of about 50 kpc (hot gas atoms, the *halo*). Our sun is located at the outer edge of one of the spiral arms, about 8.5 kpc from the galactic center. The dust limits sight toward the center to only a few kpc; without this dust the galactic center would shine as bright as our sun. The stars in our galaxy move tangentially around its center with angular velocities increasing closer to the center, indicating the existence of a heavy central object, called Sagittarius (Sgr) A*.

The Milky Way belongs to the Local Group, a cluster of some 20 galaxies that include the Large Magellanic Cloud, our nearest galaxy, 50 kpc away, and the Andromeda galaxy, 650 kpc away. The Local Group is part of the larger Virgo supercluster. The universe contains some 10^{10} galaxies. The galaxies fill only a fraction of space, less than 5%, the rest appears void of matter.

In the 1930s Hubble discovered that galaxies on the whole are equally distributed in all directions of space as observed from the earth. Thus space—on a large scale—seems to be isotropic. This idea of uniformity of the universe is called the *cosmological principle*. This information has been deduced from celestial mechanics (motion of bodies according to Newton's fundamental laws) and from spectroscopic analysis of light and other kinds of radiation. It has been found that the mass of our sun is 1.99×10^{30} kg (=1 solar mass, $M_\odot$). The mass of the Milky Way is $>2 \times 10^{11} M_\odot$; about 10% of the mass is interstellar gas, and 0.1% is dust (typically particles with diameter 0.01–0.1 μm). The interstellar gas density varies considerably in our galaxy; in our part of space it varies from about 10^9 (in dark clouds) to 10^5 atoms/m^3 (on the average $\sim$1 atom/cm^3). Though it contains mainly H and He, large rather complex molecules containing H, C (up to C_{15} molecules), N, and O (including amino acids) have also been discovered.

12.2.2 Dark matter

Astronomical models of the universe indicate that it will expand forever if the observed galaxies alone account for the total mass of the universe. Almost 90% is missing of the mass needed for a slowing down and ultimately contracting universe. Most cosmologists believe that the mass of the observed galaxies is less than 10% of the mass of the universe, the main part consisting of *dark matter*; this includes interstellar and intergalactic matter, neutron stars, "black holes," and other little-known sources of radiation, like quasars, whose masses are unknown.

From mechanics and Newton's gravitational law one can calculate the velocity needed for a body, m_x, to escape the gravitational pull of a larger mass, m, where $m_x \ll m$. For example, if m is the earth's mass (5.94×10^{24} kg), a rocket (mass m_x) must have a velocity of about 11 km/s to escape from the earth's surface (the escape velocity, v_e). Conversely, for a given velocity, v_e, one can calculate the mass and size of the large body needed to hamper such an escape. A body with our solar mass, $M_\odot$, but a radius of only 3 km, requires an escape velocity $>3 \times 10^{10}$ m/s. Thus not even light will escape such a body, which therefore is termed a *black hole*. Though we cannot observe black holes directly, some secondary effects can be observed.

Astronomical observations of star movements support the existence of black holes. For example, from movements of stars close to our galactic center, it is believed that a black hole is located at SgrA* in the center of the Milky Way, with a mass $>3 \times 10^6 M_\odot$. The radius of such a hole would be the same as that of our sun. The density of matter in the hole would be several million times that of our sun (the average value for the sun is about 1400 kg/m^3). Obviously matter cannot be in the same atomic state (nuclei surrounded by electrons) as we know on earth. Instead, we must assume that the electron shells are partly crushed; we refer to this as *degenerate matter*, because the electron quantum rules cannot be upheld. For completely crushed atoms, matter will mainly consist of compact nuclei. For example, for calcium the nuclear density is $\sim 2.5 \times 10^{17}$ kg/m^3.

Even if black holes are given a considerable portion of the missing mass, this will not be enough. However, a very recent discovery may provide the "missing" mass: detailed analysis of the variation in luminosity (a factor of about 2.5) for some 10 million double stars in the Large Magellanic Cloud gives support for the existence of nearby *gravitational microlenses*, which are believed to be unborn stars (so-called *brown dwarfs*) of sizes $\sim$10 $M_\odot$. When such a dark object passes the line of sight to a distant star it acts as a focusing lens for the light, thereby temporarily increasing that star's apparent luminosity. As these

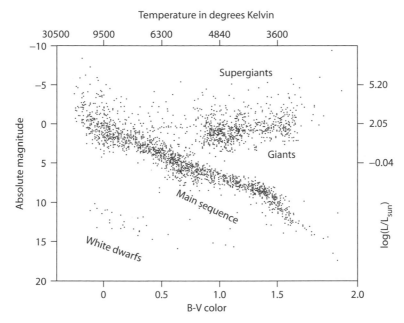

Figure 12.1 *HR* diagram showing surface temperature, spectral class, and color for stars born on the main sequence as a function of their absolute luminosity relative to the solar mass $M_\odot$.

gravitational microlenses seem to be especially abundant in the halo of our galaxy (and presumably in halos of other galaxies), they—together with neutron stars and black holes—could account for the 90% of "missing dark matter" required for an ultimately contracting universe.

12.2.3 Luminosity and Hubble's law

Spectral analysis of the light received from astronomical objects has provided us with information about their (surface) temperature (from their continuous spectrum) and outer chemical composition (from identification of spectral line frequencies), while bolometric measurements have given their luminosity (energy flux density, F/Wm^2). In 1911 Hertzsprung and Russell discovered that, if the luminosity and color (or temperature) of stars in different galaxies is compared with those of similar types of stars in the Milky Way, the stars are distributed according to a certain pattern, the so-called Zero Age Main Sequence (ZAMS); it is believed that most stars in their evolution follow the diagram beginning at the lower right side, along the main sequence into the red giant phase, then to the left and down, decreasing in size and temperature to end as blue or white dwarfs.

Hertzsprung-Russell (HR) diagrams, like the one in figure 12.1, are valid for stars of about 0.7–70 $M_\odot$: from such diagrams conclusions can be drawn about the size (or mass) and relative age of the star, as it is assumed that stars of a given mass follow the same sequence as they age. The apparent luminosity, *F*, which we observe with our telescopes, is related to the *absolute luminosity*, L^*, the total energy flux in all directions from a star, by the relation

$$F = \frac{L^*}{4Bd^2},$$ (12.1)

where d is the distance from the star. The historical classification of stars into brightness classes is now usually replaced by their relative (or apparent) magnitude, m^*, defined as $m = -2.5 \log (F/F_0)$, where F_0 is a reference flux density.

Hubble discovered that all galaxies, except for those in Virgo, show a spectral red shift (increased distance between known frequency lines). This is assumed to be a Doppler effect due to the objects moving away from us (compare the lowering of the pitch from the horn of a train moving away from us). The red shift z is

$$z = (\lambda - \lambda_0)/\lambda_0 = Hd/c ,$$ (12.2)

where H is the *Hubble constant*. For velocities $v \ll c$, the relation becomes $z = v/c$, hence

$$v_r = Hd ,$$ (12.3)

which is the common expression of *Hubble's law*; v_r is the radial velocity. If the red shift is plotted against the apparent magnitude of the brightest star in a large number of galaxies, it is seen to increase with decreasing luminosity, which is interpreted as that the more distant (faintest) galaxies move away from us faster than the closer ones. Except for the galaxies in the Local Group, all galaxies recede from us with velocities up to 20,000 km/s; hence it is concluded that the universe is expanding.

If the universe is expanding, the galaxies were once much closer to each other. If the rate of expansion has been unchanged, the inverse of the Hubble constant, H^{-1}, would represent the age of the universe. In (12.3) v_r is the radial velocity of a galaxy at distance d from us. But velocity is just distance divided by time; that is, $v_r = d/t_0$, where t_0 is the time the expansion has going on, assuming a constant speed. Thus, $d/t_0 = Hd$, or $t_0 = 1/H$. t_0 is only an upper limit of the age of the universe, because we have reasons to believe that the expansion has slowed down due to gravitational pull. According to present estimates an H value of 0.05–0.1 ms^{-1} pc^{-1} corresponds to an age of 10–20 Gy (gigayears, 10^9 years). Cosmologists also give the age in "scale factor" $(1 + z)$-values (red shift values); for example, we would observe a z-value of 10 for an object about one billion years old from the formation of the universe.

12.3 The Big Bang

In 1965 it was discovered that low energy microwave radiation (at 7.35 cm uncorrected) reaches us from all directions in space (about 400 photons cm^{-3}). This is referred to as the cosmic background radiation, whose wavelength corresponds to radiation from a black body of temperature 2.7 K (about 0.0003 eV). Thermodynamic calculations show that this is the temperature reached after adiabatic expansion of a very hot cloud for some 10 billion years. The existence of such background radiation was predicted by Gamow decades before in a cosmological hypothesis referred to as the *Big Bang model.*

The Big Bang hypothesis requires an instantaneous beginning of our universe at a point at which all energy is concentrated. Ordinary nuclear reactions cannot model this beginning, and we must turn to particle physics.

In the following we describe the formation of the universe and the elements according to the so-called *standard model of stellar evolution,* based on the models originally developed by Bethe and Weizsaecker in the 1930s for the reactions in the sun, and the Big Bang hypothesis for the formation of the universe as originally suggested in 1948 by Gamow, Alpher, and Herman, and later developed by the B^{2}FH group, Weinberg, and others.

Around "time zero," the universe consisted of an immensely dense, hot sphere of photons, quarks, leptons, and their antiparticles, in thermal equilibrium, particles being created by photons and photons by annihilation of particles. The temperature must have been $\geq 10^{13}$ K, but no light was emitted, because the enormous gravitational force pulled the photons back. The system was supposed to be in a unique state with no repulsion forces. However, just as a bottle of supercritical (overheated) water can explode by a phase transition, so did the universe, and time began. The universe expanded violently in all directions, and as age and size grew, density and temperature fell.

One-hundredth of a second later all the quarks were gone, and the universe consisted of an approximately equal number of electrons, positrons, neutrinos, and photons, with a small number of protons and neutrons; the ratio of protons to photons is assumed to have been about 10^{-9}. The temperature was about 10^{11} K and the density so high, about 4×10^6 kg m^{-3}, that even the unreactive neutrinos were hindered in escaping. The conditions can be partly understood by considering the relations

$$E(MeV) = mc^2 = 931.5 \Delta M, \tag{12.4}$$

which gives the energy required to create a particle of mass ΔM (E in MeV, ΔM in atomic mass units, u), and

$$E(MeV) = kT = 8.61 \times 10^{-11} T, \tag{12.5}$$

which gives the average kinetic energy of a particle at temperature T (K, Kelvin).

As the photon energy of E (eV) corresponds to the wavelength λ (m) according to

$$E(eV) = h\nu = hc/\lambda = 1.240 \times 10^{-6}/\lambda(m), \tag{12.6}$$

one can estimate that the creation of a proton or a neutron (rest mass 940 MeV) from radiation requires a temperature of 1.1×10^{13} K, corresponding to a photon wave length of about 10^{-15} m, the size of a nucleon. At these temperatures nucleons are formed out of radiation, but are also disrupted by photons, leading to an equilibrium with about an equal number of protons and neutrons. At temperatures below the threshold formation energy, no nucleons are formed. However, it should be remembered that particles and radiation are distributed over a range of energies, according to the Boltzmann and Planck distribution laws. Thus some formation (and disruption) of nucleons occurs even at lower temperatures.

In about 0.1 s the temperature is assumed to have decreased to 3×10^{10} K (corresponding to 2.6 MeV). Now, the equilibrium between protons, neutrons, electrons, and neutrinos can be written

$$p + \bar{\nu} = n + e^+ \; (Q = -1.80 \, \text{MeV}) \quad \text{and} \quad n + \nu = p + e^- \; (Q = 0.78 \, \text{MeV}) . \tag{12.7}$$

The mass of the neutron exceeds that of the proton by a small margin of 0.00013885 u, corresponding to 1.29 MeV; thus reaction on the left of (12.7) requires energy, while reaction on the right releases energy. The formation of protons was therefore favored over that of neutrons, leading to 38% neutrons and 62% protons.

As temperature and density further decreased, the neutrinos began to behave like free particles, and below 10^{10} K they ceased to play any active role in the formation sequence (matter became transparent to the neutrinos). The temperature corresponded then to ~1 MeV, about the threshold energy for formation of positron/electron pairs. Consequently they began to annihilate each other, leaving, for some reason, a small excess of electrons. Though neutrons and protons may react at this temperature, the thermal energies were still high enough to destroy any heavier nuclides eventually formed.

After 14 s the temperature had decreased to 3×10^9 K (0.27 MeV), and 3 min later it had reached about 10^9 K (<0.1 MeV). Now, with the number of electrons, protons, and neutrons about equal (though the universe mostly consisted of photons and neutrinos), some protons and neutrons reacted to form stable nuclides like deuterium and helium. However, for the deuteron to be stable, the temperature must decrease below the Q-value for its formation, about 10^{10} K, and, in reality, to the much lower value of about 10^9 K, because of the high photon flux, which may dissociate the deuteron into two protons. Two deuterium atoms then fuse, probably in several steps as discussed below, to form He. He is an extremely stable nucleus, not easily destroyed, as compared to nuclides with masses >4, whose binding energies (per nucleon) are only a few MeV (figure 4.3).

As the universe expanded, the probability for particle collisions decreased, while the kinetic energy available for fusion reactions was reduced. Therefore, nucleon build-up in practice stopped with ^{4}He, leading to an average universal composition of 73% hydrogen and 27% helium. A very small amount of deuterium was still left, as well as a minute fraction of heavier atoms, formed by the effects of the "Boltzmann tail" and "quantum tunneling." The remaining free neutrons (half-life 10.4 min) now decayed to protons. The situation about 35 minutes after time zero was then the following: temperature 3×10^8 K, density about 10^{-4} kg m^{-3}. The universe consisted of 69% photons, 31% neutrinos, and a fraction of 10^{-9} of particles consisting of 72–78% hydrogen, 28–22% helium, and an equivalent number of free electrons, all rapidly expanding in all directions of space.

It was still too hot for the electrons to join the hydrogen and helium ions to form neutral atoms. This did not occur until about 500,000 years later, when the temperature had dropped to a few 1,000 K. The disappearance of free electrons broke the thermal contact between radiation and matter, and radiation continued to expand freely. An outside spectator would have observed this as a huge flash and a rapidly expanding fireball. In the adiabatic expansion the radiation cooled further to the cosmic background radiation level of

2.7 K measured today. The recent observation that the cosmic background radiation shows ripples in intensity in various directions of space indicates a slightly uneven ejection of matter into space, allowing gravitational forces to act, condensing the denser cloud parts into even more dense regions, or "islands," which with time separated from each other, leaving seemingly empty space in between. Within these clouds, or proto-galaxies, local higher densities led to the formation of stars.

12.4 Stellar Evolution

12.4.1 Stars burn slowly

Energy production in the stars is a well-known process. The initial energy that ignites the process arises from the gravitational contraction of a mass of gas. The contraction increases the pressure, temperature, and density at the center of the star until values are achieved that are able to start thermonuclear reactions, initiating the star lifetime. The energy liberated in these reactions yields a pressure in the plasma, which opposes compression due to gravitation. Thus, an equilibrium is reached for the energy produced, the energy liberated by radiation, temperature, and pressure.

The sun is a star in its initial phase of evolution. The temperature at its surface is 6000° C, while in the interior it reaches 1.5×10^7 K, with a pressure of 6×10^{11} atm and density 150 g/cm^3. The present mass of the sun is $M_\odot = 2 \times 10^{33}$ g and its main composition is hydrogen (70%), helium (29%), and less than 1% heavier elements, like carbon, and oxygen.

For reactions involving charged particles, nuclear physicists often encounter cross sections near the Coulomb barrier of the order of millibarns. One can obtain a characteristic luminosity L_C based on this cross section and the nuclear energy released per reaction [Bah89]:

$$L_C \sim \epsilon N \Delta E / \tau_C, \tag{12.8}$$

where $\epsilon \approx 10^{-2}$ is the fraction of the total number of solar nuclei $N \sim 10^{57}$ that take part in nuclear fusion reactions, generating typically $\Delta E \sim 25$ MeV in hydrogen to helium conversion. Here, the τ_C is the characteristic timescale for reactions, which becomes minuscule for the cross sections at the Coulomb barrier, the ambient density and relative speed of the reactants, and so on:

$$\tau_C \sim \frac{1}{n\sigma v} = \frac{10^{-8}\text{s}}{[n/(10^{26}\text{cm}^{-3})][\sigma/mb][v/10^9 \text{cm s}^{-1}]}. \tag{12.9}$$

This would imply a characteristic luminosity of $L_C \approx 10^{20} L_\odot$, even for a small fraction of the solar material taking part in the reactions ($\epsilon \sim 10^{-2}$). If this were really the appropriate cross section for the reaction, the sun would have been gone very quickly indeed. Instead, the cross sections are much less than that at the Coulomb barrier penetration energy (say, at proton energies of 1 MeV), to allow for a long lifetime of the sun (in addition the weak interaction process gives a smaller cross section for some reactions than the electromagnetic process.

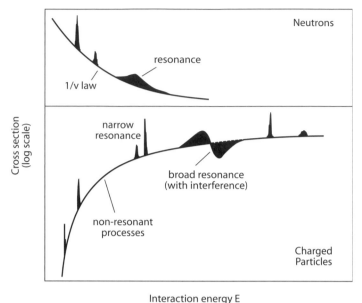

Figure 12.2 Dependence of total cross sections on the interaction energy for neutrons (top panel) and charged particles (bottom panel). Note the presence of resonances (narrow or broad) superimposed on a slowly varying nonresonant cross section.

Stellar nuclear reactions can be either charged particle reactions (both target and projectile are nuclei) or neutral particle (neutron)-induced reactions. Both reactions can go through a resonant state of an intermediate nucleus or can be nonresonant. In the former case, the intermediate state can be a narrow unstable state, which decays into other particles or nuclei. In general, a given reaction can involve both types of reaction channels. In charged particle-induced reactions, the cross section for both reaction mechanisms drops rapidly with decreasing energy, due to the effect of the Coulomb barrier (and thus it becomes more difficult to measure stellar reaction cross sections accurately). In contrast, the neutron induced reaction cross-section is very large and increases with decreasing energy (here, resonances may be superposed on a smooth nonresonant yield which follows the $1/v \sim 1/\sqrt{E}$ dependence). These reaction rates and cross sections can be then directly measured at stellar energies that are relevant (if such nuclei are long lived or can be generated). The schematic dependence of the cross sections is shown in figure 12.2.

12.4.2 Gamow peak and astrophysical S-factor

The sun and other "main sequence" stars (burning hydrogen in their core quiescently) evolve very slowly by adjusting their central temperature such that the average thermal energy of a nucleus is small compared to the Coulomb repulsion an ion-ion pair encounters. This is how stars can live for astronomically long times. A central temperature $T \geq 10^7\,\text{K}$ (or $T_7 \geq 1$; hereafter a subscript x to temperature or density indicates a temperature in units of 10^x) is required for sufficient kinetic energy of the reactants to overcome the Coulomb barrier and for thermonuclear reactions involving hydrogen to proceed at an effective rate, even though fusion reactions have positive Q-values, that is, net energy is liberated out of the reactions.

The classical turning point radius for a projectile of charge Z_2 and kinetic energy E_p (in a Coulomb potential $V_C = Z_1 Z_2 e^2/r$, and an effective height of the Coulomb barrier $E_C = Z_1 Z_2 e^2/R_n = 550$ keV for a p+p reaction), is $r_{cl} = Z_1 Z_2 e^2/E_p$. Thus, classically a p+p reaction would proceed only when the kinetic energy exceeds 550 keV. Since the number of particles traveling at a given speed is given by the *Maxwell-Boltzmann* (MB) distribution $\phi(E)$, only the tail of the MB distribution above 550 keV is effective when the typical thermal energy is 0.86 keV ($T_9 = 0.01$). The ratio of the tails of the MB distributions: $\phi(550 \text{ keV})/\phi(0.86 \text{ keV})$ is quite minuscule, and thus classically at typical stellar temperatures this reaction will be virtually absent.

Although classically a particle with projectile energy E_p cannot penetrate beyond the classical turning point, quantum mechanically, one has a finite value of the square wave function at the nuclear radius $R_n : |\psi(R_n)|^2$. The probability that the incoming particle penetrates the barrier is

$$P = \frac{|\psi(R_n)|^2}{|\psi(R_c)|^2}, \tag{12.10}$$

where $\psi(r)$ are the wavefunctions at corresponding points.

Bethe [Bet37] solved the Schröedinger equation for the Coulomb potential and obtained the transmission probability

$$P = \exp\left(-2KR_c\left[\frac{\tan^{-1}(R_c/R_n - 1)^{1/2}}{(R_c/R_n - 1)^{1/2}} - \frac{R_n}{R_c}\right]\right), \tag{12.11}$$

with $K = [2\mu/\hbar^2(E_c - E)]^{1/2}$. This probability reduces to a much simpler relation at the low energy limit $E \ll E_c$, which is equivalent to the classical turning point R_c being much larger than the nuclear radius R_n. The probability is

$$P = \exp(-2\pi\eta) = \exp[-2\pi Z_1 Z_2 e^2/(\hbar v)] = \exp\left[-31.3 Z_1 Z_2 \left(\frac{\mu}{E}\right)^{1/2}\right], \tag{12.12}$$

where in the second equality μ is the reduced mass in u and E is the center of mass energy in keV.

The exponential quantity involving the square brackets in the second expression is called the *Gamow factor*. The reaction cross section between particles of charge Z_1 and Z_2 has this exponential dependence due to the Gamow factor. In addition, because the cross sections are essentially "areas" proportional to $\pi(\lambda/2\pi\hbar)^2 \propto 1/E$, it is customary to write the cross section with these two energy dependencies filtered out,

$$\sigma(E) = \frac{\exp(-2\pi\eta)}{E} S(E), \tag{12.13}$$

where the factor $S(E)$ is called the astrophysical (or nuclear) S-factor.

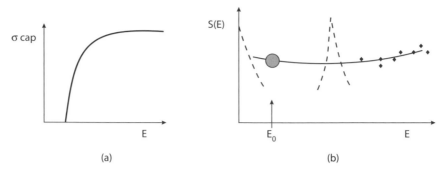

Figure 12.3 Cross section and astrophysical S-factor for charged particle reactions as a function of beam energy. The effective range of energy in stellar interiors is usually far less than the Coulomb barrier energy E_C or the lower limit E_L, where laboratory measurements can be carried out. The cross section drops sharply in the region of astrophysical interest, whereas the change is much less severe for the S-factor. Therefore, necessary extrapolation of laboratory data to lower energies relevant for astrophysical situations is more reliable in the case of the S-factor.

The S-factor may contain degeneracy factors due to spin, for example, $[(2J+1)/(2J_1+1)(2J_2+1)]$, as reaction cross sections are summed over final states and averaged over initial states. Because the rapidly varying parts of the cross section (with energy) are thus filtered out, the S-factor is a slowly varying function of center of mass energy, at least for the nonresonant reactions. It is thus much safer to extrapolate $S(E)$ to the energies relevant for astrophysical environments from the laboratory data, which are usually generated at higher energies (due to difficulties of measuring small cross sections), than directly to extrapolate the $\sigma(E)$, which contains the Gamow transmission factor (see figure 12.3). Additionally, in order to relate $\sigma(E)$ and $S(E)$, quantities measured in the laboratory to these relevant quantities in the solar interior, a correction factor f_0 due to the effects of electron screening needs to be taken into account [Sal54].

In the stellar core with a temperature T, reacting particles have many different velocities (energies) according to a Maxwell-Boltzmann distribution

$$\phi(v) = 4\pi v^2 \left(\frac{\mu}{2\pi kT}\right)^{3/2} \exp\left[-\frac{\mu v^2}{2kT}\right] \propto E \exp\left(-E/kT\right). \tag{12.14}$$

Nuclear cross sections or the reaction rates which also depend upon the relative velocity (or equivalently the center of mass energy) therefore need to be averaged over the thermal velocity (energy) distribution. Therefore, the thermally averaged reaction rate per particle pair is

$$\langle \sigma v \rangle = \int_0^\infty \phi(v)\sigma(v)v\,dv = \left(\frac{8}{\pi\mu}\right)^{1/2} \frac{1}{(kT)^{3/2}} \int_0^\infty \sigma(E)E \, \exp\left(-E/kT\right)dE. \tag{12.15}$$

The thermally averaged reaction rate per pair is, utilizing the astrophysical S-factor and the energy dependence of the Gamow factor,

$$\langle \sigma v \rangle = \left(\frac{8}{\pi\mu}\right)^{1/2} \frac{1}{(kT)^{3/2}} \int_0^\infty S(E) \exp\left[-\frac{E}{kT} - \frac{b}{\sqrt{E}}\right]dE, \tag{12.16}$$

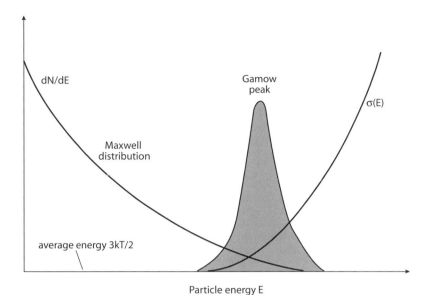

Figure 12.4 The Gamow peak is a convolution of the energy distribution of the Maxwell-Boltzmann probability and the quantum mechanical Coulomb barrier transmission probability. The peak in the shaded region near energy E_0 is the Gamow peak that gives the highest probability for charged particle reactions to take place. Usually the Gamow peak is at a much higher energy than kT, and in the figure the ordinate scale (for the Gamow peak) is magnified with respect to those of the Maxwell-Boltzmann and barrier penetration factors.

with $b^2 = E_G = 2\mu(\pi e^2 Z_1 Z_2/\hbar)^2 = 0.978\mu Z_1^2 Z_2^2$ MeV, E_G being called the Gamow energy. Note that in the expression for the reaction rate above, at low energies, the exponential term $\exp(-b/\sqrt{E}) = \exp(-\sqrt{(E_G/E)})$ becomes very small, whereas at high energies the Maxwell-Boltzmann factor $\exp(-E/kT)$ vanishes. Hence there would be a peak (at energy, say, F_0) of the integrand for the thermally averaged reaction rate per pair (see figure 12.4).

The exponential part of the energy integrand can be approximated as

$$\exp\left[-\frac{E}{kT} - bE^{-1/2}\right] \sim C\exp\left[-\left(\frac{E - E_0}{\Delta/2}\right)^2\right],\tag{12.17}$$

where

$$C = \exp(-E_0/kT - bE_0^{-1/2}) = \exp(-3E_0/kT) = \exp(-\tau),\tag{12.18}$$

with

$$E_0 = (bkT/2)^{\frac{2}{3}} = 1.22\text{keV}(Z_1^2 Z_2^2 \mu T_6^2)^{\frac{1}{3}},\tag{12.19}$$

and

$$\Delta = 4(E_0 kT/3)^{\frac{1}{2}} = 0.75\text{keV}(Z_1^2 Z_2^2 A T_6^5)^{\frac{1}{6}},\tag{12.20}$$

Table 12.1 Parameters of the thermally averaged reaction rates at $T_6 = 15$.

Reaction	Coulomb barrier (MeV)	Gamow peak (E_0) (keV)	I_{max} ($e^{-3E_0/kT}$)	Δ (keV)	$(\Delta)I_{max}$
p + p	0.55	5.9	1.1×10^{-6}	6.4	7×10^{-6}
p + N	2.27	26.5	1.8×10^{-27}	13.6	2.5×10^{-26}
$\alpha + C^{12}$	3.43	56	3×10^{-57}	19.4	5.9×10^{-56}
$O^{16} + O^{16}$	14.07	237	6.2×10^{-239}	40.4	2.5×10^{-237}

Since most stellar reactions happen in a fairly narrow band of energies, $S(E)$ will have a nearly constant value over this band averaging to S_0. With this, the reaction rate per pair of particles turns out to be

$$\langle \sigma v \rangle = \left[\frac{8}{\pi \mu (kT)^3} \right]^{1/2} S_0 \int_0^\infty \exp \left(-\tau - 4 \left(\frac{E - E_0}{\Delta} \right)^2 \right) dE$$

$$= 4.5 \times 10^{14} \frac{S_0}{AZ_1 Z_2} \tau^2 e^{-\tau} \text{cm}^3 s^{-1}. \tag{12.21}$$

Here,

$$\tau = 3E_0/kT = 42.5 (Z_1^2 Z_2^2 \mu / T_6)^{\frac{1}{3}}. \tag{12.22}$$

The maximum value of the integrand in the above equation is

$$I_{max} = \exp(-\tau). \tag{12.23}$$

The values of E_0, I_{max}, Δ, etc., apart from the Coulomb barrier for several reactions are tabulated in table 12.1 for $T_6 = 15$.

As the nuclear charge increases, the Coulomb barrier increases, and the Gamow peak E_0 also shifts toward higher energies. Note how rapidly the maximum of the integrand I_{max} decreases with the nuclear charge and the Coulomb barriers. The effective width Δ is a geometric mean of E_0 and kT, and $\Delta/2$ is much less rapidly varying between reactions (for $kT \ll E_0$). The rapid variation of I_{max} indicates that, of several nuclei present in the stellar core, the nuclear pairs with the smallest Coulomb barrier will have the largest reaction rates. The relevant nuclei will be consumed most rapidly at that stage. (Note, however, that for the p + p reaction, apart from the Coulomb barrier, the strength of the weak force, which transforms a proton to a neutron also comes into play.)

When nuclei of the smallest Coulomb barrier are consumed, there is a temporary dip in the nuclear generation rate, and the star contracts gravitationally until the temperature rises to a point where nuclei with the next lowest Coulomb barrier will start burning. At that stage, further contraction is halted. The star therefore goes through well-defined stages of different nuclear burning phases in its core at later epochs dictated by the height of the Coulomb barriers of the fuels. Note also from table 12.1, how far E_0, the effective mean energy of reaction is below the Coulomb barrier at the relevant temperature. Stellar burning is so slow because the reactions are taking place at such a far sub-Coulomb region, and this is why the stars can last so long.

The above discussion assumes that a bare nuclear Coulomb potential is seen by the charged projectile. For nuclear reactions measured in the laboratory, the target nuclei are in the form of atoms with electron clouds surrounding the nucleus and giving rise to a screened potential—the total potential then goes to zero outside the atomic radius. The effect of the screening is to reduce the effective height of the Coulomb barrier. Atoms in the stellar interiors are in most cases in a highly stripped state, and nuclei are immersed in a sea of free electrons that tend to cluster near the nucleus. When the stellar density increases, the so called *Debye-Huckel radius* $R_D = (kT/4\pi e^2 \rho N_A \xi)^{1/2}$ (here $\xi = \sum_i (Z_i^2 + Z_i)X_i/A_i$), which is a measure of this cluster "radius," decreases, and the effect of shielding upon the reaction cross section becomes more important. This shielding effect enhances thermonuclear reactions inside the star. The enhancement factor $f_0 = \exp(0.188Z_1Z_2\xi\rho^{1/2}T_6^{-3/2}$ varies between 1 and 2 for typical densities and compositions [Sal54] but can be large at high densities.

Therefore, the important ingredients to nucleosynthesis calculations are decay half-lives, electron and positron capture rates, photodisintegrations, neutrino-induced reaction rates, and strong interaction cross sections.

The solution of the above group of equations allows one to deduce the path for the r-process until reaching the heavier elements (see figure 12.9 below). The relative abundances of elements are also obtained theoretically by means of these equations using stellar models for the initial conditions, as the neutron density and the temperature vary.

12.5 The Sun

What are the nuclear processes that give rise to the huge thermonuclear energy of the sun, the latter having lasted 4.6×10^9 years (its assumed age)? It cannot be the simple fusion of two protons, or of α-particles, or even of protons with α-particles, since neither $^2_2$He, $^8_4$Be, nor $^5_3$Li, is stable. The only possibility is proton-proton fusion in the form

$$p + p \rightarrow d + e^+ + \nu_e, \tag{12.24}$$

which occurs via β-decay, that is, due to the weak interaction. The cross section for this reaction for protons of energy around 1 MeV is very small, of the order of 10^{-23} b. The average lifetime of protons in the sun due to the transformation to deuterons by means of (12.24) is about 10^{10}y. This explains why the energy radiated from the sun is approximately constant in time, and not an explosive process.

Because of the low Coulomb barrier, in the $p + p$ reaction ($E_c = 0.55$ MeV), a star like the sun would have consumed all its hydrogen quickly (note the relatively large value of $(\Delta)I_{max}$ in table 12.1), were it not slowed down by the weakness of the weak interactions. The probability calculation for deuteron formation consists of two separate considerations: 1) penetration of a mutual potential barrier in a collision of two protons in a thermal bath and 2) β-decay and positron and neutrino emission. Bethe and Critchfield [Bet38] used the original Fermi theory (point interaction) for the second part, which is adequate for the low energy process.

12.5.1 Deuterium formation

The total Hamiltonian H for the p-p interaction can be written as a sum of a nuclear term H_n and a weak interaction term H_w. As the weak interaction term is small compared to the nuclear term, first order perturbation theory can be applied, and Fermi's golden rule (chapter 6) gives the differential cross section as

$$d\sigma = \frac{2\pi\rho(E)}{hv_i}|\langle f|H_w|i\rangle|^2. \tag{12.25}$$

Here $\rho(E) = dN/dE$, is the density of final states in the interval dE and v_i is the relative velocity of the incoming particles.

For a given volume V, the number of states dN between p and $p + dp$ is

$$dN = dn_e dn_\nu = \left(V\frac{4\pi p_e^2 dp_e}{h^3}\right)\left(V\frac{4\pi p_\nu^2 dp_\nu}{h^3}\right). \tag{12.26}$$

By neglecting the recoil energy of deuterium (since this is much heavier than the outgoing positron in the final state) and neglecting the mass of the electron neutrino, we have $E = E_e + E_\nu = E_e + cp_\nu$ and $dE = dE_\nu = cp_\nu$, for a given E_e, and,

$$\rho(E) = dN(E)/dE = dn_e(dn_\nu/dE) = 16\pi^2 V^2/(c^3 h^6) p_e^2(E - E_e)^2 dp_e = \rho(E_e)dp_e. \tag{12.27}$$

The matrix element that appears in the differential cross section may be written in terms of the initial state wavefunction Ψ_i of the two protons in the entrance channel and the final state wavefunction[1] Ψ_f as

$$H_{if} = \int [\Psi_d \Psi_e \Psi_\nu]^* H_\beta \Psi_i d\tau. \tag{12.28}$$

If the energy of the electron is large compared to $Z \times$ Rydberg (Rydberg $R_\infty = 2\pi^2 me^4/ch^3$), then a plane wave approximation is a good one, $\Psi_e = 1/(\sqrt{V})\exp(i\mathbf{k_e} \cdot \mathbf{r})$, where the wavefunction is normalized over volume V. For lower energies, typically 200 keV or less, the electron wavefunction could be strongly affected by nuclear charge. Apart from this, the final state wavefunction $[\Psi_d \Psi_e \Psi_\nu]$ has a deuteron part Ψ_d whose radial part rapidly vanishes outside the nuclear domain R_0, so that the integration need not extend much beyond $r \simeq R_0$ (for example, the deuteron radius $R_d = 1.7$ fm). Note that because of the Q-value of 0.42 MeV for the reaction, the kinetic energy of the electron ($K_e \leq 0.42$ MeV) and the average energy of the neutrinos ($\bar{E}_\nu = 0.26$ MeV) are low enough that, for both electrons and neutrino wavefunctions, the product $kR_0 \leq 2.2 \times 10^{-3}$, and the exponential can be approximated by the first term of the Taylor expansion

$$\Psi_e = 1/(\sqrt{V})[1 + i(\mathbf{k_e} \cdot \mathbf{r})] \sim 1/(\sqrt{V}), \tag{12.29}$$

---

[1] The same arguments were used in chapter 8 in connection with Fermi's theory for β-decay.

and

$$\Psi_\nu \sim 1/(\sqrt{V}). \tag{12.30}$$

Then the expectation value of the Hamiltonian, given a strength of interaction governed by the coupling constant g, is

$$H_{if} = \int [\Psi_d \Psi_e \Psi_\nu]^* H_\beta \Psi_i d\tau = \frac{g}{V} \int [\Psi_d]^* \Psi_i d\tau. \tag{12.31}$$

The integration over $d\tau$ can be broken into a space part M_{space} and a spin part M_{spin}, so that the differential cross section is

$$d\sigma = \frac{2\pi}{h\nu_i} \frac{16\pi^2}{c^3 h^6} g^2 M_{\text{spin}}^2 M_{\text{space}}^2 p_e^2 (E - E_e)^2 dp_e. \tag{12.32}$$

Thus the total cross section up to an electron energy of E can be obtained by integration as proportional to

$$\int_0^E p_e^2 (E - E_e)^2 dp_e = \frac{(m_e c^2)^5}{c^3} \int_1^W (W_e^2 - 1)^{1/2} (W - W_e)^2 W_e dW_e, \tag{12.33}$$

where $W = (E + m_e c^2)/m_e c^2$.

The integral over W becomes

$$f(W) = (W^2 - 1)^{1/2} \left[\frac{W^4}{30} - \frac{3W^2}{20} - \frac{2}{15} \right] + \frac{W}{4} \ln \left[W + (W^2 - 1)^{1/2} \right], \tag{12.34}$$

so that

$$\sigma = \frac{m_e^5 c^4}{2\pi^3 \hbar^7} f(W) g^2 M_{\text{space}}^2 M_{\text{spin}}^2. \tag{12.35}$$

At large energies, the factor $f(W)$ behaves as

$$f(W) \propto W^5 \propto \frac{1}{30} E^5. \tag{12.36}$$

In the process that we are considering, $p + p \rightarrow d + e^+ + \nu_e$, the final state nucleus (deuterium in its ground state) has $J_f^\pi = 1^+$, with a predominant relative orbital angular momentum $l_f = 0$ and $S_f = 1$ (triplet S-state). For a maximum probability of the process, called the super-allowed transition, there are no changes in the *orbital* angular momentum between the initial and final states of the nuclei. Hence for super-allowed transitions, the initial two interacting protons in the $p + p$ reaction that we are considering must have $l_i = 0$. Since the two protons are identical particles, the Pauli principle requires $S_i = 0$, so that the total wavefunction will be antisymmetric in space and spin coordinates. Thus, we have a process:

$$|S_i = 0, l_i = 0\,\rangle \rightarrow |S_f = 1, l_f = 0\rangle. \tag{12.37}$$

This is a pure Gamow-Teller transition with coupling constant $g = C_A$ due to the axial vector component.

The spin matrix element in the above expression for the energy-integrated cross section σ is obtained from summing over the final states, averaging over the initial states, and dividing by 2 to take into account that we have two identical particles in the initial state. Thus,

$$\lambda = \frac{1}{\tau} = \frac{m^5 c^4}{2\pi^3 \hbar^7 v_i} f(W) g^2 \frac{M_{\text{space}}^2 M_{\text{spin}}^2}{2}, \tag{12.38}$$

where, $M_{\text{spin}}^2 = \frac{(2J+1)}{(2J_1+1)(2J_2+1)} = 3$. And the space matrix element is

$$M_{\text{space}} = \int_0^\infty \chi_f(r) \chi_i(r) r^2 dr, \tag{12.39}$$

in units of $\text{cm}^{3/2}$.

The above integral contains the radial parts of the nuclear wavefunctions $\chi(r)$, and involves Coulomb wavefunctions for barrier penetration at (low) stellar energies. The integral can be evaluated by numerical methods. In the overlap integral one needs only the S-wave part for the wavefunction of the deuteron ψ_d, as the D-wave part makes no contribution to the matrix element [Fri51], although its contribution to the normalization has to be accounted for. The wavefunction of the initial two-proton system ψ_p is normalized to a plane wave of unit amplitude, and again only the S-wave part is needed. The asymptotic form of ψ_p (well outside the range of nuclear forces) is given in terms of regular and irregular Coulomb functions and has to be defined through quantities related to the S-wave phase shifts in p-p scattering data). The result is a minuscule total cross section of $\sigma = 10^{-47} \text{cm}^2$ at a laboratory beam energy of $E_p = 1$ MeV, which cannot be measured experimentally even with milliampere beam currents.

The reaction $p + p \to d + e^+ + \nu_e$ is a nonresonant reaction, and at all energies the rate varies smoothly with energy (and with stellar temperatures), with $S(0) = 3.8 \times 10^{-22}$ keV barn and $dS(0)/dE = 4.2 \times 10^{-24}$ barn. At, for example, the central temperature of the sun $T_6 = 15$, this gives $\langle \sigma v \rangle_{pp} = 1.2 \times 10^{-43} \text{ cm}^3 \text{ s}^{-1}$. For density in the center of the sun $\rho = 100$ gm cm^{-3} and equal mixture of hydrogen and helium ($X_H = X_{He} = 0.5$), the mean life of a hydrogen nucleus against conversion to deuterium is $\tau_H(H) = 1/N_H \langle \sigma v \rangle_{pp} \sim 10^{10}$yr. This is comparable to the age of the old stars. The reaction is so slow primarily because of weak interactions and to a lesser extent because of the smallness of the Coulomb barrier penetration factor (which contributes a factor $\sim 10^{-2}$ in the rate), and is the primary reason why stars consume their nuclear fuel of hydrogen so slowly.

12.5.2 Deuterium burning

Once deuterium is produced in the weak interaction-mediated $p + p$ reaction, the main way this is burned in the sun turns out to be

$$d + p \to {}^3\text{He} + \gamma. \tag{12.40}$$

This is a nonresonant direct capture reaction to the ^{3}He ground state with a Q-value of 5.497 MeV and $S(0) = 2.5 \times 10^{-3}$keV barn. The angle averaged cross sections measured as a function of proton + deuterium center of mass energy, where the capture transitions were observed in γ-ray detectors at several angles to the incident proton beam direction, are well explained by the direct capture model.

The reactions comprising the rest of the (three) pp chains start out with the predominant product of deuterium burning: ^{3}He (manufactured from $d + p$ reaction) as the starting point. The only other reactions with a $S(0)$ greater than the above are $d(d, p)t$, $d(d, n)^3$He, $d(^3$He, p)^{4}He, and $d(^3$He, $\gamma)^5$Li. However, because of the overwhelmingly large number of protons in the stellar thermonuclear reactors, the process involving protons on deuterium dominates. The rate of this reaction is so fast compared to its precursor, $p + p \rightarrow d + e^+ \nu_e$, that the overall rate of the pp chain is not determined by this reaction.

One can show that the abundance ratio of deuterium to hydrogen in a quasi-equilibrium has an extremely small value, signifying that deuterium is destroyed in thermonuclear burning. The time dependence of deuterium abundance D is

$$\frac{dD}{dt} = r_{pp} - r_{pd} = \frac{H^2}{2} \langle \sigma v \rangle_{pp} - HD \langle \sigma v \rangle_{pd}. \tag{12.41}$$

The self-regulating system eventually reaches a state of quasi-equilibrium and has

$$(D/H) = \langle \sigma v \rangle_{pp} / (2 \langle \sigma v \rangle_{pd}) = 5.6 \times 10^{-18}, \tag{12.42}$$

at $T_6 = 5$ and 1.7×10^{-18} at $T_6 = 40$. For the solar system however, this ratio is 1.5×10^{-4} and the observed $(D/H)_{obs}$ ratio in the cosmos is $\sim 10^{-5}$. The higher cosmic ratio is due to primordial nucleosynthesis in the early phase of the universe before the stars formed. (The primordial deuterium abundance is a key quantity used to determine the baryon density in the universe). Stars only destroy the deuterium in their core due to the above reaction.

12.5.3 ^{3}He burning

The ppI chain is completed (see figure 12.5) through the burning of ^{3}He via the reaction

$$^3\text{He} + {}^3\text{He} \rightarrow p + p + {}^4\text{He}, \tag{12.43}$$

with an S-factor $S(0) = 5500$ keV barn and Q-value = 12.86 MeV. In addition, the reaction

$$^3\text{He} + D \rightarrow {}^4\text{He} + p \tag{12.44}$$

has an S-factor $S(0) = 6240$ keV barn, but since the deuterium concentration is very small as argued above, the first reaction dominates the destruction of ^{3}He even though both reactions have comparable $S(0)$ factors.

^{3}He can also be consumed by reactions with ^{4}He (the latter is preexisting from the gas cloud from which the star formed and is synthesized in the early universe and in population III objects). These reactions proceed through direct captures and lead to the ppII and ppIII parts of the chain (happening 15% of the time). Note that the reaction ^{3}He$(\alpha, \gamma)^7$Be and the subsequent reaction ^{7}Be$(p, \gamma)^8$B control the production of high energy neutrinos in the sun and are particularly important for the ^{37}Cl solar neutrino detector constructed by Ray Davis and collaborators [Bah89].

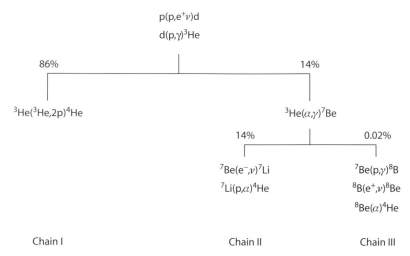

Figure 12.5 The p-p chain reaction (p-p cycle). The percentages for the several branches are calculated for the center of the sun [Bah89].

12.5.4 Reactions involving ^{7}Be

As shown in figure 12.5, about 14% of the time, ^{3}He is burned with ^{4}He radiatively to ^{7}Be. Subsequent reactions involving ^{7}Be as a first step in alternate ways complete the fusion process: 4H → ^{4}He in the ppII and ppIII chains.

Electron capture

The first step of the ppII chain is the electron capture reaction on ^{7}Be: ^{7}Be+e$^-$ → ^{7}Li + ν_e. This decay goes both to the ground state of ^{7}Li and to its first excited state at $E_X = 0.478$ keV, $J^\pi = \frac{1}{2}^-$)—the percentage of decays to the excited state being 10.4 % in the laboratory. The energy released in the reaction with a Q-value of 0.862 keV is carried away by escaping monoenergetic neutrinos with energies $E_\nu = 862$ and 384 keV. The measured laboratory mean life of the decay is $\tau = 76.9$d.

The capture rate in the laboratory can be obtained from Fermi's golden rule and utilizing the fact that the wavefunctions of both the initial and the final nucleus vanish rapidly outside the nuclear domain. The electron wavefunction in that domain. The approximated as its value at $r = 0$ and the neutrino wavefunction by a plane wave normalized to volume V, so that $H_{if} = \Psi_e(0)g/\sqrt{V} \int \Psi^*_{^7Li}\Psi_{^7Be}d\tau = \Psi_e(0)gM_n/\sqrt{V}$, where M_n represents the nuclear matrix element and the resultant capture rate is

$$\lambda_{EC} = \frac{1}{\tau_{EC}} = \frac{g^2 M_n^2}{\pi c^3 \hbar^4} E_\nu^2 |\Psi_e(0)|^2. \tag{12.45}$$

In the laboratory capture process, any of the various electron shells contribute to the capture rate; however the K shell gives the dominant contribution. At temperatures inside the sun, for example, $T_6 = 15$, nuclei such as ^{7}Be are largely ionized. The nuclei, however, are immersed in a sea of free electrons resulting from the ionized process, and therefore electron capture from continuum states is possible. Since all factors in the capture of continuum electrons in the sun are approximately the same as those in the case of atomic electron capture, except for the respective electron densities, the ^{7}Be lifetime in a star, τ_{fr} is related to the terrestrial lifetime τ_t by

$$\frac{\tau_{fr}}{\tau_t} \sim \frac{2|\Psi_t(0)|^2}{|\Psi_{fr}(0)|^2}, \tag{12.46}$$

where $|\Psi_{fr}(0)|^2$ is the density of the free electrons $n_e = \rho/m_H$ at the nucleus, ρ being the stellar density. The factor of 2 in the numerator takes care of the two spin states of calculation of the λ_t, whereas the corresponding λ_{fr} is calculated by averaging over these two orientations.

Taking account of distortions of the electron wavefunctions due to the thermally averaged Coulomb interaction with nuclei of charge Z and contribution due to hydrogen (of mass fraction X_H) and heavier nuclei, one gets the continuum capture rate as

$$\tau_{fr} = \frac{2|\Psi_t(0)|^2 \tau_t}{(\rho/M_H)[(1+X_H)/2]2\pi Z\alpha(m_e c^2/3kT)^{1/2}}, \tag{12.47}$$

with $|\Psi_e(0)|^2 \sim (Z/a_0)^3/\pi$. Bahcall et al. [Bah69] obtained for the ^{7}Be nucleus a lifetime

$$\tau_{fr}(^7\text{Be}) = 4.72 \times 10^8 \frac{T_6^{1/2}}{\rho(1+X_H)}\text{s.} \tag{12.48}$$

The temperature dependence comes from the nuclear Coulomb field corrections to the electron wavefunction, which are thermally averaged. For solar conditions the above rate [Bah69b] gives a continuum capture rate of $\tau_{fr}(^7\text{Be}) = 140$ d as compared to the terrestrial mean life of $\tau_t = 76.9$ d. Actually, under stellar conditions, there is a partial contribution from some ^{7}Be atoms which are only partially ionized, leaving electrons in the inner K shell. So the contributions of such partially ionized atoms have to be taken into account. Under solar conditions the K shell electrons from partially ionized atoms give another 21% increase in the total decay rate. Including this, gives the solar lifetime of a ^{7}Be nucleus as $\tau_\odot(^7\text{Be}) = 120$ d. In addition, the solar fusion reactions have to be corrected for plasma electrostatic screening enhancement effects.

Formation of ^{8}B

Apart from the electron capture reaction, the ^{7}Be that is produced is partly consumed by proton capture via $^7\text{Be}(p,\alpha)^8\text{B}$ reaction. Under solar conditions, this reaction happens only 0.02% of the time. The proton capture on ^{7}Be proceeds at energies away from the 640 keV resonance via the direct capture process. Since the product ^{7}Li nucleus emits an intense γ-ray flux of 478 keV, this prevents the direct measurement of the direct capture to ground state γ-ray yield. The process is studied indirectly by either the delayed positron or the breakup of the product ^{8}B nucleus into two α-particles. This reaction has a weighted average $S(0) = 0.0238$ keV barn [Fil83].

The product ^{8}B is a radioactive nucleus that decays with a lifetime $\tau = 1.1$ s

$$^8\text{B} \rightarrow {}^8\text{Be} + e^+ + \nu_e. \tag{12.49}$$

The positron decay of ^{8}B($J^\pi = 2^+$) goes mainly to the $\Gamma = 1.6$ MeV broad excited state in ^{8}Be at excitation energy $E_x = 2.94$ MeV ($J^\pi = 2^+$) due to the selection rules. This excited state has a very short lifetime and quickly decays into two α-particles. This completes the ppIII part of the pp chain. The average energy of the neutrinos from ^{8}B reactions is $\bar{E}_\nu(^8\text{B}) = 7.3$ MeV. These neutrinos, having relatively high energy, play an important role in several solar neutrino experiments.

12.6 The CNO Cycle

The sun gets most of its energy generation through the pp chain reactions (see figure 12.7). However, as the central temperature (in stars more massive than the sun) gets higher, the CNO cycle (see below for reaction sequence) comes to dominate over the pp chain at T_6 near 20 (this changeover assumes the solar CNO abundance; the transition temperature depends upon CNO abundance in the star).

The early generation of stars (usually referred to as the Population II or Pop II stars, although there is an even earlier generation of Pop III stars) generated energy primarily through the pp chain. These stars are still shining in globular clusters, and being of mass lower than that of the sun, are very old. Most other stars that we see today are later generation stars formed from the debris of heavier stars that contained heavy elements apart from (the most abundant) hydrogen. Thus in second and third generation stars (which are slightly heavier than the sun), where higher central temperatures are possible because of higher gravity, hydrogen burning can take place through faster chains of reactions involving heavy elements C, N, and O which have some reasonable abundance (exceeding 1%)) compared to other heavy elements like Li, Be, B, which are extremely low in abundance. The favored reactions involve heavier elements (than those of the pp chain), which have the smallest Coulomb barriers but with reasonably high abundance. Even though the Coulomb barriers of Li, Be, B are smaller than those of C, N, O (when protons are the lighter reactants (projectiles)), they lose out due to their lower abundance.

In 1937–1938, Bethe and von Weizsäcker independently suggested the CN part of the cycle, which goes as (see figure 12.6)

$$^{12}\text{C}(p,\gamma)^{13}\text{N}(e^+\nu_e)^{13}\text{C}(p,\gamma)^{14}\text{N}(p,\gamma)^{15}\text{O}(e^+\nu)^{15}\text{N}(p,\alpha)^{12}\text{C}. \tag{12.50}$$

This has the net result, as before, $4p \rightarrow {}^4\text{He} + 2e^+ + 2\nu_e$ with $Q = 26.73$. In these reactions, the ^{12}C and ^{14}N act merely as catalysts as their nuclei are "returned" at the end of the cycle. Therefore the ^{12}C nuclei act as seeds that can be used over and over again, even though the abundance of the seed material is minuscule compared to that of the hydrogen. Note that a loss of the catalytic material from the CN cycle takes place through the ^{15}N(p,γ)^{16}O reactions. However, the catalytic material is subsequently returned to the CN cycle by the reaction ^{16}O(p,γ)^{17}F($e^+\nu_e$)^{17}O(p,α)^{14}N.

In the CN cycle (see figure 12.6), the two neutrinos involved in the β-decays (of ^{13}N ($t_{1/2} = 9.97$min) and ^{15}O ($t_{1/2} = 122.24$s)) are of relatively low energy, and most of the total energy $Q = 26.73$ MeV from the conversion of four protons into helium is deposited in the

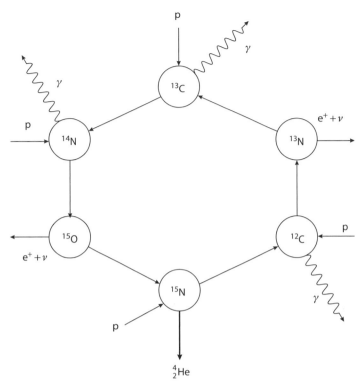

Figure 12.6 The CNO cycle [Bet38].

stellar thermonuclear reactor. The rate of the energy production is governed by the slowest thermonuclear reaction in the cycle. Here nitrogen isotopes have the highest Coulomb barriers in charged particle reactions, because their $Z = 7$. Among them $^{14}N(p, \gamma)^{15}O$ is the slowest because this reaction, having a final state photon, is governed by electromagnetic forces while that involving the other nitrogen isotope: $^{15}N(p, \alpha)^{12}C$ is governed by strong forces and is therefore faster.

From the CN cycle, there is actually a branching off from ^{15}N by the reaction $^{15}N(p, \gamma)^{16}O$ mentioned above. This involves isotopes of oxygen and is called the ON cycle; finally the nitrogen is returned to the CN cycle through ^{14}N. Together, the CN and ON cycles constitutes the CNO bi-cycle. The two cycles differ considerably in their relative cycle rates: the ON cycle operates only once for every 1000 cycles of the main CN cycle. This can be gauged from the $S(0)$ factors of the two sets of reactions branching off from ^{15}N: for the $^{15}N(p, \alpha)^{12}C$ reaction $S(0) = 65$ MeV b, whereas for $^{15}N(p, \gamma)^{16}O$, it is 64 keV b, a factor of 1000 smaller.

12.6.1 Hot CNO and rp process

The above discussion of the CNO cycle is relevant for typical temperatures $T_6 \geq 20$. These are found in quiescently hydrogen burning stars with solar composition which are only slightly more massive than the sun. There are situations where the hydrogen burning takes

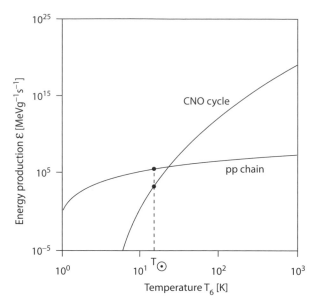

Figure 12.7 Comparison of the energy production in the pp and in the CNO cycle as a function of the star temperature [Ib65].

place at temperatures ($T \sim 10^8$–10^9 K), far in excess of those found in the interiors of the ordinary "main sequence" stars. Examples of these are hydrogen burning at the accreting surface of a neutron star or in the explosive burning on the surface of a white dwarf, that is, novae, or the outer layers of a supernova shock heated material in the stellar mantle.

These hot CNO cycles operate under such conditions on a rapid enough timescale (a few seconds) so that even "normally" β-unstable nuclei like ^{13}N will live long enough to be burned by thermonuclear charged particle reactions, before they are able to β-decay. So, unlike the normal CNO the amount of hydrogen to helium conversion in hot CNO is limited by the β-decay lifetimes of the proton-rich nuclei like ^{14}O and ^{15}O rather than the proton capture rate of ^{14}N. For temperatures, $T \geq 5 \times 10^8$K, nucleosynthesised material can leak out of the cycles. This leads to a diversion from lighter to heavier nuclei and is known as the rapid proton capture or rp process.

The nucleosynthesis path of the rp process of rapid proton addition is analogous to the r process of neutron addition. The hot hydrogen bath converts CNO nuclei into isotopes near the region of proton unbound nuclei (the proton drip line). For each neutron number, a maximum mass number A is reached where the proton capture must wait until β^+-decay takes place before the buildup of heavier nuclei (for an increased neutron number) can take place. Unlike the r process the rate of the rp process is increasingly hindered due to the increasing Coulomb barrier of heavier and higher-Z nuclei to proton projectiles. Thus the rp process does not extend all the way to the proton drip line but runs close to the β-stability valley and runs through where the β^+-decay rate compares favorably with the proton captures. A comparison of the reaction paths of rp and r processes in the (N, Z) plane is given in figure 12.9.

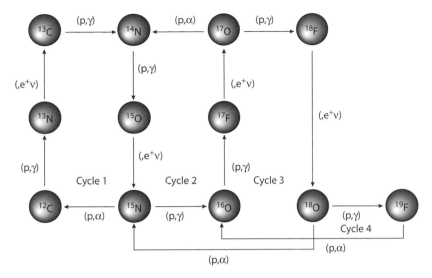

Figure 12.8 The various CNO cycles. The left part is the CN cycle where only C and N serve as catalysts for the conversion of four protons into ^{4}He. Here the slowest fusion reaction is the (p,γ) reaction on ^{14}N, whereas the slower β-decay has a half-life of 9.97m. In the CNO cycle 2 (middle), there is leakage from the CN cycle to the ON cycle through the branching at ^{15}N. The flow is returned to the CN cycle (which cycles 1000 times for each ON cycle) through $^{17}O(p,\alpha)^{14}$N. The right part represents additional cycles linking into the CNO cycle through the $^{17}O(p,\gamma)^{18}$F reaction [courtesy of Frank Timmes].

12.7 Helium Burning

After hydrogen burning in the core of the star has exhausted its fuel, the helium core contracts slowly. Its density and temperature go up as gravitational energy released is converted to internal kinetic energy. The contraction also heats hydrogen at the edge of the helium core, igniting the hydrogen to burn in a shell. At a still later stage in the star's evolution, the core has contracted enough to reach central temperature density conditions: $T_6 = 100$–200 and $\rho_c = 10^2$–10^5 gm cm^{-3} when the stellar core settles down to burn ^{4}He in a stable manner. The product of helium burning is ^{12}C. Since in nature the $A = 5$ and $A = 8$ nuclei are not stable, the question arises as to how helium burning bridges this gap. A direct interaction of three α-particles to produce a ^{12}C nucleus ($\alpha + \alpha + \alpha \rightarrow ^{12}C$) would seem at first sight, to be too improbable (as was mentioned, for example, in Bethe's 1939 paper [Bet39], which was primarily on the CN cycle). However, Öpik [Opi51] and Salpeter [Sal52, Sal57] independently proposed a two-step process where in the first step, two α-particles interact to produce a ^{8}Be nucleus in its ground state (which is unstable to α breakup), followed by the unstable nucleus interacting with another α-particle process to produce a ^{12}C nucleus.

Thus the triple-α reaction begins with the formation of ^{8}Be that has a lifetime of only 1×10^{-16} s (this is found from the width $\Gamma = 6.8$ eV of the ground state and is the cause of the $A = 8$ mass gap). This is, however, long compared to the transit time 1×10^{-19} s of two α-particles to scatter past each other nonresonantly with kinetic energies comparable to the Q-value of the reaction namely, $Q = -92.1$ keV. So it is possible to have an equilibrium

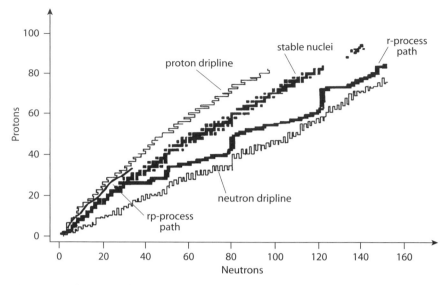

Figure 12.9 Schematic paths of the r process and rp process in the N, Z-plane with respect to the valley of β stability and the neutron drip and proton drip lines.

build-up of a small quantity of ^{8}Be in equilibrium with its decay or reaction products $\alpha + \alpha \rightarrow {}^8$Be. The equilibrium concentration of the ^{8}Be nucleus can be calculated through the Saha equation

$$N_{12} = \frac{N_1 N_2}{2} \left(\frac{2\pi}{\mu kT} \right)^{3/2} \hbar^3 \frac{(2J+1)}{(2J_1+1)(2J_2+1)} \exp\left(-\frac{E_R}{kT} \right), \tag{12.51}$$

at the relevant temperature $T_6 = 11$ and $\rho = 10^5$ gm cm^{-3} to be

$$\frac{N(^8\text{Be})}{N(^4\text{He})} = 5.2 \times 10^{-10}. \tag{12.52}$$

Salpeter suggested that this small quantity of ^{8}Be serves as the seed for the second stage of the triple-α capture into the ^{12}C nucleus. It was, however, shown by Hoyle [Hoy53] that the amount of ^{12}C produced for the conditions inside a star at the tip of the red-giant branch is insufficient to explain the observed abundance, *unless* the reaction proceeds through a resonance process [Hoy54]. The presence of such a resonance greatly speeds up the rate of the triple-α process, which then proceeds through an s-wave ($l = 0$) resonance in ^{12}C near the threshold of the ^{8}Be $+ \alpha$ reaction. Since ^{8}Be and ^{4}He both have $J^{\pi} = 0^+$, an s-wave resonance would imply that the resonant state in question has to be 0^+ in the ^{12}C nucleus.

Hoyle suggested the excitation energy to be: $E_X \sim 7.68$ MeV in the ^{12}C nucleus, and this state was experimentally found by W.A. Fowler's group [Coo57] with spin-parity $J^{\pi} = 0^+$. This state has a total width [RR88] $\Gamma = 8.9 \pm 1.08$ eV, most of which lies in Γ_{α}, due to the major propensity of the ^{12}C nucleus to break up through α-decay. (The decay of the excited state of ^{12}C by γ-rays cannot go directly to the ground state, since the resonance state as well as the ground state of the ^{12}C nucleus have both $J^{\pi} = 0^+$ and $0^+ \rightarrow 0^+$ decays forbidden. This partial width due to γ-decay is several thousand times smaller than that due to α-decay). So, $\Gamma = \Gamma_\alpha + \Gamma_{rad} \sim \Gamma_\alpha$ and $\Gamma_{rad} = \Gamma_\gamma + \Gamma_{e^+e^-} = 3.67 \pm 0.50\,\text{meV}$. Again the radiative width Γ_{rad} is dominated by the width due to photon width deexcitation: $\Gamma_\gamma = 3.58 \pm 0.46\,\text{meV}$. (Note the scales of *millielectron* volts.)

The reaction rate for the ^{12}C formation can be calculated by using the properties of the resonant state and the thermally averaged cross section

$$r_{3\alpha} = N_{^8\text{Be}} N_\alpha \langle \sigma v \rangle_{^8\text{Be}+\alpha}. \tag{12.53}$$

Here $N_{^8\text{Be}}$ and N_α are the number densities of interacting ^8Be and ^4He nuclei and the angular brackets denote thermal averaging over a Maxwell-Boltzmann distribution $\psi(E)$. This averaging leads to

$$r_{3\alpha} = N_{^8\text{Be}} N_\alpha \int_0^\infty \psi(E) v(E) \sigma(E) dE, \tag{12.54}$$

with

$$\psi(E) = \frac{2}{\sqrt{\pi}} \frac{E}{kT} \exp\left(-E/kT\right) \frac{dE}{(kTE)^{1/2}}, \tag{12.55}$$

and

$$\sigma(E) = \pi \left(\frac{\lambda}{2\pi}\right)^2 \frac{2J+1}{(2J_1+1)(2J_2+1)} \frac{\Gamma_1 \Gamma_2}{(E-E_R)^2 + (\Gamma/2)^2} \tag{12.56}$$

is the Breit-Wigner resonant reaction cross section with the resonant energy centroid at $E = E_R$. The total width Γ is a sum of all decay channel widths such as $\Gamma_1 = \Gamma_\alpha$ and $\Gamma_2 = \Gamma_\gamma$.

If the width Γ is only a few eV, then the functions $\psi(E)$ and $v(E)$ can be pulled out of the integral. Then the reaction rate will contain an integral like $\int_0^\infty \sigma_{BW}(E)dE = 2\pi (\lambda/2\pi\hbar)^2 \omega \Gamma_1 \Gamma_2 / \Gamma$, where $\omega = (2J+1)/[(2J_1+1)(2J_2+1)]$, and the functions pulled out of the integral need to be evaluated at $E = E_R$. Since most of the time the excited state of the $^{12}\text{C}^*$ breaks up into α-particles, we have $\Gamma_1 = \Gamma_\alpha$ dominating over Γ_γ and $(\Gamma_1\Gamma_2/\Gamma) \sim \Gamma_2$. This limit usually holds for resonances of energy sufficiently high that the incident particle width (Γ_1) dominates the natural width of the state (Γ_2). In that case, we can use the number density of the ^8Be nuclei in equilibrium with the α-particle nuclei bath as described by the Saha equilibrium condition

$$N(^8\text{Be}) = N_\alpha^2 \omega f \frac{h^3}{(2\pi\mu kT)^{3/2}} \exp\left(-E_r/kT\right), \tag{12.57}$$

where f is the screening factor.

It is possible to get the overall triple-α reaction rate by calculating the equilibrium concentration of the excited (resonant) state of ^{12}C reached by the $^8\text{Be} + \alpha \to {}^{12}\text{C}^*$ reaction and then multiplying that concentration by the γ-decay rate $\Gamma_\gamma/\hbar$ which leads to the final product of ^{12}C. So the reaction rate for the final step of the triple-α reaction turns out to be

$$r_{3\alpha} = N_{^8\text{Be}} N_\alpha \hbar^2 \left(\frac{2\pi}{\mu kT}\right)^{3/2} \omega f \Gamma_2 \exp\left(-E_r'/kT\right), \tag{12.58}$$

where μ is the reduced mass of the reactants ^{8}Be and α-particle. This further reduces by the above argument to

$$r_{3\alpha \to {}^{12}C} = \frac{N_\alpha^3}{2} 3^{3/2} \left(\frac{2\pi\hbar^2}{M_\alpha kT} \right)^3 f \frac{\Gamma_\alpha \Gamma_\gamma}{\Gamma\hbar} \exp\left(-\frac{Q}{kT} \right). \tag{12.59}$$

The Q-value of the reaction is the sum of $E_R(^8\text{Be} + \alpha) = 287$ keV and $E_R(\alpha + \alpha) = |Q| = 92$ keV and turns out to be $Q_{3\alpha} = (M_{{}^{12}C} - 3M_\alpha)c^2 = 379.38 \pm 0.20$ keV. Numerically, the energy generation rate for the triple-α reaction is

$$\epsilon_{3\alpha} = \frac{r_{3\alpha} Q_{3\alpha}}{\rho} = 3.9 \times 10^{11} \frac{\rho^2 X_\alpha^3}{T_8^3} f \exp\left(-42.94/T_8 \right) \text{ erg gm}^{-1} \text{ s}^{-1}. \tag{12.60}$$

The triple-α reaction has a very strong temperature dependence near a value of temperature T_0, and one can show that the energy generation rate is

$$\epsilon(T) = \epsilon(T_0) \left(\frac{T}{T_0} \right)^n, \tag{12.61}$$

where $n = 42.9/T_8 - 3$. Thus at a sufficiently high temperature and density, the helium gas is very highly explosive, so that a small temperature rise gives rise to greatly accelerated reaction rate and energy liberation. When helium thermonuclear burning is ignited in the stellar core under degenerate conditions, an unstable and sometimes explosive condition develops.

12.8 Red Giants

The product of the triple-α reactions, ^{12}C, is burned into ^{16}O by α-capture reactions

$$^{12}\text{C} + \alpha \to {}^{16}\text{O} + \gamma. \tag{12.62}$$

If this reaction proceeds too efficiently, then all the carbon will be burned up to oxygen. Carbon is, however, the most abundant element in the universe after hydrogen, helium, and oxygen, and the cosmic C/O ratio is about 0.6. In fact, the O and C burning reactions and the conversion of He into C and O take place in similar stellar core temperature and density conditions. Major ashes of He burning in red giant stars are C and O. Red giants are the source of the galactic supply of ^{12}C and ^{16}O. Fortuitous circumstances of the energy level structures of these α-particle nuclei are in fact important for the observed abundance of oxygen and carbon.

For example, if as in the case of the triple-α reaction, there were a resonance in the $^{12}\text{C}(\alpha, \gamma)^{16}\text{O}$ reaction near the Gamow window for He burning conditions ($T_9 \sim 0.1 - 0.2$), then the conversion $^{12}\text{C} \to {}^{16}\text{O}$ would proceed at a very rapid rate. However, the energy level diagram of ^{16}O shows that for temperatures up to about $T_9 \sim 2$, there is no level available in ^{16}O to foster a resonant reaction behavior. But since this nucleus is found in nature, its production must go through either: 1) a nonresonant direct capture reaction or 2) nonresonant captures into the tails of nearby resonances (subthreshold reactions).

Figure 12.10 shows, on the right of the ^{16}O energy levels, the threshold for the $^{12}\text{C} + {}^4\text{He}$ reaction, drawn at the appropriate level with respect to the ground state of the ^{16}O nucleus.

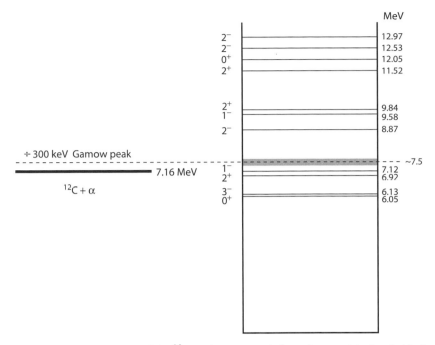

Figure 12.10 Energy levels of the ^{16}O nucleus near and above the α-particle threshold of capture on ^{12}C. The reaction rate is influenced mainly by the high energy tails of two sub-threshold resonances in ^{16}O at $E_R = -45$ keV and $E_R = -245$ keV, plus the low energy tail of another high-lying broad resonance at 9580 keV.

The Gamow energy for temperatures $T_9 = 0.1$ and above indicates that for the expected central temperatures, the effective stellar (center of mass) energy region is near $E_0 = 0.3$ MeV. This energy region is reached by the low energy tail of a broad resonance centered at $E_{CM} = 2.42$ MeV above the threshold (the $J^\pi = 1^-$ state at 9.58 MeV above the ground state of ^{16}O) with a (relatively large) resonance width of 400 keV.

On the other hand, there are two subthreshold resonances in ^{16}O (at $E_X = 7.12$ MeV and $E_X = 6.92$ MeV), that is, -45 keV and -245 keV *below* the α-particle threshold that have $J^\pi = 1^-$ and $J^\pi = 2^+$, which contribute to the stellar burning rate by their high energy tails. However, electric dipole (E1) γ-decay of the 7.12 MeV state is inhibited by isospin selection rules. Had this not been the case, the ^{12}C$(\alpha, \gamma)^{16}$O reaction would have proceeded fast and ^{12}C would have been consumed during helium burning itself. The two subthreshold states at -45 keV and -245 keV give contributions to the astrophysical S factor of $S_{1-}(E_0) = 0.1$ MeV barn and $S_{2+}(E_0) = 0.2$ MeV barn, respectively, at the relevant stellar energy $E_0 = 0.3$ MeV. The state at $E_{CM} = 2.42$ MeV ($J^\pi = 1^-$ state at 9.58 MeV) gives a contribution $S_{1-}(E_0) = 1.5 \times 10^{-3}$ MeV barn. The total S-factor at $E_0 = 0.3$ MeV is therefore close to 0.3 MeV barn. These then provide low enough S or cross section not to burn the ^{12}C away entirely to ^{16}O, so that C/O ~ 0.1 at the least.

Additionally, ^{16}O nuclei are not burned away by further α-capture in the reaction

$$^{16}\text{O} + {}^4\text{He} \rightarrow {}^{20}\text{Ne} + \gamma. \tag{12.63}$$

A look at the level schemes of ^{20}Ne (see figure 12.11) shows the existence of an $E_X = 4.97$ MeV state ($J^\pi = 2^-$) in the Gamow window. However, this state cannot form in the resonance reaction due to considerations of parity conservation (unnatural parity of the resonant state). The lower 4.25 MeV state ($J^\pi = 4^+$) in ^{20}Ne also cannot act as a subthreshold resonance as it lies too far below threshold and is formed in the g-wave state. Therefore only direct capture reactions seem to be operative, which for (α, γ) reactions lead to cross sections in the range of nanobarns or below. Thus the destruction of the ^{16}O via ^{16}O$(\alpha, \gamma)^{20}$Ne reaction proceeds at a very slow rate during the stage of helium burning in red giant stars, for which the major ashes are carbon and oxygen.

To summarize, the synthesis of two important elements for the evolution of life as we know it on the earth depended on fortuitous circumstances of nuclear properties and selection rules for nuclear reactions. These are: 1) the mass of the unstable lowest (ground) state of ^{8}Be is close to the combined mass of two α-particles; 2) the resonance in ^{12}C at 7.65 MeV, enhances the α addition reaction (the second step); and 3) parity conservation has protected ^{16}O from being destroyed in the ^{16}O$(\alpha, \gamma)^{20}$Ne reactions by making the 4.97 MeV excited state in ^{20}Ne of unnatural parity.

12.9 Advanced Burning Stages

As helium burning progresses, the stellar core is increasingly made up of C and O. At the end of helium burning, all hydrogen and helium is converted into a mixture[2] of C and O, and since H, He are the most abundant elements in the original gas from which the star formed, the amounts of C and O are far more than the traces of heavy elements in the gas cloud.

Between these two products, the Coulomb barrier for further thermonuclear reaction involving the products is lower for C nuclei. At first the C + O rich core is surrounded by He burning shells and a helium rich layer, which in turn may be surrounded by a hydrogen burning shell and the unignited hydrogen rich envelope. When the helium burning ceases to provide sufficient power, the star begins to contract again under its own gravity and, as implied by the Virial theorem, the temperature of the helium exhausted core rises. The contraction continues until either the next nuclear fuel begins to burn at a rapid enough rate or electron degeneracy pressure halts the infall.

12.9.1 Carbon burning

Stars somewhat more massive than about 0.7 $M_\odot$ contract until the temperature is large enough for carbon to interact with itself (stars less massive may settle as degenerate helium *white dwarfs*). For stars more massive than $M \geq 8 - 10$ $M_\odot$ (mass on the main sequence— *not* the mass of the C+O core), the contracting C+O core remains nondegenerate until C

---

[2] Note, however, the caveat: if the amount of ^{12}C is little due to a long stellar lifetime of He burning or to a larger rate of the ^{12}C $+ \alpha \rightarrow {}^{16}$O $+ \gamma$ reaction whose estimate outlined in the earlier section is somewhat uncertain), then the star may directly go from the He burning stage to the O burning or Ne burning stage, skipping C burning altogether [Woo86].

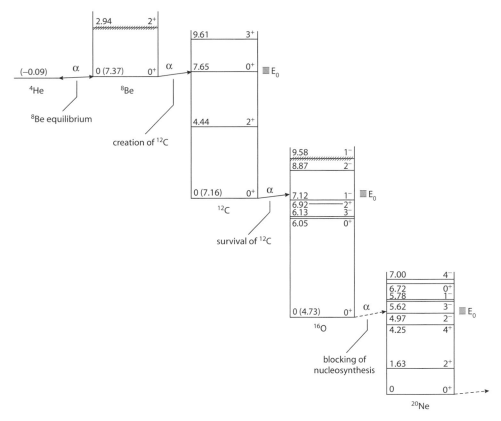

Figure 12.11 Energy levels of nuclei participating in thermonuclear reactions during the helium burning stage in red giant stars (adapted from [RR88]). The survival of both ^{12}C and ^{16}O in red giants, believed to be the source of terrestrial abundances depends upon fortuitous circumstances of nuclear level structures and other properties in these nuclei.

starts burning at $T \sim 5 \times 10^8 \, K$ and $\rho = 3 \times 10^6$ g cm^{-3}. Thereafter, sufficient power is generated and the contraction stops and quiescent (hydrostatic, not explosive) C burning proceeds.

The combined mass of two reacting ^{12}C nuclei falls at an excitation energy of 14 MeV in the compound nucleus of ^{24}Mg. At this energy there are many compound nuclear states, and the most effective range of stellar energies (the Gamow window) at the relevant temperature is about 1 MeV; hence a number of resonant states can contribute to the decay of the compound nucleus, and even the large angular momentum resonances may be important because the penetration factors in massive nuclei are not affected by centrifugal barriers. The carbon on carbon burning can proceed through multiple energetically allowed reaction channels:

$$^{12}\text{C} + {}^{12}\text{C} \rightarrow {}^{20}\text{Ne} + {}^{4}\text{He} \ \ (Q = 4.62 \text{ MeV}) \tag{12.64}$$

$$\rightarrow {}^{23}\text{Na} + \text{p} \ \ (Q = 2.24 \text{ MeV}) \tag{12.65}$$

$$\rightarrow {}^{23}\text{Mg} + \text{n} \ \ (Q = -2.62 \text{ MeV}). \tag{12.66}$$

At the temperatures where carbon burning starts, the neutron liberating reactions require too much particle kinetic energy to be effective. In addition, based on laboratory measurements at higher energies compared to the stellar energies, the electromagnetic decay channel (^{24}Mg + γ) and the three-particle channel (^{16}O + 2α) have lower probability compared to the two-particle channels ^{23}Na + p and ^{20}Ne + α. The latter two channels have nearly equal probabilities (see [Cla84]); at the lowest center of mass energies for which cross sections are measured in the laboratory for the proton and α channels (about 2.45 MeV), the branching ratios were $b_p \sim 0.6$ and $b_\alpha \sim 0.4$), and therefore the direct products of carbon burning are likely to be ^{23}Na, ^{20}Ne, protons, and α-particles. The rate for this reaction per pair of ^{12}C nuclei is [Ree59]

$$\log\lambda_{12,12} = \log f_{12,12} + 4.3 - \frac{36.55(1 + 0.1T_9)^{1/3}}{T_9^{1/3}} - \frac{2}{3}\log T_9, \tag{12.67}$$

where the factor $f_{12,12}$ is a screening factor. Now, at the temperatures of ^{12}C burning, the liberated protons and alpha particles can be quickly consumed through the reaction chain ^{12}C(p, γ)^{13}N(e$^+$$\nu_e$)^{13}C($\alpha$, n)^{16}O. Thus, the net effect is that the free proton is converted into a free neutron (which may be further captured) and the α-particle is consumed with ^{12}C into ^{16}O.

The α-particles are also captured by other α-particle nuclei, resulting in, at the end of carbon burning, nuclei like ^{16}O, ^{20}Ne, ^{24}Mg, and ^{28}Si. These secondary reactions augment the energy released by the initial carbon reaction, and Reeves [Ree59] estimated that each pair of ^{12}C nuclei releases about 13 MeV of energy. Toward the end of the carbon burning phase, other reactions such as ^{12}C+^{16}O and ^{12}C+^{20}Ne also take place. But these are less rapid and are not expected to play major roles compared to the ^{12}C+^{12}C reactions, due to their increased Coulomb barriers.

During the carbon burning and subsequent stages, the dominant energy loss from the star is due to neutrinos streaming out directly from the stellar thermonuclear furnace, rather than by photons from the surface. The neutrino luminosity is a sensitive function of core temperature and quickly outshines the surface photon luminosity of the star at the carbon burning stage. The (thermal) evolutionary timescale of the star, due to the neutrino emission, becomes very short and the core evolves rapidly—so rapidly (compared to the "cooling" timescale Kelvin-Helmholtz time $\tau_{KH} \sim GM^2/RL_{ph}$) that the conditions in the core are "not communicated" to the surface, since this communication happens by photon diffusion. The surface conditions (e.g., the temperature) of the star then do not markedly evolve as the core goes beyond the carbon burning stage, and it may not be possible just by looking at a star's surface conditions to understand whether the core is close to a supernova stage or has many thousands of years of hydrostatic thermonuclear burning to go.

12.9.2 Neon burning

The result of carbon burning is mainly neon, sodium, and magnesium, but aluminum and silicon are also produced in small quantities by the capture of α, p, and n released during carbon burning. When carbon fuel is exhausted, again the core contracts and its temperature T_c goes up. At approximately $T_9 \sim 1$, energetic photons from the high energy tail of the Planck distribution function can begin to disintegrate the ^{20}Ne ash (see figure 12.11), so that one has the reaction ^{20}Ne $+ \gamma \rightarrow$ ^{16}O $+^4$ He.

Nucleons in a nucleus are bound with typical binding energy of several to 8 MeV. An energetic γ-ray photon would be required to photo-eject a nucleon. Two-nucleon ejection would require more energy. Alpha-particles are, however, released at approximately the same energy as a nucleon due to the low separation energy of an α-particle in the nucleus. For example, the α separation energy in ^{20}Ne is 4.73 MeV. Thus, the major photonuclear reactions are (γ, n), (γ, p), and (γ, α) processes. For a photodisintegration reaction to proceed through an excited state E_X in the parent, the decay rate is

$$\lambda(\gamma, \alpha) = \left[\exp\left(-\frac{E_X}{kT} \right) \frac{2J_R + 1}{2J_0 + 1} \frac{\Gamma_\gamma}{\Gamma} \right] \times \frac{\Gamma_\alpha}{\hbar}. \tag{12.68}$$

In the above equation, the first factor in square brackets on the right-hand side is the probability of finding the nucleus in the excited state E_X and spin J_R (with J_0 being the ground state spin), while the second factor $\Gamma_\alpha/\hbar$ is the decay rate of the excited state with an α-particle emission. Now since $E_X = E_R + Q$, we have

$$\lambda(\gamma, \alpha) = \frac{\exp(-Q/kT)}{\hbar(2J_0 + 1)} (2J_R + 1) \frac{\Gamma_\alpha \Gamma_\gamma}{\Gamma} \exp\left(-\frac{E_R}{kT} \right). \tag{12.69}$$

At $T_9 \geq 1$, the photodisintegration is dominated by the 5.63 MeV level in ^{20}Ne (see figure 12.11). At approximately $T_9 \sim 1.5$, the photodissociation rate becomes greater than the rate for α capture on ^{16}O to produce ^{20}Ne (the reverse reaction), thus leading effectively to the net dissociation of ^{20}Ne. The released ^{4}He reacts with the unspent ^{20}Ne and leads to ^{4}He $+ ^{20}$Ne $\rightarrow ^{24}$Mg $+ \gamma$. Thus the net result of the photodissociation of two ^{20}Ne nuclei is $2 \times^{20}$ Ne $\rightarrow ^{16}$O $+^{24}$Mg with a net Q value of 4.58 MeV. The brief neon burning phase concludes at T_9 close to ~ 1.

12.9.3 Oxygen burning

At the end of the neon burning the core is left with a mixture of α-particle nuclei ^{16}O and ^{24}Mg. After this another core contraction phase ensues and the core heats up, until at $T_9 \sim 2$, ^{16}O begins to react with itself,

$$^{16}O +^{16} O \rightarrow ^{28}Si + ^4He \tag{12.70}$$

$$\rightarrow ^{32}S + \gamma. \tag{12.71}$$

The first reaction takes place approximately 45% of the time with a Q value of 9.593 MeV. In addition to Si and S, the O burning phase also produces Ar, Ca, and trace amounts of Cl, K, etc. up to Sc. Then at $T_9 \sim 3$, the produced ^{28}Si begins to burn in what is known as the Si burning phase.

12.9.4 Silicon burning

As we have seen, most of the stages of stellar burning involve thermonuclear fusion of nuclei to produce higher Z and A nuclei. The first exception to this is neon burning, where the photon field is sufficiently energetic to photodissociate neon before the temperature rises sufficiently to allow fusion reactions among oxygen nuclei to overcome their Coulomb repulsion. Processing in the neon burning phase takes place with the addition of helium nuclei to the undissociated neon rather than overcoming the Coulomb barrier of two neon nuclei. This trend continues in the silicon burning phase. In general, a photodisintegration channel becomes important when the temperature rises to the point that the Q-value, that is, the energy difference between the fuel and the products is smaller than approximately $30k_BT$.

With typical Q-values for reactions among stable nuclei above silicon being 8–12 MeV, photodisintegration of the nuclear products of neon and oxygen burning begins to play an important role once the temperature exceeds $T_9 \geq 3$. Then nuclei with smaller binding energies are destroyed by photodissociation in favor of their more tightly bound neighbors, and many nuclear reactions involving α-particles, protons, and neutrons interacting with all the nuclei in the mass range $A = 28$–65 take place. In contrast to the previous burning stages, where only a few nuclei underwent thermonuclear reactions among themselves, here the nuclear reactions are primarily of a rearrangement type, in which a particle is photoejected from one nucleus and captured by another and a given fuel nucleus is linked to a product nucleus by a multitude of reaction chains and cycles, so it is necessary to keep track of many more nuclei (and many reaction processes involving these) than for previous burning stages. More and more stable forms of the nuclei form in a nuclear reaction network as the rearrangement proceeds. Since there exists a maximum in the binding energy per nucleon at the ^{56}Fe nucleus, the rearrangements lead to nuclei in the vicinity of this nucleus (iron-group nuclei).

In the mass range $A = 28$–65, the levels in the compound nuclei that form in the reactions during silicon burning are so dense that they overlap. Moreover, at the high temperatures that are involved ($T_9 = 3$–5), the net reaction flux may be small compared to the large forward and backward reactions involving a particular nucleus and a quasi-equilibrium may ensue between groups of nuclei which are connected between separate groups by a few, slow, rate-limiting reactions ("bottlenecks"). However, as the available nuclear fuel(s) are consumed and thermal energy is removed due to escaping neutrinos, various nuclear reactions may no longer occur substantially rapidly ("freeze-out").

Weak interaction processes such as electron capture and β-decay of nuclei are important, by influencing the Y_e and thereby the reaction flow. These ultimately affect both the stellar core density and entropy structures, and it is important to track and include the changing Y_e (the number of electrons per nucleon) of the core material, not only in the silicon burning phase, but even from earlier oxygen burning phases.

For temperatures above 3×10^9 K, more photonuclear processes appear. These yield more nuclei to be burned and heavier nuclei are produced:

$$\gamma + {}^{28}_{14}\text{Si} \rightarrow {}^{24}_{12}\text{Mg} + {}^{4}_{2}\text{He}, \quad {}^{4}_{2}\text{He} + {}^{28}_{14}\text{O} \rightarrow {}^{32}_{16}\text{S} + \gamma, \text{ etc.} \tag{12.72}$$

Due to the large number of free neutrons, many (n,γ) reactions (radiative neutron capture) elements in the mass range $A = 28$–57 are formed. This leads to a large abundance of elements in the iron mass region, which have the largest binding energy per nucleon. For elements heavier than iron the nuclear fusion processes do not generate energy.

For $A > 100$ the distribution of nuclei cannot be explained in terms of fusion reactions with charged particles. They are formed by the successive capture of slow neutrons and of β^-decay. The maxima of the element distribution in $N = 50, 82, 126$ are due to the small capture cross sections corresponding to the magic numbers. This yields a trash of isotopes at the observed element distribution.

12.10 Synthesis of Heaviest Elements

So far we have been dealing primarily with charged particle reactions and photodisintegration which lead to the production of lighter elements ($A = 1$–40) and the recombination reactions for the production of elements, $A = 40$–65. However, the heavier elements ($A \geq 65$), because of their high charge and relatively weak stability, cannot be produced by these two processes. It was therefore natural to investigate the hypothesis of neutron-induced reactions on the elements that are formed already in the various thermonuclear burning stages, and in particular on the iron group elements.

The study of the nuclear reaction chains in stellar evolution shows that during certain phases large neutron fluxes are released in the core of a star. On the other hand, the analysis of the relative abundance of elements shows certain patterns that can be explained in terms of the neutron absorption cross sections of these elements. If the heavier elements above the iron peak were to be synthesized during, for example, charged particle thermonuclear reactions during silicon burning, their abundance would drop much more steeply with increasing mass (larger and larger Coulomb barriers) than the observed behavior of abundance curves, which shows a much lesser than expected decrease. Based on the abundance data, Suess and Urey [Sue56] and Burbidge et al. [Bur57] (hereafter B^2FH) argued that heavy elements are made instead by thermal neutron capture.

Two distinct neutron processes are required to make the heavier elements. The slow neutron capture process (s process) has a lifetime for β-decay τ_β shorter than the competing neutron capture time τ_n ($\tau_\beta \leq \tau_n$). This makes the s process nucleosynthesis run through the valley of β-stability. The rapid neutron capture r process, on the other hand, requires $\tau_n \ll \tau_\beta$. This process takes place in extremely neutron rich environments, for the neutron capture timescale is inversely proportional to the ambient neutron density. The r process, in contrast to the s process, goes through very neutron rich and unstable nuclei that are far off the valley of stability. The relevant properties of such nuclei are most often not known experimentally, and are usually estimated theoretically. Some of the key parameters are the half-lives of the β-unstable nuclei along the s process path. But the nuclear half-life in the stellar environment can change not just due to transitions from the ground state of the parent nucleus, but also because its excited states are thermally populated.

In the r process, the β-decay properties of the nuclei regulate the reaction flow to larger charge numbers and determine the resultant abundance pattern and the duration of the process. The r process lasts for typically a few seconds, in an intense neutron density environment $n_n \sim 10^{20}$–10^{25}cm^{-3}. In comparison, the neutron densities in the s process are much more modest, say $n_n \sim 10^8 \text{cm}^{-3}$; these neutron irradiation can take place, for example, in the helium burning phase of red giant stars. Nuclei above the iron group up to about $A = 90$ are produced in massive stars mainly by the s process. Above $A = 100$ the s process does very little in massive stars, although there are redistributions of some of the heavy nuclei. Most of the s process above mass 90 is believed to come from asymptotic giant branch (AGB) stars.

12.11 White Dwarfs and Neutron Stars

If the thermonuclear processes in massive stars achieve the production of iron, there are the following possibilities for the star evolution.

(a) For stars with masses $<1.2\,M_\odot$ the internal pressure of the degenerated electron gas (when the electrons occupy all states allowed by the Pauli principle) does not allow star compression due to the gravitational attraction continuing indefinitely. For a free electron gas at temperature $T = 0$ (lowest energy state), the electrons occupy all energy states up to the Fermi energy. The total density of the star can be calculated by adding up the individual electronic energies. Since each phase-space cell $d^3p \cdot V$ (where V is the volume occupied by the electrons) contains $d^3p \cdot V/(2\pi\hbar)^3$ states, we get

$$\frac{E}{V} = 2 \int_0^{p_F} \frac{d^3p}{(2\pi\hbar)^3}\, E(p) = 2 \int_0^{p_F} \frac{d^3p}{(2\pi\hbar)^3} \sqrt{p^2c^2 + m_e^2 c^4} = n_0 m_e c^2 x^3 \epsilon(x),$$

$$\epsilon(x) = \frac{3}{8x^3}\left\{ x(1 + 2x^2)(1 + x^2)^{1/2} - \log[x + (1 + x)^{1/2}] \right\}, \tag{12.73}$$

where the factor 2 is due to the electron spin, and

$$x = \frac{p_F c}{m_e c^2} = \left(\frac{n}{n_0}\right)^{1/3} = \left(\frac{\rho}{\rho_0}\right)^{1/3}, \tag{12.74}$$

where

$$n_0 = \frac{m e^3 c^3}{\hbar^3} \quad \text{and} \quad \rho_0 = \frac{m_N n_0}{Y_e} = 9.79 \times 10^5\, Y_e^{-1}\, \text{g/cm}^3. \tag{12.75}$$

In the above relations p_F is the Fermi momentum of the electrons, m_e (m_N) is the electron (nucleon) mass, n is the density of electrons, and ρ is the mass density in the star. Y_e is the number of electrons per nucleon.

The variable x characterizes the electron density in terms of

$$n_0 = 5.89 \times 10^{29}\ \text{cm}^{-3}. \tag{12.76}$$

At this density the Fermi momentum is equal to the inverse of the Compton wavelength of the electron.

Using traditional methods of thermodynamics, the pressure is related to the energy variation by

$$P = -\frac{\partial E}{\partial V} = -\frac{\partial E}{\partial x}\frac{\partial x}{\partial V} = -\frac{\partial E}{\partial x}\left(-\frac{x}{3V}\right) = \frac{1}{3}\, n_0 m_e c^2 x^4 \frac{d\epsilon}{dx}. \tag{12.77}$$

This model allows us to calculate the pressure in the electron gas in a very simple form. Since the pressure increases with the electron density, which increases with the decreasing volume of the star, we expect that the gravitational collapse stops when the electronic pressure equals the gravitational pressure. When this occurs the star cools slowly and its luminosity decreases. The star becomes a *white dwarf* and in some cases its diameter can become smaller than that of the moon.

(b) For stars with masses in the interval 1.2–1.6 $M_\odot$, the electron pressure is not sufficient to balance the gravitational attraction. The density increases to $2 \times 10^{14} g\ cm^{-3}$ and the matter "neutronizes." This occurs via electron capture by the nuclei (inverse beta decay), transforming protons into neutrons. The final product is a *neutron star*, with a small radius (see figure 12.12). For example, if it were possible to form a neutron star from the sun it would have a radius given by

$$\left(\frac{M_\odot}{\frac{4\pi}{3}\rho}\right)^{1/3} = \left(\frac{2 \times 10^{33}\ g}{\frac{4\pi}{3} \times 2 \times 10^{14}\ g\ cm^{-3}}\right)^{1/3} \simeq 14\ km.$$

The process of transformation of iron nuclei into neutron matter occurs as follows. For densities of the order of $1.15 \times 10^9\ g\ cm^{-3}$ the Fermi energy of the electron gas is larger than the upper energy of the energy spectrum for the β-decay of the isotope $^{56}_{25}Mn$. The decay of this isotope can be inverted and two neutron rich isotopes of $^{56}_{25}Mn$ are formed, that is,

$$^{56}_{26}Fe + e^- \rightarrow {}^{56}_{25}Mn + \nu_e .\qquad(12.78)$$

These nuclei transform in $^{56}_{24}Cr$ by means of the reaction

$$^{56}_{25}Mn + e^- \rightarrow {}^{56}_{24}Cr + \nu_e.\qquad(12.79)$$

With the increasing of the pressure more isotopes can be formed, until neutrons start being emitted

$$^{A}_{Z}X + e^- \rightarrow {}^{A-1}_{Z-1}X + n + \nu_e.\qquad(12.80)$$

For $^{56}_{26}Fe$ this reaction network starts to occur at an energy of 22 MeV, which corresponds to a density of $4 \times 10^{11}\ g\ cm^{-3}$. With increasing density, the number of free neutrons increases and, when the density reaches $2 \times 10^{14}\ g\ cm^{-3}$, the density of free neutrons is 100 times larger than the density of the remaining electrons.

A *pulsar* is a rapidly rotating neutron star. Like a black hole, it is an endpoint to stellar evolution. The "pulses" of high energy radiation we see from a pulsar are due to a misalignment of the neutron star's rotation axis and its magnetic poles (see figure 12.13). Neutron stars for which we see such pulses are called "pulsars." They have a mass 40% larger than the sun and their radius is just 20 kilometers. This means that a cubic centimeter of this matter weighs 100 million tons! The neutron stars are at the limit of density that matter can have, the subsequent step being a black hole. Today we know more than 600 pulsars, and they are formidable astrophysics laboratories, since (a) their density is comparable to that of an atomic nucleus; (b) their mass and size give place to gravitational fields smaller than those of the black holes, but easier to measure; (c) the fastest of the pulsars has 600 turns about its axis in one second, so, its surface rotates at 36,000 kilometers a second;

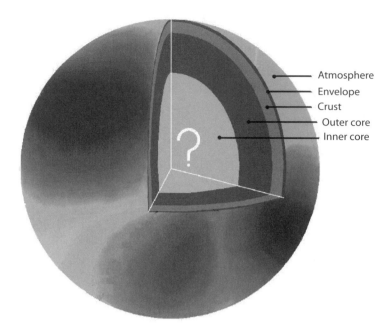

Figure 12.12 The structure of a neutron star. The outer regions of a neutron star may consist of thin layers of various elements that were produced by nuclear reactions during the star's lifetime. These outer layers are thought to have a rigid crystalline structure because of the intense gravitational field of the neutron star. The composition of the inner core is unknown.

(d) neutron stars have more intense magnetic fields than any other known object in the universe, million of times stronger than those produced in any terrestrial laboratory; (e) in some cases the regularity of their pulsations is the same or greater than the precision of the atomic clock, the latter being otherwise the best we have.

Figure 12.14 shows the relative distribution of elements in our galaxy. It has two distinct regions: in the region $A < 100$ it decreases with A approximately like an exponential, whereas for $A > 100$ it is approximately constant, except for the peaks in the region of the magic numbers $Z = 50$ and $N = 50, 82, 126$.

12.12 Supernova Explosions

It has long been observed that, occasionally, a new star appears in the sky, increases in brightness to a maximum value, and decays afterward until its visual disappearance. Such stars were called *novae*. Among the novae some stars present an exceptional variation in brightness and are called *supernovae*.

Schematically a pre-supernova has the onion structure presented in figure 12.15. Starting from the center of the star, we first find a core of iron, the remnant of silicon burning. After that we pass successive regions where ^{28}Si, ^{16}O, ^{12}C, ^{4}He, and ^{1}H form the dominant fraction. In the interfaces, nuclear burning continues to happen.

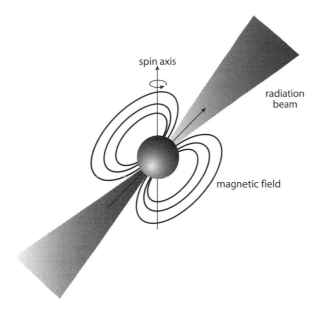

Figure 12.13 A rapidly rotating neutron star, or pulsar. At the magnetic poles, particles can escape and give rise to radio emission. If the magnetic axis is misaligned with the rotation axis of the neutron star (as shown), the star's rotation sweeps the beams over the observer as it rotates like a lighthouse, and one sees regular, sharp pulses of light (optical, radio, X-ray, etc.).

Silicon burning exhausts the nuclear fuel. As we mentioned previously, the gravitational collapse of the iron core cannot be held by means of pressure heat from nuclear reactions. However, Chandrasekhar [Ch31] showed that a total collapse can be avoided by electronic pressure. In this situation, the core is stabilized due to the pressure of the degenerated electron gas, $P(r)$, and the inward gravitational pressure. This means that for a given point inside the star,

$$-\frac{Gm(r)}{r^2}\,\rho(r) = \frac{dP(r)}{dr} = \frac{d\rho}{dr}\frac{dP}{d\rho},$$
$$\frac{dm}{dr} = 4\pi r^2 \rho(r),$$
$$\frac{dP}{d\rho} = Y_e\,\frac{m_e}{m_N}\,\frac{x^2}{3\sqrt{1+x^2}}. \tag{12.81}$$

where x and Y_e are defined following (12.73).

This model is appropriate for a nonrotating white dwarf. With the boundary conditions $m(r=0)=0$ and $\rho(r=0)=\rho_c$ (the central density), these equations can be solved easily [Ko86]. For a given Y_e, the model is totally determined by ρ_c. Figure 12.16 shows the mass density of a white dwarf. We observe that the total mass of a white dwarf (of the order of a solar mass, $M_\odot = 1.98 \times 10^{33}$ g) increases with ρ_c. Nonetheless, and perhaps most important, it cannot exceed the finite value of

$$M \le M_{Ch} \simeq 1.45\,(2Y_e)^2\,M_\odot, \tag{12.82}$$

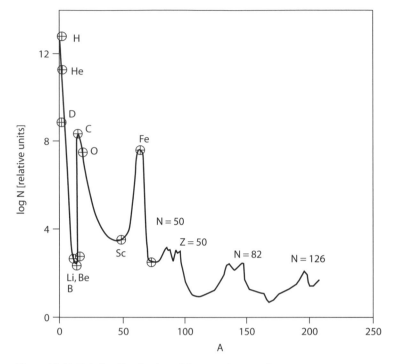

Figure 12.14 Relative distribution of elements in our galaxy.

known as the *Chandrasekhar mass* [Ch31]. Applying these results to the nucleus of a star with any mass, we get from (12.81) that stars with mass $M > M_{Ch}$ cannot be stable against gravitational collapse by the pressure of the degenerate electron gas. The collapse occurs inevitably for a massive star, since the silicon burning adds more and more material to the stellar core.

At the beginning of the collapse the temperature and density are of the order of $T \sim 10^{10}$ K and $\rho \sim 3 \times 10^9$ g/cm^3. The core is made of ^{56}Fe and of electrons. There are two possibilities, both accelerating the collapse.

1. At conditions present in the collapse the strong reactions and the electromagnetic reactions between the nuclei are in inverse equilibrium,

$$\gamma + {}^{56}_{26}\text{Fe} \Longleftrightarrow 13({}^4\text{He}) + 4n - 124 \text{ MeV}. \tag{12.83}$$

For example, with $\rho = 3 \times 10^9$ g/cm^3 and $T = 11 \times 10^9$ K, half of ^{56}Fe is dissociated. This dissociation takes energy from the core and causes pressure loss. The collapse is thus accelerated.

2. If the mass of the core exceeds M_{Ch}, electrons are captured by the nuclei to avoid violation of the Pauli principle

$$e^- + (Z, A) \rightarrow (Z - 1, A) + \nu_e. \tag{12.84}$$

The neutrinos can escape the core, taking away energy. This is again accompanied by a pressure loss due to the decrease of the free electrons (this also decreases M_{Ch}). The collapse is again accelerated.

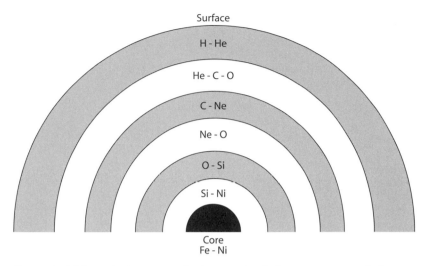

Figure 12.15 The "onion" structure of a $20M_\odot$ star just before a supernova explosion.

The gravitational contraction increases the temperature and density of the core. An important change in the physics of the collapse occurs when the density reaches $\rho_{\text{trap}} \simeq 4 \times 10^{11}$ g/cm^3. The neutrinos become essentially confined to the core, since their diffusion time in the core is larger than the collapse time. After the neutrino confinement no energy is taken out of the core. Also, all reactions are in equilibrium, including the capture process (12.84). The degeneracy of the neutrino Fermi gas avoids a complete neutronization, directing the reaction (12.84) to the left. As a consequence, Y_e remains large during the collapse ($Y_e \approx 0.3$–0.4 [Be79]). To equilibrate the charge, the number of protons must also be large. To reach $Z/A = Y_e \approx 0.3$–0.4, the protons must be inside heavy nuclei that will therefore survive the collapse.

Two consequences follow

1. The pressure is given by the degenerate electron gas that controls the whole collapse; the collapse is thus adiabatic, with the important consequence that the collapse of the most internal part of the core is *homologous*, that is, the position $r(t)$ and the velocity $v(t)$ of a given element of mass of the core are related by

$$r(t) = \alpha(t)r_0; \quad v(t) = \frac{\dot{\alpha}}{\alpha}r(t), \tag{12.85}$$

where r_0 is the initial position.

2. Since the nuclei remain in the core of the star, the collapse has a reasonably large order and the entropy remains small during the collapse [Be79] ($S \approx 1.5\ k$ per nucleon, where k is the Boltzmann constant).

The collapse continues homologously until nuclear densities of the order of $\rho_N \approx 10^{14}$ g/cm^3 are reached, when the matter can be thought as approximately a degenerate Fermi gas of nucleons. Since the nuclear matter has a finite compressibility, the homologous core decelerates and starts to increase again as a response to the increase of the nuclear

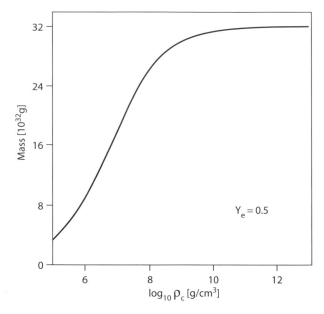

Figure 12.16 Masses of white dwarfs calculated as a function of ρ_c, the central density. With increasing ρ_c, the mass reaches a limiting value, the Chandrasekhar mass.

matter. This eventually leads to a *shock wave* which propagates to the external core (the iron core outside the homologous core), which, during the collapse time, continues to contract reaching the supersonic velocity. The collapse break followed by the shock wave is the mechanism which creates the supernova explosion. Nonetheless, several ingredients of this scenario are still unknown, including the equation of state of the nuclear matter. The compressibility influences the available energy for the shock wave, which must be of the order of 10^{51} erg.

The exact mechanism for the explosion of a supernova is still controversial.

1. In the *direct mechanism*, the shock wave is not only strong enough to stop the collapse, but also to explode the exterior stellar shells.

2. If the energy in the shock wave is insufficient for a direct explosion, the wave will deposit its energy in the exterior of the core, e.g., by excitation of the nuclei, being frequently followed by electronic capture and emission of neutrinos (*neutrino eruption*). Additionally, neutrinos of the all three species are generated by the production of pairs in the hot environment. A new shock wave can be generated by the outward diffusion of neutrinos, indeed carrying the most part of the energy liberated in the gravitational collapse of the core ($\approx 10^{53}$ erg). If about 1% of the energy of the neutrinos is converted into kinetic energy due to the coherent neutrino-nucleus scattering, a new shock wave arises. This will be strong enough to explode the star. This process is known as the *retarded mechanism* for supernova explosion.

To know which of the above mechanism is responsible for the supernova explosion, one needs to know the rate of electron capture, the nuclear compressibility, and the way neutrinos are transported. The iron core, the remnant of the explosion (the homologous core and part of the external core) will not explode and will become either a neutron star, and possibly later a *pulsar* (rotating neutron star), or a *black hole*, as in the case of more massive stars, with $M \geq 25-35 M_\odot$.

Type II supernovae are defined as those showing H-lines in their spectra. It is likely that most, if not all, of the exploding massive stars still have some H-envelope left, and thus exhibit such a feature. In contrast, Type I supernovae lack H in their ejecta.

12.13 Nuclear Reaction Models

Explosive nuclear burning in astrophysical environments produces unstable nuclei, which again can be targets for subsequent reactions. In addition, it involves a very large number of stable nuclei, which are not fully explored by experiments. Thus, it is necessary to be able to predict reaction cross sections and thermonuclear rates with the aid of theoretical models. Especially during the hydrostatic burning stages of stars, charged particle induced reactions proceed at such low energies that a direct cross section measurement is often not possible with existing techniques. Hence extrapolations down to the stellar energies of the cross sections measured at the lowest possible energies in the laboratory are the usual procedures to apply. To be trustworthy, such extrapolations should have as strong a theoretical foundation as possible. Theory is even more mandatory when excited nuclei are involved in the entrance channel, or when unstable very neutron rich or neutron deficient nuclides (many of them being even impossible to produce with present-day experimental techniques) have to be considered. Such situations are often encountered in the modeling of explosive astrophysical scenarios.

Various models have been developed in order to complement the experimental information.

12.13.1 Microscopic models

In this model, the nucleons are grouped into clusters, as was explained in section 3.12. Keeping the internal cluster degrees of freedom fixed, the totally antisymmetrized relative wavefunctions between the various clusters are determined by solving the Schrödinger equation for a many-body Hamiltonian with an effective nucleon-nucleon interaction. When compared with most others, this approach has the major advantage of providing a consistent, unified and successful description of the bound, resonant, and scattering states of a nuclear system. Various improvements of the model have been made [Des98].

The microscopic model has been applied to many important reactions involving light systems, and in particular to the various p-p chain reactions [Lan96]. The available experimental data can generally be well reproduced. The microscopic cluster model (or its variant, the microscopic potential model) has also made an important contribution to the understanding of the key $^{12}C(\alpha, \gamma)^{16}O$ reaction rate [Des93a].

12.13.2 *Potential and DWBA models*

The potential model has been known for a long time to be a useful tool in the description of radiative capture reactions. It assumes that the physically important degrees of freedom are the relative motion between the (structureless) nuclei in the entrance and exit channels, and by the introduction of spectroscopic factors and strength factors in the optical potential. The associated drawbacks are that the nucleus-nucleus potentials adopted for calculating the initial and final wavefunctions from the Schrödinger equation cannot be unambiguously defined, and that the spectroscopic factors cannot be derived from first principles. They have instead to be obtained from more or less rough "educated guesses."

In the *potential model* the bound state wavefunctions of $c = a + b$ are specified by

$$\Psi_{JM}(\mathbf{r}) = \frac{u_{lj}^J(r)}{r} \mathcal{Y}_{JM}^l , \tag{12.86}$$

where $\mathbf{r}$ is the relative coordinate of a and b, $u_{lj}^J(r)$ is the radial wavefunction and $\mathcal{Y}_{JM}^l$ is the spin-angle wavefunction

$$\mathcal{Y}_{JM}^l = \sum_{m, M_a} \langle jm I_a M_a | JM \rangle |jm\rangle |I_a M_a\rangle , \quad \text{with} \quad |jm\rangle = \sum_{m_l, M_b} Y_{lm_l}(\hat{\mathbf{r}}) \chi_{M_b}, \tag{12.87}$$

where χ_{M_b} is the spinor wavefunction of particle b and $\langle jm I_a M_a | JM \rangle$ is a Clebsch-Gordan coefficient.

As described in chapter 9, the operators for electric transitions of multipolarity $\lambda\pi$ are given by

$$\mathcal{M}_{E\lambda\mu} = e_\lambda \, r^\lambda Y_{\lambda\mu}(\hat{\mathbf{r}}), \tag{12.88}$$

where the effective charge, which takes into account the displacement of the center of mass, is

$$e_\lambda = Z_b e \left(-\frac{m_a}{m_c} \right)^\lambda + Z_a e \left(\frac{m_b}{m_c} \right)^\lambda . \tag{12.89}$$

For magnetic dipole transitions,

$$\mathcal{M}_{M1\mu} = \sqrt{\frac{3}{4\pi}} \mu_N \left[e_M l_\mu + \sum_{i=a,b} g_i (s_i)_\mu \right], \quad e_M = \left(\frac{m_a^2 Z_a}{m_c^2} + \frac{m_b^2 Z_b}{m_c^2} \right), \tag{12.90}$$

where l_μ and s_μ are the spherical components of order μ ($\mu = -1, 0, 1$) of the orbital and spin angular momentum ($\mathbf{l} = -i\mathbf{r} \times \nabla$, and $\mathbf{s} = \sigma/2$) and g_i are the gyromagnetic factors of particles a and b. The nuclear magneton is given by $\mu_N = e\hbar/2m_N c$.

The matrix element for the transition $J_0 M_0 \rightarrow JM$, using the convention of [BM69] is given by

$$\langle JM | \mathcal{M}_{E\lambda\mu} | J_0 M_0 \rangle = \langle J_0 M_0 \lambda\mu | JM \rangle \frac{\langle J \| \mathcal{M}_{E\lambda} \| J_0 \rangle}{\sqrt{2J+1}} . \tag{12.91}$$

The multipole strength, or response function, for a particular partial wave, summed over final channel spins, is defined by

$$\frac{dB\left(\pi\lambda; l_0 j_0 \to klj\right)}{dk} = \sum_J \frac{\left|\langle kJ \|\mathcal{M}_{\pi\lambda}\| J_0\rangle\right|^2}{2J_0 + 1}, \tag{12.92}$$

where $\pi = E$, or M. In this equation $l_0 j_0$ (lj) are the ground (continuum) state angular momentum quantum numbers and k denotes the relative momentum of the final (continuum) state (the relative energy is $E = \hbar^2 k^2 / 2\mu$, where μ is the reduced mass of $a + b$).

The photo-absorption cross section for the reaction $\gamma + c \to a + b$ is given in terms of the response function by

$$\sigma_\gamma^{(\lambda)}(E_\gamma) = \frac{(2\pi)^3(\lambda + 1)}{\lambda\left[(2\lambda + 1)!!\right]^2} \left(\frac{m_{ab}}{\hbar^2 k}\right) \left(\frac{E_\gamma}{\hbar c}\right)^{2\lambda - 1} \frac{dB(\pi\lambda)}{dE}, \tag{12.93}$$

where $E_\gamma = E + |E_B|$, with $|E_B|$ being the binding energy of the $a + b$ system.

The cross section for the radiative capture process $a + b \to c + \gamma$ can be related by detailed balance to (12.93), that is,

$$\sigma_{(\pi\lambda)}^{(rc)}(E) = \left(\frac{E_\gamma}{\hbar c}\right)^{2\lambda - 1} \frac{2(2I_c + 1)}{(2I_a + 1)(2I_b + 1)} \sigma_\gamma^{(\lambda)}(E_\gamma). \tag{12.94}$$

The total capture cross section σ_{nr} is determined by the capture to all bound states with the single particle spectroscopic factors S_i in the final nucleus

$$\sigma_{\mathrm{nr}}(E) = \sum_{i,\pi,\lambda} S_i\, \sigma_{(\pi\lambda),i}^{(rc)}(E). \tag{12.95}$$

Experimental information or detailed shell model calculations have to be performed to obtain the spectroscopic factors S_i.

12.13.3 Parameter fit

Reaction rates dominated by the contributions from a few resonant or bound states are often extrapolated in terms of R- or K-matrix fits, which rely on quite similar strategies. A sketch of R-matrix theory was presented in section 4.9. The appeal of these methods rests on the fact that analytical expressions which allow for a rather simple parametrization of the data can be derived from underlying formal reaction theories. However, the link between the parameters of the R-matrix model and the experimental data (resonance energies and widths) is only quite indirect. The K-matrix formalism solves this problem, but suffers from other drawbacks [Bar94].

The R- and K-matrix models have been applied to a variety of reactions, in particular to the analysis of the $^{12}C(\alpha, \gamma)^{16}O$ reaction rate [Az95].

12.13.4 Statistical models

Many astrophysical scenarios involve a wealth of reactions on intermediate mass or heavy nuclei. This concerns the nonexplosive or explosive burning of C, Ne, O, and Si, as well as the s-, r- and p-process nucleosynthesis. Fortunately, a large fraction of the reactions of interest proceed through compound systems that exhibit high enough level densities for statistical methods to provide a reliable description of the reaction mechanism. In this respect, the *Hauser-Feshbach (HF) model* has been widely used with considerable success. Explosive burning in supernovae involves in general intermediate mass and heavy nuclei. Due to a large nucleon number, they have an intrinsically high density of excited states. A high density in the compound nucleus at the appropriate excitation energy allows one to make use of the statistical model approach for compound nuclear reactions [HF52] which averages over resonances. Averaged transmission coefficients T, which do not reflect a resonance behavior, but rather describe absorption via an imaginary part in the (optical) nucleon-nucleus potential as described in [MW79]. This leads to the expression derived in section 4.14,

$$
\sigma_i^{\mu\nu}(j,o;E_{ij}) = \frac{\pi\hbar^2/(2\mu_{ij}E_{ij})}{(2J_i^\mu+1)(2J_j+1)}
$$
$$
\times \sum_{J,\pi}(2J+1)\frac{T_j^\mu(E,J,\pi,E_i^\mu,J_i^\mu,\pi_i^\mu)T_o^\nu(E,J,\pi,E_m^\nu,J_m^\nu,\pi_m^\nu)}{T_{tot}(E,J,\pi)} \tag{12.96}
$$

for the reaction $i^\mu(j,o)m^\nu$ from the target state i^μ to the excited state m^ν of the final nucleus, with a center of mass energy E_{ij} and reduced mass μ_{ij}. J denotes the spin, E the corresponding excitation energy in the compound nucleus, and π the parity of excited states. When these properties are used without subscripts they describe the compound nucleus, while subscripts refer to states of the participating nuclei in the reaction $i^\mu(j,o)m^\nu$ and superscripts indicate the specific excited states. Experiments measure $\sum_\nu \sigma_i^{0\nu}(j,o;E_{ij})$, summed over all excited states of the final nucleus, with the target in the ground state. Target states μ in an astrophysical plasma are thermally populated and the astrophysical cross section $\sigma_i^*(j,o)$ is given by

$$
\sigma_i^*(j,o;E_{ij}) = \frac{\sum_\mu(2J_i^\mu+1)\exp(-E_i^\mu/kT)\sum_\nu \sigma_i^{\mu\nu}(j,o;E_{ij})}{\sum_\mu(2J_i^\mu+1)\exp(-E_i^\mu/kT)}. \tag{12.97}
$$

The summation over ν replaces $T_o^\nu(E,J,\pi)$ in (12.96) by the total transmission coefficient

$$
T_o(E,J,\pi) = \sum_{\nu=0}^{\nu_m} T_o^\nu(E,J,\pi,E_m^\nu,J_m^\nu,\pi_m^\nu)
$$
$$
+ \int_{E_m^{\nu_m}}^{E-S_{m,o}} \sum_{J_m,\pi_m} T_o(E,J,\pi,E_m,J_m,\pi_m)\rho(E_m,J_m,\pi_m)dE_m. \tag{12.98}
$$

Here $S_{m,o}$ is the channel separation energy, and the summation over excited states above the highest experimentally known state ν_m is changed to an integration over the level density ρ. The summation over target states μ in (12.97) has to be generalized accordingly.

The important ingredients of statistical model calculations as indicated in the above equations are the particle and γ transmission coefficients T and the level density of excited states ρ. Therefore, the reliability of such calculations is determined by the accuracy with which these components can be evaluated (often for unstable nuclei).

The gamma transmission coefficients have to include the dominant gamma transitions (E1 and M1) in the calculation of the total photon width. The smaller, and therefore less important, M1 transitions have usually been treated with the simple single particle approach $T \propto E^3$ of [BW52]. The E1 transitions are usually calculated on the basis of the Lorentzian representation of the giant dipole resonance (see section 6.8). Within this model, the E1 transmission coefficient for the transition emitting a photon of energy E_γ in a nucleus $^A_N Z$ is given by

$$T_{E1}(E_\gamma) = \frac{8}{3} \frac{NZ}{A} \frac{e^2}{\hbar c} \frac{1+\chi}{mc^2} \sum_{i=1}^{2} \frac{i}{3} \frac{\Gamma_{G,i} E_\gamma^4}{(E_\gamma^2 - E_{G,i}^2)^2 + \Gamma_{G,i}^2 E_\gamma^2}. \tag{12.99}$$

Here $\chi (= 0.2)$ accounts for the neutron-proton exchange contribution, and the summation over i includes two terms that correspond to the split of the GDR in statically deformed nuclei, with oscillations along $(i = 1)$ and perpendicular $(i = 2)$ to the axis of rotational symmetry.

12.14 Exercises

1. (a) What is the most probable kinetic energy of a hydrogen atom at the interior of the sun $(T = 1.5 \times 10^7$ K)? (b) What fraction of these particles would have kinetic energy in excess of 100 keV?

2. Suppose the iron core of a supernova star has a mass of 1.4 $M_\odot$ (the sun's mass is $M_\odot = 1.99 \times 10^{30}$ kg) and a radius of 100 km, and that it collapses to a uniform sphere of neutrons of radius 10 km. Assume that the virial theorem

$$2 \langle T \rangle + \langle V \rangle = 0$$

holds, where $\langle T \rangle$ is the average of the internal kinetic energy and $\langle V \rangle$ is the average of the gravitational potential energy. $E = \langle T \rangle + \langle V \rangle$ is the total mechanical energy of the system. Calculate the energy consumed in neutronization and the number of electron neutrinos produced. Given that the remaining energy is radiated as neutrino-antineutrino pairs of all kinds of average energy 12 + 12 MeV, calculate the total number of neutrinos radiated.

3. About 3 s after the onset of the Big Bang, the neutron-proton ratio became frozen when the temperature was still as high as 10^{10} K $(kT \simeq 0.8$ MeV). About 250 s later, fusion reactions took place converting neutrons and protons into ^{4}He nuclei. Show that the resulting ratio of the masses of hydrogen and helium in the universe was close to 3. The neutron half-life = 10.24 min and the neutron-proton mass difference is 1.29 MeV.

4. Given that the supernova of exercise 2 is at a distance of 163,000 light years, calculate the total number of neutrinos of all types arriving at each square meter of the earth. Also estimate the number of reactions

$$\bar{\nu}_e + p \longrightarrow n + e^+$$

that will occur in 1000 metric tons of water. Assume that the cross section is given by

$$\sigma = \frac{4p_e E_e G_F^2}{\pi \hbar^4 c^3}$$

where p_e and E_e are the positron momentum and energy, respectively, and G_F is the Fermi coupling constant. Assume that only one-sixth of the neutrinos are electron neutrinos.

5. Evaluate the radius of a neutron star with mass of 1.2 $M_\odot$.

6. The rate of energy delivered by the sun to the earth is known as the solar constant and equals 1.4×10^6 erg cm^{-2} s^{-1}. Knowing that the distance between the sun and the earth is of the order of 1.5×10^8 km, give a lower estimate of the rate at which the sun is losing mass to supply the radiated energy.

7. Calculate the energy radiated during the contraction of the primordial gas into the sun (the sun's diameter is 1.4×10^6 km). What energy would be released if the solar diameter were suddenly to shrink by 10%?

8. Given that the sun was originally composed of 71% hydrogen by weight and assuming it has generated energy at its present rate (3.86×10^{36} W) for about 5×10^9 years by converting hydrogen into helium, estimate the time it will take to burn 10% of its remaining hydrogen. Take the energy release per helium nucleus created to be 26 MeV.

9. The CNO cycle that may contribute to energy production in stars similar to the sun begins with the reaction p + ^{12}C → ^{13}N + γ. Assuming the temperature near the center of the sun to be 15×10^6 K, find the peak energy and width of the reaction rate.

10. Assume that we know the bound state wavefunction for the relative motion of particles a and b. Initially, the particles are thought to be in a quasi-stationary excited state of the compound nucleus. There is an exponential decrease of the wavefunction through the potential barrier which turns into an outgoing wave at infinity. We define the decay rate of that state (for particle emission) as

$$\lambda = \frac{1}{\tau} = \text{probability/sec for a decay through a large spherical shell.}$$

(a) If the wavefunction is written as

$$\Psi_{nlm} = \frac{u_l(r)}{r} Y_{lm}(\theta, \phi),$$

show that $\lambda = v \left| u_l(\infty) \right|^2$.

We define the penetration factor for particles of relative angular momentum l as

$$P_l = \frac{|u_l(\infty)|^2}{|u_l(R)|^2},$$

where $r = R$ is a position where the nuclear potential is close to zero (see figure 12.17 below).

Thus the decay rate can be written as $\lambda = v P_l |u_l(R)|^2$. For a uniform probability density inside the compound nucleus

$$|u_l(R)|^2 dr = \frac{4\pi R^2 dr}{4\pi R^3/3} = \frac{3}{R} dr .$$

One defines the reduced width θ_l by means of

$$|u_l(R)|^2 = \theta_l^2 \frac{3}{R}.$$

For realistic nuclear states

$$0.01 < \theta_l^2 < 1,$$

and θ_l^2 gives a measure of the degree to which a quasi-stationary nuclear state can be described by a relative motion of a an b in a potential.

(b) Show that the partial width of that state is given by

$$\Gamma_l = \frac{3\hbar v}{R} P_l \theta_l^2.$$

(c) Show that for charged particles, with

$$V_l(r) = \begin{cases} l(l+1)\hbar^2/2\mu r^2 + Z_1 Z_2 e^2/r & (r > R), \\ l(l+1)\hbar^2/2\mu r^2 + V_N & (r < R), \end{cases}$$

we get

$$P_l = \frac{1}{F_l^2(R) + G_l^2(R)},$$

where F_l and G_l are the regular and irregular Coulomb functions.

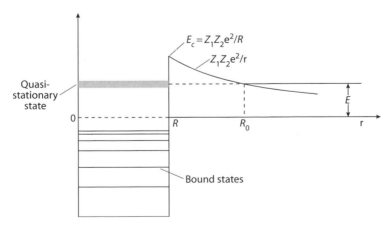

Figure 12.17 Potential barrier for charged particles.

(d) Show that using WKB wavefunctions for $u_l(r)$, one gets

$$P_l = \left[\frac{V_l(R) - E}{E} \right]^{1/2} \exp \left\{ -\frac{2\sqrt{2\mu}}{\hbar} \int_R^{R_0} \left[V_l(r) - E \right]^{1/2} dr \right\},$$ (12.100)

with $l(l+1) \rightarrow (l+1/2)^2$.

11. A numerical computation of the above equation shows that a good approximation consists in replacing $V_l(R) - E \cong E_c = Z_1 Z_2 e^2 / R$ in the factor preceding the exponential. This approximation is justified since at astrophysical energies $E \ll E_c$.

(a) Calling the exponent in (12.100) W_l, show that, to lowest order in E/E_c,

$$W_0 = 2\pi \frac{Z_1 Z_2 e^2}{\hbar v} \left[1 - \frac{4}{\pi} \left(\frac{E}{E_c} \right)^{1/2} + \frac{2}{3\pi} \left(\frac{E}{E_c} \right)^{3/2} - \cdots \right].$$ (12.101)

(b) Write the expression for the lifetime, Γ_0, of a state with $l = 0$ in terms of the reduced width θ_0^2 (see exercise 10) and W_0. Show that, to first order in E/E_c,

$$\Gamma_0 \propto \exp\left(-bE^{-1/2} \right).$$

12. For $l \neq 0$,

$$W_l = \frac{2\sqrt{2\mu}}{\hbar} \int_R^{R_0} \left[E_c \frac{R}{r} + E_l \left(\frac{R}{r} \right)^2 - E \right]^{1/2} dr.$$

where $E_l = \left(l + \frac{1}{2}\right)^2 \hbar^2 / 2\mu R^2$. For astrophysically relevant cases, $R/R_0 \lesssim 10^{-3}$ and the ratio R/r is quite small for most range of the integration. As a consequence the second term in the square root bracket never dominates, and the integrand may be expanded, whereupon the leading terms become

$$W_l \simeq \frac{2\sqrt{2\mu}}{\hbar} \int_R^{R_0} \left(E_c \frac{R}{r} - E \right)^{1/2} dr + \frac{\sqrt{2\mu}}{\hbar} \int_R^{R_0} \frac{E_l R^{3/2}}{E_c^{1/2}} \frac{dr}{r^{3/2}}.$$ (12.102)

The first term is just equal to W_0, whereas the second term reflects the additional effects of the centrifugal barrier.

(a) Show that (12.102) becomes

$$W_l = W_0 + 2 \left[\frac{(l+1/2)^2 E_l}{E_c} \right]^{1/2} \left[1 - \left(\frac{E}{E_c} \right)^{1/2} \right].$$

(b) Neglecting the correction in $(E/E_c)^{1/2}$ in the above equation, show that

$$P_l \approx \left(\frac{E_c}{E} \right)^{1/2} \exp \left[-2\pi \frac{Z_1 Z_2 e^2}{\hbar v} + 4 \left(\frac{2\mu R^2 E_c}{\hbar^2} \right)^{1/2} - 2(l+1/2)^2 \left(\frac{\hbar^2}{2\mu R^2 E_c} \right)^{1/2} \right],$$

where the correction of order $(E/E_c)^{3/2}$ in W_0 has also been dropped.

(c) Show that

$$\Gamma_l = 6\theta_l^2 \left(\frac{\hbar^2 E_c}{2\mu R^2} \right)^{1/2} \exp\left(-W_l \right).$$ (12.103)

(d) The reaction $^{12}C(p, \gamma)^{13}N$ has a peak at 424 keV center of mass energy, corresponding to a $J^\pi = \frac{1}{2}^+$ resonance. The resonance has a full width at half maximum $\Gamma = 40$ keV. This width is essentially the proton width, since the only other channel is Γ_γ, which is much smaller than Γ_p. What is the value of the dimensionless reduced width θ_l^2 for that state?

13. (a) Show that close to a resonance the astrophysical S-factor is given by

$$S(E) = \frac{\pi \hbar^2}{2\mu} \frac{g\Gamma_\alpha(E)\Gamma_\beta(E)}{(E - E_r)^2 + \Gamma^2/4} \exp\left(2\pi \frac{Z_1 Z_2 e^2}{\hbar v} \right),$$

where $g = (2J + 1)/(2J_\alpha + 1)(2J_\beta + 1)$ is the statistical spin factor. The decay widths for the entrance and decay channels, $\Gamma_\alpha(E)$ and $\Gamma_\beta(E)$, respectively, also depend on the energy for reactions of astrophysical interest.

(b) In some situations, Γ_β is approximately constant, that is, when the final channel is a γ- or α- decay. In this case, use (12.103) for $\Gamma_\alpha^{(l)}(E)$ and write an expression for $S(E)$ in terms of $l, E, E_c, \Gamma_\beta^{(l)}$, and Γ.

14. In stellar interiors, and also in laboratory experiments, the nuclear fusion cross sections for naked nuclei are modified due to the presence of electrons. The electron shielding around the nuclei is equivalent to a constant (negative) potential U_e that is usually much smaller than the energy E.

(a) Show that this potential modifies the fusion cross section so that

$$\sigma_{\text{screened}}(E) = \exp\left(\pi \eta \frac{U_e}{E} \right) \sigma_{\text{bare}}(E).$$

Find in the literature the experimental values of U_e (screening by electrons in the target) for five reactions of astrophysical interest. Comment about the comparison with theoretical values.

15. Show that for a resonant (p, γ) cross section reaction, for example, $^{12}C(p, \gamma)^{13}N$ at 424 keV, one finds

$$\langle \sigma v \rangle_{\text{resonant}} = \left(\frac{2\pi}{\mu kT} \right)^{3/2} \frac{\Gamma_p \Gamma_\gamma}{\Gamma} e^{-E_r/kT}.$$

16. The nonexistence of a bound nucleus with $A = 8$ was one of the major puzzles in nuclear astrophysics. How could heavier elements than $A = 8$ be formed? Using typical values of concentration of α-particles in the core of a heavy star, $n_\alpha \sim 1.5 \cdot 10^{28}/cm^3$ (corresponding to $\rho_\alpha \sim 10^5$ g/cm^3) and $T_8 \sim 1$, one obtains

$$\frac{n(^8\text{Be})}{n(\alpha)} \sim 3.2 \times 10^{-10}.$$

Salpeter suggested that this concentration would then allow $\alpha + {}^8\text{Be}(\alpha + \alpha) \rightarrow {}^{12}\text{C}$ to take place. Hoyle then argued that this reaction would not be fast enough to produce significant burning unless it was also resonant. Now the mass of $^8\text{Be} + \alpha$ is 7.366 MeV, and each nucleus has $J^\pi = 0^+$. Thus s wave capture would require a 0^+ resonance in ^{12}C at ~ 7.4 MeV. No such state was then known, but an experimental search revealed a 0^+ level at 7.644 MeV, with decay channels $^8\text{Be} + \alpha$ and γ-decay to the 2^+ 4.433 level in ^{12}C. The parameters are

$$\Gamma_\alpha \sim 8.9\text{eV},$$

$$\Gamma_\gamma \sim 3.6 \cdot 10^{-3}\text{eV}.$$

(a) Show that

$$r_{48} = n_\alpha^3 \, T_8^{-3} \, \exp\left(-\frac{42.9}{T_8}\right) (6.3 \cdot 10^{-54}\text{cm}^6/\text{sec}).$$

If we denote by $\omega_{3\alpha}$ the decay rate of an α in the plasma, then

$$\omega_{3\alpha} = 3 \, n_\alpha^2 \, T_8^{-3} \, \exp\left(-\frac{42.9}{T_8}\right) (6.3 \cdot 10^{-54}\text{cm}^6/\text{sec})$$

$$= \left(\frac{n_\alpha}{1.5 \cdot 10^{28}/\text{cm}^3}\right)^2 (4.3 \cdot 10^3/\text{sec}) \, T_8^{-3} \, \exp\left(-\frac{42.9}{T_8}\right).$$

(b) Since the energy release per reaction is 7.27 MeV show that the energy produced per gram, ϵ, is

$$\epsilon = (2.5 \cdot 10^{21}\text{erg/g sec}) \left(\frac{n_\alpha}{1.5 \cdot 10^{28}/\text{cm}^3}\right)^2 T_8^{-3} \, \exp\left(-\frac{42.9}{T_8}\right).$$

13 | Rare Nuclear Isotopes

13.1 Introduction

The study of nuclear physics demands beams of energetic particles to induce nuclear reactions on the nuclei of target atoms. It was from this need that accelerators were born. Over the years nuclear physicists have devised many ways of accelerating charged particles to ever increasing energies. Today we have beams of all nuclei from protons to uranium ions available at energies well beyond those needed for the study of atomic nuclei. This basic research activity, driven by the desire to understand the forces that dictate the properties of nuclei, has spawned a large number of beneficial applications. Among its many progeny we can count reactor- and spallation-based neutron sources, synchrotron radiation sources, particle physics, materials modification by implantation, carbon dating, and much more. It is an excellent example of the return to society of investment in basic research.

All these achievements have been realized by accelerating the 283 stable or long-lived nuclear species we find here on Earth. In recent years, however, it has become evident that it is now technically possible to create and accelerate unstable nuclei, and there are some 6,000–7,000 distinct nuclear species that live long enough to be candidates for acceleration. They are the nuclei within the so-called drip lines, the point where the nucleus can no longer hold another particle. It needs little imagination to see that this development might not only transform nuclear physics but also lead to many new, undreamed of opportunities in industry, medicine, material studies, and the environment.

Figure 13.1 shows schematically the two main methods of radioactive beam production that have been proposed. They are commonly known as the ISOL-isotope separation on line and in-flight techniques. In the ISOL method, we must first make the radioactive nuclei in a target/ion source, extract them in the form of ions, and, after selection of mass by an electromagnetic device, accelerate them to the energy required for the experiments. In contrast, the in-flight method relies on energetic beams of heavy ions impinging on a thin target. Interactions with the target nuclei can result in fission or fragmentation, with the nuclei produced leaving the target with velocities close to those of the projectiles.

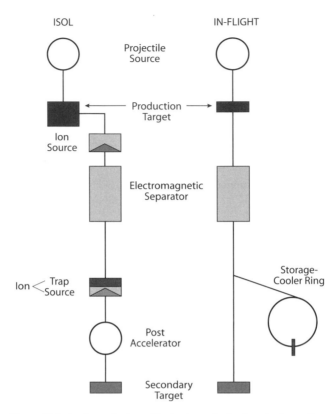

Figure 13.1 A schematic view of the basic methods of producing radioactive nuclear beams. At the left is the ISOL method with and without a post-accelerator. At the right is the in-flight method and the proposed hybrid in which fragments are caught in a gas cell and then re-accelerated.

A cocktail of many different species is produced which, since the ions have high velocities, does not need further acceleration to transport it to the secondary target. En route to the target the reaction products can be identified by mass, charge, and momentum in a spectrometer (fragment separator). Thus a pure beam is not separated out from the cocktail. Instead, each ion is tagged and identified by these primary characteristics and the secondary reactions are studied on an event-by-event basis. Another possibility is a combination of the two methods in which the in-flight reaction products are brought to rest in a gas cell, sucked out and separated by mass, and then re-accelerated to the required energy.

The ISOL and in-flight methods are complementary in almost every respect. With the ISOL technique one can produce beams of high quality, comparable to that of stable beams. Since we start with ions at the temperature of the target/ion source, the process is similar to the way the beam is generated in a stable beam accelerator, so we can produce beams of similar quality. Strong ISOL beams can be produced, but the intensity varies markedly according to the chemical species involved and how far from stability they are. Refractory elements such as zirconium and molybdenum are extremely difficult to ionize and are not suitable for the method at present. This technique also relies on the diffusion and

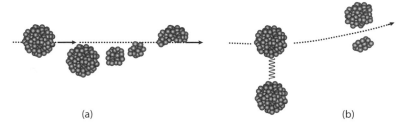

(a)   (b)

Figure 13.2 (a) Schematic description of a nuclear fragmentation reaction producing rare isotopes. The lower fragments are called participants, while the upper one is called the spectator. Using uranium projectiles ($N/Z \sim 1.6$) one expects to produce (light) spectator nuclei of about the same N/Z ratio. (b) Coulomb fission of relativistic projectiles leading to the production of rare isotopes. For a heavy unstable projectile an exchanged photon with the target can give it enough energy to fragment into several types of isotopes.

effusion of the radioactive atoms in the target, which is maintained at high temperatures ($\sim$2500 $^\circ$ C) to speed the process up. Such diffusion processes vary a great deal in speed. For short-lived nuclear species, with half-lives of milliseconds or less, this is often the limiting factor in intensity because the atoms decay before they reach the final target.

In contrast, in-flight facilities can produce all chemical species with half-lives greater than about 150ns, the time of transit through the fragment separator, and since the beams are produced at high energy they do not need re-acceleration. The main drawbacks of this method are that a) the beams are weak; b) they are not separated physically—the individual ions are simply tagged electronically by A, Z, and momentum; and c) they are of poor quality in terms of energy and focusing.

Assume that a highly energetic uranium projectile ($N/Z \sim 1.6$) hits a target nucleus in an almost central collision, as shown in figure 13.2. A part of the projectile (participant) is scrapped off and forms a highly excited mixture of nucleons with a part of the target. A piece of the projectile (spectator) flies away with nearly the same velocity of the beam. The neutron-to-proton ratio of the spectator is nearly equal to that of the projectile. Since the N/Z ratio of light nuclei (stable) is close to one, the fragment is far from the stability line. Statistically, a large number of fragments with different N/Z ratios are created and several new exotic nuclei have been discovered in this way.

Experiments with secondary beams are limited by reaction cross section and luminosity. The luminosity L is defined as the product of beam intensity i and target thickness t:

$$L = i \cdot t. \tag{13.1}$$

The reaction rate N is the product of luminosity and reaction cross section σ_r:

$$N = \sigma_r \cdot L. \tag{13.2}$$

In most of the reactions the usable target thickness is limited by the width of the excitation function (the cross section as a function of the excitation energy). Production reactions with a wide excitation function covering a broad energy range can profit in luminosity by the use of thick targets.

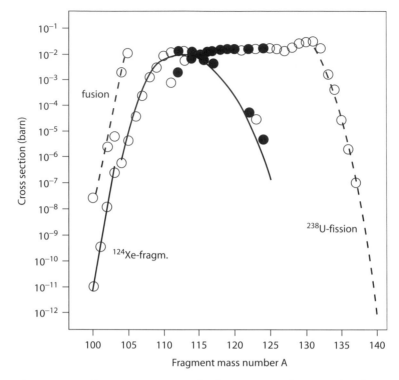

Figure 13.3 Production cross sections for the tin isotopes from complete fusion (dotted line), fragmentation (solid line), and projectile fission of ^{238}U (dashed line). The symbols represent experimental data. The fragmentation cross sections (solid line) have been calculated with a semi-empirical code [Sue00].

The condition for fragmentation of heavy ions is that the projectile should move faster than the nucleons move inside the nucleus. The projectile energy should be sufficiently above the Fermi domain, for example, above 100 A MeV. The usable target thickness for these high energies is of the order of several grams per square centimeter, corresponding to 10^{23} atoms/cm^2. The excitation function for complete fusion of heavy ions, however, has a width of only 10 MeV. This corresponds to a usable target thickness of the order of one milligram per square centimeter or 10^{18} atoms/cm^2. Consequently beam intensities for the investigation of complete fusion reactions must be four to five orders higher to achieve the same luminosity as for fragmentation.

Figure 13.3 shows as an example the production cross sections for the tin isotopes from complete fusion (dotted line), nuclear fragmentation (solid line), and Coulomb fission of ^{238}U (dashed line). The symbols represent experimental data. The fragmentation cross sections (solid line) have been calculated with a semi-empirical code [Sue00].

13.2 Light Exotic Nuclei

The first experiments with unstable nuclear beams were designed to measure the nuclear sizes, namely, the matter distribution of protons and neutrons. For stable nuclei such

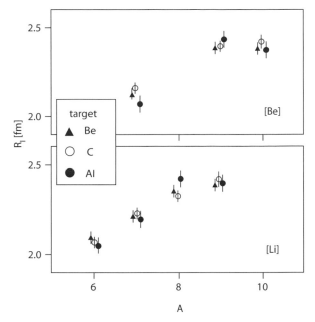

Figure 13.4 Interaction radii for Li and Be isotopes using three different targets [Ta85].

experiments are best accomplished with electron beams, which probe the nuclear charge (proton) distribution. Electron scattering experiments with unstable beams can only be performed in an electron-nucleus collider. Such machines are not yet available. The easiest solution is to measure the *interaction cross section* in collisions of unstable beams with a fixed target nucleus.

The interaction cross section is defined as the cross section for the change of proton and/or neutron number in the incident nucleus. To extract the *interaction radii* of the radioactive secondary beam nuclei, it is assumed that it can be expressed as [Ta85]

$$\sigma_I(P, T) = \pi \left[R_I(P) + R_I(T) \right]^2, \tag{13.3}$$

where $R_I(P)$ and $R_I(T)$ are the interaction radii of the projectile and the target nuclei, respectively. $R_I(T)$ can be obtained from σ_I in collisions between identical nuclei, while $R_I(P)$ can be obtained by measuring σ_I for different targets T [Ta85].

The above equation assumes a separability of the projectile and target radius. This hypothesis has been tested by Tanihata and collaborators [Ta85]. As an example, the interaction radii R_I for Li and Be isotopes have been obtained using three different targets. The results are shown in figure 13.4.

The reaction cross section in high energy collisions is given by

$$\sigma_R = 2\pi \int [1 - T(b)]b \, db, \tag{13.4}$$

where

$$T(b) = \exp\left\{ -\sigma_{NN} \int_{-\infty}^{\infty} dz \int \rho_P(\mathbf{r})\rho_T(\mathbf{R} + \mathbf{r})d^3r \right\}, \tag{13.5}$$

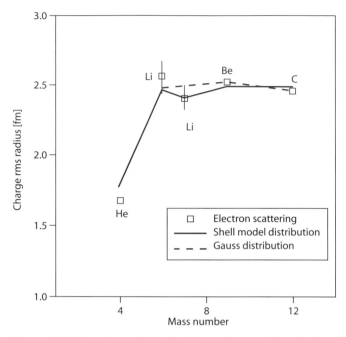

Figure 13.5 Comparison of calculated rms charge radii with those obtained by electron scattering experiments with stable nuclei. The calculations using harmonic oscillator distributions are shown by the solid line. Calculations using Gaussian distributions (which are the same for protons as for neutrons) is shown by the dashed line.

with $\mathbf{R} = (\mathbf{b}, z)$. σ_{NN} is the nucleon-nucleon cross section at the corresponding bombarding energy, and $\rho_{P(T)}$ is the projectile (target) matter density distribution. $T(b)$ is known as the *transparency function*. It is the probability that a reaction occurs for a collision with impact parameter b. The exponent $\left(\sigma_{NN} \int \rho_P(\mathbf{r})\rho_T(\mathbf{R} + \mathbf{r})d^3r\right)^{-1}$ is interpreted as the mean free path for a nucleon-nucleon collision. The reaction cross sections can be calculated using the matter distributions of the target in eq. 13.4.

13.2.1 Halo nuclei

In order to show that the rms (root-mean-square) radii obtained by a comparison of reaction cross section calculations with the experimentally determined σ_I are equal, figure 13.5 shows the calculated rms charge radii and those obtained by electron-scattering experiments for stable nuclei. Even the difference between the radii of ^{6}Li and ^{7}Li because of the occupation number difference between protons and neutrons is reproduced by the harmonic oscillator distribution (solid line). The rms radii obtained with Gaussian distributions (which are the same for protons as for neutrons) is shown by the dashed line.

The calculations also show that R_I represents the radius where the matter density is about 0.05 fm^{-3} for $A \geq 6$ nuclei. Now we can understand why the rms radii and R_I behave differently with A. While the rms radii stay constant, the absolute density increases with

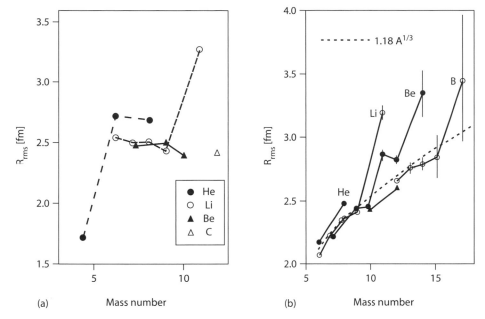

Figure 13.6 (a) Rms radii for the neutron rich isotopes He, Li, Be, and C. (b) Matter density radii of several light nuclei compared to the trend $R \sim 1.18\ A^{1/3}$ fm (dashed line) for normal nuclei. The solid lines are guides to the eyes.

A. Therefore R_I, which represents constant density, increases with A. These interesting results are presented in figure 13.6a, where the rms radii of He, Li, Be, and C isotopes are shown [Ta89]. The curves are guides to the eyes.

We observe a great increase in the rms radii for the neutron rich isotopes ^{6}He, ^{8}He, and ^{11}Li. Thus, the addition of the neutrons to ^{4}He and ^{9}Li nuclei increase their radii considerably. This might be understood in terms of the binding energy of the outer nucleons. The large matter radii of these nuclei have lead the experimentalists to call them "halo nuclei." The binding energy of the last two neutrons in ^{11}Li is about 300 keV. In ^{6}He it is 0.97 MeV. These are very small values and should be compared with $S_n = 6$–8 MeV, the average binding of nucleons in stable nuclei.

The wavefunction of a loosely bound nucleon (as in the case of the deuteron) extends far beyond the nuclear potential. For large distances the wavefunction behaves like a Yukawa function,

$$R(r) \sim \frac{e^{-\eta r}}{r}, \tag{13.6}$$

where $(\hbar \eta)^2 = 2mB$, with B the binding energy and m the nucleon mass. Thus, the smaller the value of B, the more the wavefunction extends to larger $r's$. Thus the "halo" in an exotic unstable nuclei, like ^{11}Li, is a simple manifestation of the weak binding energy of the last nucleons. What is not so trivial is to know why ^{6}He and ^{11}Li are bound while ^{5}He and ^{10}Li are not. We will continue this discussion later.

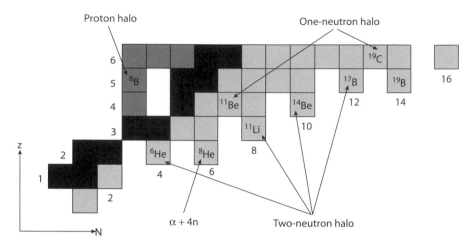

Figure 13.7 Nuclear chart showing the halo nuclei (courtesy of Haik Simon).

Abnormally large radii were also found for other light neutron-rich nuclei [Ta89] as shown in figure 13.6b.

The matter density radii of these nuclei do not follow the commonly observed trend $R \sim 1.18 \; A^{1/3}$ fm of normal nuclei. Thus the halo seems to be a common feature of loosely bound neutron rich nuclei. In fact, it has been observed experimentally that the separation energy of one neutron (S_n) and of two neutrons (S_{2n}) decreases with increasing matter radius. The two-neutron separation energies of ^{11}Li, ^{14}Be, and ^{17}B are very small and are responsible for large matter radii of these nuclei, as seen in figure 13.6b. A nuclear chart with the halo nuclei is shown in figure 13.7.

An RMF calculation was done in [Ta92] for the matter density distribution. As shown in figure 13.8c, d the agreement, with the "experimental" distributions is reasonable.

Since the number of neutrons in the outer orbits in ^{8}He is large (4) but ΔR^{rms} is not much larger than that for ^{6}He, the term "skin" nucleus was coined for ^{8}He [Ta92]. Neutron skins appear in many nuclei away from the stability line (close to the drip line). While a considerable number of neutrons can be included in a neutron skin, a neutron halo is expected to include only a few neutrons in the last orbital.

The fact emerging from the experiments is that the long tail of the matter distribution in some light neutron rich nuclei is due to the small binding of the last neutrons. The contribution from the nucleons in the core and in the valence orbitals to the total matter distribution yields different distribution shapes. This fact is well displayed in figure 13.9a, where a phenomenological density for ^{11}Be was used to describe its reaction cross section with Al and C targets at two different energies, 33 MeV/nucleon and 790 MeV/nucleon, respectively. The ^{11}Be was described as a ^{10}Be-core + a valence neutron. The density distribution of ^{10}Be was assumed to be Gaussian, which describes very well the matter distribution of light stable nuclei (see figure 13.5). The wavefunction of a nucleon with separation energy ϵ has a tail of the form $e^{-\eta r/r}$, where $\hbar^2 \eta^2 = 2\mu\epsilon$ and μ is the reduced mass. Thus, the valence neutron contributes with a density which is proportional to the

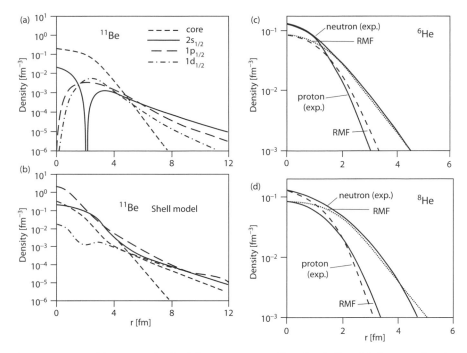

Figure 13.8 (a and b) Core and orbital densities (a), as well as the shell model densities (b). The core and orbital densities are multiplied by the occupation probabilities. The figures are from [Sa92]. (c and d) Relativistic mean field (RMF—see section 5.10.3) calculations for the ground state densities of ^{6}He and ^{8}He (solid lines).

square of this distribution. The summed distribution was used in (13.4) and (13.5), and the reaction cross section was compared with the experiment. One sees that a single Gaussian distribution is unable to describe both set of data simultaneously. But a Gaussian plus a distribution obtained from a Yukawa function does it quite well. Also shown is the result of a calculation (solid line) using the modified Hartree-Fock method (see section 5.10).

The material presented in this chapter leads us to conclude that the matter distribution in ^{11}Li and ^{11}Be is much like what one sees in figure 13.10. The calculations support the idea that the nuclei possess a "halo" generated by the loosely bound neutrons in the last orbit.

13.2.2 Borromean nuclei

The nucleus ^{11}Li is a prototype of a halo nucleus with fascinating properties. The reason is its unusually large radius and the difficulty in explaining its properties by means of traditional theoretical tools. The nucleus ^{10}Li is not bound, having a resonant state at $\sim$500 keV in the continuum. However, ^{11}Li is bound by $\sim$350 keV. The valence neutrons in ^{11}Li have to be treated separately from the core nucleons in order to explain its properties (i.e., radius) with traditional techniques. Thus, it seems that ^{11}Li should be treated theoretically as a three-body problem, ^{9}Li + n + n, in which to a good approximation ^{9}Li is treated as an inert core. This assumption is supported by the fact that the valence neutrons are distributed in a spatial region which is much larger than the core.

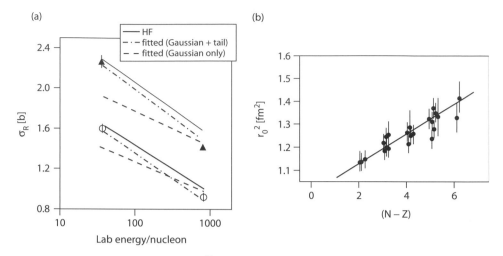

Figure 13.9 (a) Reaction cross section for ^{11}Be with Al (upper points) and C (lower points) targets at two different energies, 33 MeV/nucleon and 790 MeV/nucleon. Hartree-Fock calculations and phenomenological Gaussian densities are also shown and compared with the experiments. (b) Radius parameter used to fit the experimental reaction cross section, as a function of the neutron excess in several nuclei.

Experimentally, $r_{ms}(^{11}\text{Li}) \cong 3.1\,\text{fm}$ while $r_{ms}(^{9}\text{Li}) \cong 2.4\,\text{fm}$. Calling $r_{ms}(n)$ the root mean square radius of the distribution of a valence neutron, one can write

$$r_{ms}\left(^{11}\text{Li}\right) \cong \frac{2}{11}\, r_{ms}(n) + \frac{9}{11}\left(^{9}\text{Li}\right) \cong \frac{2}{11}(6\,\text{fm}) + \frac{9}{11}(2.4\,\text{fm}). \tag{13.7}$$

Thus, the r_{ms} of the valence distribution is of order of 6 fm, which is much larger than $r_{ms}(^{9}\text{Li})$. In other words, the valence neutrons in ^{11}Li "see" the core (^{9}Li) almost like a point particle.

The properties of the three-body halo nuclei were so intriguing that a new name was coined for them: *Borromean nuclei* [Zh93]. The name recalls the Borromean rings, heraldic symbol of the princes of Borromeo, which are carved in the stone of their castle on an

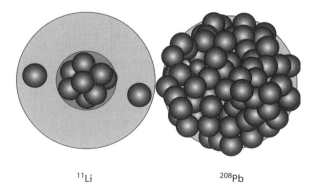

^{11}Li $\qquad\qquad$ ^{208}Pb

Figure 13.10 Light exotic nucleus (left), with a halo formed by loosely bound nucleons and a tight core (^{11}Li), compared with a normal heavy nucleus (^{208}Pb).

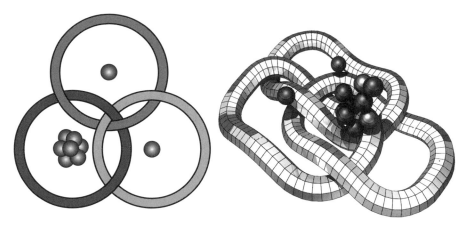

Figure 13.11 The ^{11}Li nucleus consists of a core ^{9}Li nucleus orbited by two neutrons. If any one of the three bodies is removed the remaining two would be unbound, rather like heraldic Borromean rings.

island in Lago Maggiore in Northern Italy. The three rings are interlocked in such a way that if any of them were removed, the other two would also fall apart. The three-body quantum analog is one where the three-body system is bound, but where none of the two-body subsystems are bound. Figure 13.11 shows an indication of such character.

Nature has realized Borromean systems in loosely bound halo-like nuclei such as ^{11}Li, which are now produced as secondary beams. To understand these systems one needs to go beyond the mean field approach, to few-body theoretical procedures such as expansion on *hyperspherical harmonics* or the coordinate space *Fadeev approach* [Zh93]. Nuclei like ^{6}He and ^{11}Li can be modeled as a core surrounded by two loosely bound valence neutrons. Not only the bound state (^{6}He and ^{11}Li both have only one bound state!), but also the structure of the continuum and the details of reaction mechanisms can be described in this way.

Besides the three-body halos (e.g., ^{6}He, ^{11}Li) the two-body halos (e.g., ^{8}Be, ^{11}Be, ^{19}C) can also be treated in terms of a core and a proton, or neutron, outside the core. Some characteristics of these nuclei can be reproduced in this way. However, to obtain the correct assignments of spin and parity the model has to be improved.

Several other methods have been devised to probe the structure of nuclei far from the valley of stability (see ref. [BHM02]). Among these are a) Coulomb dissociation [BBR86]; b) Trojan horse method [Bau86]; c) asymptotic normalization coefficients [Xu04]; d) heavy ion charge-exchange [St96]; e) knockout [HT03]; and f) fusion reactions [Can05]. These methods yield different insights into the structure of exotic nuclei and comprise most of experiments in radioactive beam facilities. Next we shall briefly describe how fusion reactions have been used to study the other extreme of the nuclear chart: the superheavy elements.

13.3 Superheavy Elements

The heaviest element found in nature is $^{238}_{92}$U. It is radioactive, but it survived from its formation in the supernova explosions since it has a decay half-life of the order of the age of the earth. Elements with larger atomic number (transuranic) have shorter half-lives and have disappeared. They are created artificially through nuclear reactions using as targets heavy elements or transuranic elements obtained previously. The projectiles were initially light particles: protons, deuterons, α-particles and, neutrons. Use of the neutron is justified because the β^--emission of the compound nucleus increases the value of Z and it was in this way [MA40] that the first transuranic element, neptunium, was obtained:

$$n + {}^{238}U \rightarrow {}^{239}U \rightarrow {}^{239}Np + \beta^-. \tag{13.8}$$

Reactions with light particles could produce isotopes up to mendelevium ($Z = 101$), but with this way of production it is not possible to go beyond that; the half-lives for α-emission or spontaneous fission become extremely short when this region of the periodic table is reached, making the preparation of a target impracticable. The alternative is to place a heavy element under a flux of very intense neutrons. This can be done using special reactors of high flux or using the residual material of nuclear explosions. The elements einsteinium ($Z = 99$) and fermium ($Z = 100$) were discovered in this way in 1955, but the increasing competition the β-decay has with α-decay and with spontaneous fission prevents the formation of elements with larger Z.

Starting in 1955 heavy ion accelerators began to deliver beams with high enough intensity and energy to compete in the production of transuranic isotopes. The first positive result was the production of two californium isotopes ($Z = 98$) in the fusion of carbon and uranium nuclei:

$$\begin{align} {}^{12}_{6}C + {}^{238}_{92}U &\rightarrow {}^{244}_{98}Cf + 6n, \tag{13.9}\\ {}^{12}_{6}C + {}^{238}_{92}U &\rightarrow {}^{246}_{98}Cf + 4n. \tag{13.10} \end{align}$$

This opened the possibility of reaching directly the nucleus one wants to create from the fusion of two appropriately chosen smaller nuclei. The difficulty of such a task is that the cross sections for the production of heavy isotopes are extremely low. For example, the reaction ${}^{50}_{22}Ti + {}^{208}_{82}Pb \rightarrow {}^{257}_{104}Rf + n$, which produces the element rutherfordium, has a cross section of 5 nb. A small increase in the charge drastically reduces that value: the cross section for the fusion reaction ${}^{58}_{26}Fe + {}^{208}_{82}Pb \rightarrow {}^{265}_{108}Hs + n$ is 4 pb. In comparison, the typical cross sections of DIC (deep inelastic collisions; see [BD04]). For heavy nuclei are in the range 1–2 b.

In spite of the experimental refinement these low cross sections demand, one is able to produce an isotope as heavy as ${}^{277}112$ (not yet named superheavy elements are represented by their mass number). The understanding of the mechanisms that lead to fusion is, however, not fully understood. According to the traditionally accepted model, the fusion of two nuclei proceeds in two stages: the formation of a compound nucleus and the de-excitation of the compound nucleus by evaporation of particles, preferentially neutrons. The difficulties for the materialization of the process in very heavy nuclei reside in both stages and they will be discussed next.

When two light nuclei are in contact the attractive nuclear force overcomes the repulsive Coulomb force and the final fate of the system is fusion. In heavier systems the opposite

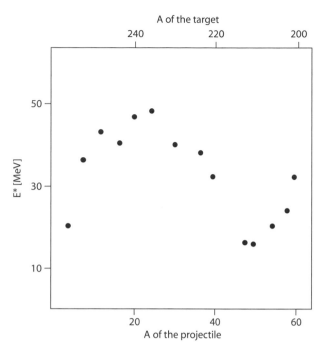

Figure 13.12 Minimum excitation energy of the compound nucleus $^{258}_{104}$Rf formed by several projectile/target combinations.

happens and the projectile should have a certain minimum kinetic energy to penetrate the target. At this stage an appreciable amount of the kinetic energy is transferred to the internal excitation energy and this has an effect equivalent to the increase of the Coulomb barrier. An increment in the incident energy, or as is said, an "extra push," is necessary to take the fusion ahead. Effects of nuclear structure can also be present, favoring or hindering the process.

But an increase in the kinetic energy of the projectile is conditioned to the survival of the nucleus to the de-excitation process. In each stage of the neutron emission there is a possibility of fission. If the excitation energy is very high, say, about 50 MeV, it will be much larger than the fission barrier. The fission will very probably win the competition and the final stage will not be a cold residual nucleus of atomic number $Z = Z_1 + Z_2$ as one wants. One should also keep in mind the limitation, already mentioned, of the value of the angular momentum l transferred to the nucleus: it cannot surpass l_{crit}, so that the allowed values of l are located inside a small window around $l = 0$. In other words, only central collisions have some chance of producing a compound nucleus that de-excites without fission.

A guide for the choice of more appropriate projectiles and targets to produce a given nucleus should come from an examination of a graph such as that of figure 13.12, where we show, as an example, the minimum value of the excitation energy of the compound nucleus $^{258}_{104}$Rf* created from several projectile/target configurations. This value was calcu-

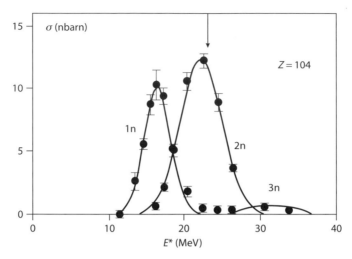

Figure 13.13 Excitation function for the production of $_{104}$Rf from the fusion of $^{208}_{82}$Pb with $^{50}_{22}$Ti. The arrow indicates the value of the interaction barrier, as calculated by Bass [Ba74]. The abscissa is the dissipated energy, calculated starting from the energy in the CMS at half the thickness of the target.

lated [Og75] adding the negative value of Q of the reaction to the value of the interaction barrier $B = Z_1 Z_2 e^2 / [1.45(A_1^{1/3} + A_2^{1/3})]$. It is evident that projectiles around $A = 50$ hitting targets of Pb and Bi produce the smallest excitation energy. This "cold fusion" has a larger probability of happening. The excitation function for a reaction of this type is seen in figure 13.13, where $_{104}$Rf is formed by the irradiation of $^{208}_{82}$Pb with $^{50}_{22}$Ti [Ho96]. The maximum probability of fusion with emission of one or no neutrons happens for a center of mass energy smaller than the Coulomb barrier! In a model that one can envisage for an event of this type, the projectile has a central collision with the target and stops before reaching the maximum of the Coulomb barrier. The idea is that at this point the external shells are in contact and the nucleon transfer between projectile and target can decrease the Coulomb barrier and allow the nuclei to overcome it.

The favorable conditions provided by the method of cold fusion imply that, for the creation of the heaviest elements, it is preferable to do reactions of this kind than with actinide targets and lighter projectiles.

In fact, it was through the reactions

$$^{54}_{24}\text{Cr} + \,^{209}_{83}\text{Bi} \rightarrow \,^{262}_{107}\text{Bh} + \text{n},$$

$$^{58}_{26}\text{Fe} + \,^{208}_{82}\text{Pb} \rightarrow \,^{265}_{108}\text{Hs} + \text{n},$$

$$^{58}_{26}\text{Fe} + \,^{209}_{83}\text{Bi} \rightarrow \,^{266}_{109}\text{Mt} + \text{n},$$

$$^{62}_{28}\text{Ni} + \,^{208}_{82}\text{Pb} \rightarrow \,^{269}110 + \text{n},$$

$$^{64}_{28}\text{Ni} + \,^{209}_{83}\text{Bi} \rightarrow \,^{272}111 + \text{n},$$

$$^{70}_{30}\text{Zn} + \,^{208}_{82}\text{Pb} \rightarrow \,^{277}112 + \text{n}, \tag{13.11}$$

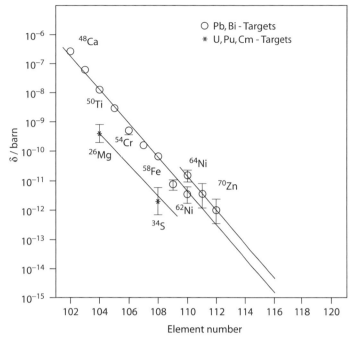

Figure 13.14 Cross section systematics for the production of heavy and superheavy elements with cold fusion and actinide based fusion.

that for the first time isotopes of the elements $Z = 107$ to $Z = 112$ [Ar85, Ho96] were synthesized, the first of the list in 1976 and the last in 1996. In spite of that, reactions with actinide targets were not abandoned. In 1996, Lazarev and collaborators [La96] synthesized the isotope $^{273}110$, bombarding ^{244}Pu with ions of ^{34}S. The isotope $^{273}110$ was obtained starting from the evaporation of 5 neutrons of a nucleus of $^{278}110$ excited with approximately 50 MeV. The cross section for the process was 0.4 pb!

Figure 13.14 shows the cross section systematic for the production of the elements from nobelium to 116. The production cross sections for elements 104 to 112 decrease by a factor of about 10 per two elements down to 1 picobarn for element 112 as shown in figure 13.14. The competing processes of barrier penetration through the Coulomb, or more precisely, the fusion barrier, and the survival of the compound nucleus can be inferred from the excitation functions. For element 104 the two-neutron (2n) and the three-neutron (3n) evaporation channels are still present; the 1n and 2n channels section show similar cross sections. For element 108 the 2n channel is already rather weak. It completely disappears for element 110. Moreover, the 1n channel is shifted toward low excitation energies, and comes close to the binding energy of the last neutron. The synthesis of element 112 is the coldest fusion ever observed. The width of the excitation functions shrink toward element 110 where the half width becomes only 5 MeV. All these observations point toward a strong competition between barrier transmission and survival probability.

13.4 Exercises

1. An object of mass M confines a loosely bound light particle of mass m with binding energy ϵ. The particle is suddenly released from the system by an external perturbation, and the distribution of probabilities $W(\mathbf{p})$ of different values of the particle momentum $\mathbf{p}$ is measured.

   a. Assuming that the particle in the bound state spends most of the time outside the well, give a simple analytical expression for $W(\mathbf{p})$.

   b. In many experiments, instead of the full three-dimensional momentum distribution $W(\mathbf{p})$, only the "longitudinal" distribution $W_x(p_x)$ of the momentum component p_x along a specific axis x is measured. Using the result a, predict the function $W_x(p_x)$.

   c. Assume that the neutron binding energy in a "halo" nucleus which consists of 3 protons and 8 neutrons is only 0.3 MeV. The experiment measuring $W_x(p_x)$ shows a bell-shaped curve with the full width at half maximum equal to $\Delta p_x = 44$ MeV/c where the convenient units for the momentum are used, and c is the speed of light. Does this result agree with the predictions of the point b?

2. The table below shows some experimental values of the fusion cross-section for ^{32}S on ^{24}Mg as a function of the energy E_{lab} of the ^{32}S nuclei in the laboratory coordinate system. Use these data to test the hypothesis that $\sigma_R = \sigma_{fus}$ at these energies.

E_{lab} [MeV]	70	80	90	110
σ_{fus} [mb]	150	414	670	940

Appendix A | Angular Momentum

A.1 Orbital Momentum

In spherical coordinates the Laplace operator is a sum of the radial and angular parts,

$$\nabla^2 = \frac{1}{r^2}\frac{\partial}{\partial r}r^2\frac{\partial}{\partial r} - \frac{1}{r^2}\mathbf{l}^2, \tag{A.1}$$

$$\mathbf{l}^2 = -\frac{1}{\sin^2\theta}\frac{\partial^2}{\partial\varphi^2} - \frac{1}{\sin\theta}\frac{\partial}{\partial\theta}\sin\theta\frac{\partial}{\partial\theta}. \tag{A.2}$$

A simple transformation of variables shows that the operator in the angular part (A.1) is nothing but the square of the vector $\mathbf{l}$ of orbital momentum (in units of $\hbar$),

$$\mathbf{l} = \frac{[\mathbf{r}\times\mathbf{p}]}{\hbar} = -i[\mathbf{r}\times\nabla]. \tag{A.3}$$

The orbital momentum components in spherical polar coordinates are

$$l_\pm \equiv l_x \pm il_y = e^{\pm i\varphi}\left[\pm\frac{\partial}{\partial\theta} + i\cot\theta\frac{\partial}{\partial\varphi}\right], \tag{A.4}$$

$$l_z = -i\frac{\partial}{\partial\varphi}. \tag{A.5}$$

Obviously, the decomposition (A.1) physically means that kinetic energy can be presented as a sum of the radial and rotational part. Being rotationally invariant, it does not imply the specification of a quantization axis.

A.2 Spherical Functions

Explicitly acting on a function of angles by the operator (A.1), we obtain

$$l^2 F_l(\theta, \varphi) = l(l+1) F_l(\theta, \varphi). \tag{A.6}$$

This means that the functions $F_l(\theta, \phi)$ are the *eigenfunctions* of the orbital momentum squared with the eigenvalue $l(l+1)$, where, by construction, $l = 0, 1, 2, \ldots$. One can construct $(2l+1)$ various angular functions F that are called *spherical harmonics* of rank l,

$$Y_{lm}(\theta, \varphi) = \Theta_{lm}(\theta) \frac{1}{\sqrt{2\pi}} e^{im\varphi}, \tag{A.7}$$

with $m = -l, -l+1, \ldots, -1, 0, 1, \ldots, l-1, l$. Of course, the function (A.7) is an eigenfunction of the operator (A.5) with the eigenvalue $l_z = m$,

$$l_z Y_{lm} = m Y_{lm}. \tag{A.8}$$

By construction, all values of l and m are integer.

According to general properties of Hermitian operators, the spherical functions with different quantum numbers (l, m) are automatically orthogonal, and the total normalization will always be taken as

$$\int do\, Y_{l'm'}^*(\mathbf{n}) Y_{lm}(\mathbf{n}) \equiv \int_0^{2\pi} d\varphi \int_0^\pi \sin\theta\, d\theta\, Y_{l'm'}^*(\theta, \varphi) Y_{lm}(\theta, \varphi) = \delta_{l'l} \delta_{m'm}. \tag{A.9}$$

There can still be a phase factor in the definition of Y_{lm} that can be fixed arbitrarily. A conventional choice is given by the condition

$$Y_{l-m}(\theta, \varphi) = (-)^m Y_{lm}^*(\theta, \varphi). \tag{A.10}$$

A.3 Generation of Rotations

Consider a rotation through an infinitesimal angle $\delta\alpha$ around the axis characterized by the unit vector $\mathbf{n}$. Under this rotation, a wavefunction ψ changes by an amount proportional to $\delta\alpha$. This transformation, $R_\mathbf{n}(\delta\alpha)$, is generated via the action of the operator $(\mathbf{J} \cdot \mathbf{n})$, the projection of the angular momentum onto the rotation axis,

$$\psi \to \psi' = [1 - i(\mathbf{J} \cdot \mathbf{n})\delta\alpha]\, \psi; \tag{A.11}$$

here and always we measure all angular momentum operators in units of $\hbar$. Equation (A.11) is nothing more than a *definition* of the angular momentum operator for a given system; we should find the transformed function explicitly and compare with (A.11) in order to determine the operator $\mathbf{J}$.

A *finite rotation* by an angle α can be achieved as a limit of a large number, $N \to \infty$, of sequential small rotations by $\delta\alpha = \alpha/N$. The operator of a finite rotation is

$$R_\mathbf{n}(\alpha) = \lim_{N\to\infty} [1 - i(\mathbf{J} \cdot \mathbf{n})\alpha/N]^N = \exp[-i(\mathbf{J} \cdot \mathbf{n})\alpha]. \tag{A.12}$$

Here we used the fact that the rotations through different angles around the same axis commute. Rotations preserve relations between the state vectors in Hilbert space: amplitudes are the same before and after any rotation R of the entire space,

$$\langle \psi_2'|\psi_1' \rangle \equiv \langle R\psi_2|R\psi_1 \rangle \equiv \langle \psi_2|R^\dagger R\psi_1 \rangle = \langle \psi_2|\psi_1 \rangle. \tag{A.13}$$

Therefore the transformation operator (A.12) has to be *unitary*,

$$R^\dagger R = 1 \Rightarrow R^\dagger = R^{-1}. \tag{A.14}$$

Any unitary operator U can be expressed as

$$U = e^{iG} \equiv \sum_{n=0}^{\infty} \frac{(iG)^n}{n!}, \tag{A.15}$$

where the exponent is a symbolic representation of the infinite series and the generator G of the unitary transformation, $\mathbf{J}$ in our case, is Hermitian.

A.4 Orbital Rotations

Let us see how (A.12) goes into the definition (A.3) in the particular case of the *orbital momentum* of a particle, $\mathbf{J} \Rightarrow \mathbf{l}$. In this case it is convenient to deal with the direct coordinate representation $\psi(\mathbf{r})$ of the particle wavefunction. The result of the rotational transformation of $\psi(\mathbf{r})$ is known: after the rotation one sees at the point $\mathbf{r}$ the value of the function that before the rotation was at the point $R^{-1}\mathbf{r}$, with R^{-1} denoting the inverse rotation,

$$\psi'(\mathbf{r}) = R\psi(\mathbf{r}) = \psi(R^{-1}\mathbf{r}). \tag{A.16}$$

For example, take the rotation around the z-axis through an angle α. A vector

$$\mathbf{r} = (x, y, z) = (r \sin\theta \cos\varphi, r \sin\theta \sin\varphi, r \cos\theta) \tag{A.17}$$

goes into

$$\mathbf{r}' = (r \sin\theta \cos(\varphi + \alpha), r \sin\theta \sin(\varphi + \alpha), r \cos\theta), \tag{A.18}$$

or, in the limit of the infinitesimal $\delta\alpha$,

$$R_z(\delta\alpha)(x, y, z) = (x - \delta\alpha \, y, y + \delta\alpha \, x, z). \tag{A.19}$$

The arguments of the transformed function in (A.16) correspond to the *opposite* rotation, so that, using the linear momentum operator in the coordinate representation, $\mathbf{p} = -i\hbar\nabla$, we get

$$R_z(\delta\alpha)\psi(x, y, z) = \psi(x + \delta\alpha \, y, y - \delta\alpha \, x, z) = \psi(x, y, z) + \delta\alpha \left[y\frac{\partial\psi}{\partial x} - x\frac{\partial\psi}{\partial y} \right]$$

$$= \{1 - (i/\hbar)\delta\alpha(xp_y - yp_x)\}\psi = \{1 - i\delta\alpha l_z\}\psi. \tag{A.20}$$

It means that the generator of rotations (A.11) is here the orbital angular momentum $\mathbf{l}$, (A.3).

Our definition (A.16) of the rotation implies that we rotate a physical object ("active") picture). Rotation of a coordinate frame is equivalent, from a viewpoint of transformations, to an opposite rotation of the system ("passive" picture). The corresponding rotation operators would be conjugate to ours (A.16).

We can compare this procedure with a simpler case of the linear momentum $\mathbf{p}$ that is the generator of *translations* $D(\mathbf{a})$, when the coordinate $\mathbf{r}$ of the object is shifted by a constant vector $\mathbf{a}$, and the new function after the shift comes to the point $\mathbf{r}$ from the point $\mathbf{r} - \mathbf{a}$,

$$D(\mathbf{a})\psi(\mathbf{r}) = \psi'(\mathbf{r}) = \psi(\mathbf{r} - \mathbf{a}). \tag{A.21}$$

For an infinitesimal translation, $\mathbf{a} \to \delta\mathbf{a}$,

$$\psi'(\mathbf{r}) \approx \psi(\mathbf{r}) - (\delta\mathbf{a} \cdot \nabla)\psi(\mathbf{r}) = \left\{1 - \frac{i}{\hbar}(\delta\mathbf{a} \cdot \mathbf{p})\right\}\psi(\mathbf{r}). \tag{A.22}$$

Comparison with the rotational case (A.11) shows that the linear momentum (in units of $\hbar$) is the generator of translations. A finite translation $D(\mathbf{a})$, analogous to (A.12), is a product of an infinite number of infinitesimal translations,

$$D(\mathbf{a}) = e^{-(i/\hbar)(\mathbf{a} \cdot \mathbf{p})}. \tag{A.23}$$

Since the translations along different axes commute, here it is not necessary to take all small shifts $\delta\mathbf{a}$ along the same direction.

A.5 Spin

The result of a rotation in general cannot be reduced to the explicit coordinate transformation (A.20). The wavefunction may consist of several components that undergo a linear transformation between themselves, in addition to the transformation (A.20) of their coordinate dependence. Such components describe different possible *intrinsic* states of an object and usually are referred to as *spin* degrees of freedom. If $\mathbf{S}$ is the vector generator (A.11) of this transformation, the whole effect of the rotation onto the wavefunction of a system is described by the *total* angular momentum,

$$\mathbf{J} = \mathbf{L} + \mathbf{S}, \tag{A.24}$$

where $\mathbf{L}$ generalizes the single-particle orbital momentum $\mathbf{l}$ of (A.3) for an arbitrary system. For a many-body system, the global rotation acts on all particles in the same way so that the total momenta are additive combinations of the single-particle ones,

$$\mathbf{J} = \sum_a \mathbf{j}_a, \quad \mathbf{L} = \sum_a \mathbf{l}_a, \quad \mathbf{S} = \sum_a \mathbf{s}_a, \quad \mathbf{j}_a = \mathbf{l}_a + \mathbf{s}_a. \tag{A.25}$$

As a natural example we consider a *vector function* $\mathbf{V}(\mathbf{r})$. At each point $\mathbf{r}$ we have three functions $V_i(\mathbf{r})$, but they are components of the same vector object. Under rotations not only should each of these functions be transformed as we have seen earlier but, apart from that, the components V_i are transformed among themselves, as would occur even for a constant vector $\mathbf{V}$ with no coordinate dependence.

For an arbitrary rotation R we have

$$RV_x(x, y, z) = V'_x(R^{-1}x, R^{-1}y, R^{-1}z), \tag{A.26}$$

where the notation V' means that the components of the vector also undergo a transformation. As we have seen in (A.19), for an infinitesimal rotation $R = R_z(\delta\alpha)$ by an angle $\delta\alpha$ around the z-axis,

$$R^{-1}x = x + \delta\alpha\, y, \quad R^{-1}y = y - \delta\alpha\, x, \quad R^{-1}z = z. \tag{A.27}$$

This shows what argument is to be taken for the vector as a function of the coordinates in the right-hand side of (A.26). On the other hand, in addition to this parallel transport, the vector $\mathbf{V}$ itself rotates around the z-axis by an angle $\delta\alpha$, so its azimuthal angle ϕ_0 changes to $\phi_0 + \delta\alpha$ (the angles with subscript 0 are those of the direction of $\mathbf{V}$ rather than of the coordinate point $\mathbf{r}$). Then

$$V'_x = |\mathbf{V}| \sin\theta_0\, \cos(\phi_0 + \delta\alpha) \approx V_x - \delta\alpha\, V_y, \tag{A.28}$$

and, analogously,

$$V'_y \approx V_y + \delta\alpha\, V_x, \quad V'_z = V_z. \tag{A.29}$$

The result of the transformation of the components (note that it again has a sign corresponding to an active rotation opposite to that for the arguments of the wave function) can be expressed as an action of a 3×3 matrix S_z on a column of the components V_i,

$$R_z(\delta\alpha)\mathbf{V} = (1 - i\delta\alpha\, S_z)\mathbf{V}, \quad S_z = \begin{pmatrix} 0 & -i & 0 \\ i & 0 & 0 \\ 0 & 0 & 0 \end{pmatrix}. \tag{A.30}$$

The total infinitesimal transformation of our vector function is given by

$$\begin{aligned} R_z(\delta\alpha)\, V_x(\mathbf{r}) &= V'_x(x + y\delta\alpha, y - x\delta\alpha, z) \\ &= V_x(x + y\delta\alpha, y - x\delta\alpha, z) - \delta\alpha\, V_y(x + y\delta\alpha, y - x\delta\alpha, z), \end{aligned} \tag{A.31}$$

or, collecting all terms of the first order with respect to $\delta\alpha$,

$$R_z(\delta\alpha)\, V_x(\mathbf{r}) = V_x(x, y, z) - \delta\alpha\, V_y(x, y, z) - \delta\alpha\left(x\frac{\partial V_x}{\partial y} - \frac{\partial V_x}{\partial x} \right). \tag{A.32}$$

In operator form this means that for the vector field

$$R_z(\delta\alpha) = 1 - i\delta\alpha(S_z + L_z) = 1 - i\delta\alpha J_z, \tag{A.33}$$

where the orbital momentum $\mathbf{L}$ is, as usual, $-i[\mathbf{r} \times \nabla]$. Finite rotations require the exponentiation of the total angular momentum operator, as in (A.12). We need to stress that the operators $\mathbf{S}$ and $\mathbf{L}$ act on different variables and therefore always commute.

A.6 Ladder Operators

We now consider the one-dimensional case of a particle described by a coordinate x and conjugate momentum p, operators with the standard commutation relation

$$[x, p] = i\hbar. \tag{A.34}$$

The operators x and p are Hermitian. Instead we can introduce their non-Hermitian linear combinations

$$a = \sqrt{\frac{1}{2\nu\hbar}}(\nu x + ip), \quad a^\dagger = \sqrt{\frac{1}{2\nu\hbar}}(\nu x - ip), \tag{A.35}$$

which are Hermitian conjugate to each other. In (A.35) we introduced an arbitrary positive constant ν of dimension (mass/time), which makes the new variables a and $a^\dagger$ dimensionless.

The *Heisenberg-Weyl algebra* (A.34) is translated to the new variables (A.35) as

$$[a, a^\dagger] = 1. \tag{A.36}$$

The operator product of these operators,

$$N = a^\dagger a, \tag{A.37}$$

is Hermitian and in terms of the original variables equals

$$N = \frac{\hbar}{2\nu}(\nu x - ip)(\nu x + ip) = \frac{\nu x^2}{2\hbar} + \frac{p^2}{2\nu\hbar} - \frac{1}{2}, \tag{A.38}$$

where the commutator (A.34) was taken into account. With a specific choice of

$$\nu = m\omega, \tag{A.39}$$

we come to the Hamiltonian of a harmonic oscillator with mass m and frequency ω presented in the form

$$H = \frac{m\omega^2 x^2}{2} + \frac{p^2}{2m} = \hbar\omega \left(N + \frac{1}{2} \right), \tag{A.40}$$

although the entire construction is meaningful independently of any oscillator system.

The ladder of the eigenvalues of N can be built if one notices that

$$[a, N] = a, \quad [a^\dagger, N] = -a^\dagger. \tag{A.41}$$

Thus, N has a spectrum $\{\ldots, n-1, n, n+1, \ldots\}$, where n is an eigenvalue at the point where we have started building the ladder. The specific features of the ladder in this case are determined by the fact that the operator N is positively defined. Indeed, for any state $|\psi\rangle$ we can construct the state $|\psi_a\rangle \equiv a|\psi\rangle$ and see that the expectation value of N for the state $|\psi\rangle$ reduces to the norm of the state $|\psi_a\rangle$ and therefore it is not negative:

$$\langle\psi|N|\psi\rangle = \langle\psi|a^\dagger a|\psi\rangle = \langle\psi_a|\psi_a\rangle \geq 0. \tag{A.42}$$

Together with (A.38), this shows that for any state of a particle and any value of the positive parameter ν,

$$\left\langle \nu x^2 + \frac{p^2}{\nu} \right\rangle \geq \hbar, \tag{A.43}$$

an alternative form of the *uncertainty relation*. The minimum of this relation is achieved at the ground state of the harmonic oscillator, (A.39) and (A.40).

Equation (A.42) implies that all eigenvalues n of the operator N are non-negative, $n \geq 0$. The expectation value (A.42) can vanish if and only if the norm of $|\psi_a\rangle$ vanishes, that is, this state is the zero vector. Let us call the state $|\psi\rangle$ annihilated by the lowering operator the *vacuum* state and denote this state as $|\text{vac}\rangle$,

$$|\psi\rangle = |\text{vac}\rangle \Rightarrow |\psi_a\rangle = a|\text{vac}\rangle = 0. \tag{A.44}$$

For the vacuum state

$$\langle\text{vac}|N|\text{vac}\rangle = 0. \tag{A.45}$$

On the other hand, any state can be represented with the help of the complete set of the eigenstates of a Hermitian operator. Taking the family of the eigenstates $|n\rangle$ of the operator N with the eigenvalues n as a basis, we can write down for the vacuum state

$$|\text{vac}\rangle = \sum_n C_n^{\text{vac}} |n\rangle. \tag{A.46}$$

Then (A.45) shows that

$$\sum_{n=0}^{\infty} |C_n^{\text{vac}}|^2 n = 0. \tag{A.47}$$

Since all $n \geq 0$, this is possible only if for the vacuum state

$$C_n^{\text{vac}} = \delta_{n0} \rightarrow |\text{vac}\rangle = |n = 0\rangle \equiv |0\rangle, \tag{A.48}$$

the vacuum state is an *eigenstate* of N with eigenvalue $n = 0$. Because of (A.44), the ladder cannot continue down from the vacuum state: this would bring us to impossible negative eigenvalues of N. But applying the raising operator $a^\dagger$ we can go upstairs in steps by $\Delta n = 1$ with no restriction. Thus, the ladder is limited from below but any *integer* $n \geq 0$ is an allowed eigenvalue of the operator N.

In order to complete our algebraic consideration we have to build explicitly the set of eigenstates $|n\rangle$ that satisfy the following equations (now we can label the states by the eigenvalues of N):

$$N|n\rangle = n|n\rangle, \quad a|n\rangle = \mu_n|n-1\rangle, \quad a^\dagger|n\rangle = \tilde{\mu}_n|n+1\rangle. \tag{A.49}$$

Here we assume that the states are normalized,

$$\langle n'|n\rangle = \delta_{nn'}, \tag{A.50}$$

so that the factors μ_n and $\tilde{\mu}_n$ are unknown matrix elements interrelated by the Hermitian conjugation,

$$\mu_n = \langle n-1|a|n\rangle, \quad \tilde{\mu}_n = \langle n+1|a^\dagger|n\rangle = \langle n|a|n+1\rangle^* = \mu_{n+1}^*. \tag{A.51}$$

For a consistent determination of the matrix elements we need a *nonlinear* relation. For example, we can take the commutator (A.36) or the definition (A.37) of N,

$$\langle n|N|n\rangle = n = \langle n|a^\dagger|n+1\rangle\langle n+1|a|n\rangle = |\mu_n|^2. \tag{A.52}$$

The *phase* of the matrix elements remains arbitrary. Indeed, the commutator (A.36), which determines the Heisenberg-Weyl algebra, does not change under a phase transformation

$$a \to a' = ae^{i\alpha}, \quad a^\dagger \to a^{\dagger'} = a^\dagger e^{-i\alpha} \tag{A.53}$$

with any real value of α. The transformations preserving the operator algebra can be called *canonical* in analogy with classical mechanics. Therefore the matrix elements can be determined up to an arbitrary phase. The simplest choice is to make them *real*,

$$\mu_n = \sqrt{n}, \quad \tilde{\mu}_n = \sqrt{n+1}. \tag{A.54}$$

Now we can recurrently construct the entire ladder starting from the vacuum state $|0\rangle$ and raising n:

$$a^\dagger|0\rangle = |1\rangle, \tag{A.55}$$

$$a^\dagger|1\rangle = \sqrt{2}|2\rangle \to |2\rangle = \frac{a^\dagger}{\sqrt{2}}\left|1\right\rangle = \frac{(a^\dagger)^2}{\sqrt{2}}\left|0\right\rangle, \tag{A.56}$$

and so on. The general recipe is evident,

$$|n\rangle = \frac{(a^\dagger)^n}{\sqrt{n!}}\left|0\right\rangle. \tag{A.57}$$

Parenthetically we can mention that the structure of the spectrum we obtained is that of noninteracting but *indistinguishable* quanta, and the quantum number n can be interpreted as a number of quanta in the quantum state $|n\rangle$. This approach is used in the general procedure of *secondary quantization* applicable to the general case of identical particles treated as quanta of a *quantum field*. The quantal picture attributes an important physical meaning to formal relations (A.49) and (A.54). Now we can rename raising and lowering operators into those of *creation* and *annihilation* of quanta, respectively. The annihilation operator *a* may describe the *absorption* of quanta. The probability of this process is proportional to the square of the matrix element $|\mu_n|^2 = n$, that is, to the number of available quanta (they are indistinguishable). The probability of the inverse process of *radiation* of a quantum is proportional to $|\tilde{\mu}_n|^2 = n + 1$ and contains, along with the *spontaneous* radiation independent of the number of quanta in the system also the effect of the *induced*, or *stimulated*, emission that is proportional to *n*. This effect is at the heart of laser physics.

A.7 Angular Momentum Multiplets

As clear from elementary geometric arguments, the result of two consecutive rotations around *different* axes depend on their order—the corresponding rotation operators do not commute. Using the explicit expression of the orbital momentum components as in (A.4) and (A.5) we obtain

$$[l_j, l_k] = i\epsilon_{jkn}l_n,$$ (A.58)

Since the commutation relations reflect a geometrical connection of rotations, they should be the same for any angular momentum operator **J**, spin or orbital, single-particle or many-body,

$$[J_j, J_k] = i\epsilon_{jkn}J_n.$$ (A.59)

It is worthwhile to notice that the linear momentum components p_j commute since these operators generate the shifts of Cartesian coordinates (A.22), and the results of two consecutive translations performed in different order coincide (the *abelian* group of translations in contrast to the *non-abelian* rotation group). As follows from the algebra (A.59), different components of **J** cannot simultaneously have certain values.

The most important new element that appears in this algebra compared to the simple case of the previous section is the possibility to construct an operator *C*, the so-called *Casimir operator*, that commutes with all generators J_k. Of course, any function of *C* also satisfies this condition but a more complicated algebra can have several *independent* Casimir operators. It is easy to see that the absolute value squared of the angular momentum plays the role of the Casimir operator,

$$\left[J_k, \mathbf{J}^2\right] = \left[J_k, J_x^2 + J_y^2 + J_z^2\right] = 0.$$ (A.60)

One of the projections, let us say J_z, and $\mathbf{J}^2$ can have certain values simultaneously. As we have seen for the orbital momentum, this characterization is associated with the choice of the quantization axis and therefore with the apparent violation of rotational symmetry. The symmetry is restored by the potential possibility of rotation to another frame.

Analogously to (A.35), instead of two Hermitian components of the angular momentum in the plane transverse to the quantization axis, J_x and J_y, we introduce two new operators, Hermitian conjugate to each other,

$$J_\pm = J_x \pm iJ_y, \quad J_+ = (J_-)^\dagger. \tag{A.61}$$

According to (A.59), these operators satisfy

$$[J_\mp, J_z] = \pm J_\mp, \tag{A.62}$$

$$[J_-, J_+] = -2J_z. \tag{A.63}$$

The first relation (A.62) is of the ladder type, (A.41). Starting with a state with a certain value M of the projection J_z, the operator J_- lowers this eigenvalue, $M \to M - 1$, whereas J_+ raises M, $M \to M + 1$. Let the initial state, apart from M, have also a certain value of $\mathbf{J}^2$. The Casimir operator commutes with $J_\pm$; therefore all states encountered along the ladder of various values of M still belong to the same value of $\mathbf{J}^2$. Geometrically it means that $J_\pm$ generate small rotations around the perpendicular axis that change the orientation (projection $J_z = M$) of the angular momentum vector relative to the quantization axis but do not change the absolute value that is invariant under rotations and characterizes the ladder as a whole.

Consider the ladder family of the states with a given value of $\mathbf{J}^2$ and various values of M. Since for any state

$$C = \langle \mathbf{J}^2 \rangle = \langle J_x^2 + J_y^2 + J_z^2 \rangle \geq \langle J_z^2 \rangle = M^2, \tag{A.64}$$

the ladder cannot be infinite, it ends (in both directions) at some limiting values M_{max} and M_{min}. These values are determined by the value C of the Casimir operator for the family under consideration. At the upper (lower) end of the multiplet the action of the raising (lowering) operator should give zero, similarly to eq. (A.44) for the Heisenberg-Weyl algebra. Using the equivalent expressions for the Casimir operator that follow from (A.60) and (A.63),

$$\mathbf{J}^2 = J_z^2 + \frac{1}{2}(J_+J_- + J_-J_+)$$

$$= J_+J_- + J_z^2 - J_z = J_-J_+ + J_z^2 + J_z, \tag{A.65}$$

having in mind that the expectation value of C is the same all over the ladder, and applying two last forms of (A.65) to the states with M_{min} and M_{max}, respectively, we obtain

$$C = M_{min}^2 - M_{min} = M_{max}^2 + M_{max}. \tag{A.66}$$

The appropriate solution for M_{min} is $M_{min} = -M_{max}$. The maximum possible projection M_{max} will be denoted as J. Usually this number is simply called "angular momentum," and

$$C = J(J + 1). \tag{A.67}$$

We see that the value of the Casimir operator $\mathbf{J}^2$ in the multiplet $|JM\rangle$ is *larger* than the squared maximum projection $M_{max}^2 = J^2$. This can be interpreted as a consequence of the uncertainty relation. In the state with the certain values of $\mathbf{J}^2$ and J_z, the noncommuting with J_z transverse components $J_{x,y}$ of the angular momentum cannot have definite values which would be the case if one could align the vector $\mathbf{J}$ along the z-axis, $\mathbf{J}^2 = J_z^2 = M_{max}^2$. The difference is due to the quantum fluctuations of $J_x^2 + J_y^2$. In other words, it is impossible to construct a state with a certain value of the angular momentum and a certain orientation in space, for example, along the quantization axis. The uncertainty relation $(\Delta J_x)(\Delta J_y) \geq |\langle J_z\rangle|/2$ is similar to the relation between a coordinate x and conjugate linear momentum p_x. From classical mechanics we remember that the angular momentum components are conjugate to angular coordinates.

Starting from the lowest state $M_{min} = -M_{max} = -J$ and applying the raising operator J_+, one can construct the entire ladder. The number of steps k from $M = -J$ to $M = +J$ is always integer and equals $2J$. Therefore only *integer* and *half-integer* values of J are possible. Correspondingly, the values of the projection M along the ladder are all integer or all half-integer. They can be labeled $|JM\rangle$ by the common value J (the family name) and the individual tag M (the first name of the family member), where $-J \leq M \leq +J$. The total number of states in the family (*multiplet*) $|JM\rangle$ is $k + 1 = 2J + 1$.

Now we can find the matrix elements of the generators inside the multiplet $|JM\rangle$. Being the eigenstates of the Hermitian operators $\mathbf{J}^2$ and J_z, the states $|JM\rangle$ are orthogonal and assumed to be normalized,

$$\langle J'M'|JM\rangle = \delta_{J'J}\delta_{M'M}. \tag{A.68}$$

The operators $J_\pm$ connect the adjacent states in the multiplet,

$$J_\pm|JM\rangle = \mu_\pm(JM)|JM \pm 1\rangle, \tag{A.69}$$

where, as a result of Hermitian conjugation,

$$\mu_-(JM) = \mu_+^*(JM - 1). \tag{A.70}$$

Taking the expectation value of the Casimir operator (A.65) in an arbitrary state $|JM\rangle$, we find the absolute values of the matrix elements $\mu_\pm(JM)$. Their phases remain arbitrary, and as we did earlier in (A.54), we take them real:

$$\mu_\pm(JM) = \sqrt{(J \mp M)(J \pm M + 1)}. \tag{A.71}$$

Thus, the Cartesian components of the angular momentum have simple selection rules with respect to quantum numbers of the states in the multiplet:

$$\langle J'M'|J_x|JM\rangle = \frac{1}{2}\big(\mu_+(JM)\delta_{M',M+1} + \mu_-(JM)\delta_{M',M-1}\big)\delta_{J'J}, \tag{A.72}$$

$$\langle J'M'|J_y|JM\rangle = \frac{1}{2i}\big(\mu_+(JM)\delta_{M',M+1} - \mu_-(JM)\delta_{M',M-1}\big)\delta_{J'J}, \tag{A.73}$$

$$\langle J'M'|J_z|JM\rangle = M\delta_{M'M}\delta_{J'J}. \tag{A.74}$$

The physical image corresponding to the state $|JM\rangle$ is that of *precession* of the angular momentum vector around the quantization axis z, then $\mathbf{J}^2$ and J_z are fixed, the expectation values $\langle J_x \rangle$ and $\langle J_y \rangle$ are averaged out and vanish, and the expectation values of J_x^2 and J_y^2 are positive.

A.8 Multiplets as Irreducible Representations

Any rotation can be represented as a function (A.12) of the generators. None of them can change the magnitude J: starting from a state $|JM\rangle$ and applying various finite rotations we are always confined to the family of states with different M and the same J. Thus, any state $|JM\rangle$ transforms under rotation R into a superposition of the states belonging to the same multiplet $|JM\rangle$. This fact can be written explicitly as

$$R|JM\rangle = \sum_{M'} D^J_{M'M}(R)|JM'\rangle, \tag{A.75}$$

where

$$D^J_{M'M}(R) = \langle JM'|R|JM\rangle \tag{A.76}$$

are matrix elements of the *finite rotation* R in a given *representation*; here we take into account that the states $|JM\rangle$ with different values of M are orthogonal and assume that they are normalized (A.68). The unitarity of rotations implies the unitarity of matrices (A.75),

$$D^J(D^J)^\dagger = (D^J)^\dagger D^J = 1, \tag{A.77}$$

or, explicitly in matrix elements,

$$\sum_M D^J_{KM}(R)D^{J*}_{K'M}(R) = \delta_{KK'}, \quad \sum_K D^{J*}_{KM'}(R)D^J_{KM}(R) = \delta_{M'M}. \tag{A.78}$$

In algebraic terms, matrices (A.76) give a *unitary representation* of the rotation group of the dimension $2J+1$. It means that for a rotation performed in two steps, $R = R_2 R_1$, the corresponding matrix (A.76), $D^J(R)$, is the matrix product of the matrices representing individual rotations performed in the same order,

$$D^J(R) = D^J(R_2)D^J(R_1). \tag{A.79}$$

All geometric properties of rotations are adequately reflected in relations between the corresponding matrices. Thus, the unit matrix corresponds to the rotation by zero angle, and for the inverse rotation $D^J(R^{-1}) = (D^J(R))^{-1}$. The representation D^J is *irreducible*: the multiplet $|JM\rangle$ of dimension $2J+1$ does not contain any smaller subset of states that transform only within this subset under all rotations.

A.9 SU (2) Group and Spin $\frac{1}{2}$

The orbital momentum **l** generates multiplets with an *integer* $J = l$. The spin angular momentum *s* can take both integer and half-integer values. We can understand the physical reason for this difference. Under rotation (A.12) around the z-axis, the wavefunction of the state $|JM\rangle$ acquires a phase,

$$R_z(\alpha)|JM\rangle = e^{-iJ_z\alpha}|JM\rangle = e^{-iM\alpha}|JM\rangle. \tag{A.80}$$

This determines a particular matrix element

$$D^J_{M'M}(R_z(\alpha)) = e^{-iM\alpha}\delta_{M'M}. \tag{A.81}$$

Consider rotation through an angle $\alpha = 2\pi$. The states with integer J do not change, $\exp(-i2\pi M) = 1$, but the states with half-integer J gain a factor -1. As we have seen, the orbital momentum transforms the explicit coordinate dependence of the wavefunctions. Since the directions marked by the angles 0 and 2π physically coincide, a single-valued wavefunction has to be periodic as a function of angles with the period 2π, that is, it should have *integer* angular momentum. Spin wavefunctions are not explicit functions of coordinates, so the requirement of periodicity is absent. Since the physical predictions are given in terms of the amplitudes, which are bilinear in wavefunctions, the *double-valued* representations of the rotation corresponding to a half-integer spin are allowed.

Spin $\frac{1}{2}$ plays an exceptional role because the objects with $J = s = \frac{1}{2}$ are the most fundamental objects in nature. The main building blocks of matter, electrons and quarks, have spin $\frac{1}{2}$. Combining constituents of spin $\frac{1}{2}$ one can construct an arbitrary high angular momentum J. The spin-$\frac{1}{2}$ objects realize the lowest nontrivial representation of the SU(2) group with dimension $2s + 1 = 2$. In a general SU(n) group the *fundamental* representation of dimension n describes similar basic constituents (the simplest nontrivial set of objects which is irreducible under all group transformations).

Our canonical basis $\chi_M = |J = 1/2, M\rangle$ consists of two basis vectors, $M = \pm\frac{1}{2}$. Sometimes it is convenient to call them "spin up," $\chi_{1/2} = \chi_+ \equiv\uparrow$, and "spin down," $\chi_{-1/2} \equiv \chi_- =\downarrow$. All operators in this space are 2×2 matrices. The algebra (A.59) is satisfied with $\mathbf{s} = \frac{1}{2}\boldsymbol{\sigma}$, where the components of $\boldsymbol{\sigma}$ are *Pauli matrices*

$$\sigma_x = \begin{pmatrix} 0 & 1 \\ 1 & 0 \end{pmatrix}, \quad \sigma_y = \begin{pmatrix} 0 & -i \\ i & 0 \end{pmatrix}, \quad \sigma_z = \begin{pmatrix} 1 & 0 \\ 0 & -1 \end{pmatrix}. \tag{A.82}$$

It is easy to check the commutation relations (A.59) and the matrix elements (A.72)−(A.74); the matrices are traceless.

Together with the unit matrix, the matrices (A.82) form a complete set of four independent matrices in 2×2 space. In particular, their products are again matrices of the same set. This allows one to accumulate the entire spin algebra in the identity

$$\sigma_k\sigma_l = \delta_{kl} + i\epsilon_{kln}\sigma_n. \tag{A.83}$$

It follows that any operator function of Pauli matrices σ_k can be reduced to a linear expression. The first term of (A.83) is Hermitian and *symmetric* in vector subscripts k, l; the second one is anti-Hermitian and *antisymmetric*. Frequently one has to deal with scalar products $\mathbf{a} \cdot \vec{\sigma}$ of Pauli matrices with (nonmatrix) vectors. Then (A.83) gives

$$(\mathbf{a} \cdot \boldsymbol{\sigma})(\mathbf{b} \cdot \boldsymbol{\sigma}) = (\mathbf{a} \cdot \mathbf{b}) + i[\mathbf{a} \times \mathbf{b}] \cdot \boldsymbol{\sigma}. \tag{A.84}$$

According to (A.84),

$$(\mathbf{n} \cdot \boldsymbol{\sigma})^2 = \mathbf{n}^2 = 1 \tag{A.85}$$

for any unit vector $\mathbf{n}$.

In the representation (A.82) the matrix $\sigma_z = 2s_z$ is diagonal, and taking the basis states $\chi_\pm$ as the eigenstates of σ_z with the eigenvalues ± 1, we obtain our canonical angular momentum basis with z as the quantization axis. In the representation corresponding to the matrices (A.82), the basis states $|1/2\ m\rangle$ are two-component columns

$$\chi_+ = \begin{pmatrix} 1 \\ 0 \end{pmatrix}, \quad \chi_- = \begin{pmatrix} 0 \\ 1 \end{pmatrix}. \tag{A.86}$$

Such objects implementing the fundamental representation of the $SU(2)$ algebra are called *spinors*. Any state of spin $\frac{1}{2}$ can be represented as a superposition $a_+\chi_+ + a_-\chi_-$ of basic spinors (A.86) with the upper (lower) component $a_+(a_-)$ giving an amplitude of finding the value of s_z equal to $\frac{1}{2}$ ($\frac{1}{2}$). Starting from the state χ_+ (spin *polarized* along the z-axis) and applying various rotations $\exp -(i/2)(\boldsymbol{\sigma} \cdot \mathbf{n})\alpha$, one can get states with any spin orientation.

A.10 Properties of Spherical Harmonics

A.10.1 Explicit derivation

The spherical harmonics $Y_{lm}(\mathbf{n}) \equiv Y_{lm}(\theta, \varphi)$ are the eigenfunctions of the orbital momentum operators $\mathbf{l}^2$ and l_z and therefore they transform among themselves under rotations generated by the orbital momentum $\mathbf{l}$, (A.11), (A.16), and (A.75), according to an integer irreducible representation, $J = l$ (integer), $J_z = m = -l, \ldots, +l$.

Given the coordinate frame with the quantization axis z, we define the polar, θ, and azimuthal, φ, angles for any direction $\mathbf{n}$. Using the general recipe (A.16) and the result (A.80) for rotation around the z-axis, we obtain

$$R_z(\alpha) Y_{lm}(\theta, \varphi) = Y_{lm}(\theta, \varphi - \alpha) = e^{-im\alpha} Y_{lm}(\theta, \varphi). \tag{A.87}$$

The second equation (A.87) determines the universal periodic dependence of spherical functions on the azimuthal angle, (A.7). As it should be, the raising and lowering operators (A.4) change the φ-dependence of the functions (A.87) in an appropriate, way adding the factor $\exp(\pm i\varphi)$. The Casimir operator $\mathbf{l}^2$ is, in this representation, the angular part (A.2) of the Laplace operator.

Since the upper state with $m = l$ is annihilated by the raising operator, we should have $l_+ Y_{ll} = 0$. It gives the simple equation of the first order for $\Theta_{ll}(\theta)$ defined by (A.7),

$$\frac{d\Theta_{ll}}{d\theta} = l \cot\theta\, \Theta_{ll}. \tag{A.88}$$

The solution of (A.88) normalized in accordance with (A.9) is

$$\Theta_{ll}(\theta) = \sqrt{\frac{(2l+1)!}{2}}\, \frac{1}{2^l l!} \sin^l\theta. \tag{A.89}$$

The larger l is, the more this function becomes concentrated near the equator, $\theta = \pi/2$, which characterizes the semiclassical orbit in the plane perpendicular to the direction of the angular momentum.

Now we can act by the lowering operator l_-, (A.4), and go down to all members of the multiplet. Using the matrix elements (A.71), we obtain

$$Y_{ll-1} = \frac{1}{\sqrt{2l}} l_- Y_{ll}, \ldots, \qquad Y_{lm} = \left[\frac{(l+m)!}{(l-m)!(2l)!}\right]^{1/2} (l_-)^{l-m} Y_{ll}. \tag{A.90}$$

After simple algebra, the result can be expressed in terms of the *associated Legendre polynomials* $P_{lm}(x)$,

$$\Theta_{lm}(\theta) = (-)^m \left[\frac{2l+1}{2}\frac{(l-m)!}{(l+m)!}\right]^{1/2} P_{lm}(\cos\theta), \tag{A.91}$$

$$P_{lm}(\cos\theta) = (-)^{l-m} \frac{(l+m)!}{(l-m)!}\frac{1}{2^l l!}\frac{1}{\sin^m\theta}\frac{d^{l-m}}{(d\cos\theta)^{l-m}} \sin^{2l}\theta, \tag{A.92}$$

where the so-called normalization according to *Condon and Shortley* [CS51] was used which implies the symmetry properties (A.10). Note that the definitions by various authors can differ in phase conventions.

A.10.2 Legendre polynomials

For the forward angles, $\theta \to 0$, a regular function of angles, as Y_{lm}, cannot depend on φ since the azimuthal angle is not defined for $\theta = 0$. Therefore all Y_{lm} vanish at $\theta = 0$ except for Y_{l0}, which does not carry any φ-dependence. At $m = 0$ the associated Legendre polynomials (A.92) reduce to the ordinary Legendre polynomials,

$$P_l(x) \equiv P_{l0}(x), \tag{A.93}$$

so that

$$Y_{l0}(\mathbf{n}) = \sqrt{\frac{2l+1}{4\pi}} P_l(\cos\theta). \tag{A.94}$$

It is easy to see from (A.92) that all Legendre polynomials are equal in the forward direction,

$$P_l(1) = 1. \tag{A.95}$$

Hence, for the direction along the quantization axis,

$$Y_{lm}(\theta = 0) = \delta_{m0}\sqrt{\frac{2l+1}{4\pi}}. \tag{A.96}$$

The Legendre polynomials are orthonormalized on the segment from -1 to $+1$,

$$\int_{-1}^{1} dx\, P_{l'}(x) P_l(x) = \frac{2}{2l+1}\delta_{l'l}. \tag{A.97}$$

The first four polynomials are

$$P_0(x) = 1, \quad P_1(x) = x, \quad P_2(x) = \frac{1}{2}(3x^2 - 1), \quad P_3(x) = \frac{1}{2}(5x^3 - 3x). \tag{A.98}$$

A.10.3 Completeness

The set of spherical functions $Y_{lm}(\theta, \varphi)$ for all l and m is complete, so any regular function of angles can be expanded over Y_{lm}. The coefficients of the expansion can be readily found with the aid of the orthonormality conditions (A.9). Azimuth-independent functions of $\cos\theta$ can be expanded into a series of Legendre polynomials with the help of (A.97). For an arbitrary function of angles $F(\mathbf{n})$, the expansion is

$$F(\mathbf{n}) = \sum_{lm} F_{lm}\, Y_{lm}(\mathbf{n}). \tag{A.99}$$

According to (A.9),

$$F_{lm} = \int do\, Y_{lm}^*(\mathbf{n}) F(\mathbf{n}). \tag{A.100}$$

Inserting (A.100) back into the expansion (A.99), one obtains the identity

$$F(\mathbf{n}) = \int do'\, F(\mathbf{n}') \sum_{lm} Y_{lm}^*(\mathbf{n}') Y_{lm}(\mathbf{n}). \tag{A.101}$$

Therefore the completeness of the set of spherical harmonics can be written as

$$\sum_{lm} Y_{lm}^*(\mathbf{n}') Y_{lm}(\mathbf{n}) = \delta(\mathbf{n} - \mathbf{n}'). \tag{A.102}$$

Taking $\mathbf{n}' = \mathbf{e}_z$, the unit vector along the z-axis, and applying (A.94) and (A.96), we get another useful relation,

$$\sum_l (2l+1) P_l(x) = 4\pi\, \delta(x - 1), \tag{A.103}$$

showing that the Legendre polynomials cancel each other in all directions except for the forward one.

A.10.4 Spherical functions as matrix elements of finite rotations

The spherical harmonics $Y_{lm}(\mathbf{n})$ are the wavefunctions in the coordinate representation of the states $|lm\rangle$ (described in the frame with the fixed quantization axis),

$$Y_{lm}(\mathbf{n}) \equiv \langle \mathbf{n}|lm\rangle. \tag{A.104}$$

Let R be a rotation that brings the directional vector $\mathbf{e}_z$ of the quantization axis to a new direction $\mathbf{n}$,

$$R\mathbf{e}_z = \mathbf{n}(\theta, \varphi). \tag{A.105}$$

The rotation R^{-1}, inverse to that in (A.105), acting on the state $|lm\rangle$, transforms it into a superposition of the multiplet states according to the general rule (A.75):

$$R^{-1}|lm\rangle = \sum_{m'} D^l_{m'm}(R^{-1})|lm'\rangle. \tag{A.106}$$

The coordinate representation of this equality is obtained by the projecting on the localized state vector $\mathbf{n}_0$,

$$\langle \mathbf{n}_0|R^{-1}|lm\rangle = \sum_{m'} D^l_{m'm}(R^{-1})\langle \mathbf{n}_0|lm'\rangle = \sum_{m'} D^l_{m'm}(R^{-1}) Y_{lm'}(\mathbf{n}_0). \tag{A.107}$$

Due to unitarity of rotations, the left hand side here is

$$\langle \mathbf{n}_0|R^{-1}|lm\rangle = \langle R\mathbf{n}_0|lm\rangle = Y_{lm}(R\mathbf{n}_0). \tag{A.108}$$

The direction $\mathbf{n}_0$ is arbitrary. Taking this function in the direction of the polar axis, $\mathbf{n}_0 \to \mathbf{e}_z$, we come to $Y_{lm}(R\mathbf{e}_z)$. that is, the spherical function of the original angles θ, φ, (A.105). In the right-hand side of (A.107) we can use the result (A.96) for $Y_{lm'}(\mathbf{e}_z)$. This leads to the connection sought for,

$$Y_{lm}(\mathbf{n}) = \sqrt{\frac{2l+1}{4\pi}} D^l_{0m}(R^{-1}) = \sqrt{\frac{2l+1}{4\pi}} D^{l*}_{m0}(R), \tag{A.109}$$

where the $D(R^{-1})$ is the matrix element for the rotation that, inversely to (A.105), brings the vector $\mathbf{n}$ to the direction of the polar axis, and the second equality uses the relation between $D(R)$ and $D(R^{-1}) = D^\dagger(R)$. The Legendre polynomials, (A.94), are

$$P_l(\cos\theta) = D^l_{00}(R^{-1}) = D^l_{00}(R). \tag{A.110}$$

A.10.5 Addition theorem

Frequently one needs a scalar function of an angle γ between two directions, $\mathbf{n}(\theta, \varphi)$ and $\mathbf{n}'(\theta', \varphi')$. Being a scalar, such a function is in fact a function of the scalar product $(\mathbf{n} \cdot \mathbf{n}')$

and can be expanded with the use of the Legendre polynomials $P_l(\cos \gamma)$, where

$$\cos \gamma = (\mathbf{n} \cdot \mathbf{n}') = \sin \theta \sin \theta' \cos (\varphi - \varphi') + \cos \theta \cos \theta'. \tag{A.111}$$

Using the rotation transformation properties of the spherical harmonics one can prove the addition theorem for spherical harmonics,

$$P_l(\mathbf{n} \cdot \mathbf{n}') = \frac{4\pi}{2l+1} \sum_m Y_{lm}(\mathbf{n}) Y_{lm}^*(\mathbf{n}'). \tag{A.112}$$

As a particular case, for coinciding $\mathbf{n}$ and $\mathbf{n}'$,

$$\sum_m |Y_{lm}(\mathbf{n})|^2 = \frac{2l+1}{4\pi}. \tag{A.113}$$

Appendix B | Angular Momentum Coupling

B.1 Tensor Operators

B.1.1 Transformation of operators

If the state vectors $|\psi\rangle$ are transformed by a unitary transformation U into $|\psi'\rangle = U|\psi\rangle$, and the operators O are transformed according to

$$O \Rightarrow O' = UOU^{-1}, \tag{B.1}$$

all physical amplitudes are preserved,

$$\langle\psi_2'|O'|\psi_1'\rangle = \langle\psi_2|U^{-1}UOU^{-1}U|\psi_1\rangle = \langle\psi_2|O|\psi_1\rangle. \tag{B.2}$$

It means that the new operators O' play exactly the same role in the new conditions as the old operators O did before the transformation. In other words, when applied to rotation $(U \Rightarrow R)$, we made the operators rotate along with the system so that physical measurements by the rotated tools give the same results.

The operators can be classified by their behavior under rotations in the same way that the state vectors were subdivided into rotational multiplets according to their transformation properties. The set of $2J + 1$ operators T_{JM}, where J is an integer or half-integer and $M = -J, -J + 1, \ldots, J$, is said to form a *tensor operator* of rank J if the operators of the set are transformed under rotations according to the same rules (A.75) as the state vectors $|JM\rangle$,

$$RT_{JM}R^{-1} = \sum_{M'} D^J_{M'M}(R)T_{JM'}. \tag{B.3}$$

For integer $J = l$, the tensor operators T_{lm} have to transform as spherical functions Y_{lm}. In the case of a spinless particle, the tensor operators $T_{lm}(\mathbf{r})$, which are functions of coordinates, should have the same angular dependence as $Y_{lm}(\mathbf{n})$,

$$T_{lm}(\mathbf{r}) = t_l(r) Y_{lm}(\mathbf{n}), \tag{B.4}$$

where the radial factor $t_l(r)$ is the same for all m. In this case it is easy to check directly that the transformation rule (B.3) preserves amplitudes (B.4). Indeed, as we know from (A.16) and (A.101), the transformation (B.3) for the function of coordinates (B.4) should give $T_{lm}(R^{-1}\mathbf{n})$. (Remember that here we transform an operator; the first factor R on the left-hand side of (B.3) transforms only T_{JM} and cancels R^{-1} so that all functions placed after T_{JM} are not affected.) The transformed matrix element (B.2) differs from the original one merely by the change of angular variables $\mathbf{n} \to R^{-1}\mathbf{n}$ in the integrand, which cannot change the integral. A similar conclusion holds for the functions of the momentum, which will be proportional to $Y_{lm}(\mathbf{n_p})$, where $\mathbf{n_p} = \mathbf{p}/|\mathbf{p}|$.

B.1.2 Scalars and vectors

Any function of coordinates can be represented by a series of spherical harmonics with coefficients depending only on r. This is equivalent to an expansion over irreducible tensor operators. The lowest term, $l = 0$ or s-wave, is proportional to the spherical function

$$Y_{00} = \sqrt{\frac{1}{4\pi}} \tag{B.5}$$

and does not depend on angles. It is not affected by rotations ($T_{00} \sim Y_{00}$ is *scalar*).

There are 3 p-wave functions, $l = 1$,

$$Y_{10} = \sqrt{\frac{3}{4\pi}} \cos\theta, \quad Y_{1\pm1} = \mp\sqrt{\frac{3}{8\pi}} \sin\theta\, e^{\pm i\varphi}. \tag{B.6}$$

For any vector $\mathbf{V}$ we can introduce, instead of Cartesian components $V_i = (V_x, V_y, V_z)$, the so-called *spherical* components V_m, $m = 0, \pm1$:

$$V_0 = V_z, \quad V_{\pm1} = \mp\frac{1}{\sqrt{2}}(V_x \pm iV_y). \tag{B.7}$$

Note that the spherical components $J_{\pm1}$ of the angular momentum differ only by a factor $\mp 1/\sqrt{2}$ from the lowering and raising operators $J_\pm$, (A.61). From (B.6) and (B.7) we see that the functions $Y_{1m}(\mathbf{n})$ are essentially the spherical components of the vector $\mathbf{n}$,

$$Y_{1m}(\mathbf{n}) = \sqrt{\frac{3}{4\pi}}\, n_m. \tag{B.8}$$

The scalar product of the vectors in Cartesian coordinates can be also expressed in terms of the spherical components (B.7):

$$(\mathbf{a} \cdot \mathbf{b}) = \sum_{m=0,\pm1} (-)^m a_m b_{-m}. \tag{B.9}$$

Note, that for unit vectors (B.9) is merely a particular case of the addition theorem (A.112) for $l = 1$.

All vectors behave in the same way under rotations. Therefore we conclude that any vector operator is a tensor of rank 1. The coordinate vector $\mathbf{n}$ is an example of a *polar* vector. Its components, together with the spherical functions Y_{1m}, change sign under spatial inversion.

B.1.3 Tensors of rank 2

Let us consider 9 quantities $T_{ij} = a_i b_j$ constructed of the Cartesian components of vectors $\mathbf{a}$ and $\mathbf{b}$. They are *reducible* under rotations and can be grouped in smaller *irreducible* sets. First we separate two parts with different symmetry (submatrices which are symmetric, S, or antisymmetric, A, under a transposition of matrix indices),

$$T_{ij} = S_{ij} + A_{ij} = \tfrac{1}{2}(a_i b_j + a_j b_i) + \tfrac{1}{2}(a_i b_j - a_j b_i). \tag{B.10}$$

The symmetric part S_{ij} is further reducible since the trace $\mathrm{Tr}\, S = S_{ii} = (\mathbf{a} \cdot \mathbf{b})$ is a scalar. We can subtract the invariant scalar with a coefficient such that the rest are traceless,

$$S_{ij} = \tfrac{1}{3}(\mathbf{a} \cdot \mathbf{b})\delta_{ij} + Q_{ij}. \tag{B.11}$$

The symmetric tensor

$$Q_{ij} = \tfrac{1}{2}\left(a_i b_j + a_j b_i - \tfrac{2}{3}(\mathbf{a} \cdot \mathbf{b})\delta_{ij}\right) \tag{B.12}$$

is traceless, $\mathrm{Tr}\, Q = Q_{ii} = 0$, and irreducible. It has 5 independent components. The antisymmetric part

$$A_{ij} = \tfrac{1}{2}(a_i b_j - a_j b_i) = \tfrac{1}{2}\epsilon_{ijk}[\mathbf{a} \times \mathbf{b}]_k \tag{B.13}$$

has 3 independent components. It is equivalent (with respect to rotations) to a vector $[\mathbf{a} \times \mathbf{b}]$. If both $\mathbf{a}$ and $\mathbf{b}$ are polar vectors (previous paragraph), the components of their vector product do not change sign under inversion of spatial coordinates. Such a vector is called an *axial* or pseudovector; as an example we can recall the orbital momentum (A.3).

Summarizing, the decomposition of the reducible tensor T_{ij} into irreducible parts S_{ij}, A_{ij}, and Q_{ij} can be symbolically presented as

$$\underline{3} \times \underline{3} = \underline{1} + \underline{3} + \underline{5}, \tag{B.14}$$

where the underlined numbers designate the dimensions of representations.

In order to understand the rotational properties of the symmetric tensor (B.12), we have to compare its transformational features with those of spherical harmonics. Since all vectors transform in the same way, it is sufficient to consider the case $\mathbf{a} = \mathbf{b} = \mathbf{n}(\theta, \varphi)$. Then we can establish a one-to-one correspondence between the spherical functions of the second rank $Y_{2m}(\mathbf{n})$ and certain linear combinations of components Q_{ij},

$$Y_{20}(\mathbf{n}) = \sqrt{\frac{5}{16\pi}}(2\cos^2\theta - \sin^2\theta) \Rightarrow \sqrt{\frac{5}{4\pi}}\frac{1}{2}(2Q_{zz} - Q_{xx} - Q_{yy}), \tag{B.15}$$

$$Y_{2\pm 1}(\mathbf{n}) = \mp\sqrt{\frac{15}{8\pi}}\cos\theta\sin\theta e^{\pm i\varphi} \Rightarrow \sqrt{\frac{5}{4\pi}}\sqrt{\frac{3}{2}}(Q_{xz} \pm iQ_{yz}), \tag{B.16}$$

$$Y_{2\pm 2}(\mathbf{n}) = \sqrt{\frac{15}{32\pi}}\sin^2\theta e^{\pm 2i\varphi} \Rightarrow \sqrt{\frac{5}{4\pi}}\sqrt{\frac{3}{8}}(Q_{xx} \pm 2iQ_{xy} - Q_{yy}). \tag{B.17}$$

Inversely, the components Q_{ij} are linear combinations of $Y_{2m}(\mathbf{n})$ and therefore correspond to a tensor operator of the second rank. The five combinations on the right-hand side of (B.15–B.17) are organized in such a way that they form a spherical tensor

$$Q_{2m} \propto \sqrt{\frac{4\pi}{5}}Y_{2m}(\mathbf{n}). \tag{B.18}$$

The conclusion is that any set of 9 quantities T_{ij} which transform under rotations as a product of two vectors can be decomposed into scalar, vector, and symmetric second rank tensor parts. The procedure can be extended for any tensor $T_{ijk\ldots} \sim a_i b_j c_k \ldots$.

B.1.4 Introduction to selection rules

Tensor properties of the operators are important in calculations of physical amplitudes proportional to the matrix elements $\langle \psi'_{J_2 M_2}|T_{JM}|\psi_{J_1 M_1}\rangle$. For given initial and final multiplets of states, we have here $(2J_2 + 1)(2J + 1)(2J_1 + 1)$ different matrix elements. However, as we will see, only one number characterizes the relevant physics. The rest are completely determined by geometrical considerations. Some matrix elements vanish exclusively due to the rotational symmetry of the states and the operators, others turn out to be strongly interrelated.

The simplest selection rules can be discovered directly from the definition of tensor operators (B.3). Let us consider an infinitesimal rotation through an angle $\delta\alpha$ around the axis $\mathbf{n}$. The corresponding operator is $R = 1 - i(\mathbf{J}\cdot\mathbf{n})\delta\alpha$, (A.11). Keeping linear terms in $\delta\alpha$, the left-hand side of (B.3) is expressed via the commutator of the tensor T_{JM} with the angular momentum

$$RT_{JM}R^{-1} = T_{JM} - i\delta\alpha[(\mathbf{J}\cdot\mathbf{n}), T_{JM}]. \tag{B.19}$$

In a given representation the matrix elements (A.70) of this rotation are

$$D^J_{M'M}(R) = \delta_{M'M} - i\delta\alpha\langle JM'|(\mathbf{J}\cdot\mathbf{n})|JM\rangle. \tag{B.20}$$

Since the axis direction $\mathbf{n}$ is arbitrary, (B.19) implies the commutation relation valid for any tensor operator,

$$[J_m, T_{JM}] = \sum_{M'}\langle JM'|J_m|JM\rangle T_{JM'} \tag{B.21}$$

where J_m can be either Cartesian or spherical, (B.7).

The $J_z = J_0$ component of (B.21) reads

$$[J_z, T_{JM}] = MT_{JM}. \tag{B.22}$$

This is a typical ladder relation, recall (A.41). We conclude that, acting on a state with a certain z-projection of the total angular momentum of a system, a tensor operator T_{JM} *raises* this projection by M. We obtained the simple *selection rule*: in the transitions $\langle a_2 J_2 M_2 | T_{JM} | a_1 J_1 M_1 \rangle$, where a_1 and a_2 are symbols for all additional (nonrotational) quantum numbers, the only nonzero amplitudes are those with $\Delta J_z \equiv M_2 - M_1 = M$,

$$T_{JM}: \quad \Delta J_z = M. \tag{B.23}$$

Our notation in (B.7) for the spherical components of vectors agrees with this general rule. The result does not depend on the specific values of J_1, J_2, or other (nonrotational) quantum numbers a_1, a_2.

The raising, $m = +1$, component of (B.21) contains on the right-hand side the only term $M' = M + 1$. We see that the operator product of J_{+1}, which has a selection rule $\Delta J_z = +1$, and T_{JM} creates a new operator corresponding to $\Delta J_z = M + 1$. For the lowering component of (B.21), $m = -1$ and $\Delta J_z = M - 1$. In the product of operators, the selection rules for the projection J_z are simply added *algebraically*.

When applied to a vector operator, $T_{1M} \to V_M$, the general equation (B.22) gives $[J_z, V_{\pm 1}] = \pm V_{\pm 1}$. Obviously, the rotation around the z-axis does not change the z-component of a vector, $[J_z, V_z] = 0$. In Cartesian coordinates, such relations are equivalent to the commutation relation

$$[J_k, V_l] = i\epsilon_{kln} V_n, \tag{B.24}$$

which generalizes the angular momentum algebra (A.59) for an arbitrary vector. All such commutation rules are of pure geometrical origin and therefore universally valid regardless of specific nature, or behavior under inversion, of the tensor operator.

B.2 Angular Momentum Coupling

B.2.1 *Two subsystems*

Consider two subsystems with angular momenta j_1 and j_2. The total quantum space contains $d = (2j_1 + 1)(2j_2 + 1)$ states obtained by combinations of various members of multiplets $|j_1 m_1\rangle$ and $|j_2 m_2\rangle$ with projections $m_1 = -j_1, \ldots, j_1$ and $m_2 = -j_2, \ldots, j_2$, respectively. Those basis states can be designated as

$$|j_1 m_1; j_2 m_2\rangle. \tag{B.25}$$

If the subsystems do not interact, all four quantum numbers j_1, j_2, m_1, m_2 are conserved (we assume the rotational invariance of the whole system). Then it is convenient to use the basis states of independent subsystems. Each system can be rotated separately according to its angular momentum operators $\mathbf{j}_1$ and $\mathbf{j}_2$, generating corresponding transformations. One can imagine the picture of separate precession of the constituent angular momenta around the common axis.

We can characterize the system in a different way by probing its behavior under common rotation when the subsystems are rotated together. The generator of such rotations is the total angular momentum

$$\mathbf{J} = \mathbf{j}_1 + \mathbf{j}_2. \tag{B.26}$$

In the previous picture of separate precessions the operator $\mathbf{J}$ has no certain value because the result of the vector addition (B.26) depends on the instantaneous mutual orientation of $\mathbf{j}_1$ and $\mathbf{j}_2$. The states (B.25) are superpositions of states with definite values of $\mathbf{J}^2$. In the case of interacting subsystems the separate rotations in general violate structure, which makes states (B.25) nonstationary, whereas common rotations preserve the intrinsic structure. Then it is more convenient to describe the states by the quantum numbers J and M related to the generator (B.26) of total rotations, even if both descriptions use the complete set of states being therefore mathematically equivalent.

With respect to common rotations, when the relative orientation of the subsystems is kept intact and they rotate as a whole, the complete set of the states (B.25) is *reducible*. Any possible relative orientation will give rise to a multiplet $|JM\rangle$ of states transforming between each other under common rotations. We first define the relative orientation and corresponding total momentum $\mathbf{J}$ (the angular momenta of subsystems precess around $\mathbf{J}$) and then allow the total construction to rotate around the space-fixed quantization axis that defines the projection M. The z-projections m_1 and m_2 cease to conserve (but absolute values j_1 and j_2 still do because we do not change the internal structure of the subsystems), so that we obtain the new set of states

$$|j_1j_2;JM\rangle \tag{B.27}$$

that form multiplets *irreducible* under common rotations. For individual angular momenta of the subsystems the effective quantization axis is now that of the total vector $\mathbf{J}$. Indeed, as seen from eq. (B.26), the state (B.27) has certain projections

$$(\mathbf{j}_1 \cdot \mathbf{J}) = \frac{J(J+1) + j_1(j_1+1) - j_2(j_2+1)}{2}, \tag{B.28}$$

and similarly for $(\mathbf{j}_2 \cdot \mathbf{J})$.

The relative orientations allowed in quantum mechanics are quantized in space. Therefore the possible total momentum J, (B.26), can take only a finite discrete set of (positive) values. In any case, the new states (B.27), where each multiplet contains $2J + 1$ members, should be as complete as the old set (B.25), so their dimensions have to coincide:

$$d = \sum_J (2J + 1) = d_1 d_2 = (2j_1 + 1)(2j_2 + 1). \tag{B.29}$$

B.2.2 Decomposition of reducible representations

First we have to find all irreducible representations that taken together span the whole space (B.25). It can be done by a simple construction which is equivalent to the standard group theory procedure of finding the characters of the representations (traces of the matrix D^J).

Put all basis states (B.25) into a $d_1 \times d_2$ table that has d_1 vertical columns numbered by $m_1, -j_1 \le m_1 \le j_1$, and d_2 horizontal rows numbered by $m_2, -j_2 \le m_2 \le j_2$. For definiteness, assume $j_1 \ge j_2$. Each state (square of the table) has a certain value

$$M = m_1 + m_2 \tag{B.30}$$

of the total projection $J_z = j_{1z} + j_{2z}$, (B.26). Any state $|JM\rangle$ of the set (B.27) will be a superposition of states lying on a straight diagonal line (B.30) corresponding to a given M-value. The number of squares on this line is equal to the number of multiplets (B.27) that include this value of the projection, that is, with angular momentum $J \ge M$.

Start with the upper right corner, $M = j_1 + j_2$. This is the maximum possible total projection. It is constructed uniquely (alignment of constituent momenta). There is only one multiplet where this value of M gives the maximum projection so this state has the highest possible magnitude of $J_{max} = M_{max} = j_1 + j_2$. This highest multiplet should have all other members, $M = J - 1, J - 2, \ldots, -J = -(j_1 + j_2)$ as well.

Let us come to the next diagonal line $M = J_{max} - 1$. There are two such states. They can form two linearly independent combinations. One of them belongs to the highest multiplet as was mentioned in the paragraph above. This combination $|J_{max} \; M = J_{max} - 1\rangle$ can be obtained by the action of the lowering operator

$$J_- = j_{1-} + j_{2-} \tag{B.31}$$

on the maximum aligned state (recall that the components of $\mathbf{J}$ act only within the multiplet). According to (A.71), the result is

$$|J = j_1 + j_2 \; M = j_1 + j_2 - 1\rangle = \sqrt{2j_1}|j_1 j_1 - 1; j_2 j_2\rangle + \sqrt{2j_2}|j_1 j_1; j_2 j_2 - 1\rangle. \tag{B.32}$$

On the other hand, this should be equal to the action of the total J_-,

$$J_-|J = j_1 + j_2 \; M = j_1 + j_2\rangle = \sqrt{2(j_1 + j_2)}|J = j_1 + j_2 \; M = j_1 + j_2 - 1\rangle. \tag{B.33}$$

The comparison of the last two expressions defines

$$|J = j_1 + j_2 \; M = j_1 + j_2 - 1\rangle = \sqrt{\frac{j_1}{j_1 + j_2}}\Big|j_1 j_1 - 1; j_2 j_2\Big\rangle + \sqrt{\frac{j_2}{j_1 + j_2}}\Big|j_1 j_1; j_2 j_2 - 1\Big\rangle. \tag{B.34}$$

The second possible combination along the same short diagonal with $M = j_1 + j_2 - 1$ is the highest for the second multiplet. Therefore we open the new multiplet and see that the value of the total momentum $J = j_1 + j_2 - 1$ is also possible. This state has another J and has to be orthogonal to the state (B.33), although they have the same value of M. By orthogonality, we find

$$|J = j_1 + j_2 - 1 \; M = j_1 + j_2 - 1\rangle = \sqrt{\frac{j_2}{j_1 + j_2}}\Big|j_1 j_1 - 1; j_2 j_2\Big\rangle - \sqrt{\frac{j_1}{j_1 + j_2}}\Big|j_1 j_1; j_2 j_2 - 1\Big\rangle. \tag{B.35}$$

Here we can add an arbitrary extra phase, for example change the common sign—this is a matter of convention.

The next step on the way down shows three states with $M = j_1 + j_2 - 2$. Two of the three combinations belong to the previous multiplets while the third one opens the new multiplet with $J = j_1 + j_2 - 2$. This procedure is obvious and regular. Each step to a lower diagonal line adds a new multiplet with angular momenta steadily decreasing. This takes place for the last time when we reach the main diagonal, which corresponds to $M = j_1 - j_2$. At this step we open the multiplet with the lowest possible angular momentum $J_{\min} = j_1 - j_2$. After that the number of possible M does not increase which means that we just fill up the available multiplets. The lowest multiplet will be full at the line reaching the left upper corner of the table. Later on, each next step one completes one of the multiplets until we arrive at the left lower corner with only one state $M = -M_{\max} = -j_1 - j_2$ which completes the largest multiplet $J = j_1 + j_2$.

We can summarize the result of this exercise saying that possible values of the total angular momentum J in the vector coupling of subsystems with angular momenta j_1 and j_2 are

$$|j_1 - j_2| \leq J \leq j_1 + j_2. \tag{B.36}$$

Each value of J appears only once and it is easy to check the fulfillment of (B.29): we used all squares of our table in the rearrangement of the reducible space (B.25) into irreducible multiplets (B.27). While projections (B.30) are added algebraically, the magnitudes of angular momenta are added geometrically; inequalities (B.36) give exactly the same boundaries that would be valid for the addition of two Euclidean vectors (*triangle conditions*). However, quantum mechanics put an extra constraint of space quantization for the total angular momentum whose allowed values (B.36), in accordance with general rules for the SU(2) group, are all integer or half-integer depending on the values of j_1 and j_2.

B.2.3 Tensor operators and selection rules revisited

Electric and magnetic multipoles are typical examples of operators forming sets of $2\lambda + 1$ quantities $T_{\lambda\mu}$ that are closed with respect to the rotation group. Under rotations such quantities are transformed into linear combinations of quantities belonging to the same set, and the rule of transformation is exactly the same as for the spherical functions $Y_{\lambda\mu}$. Such a set of operators are said to form a *tensor operator* of rank λ. The physical consequences that follow from geometrical considerations are analogous for all tensor operators of the same rank, regardless of their physical nature.

In the case of the operator proportional to a spherical function $Y_{\lambda\mu}$, its action on a state $|J_1 M_1\rangle$ can be considered as a vector coupling of angular momenta of two "subsystems," $\mathbf{J}_1$ of the state and $\vec{\lambda}$ of the operator. According to the rules of the rotation group, the final angular momentum

$$\mathbf{J}_2 = \mathbf{J}_1 + \vec{\lambda} \tag{B.37}$$

can take all values J_2 which differ by one unit within the limits put by the triangle condition (B.36),

$$|J_1 - \lambda| \leq J_2 \leq J_1 + \lambda. \tag{B.38}$$

The projections of angular momenta are added algebraically, (B.30),

$$M_2 = M_1 + \mu. \tag{B.39}$$

In fact, the triangle conditions (B.38) are symmetric with respect to all three angular momenta J_1, J_2 and λ.

Equations (B.38)–(B.39) determine selection rules which are exactly the same for any tensor operator $T_{\lambda\mu}$: matrix elements $\langle a_2 J_2 M_2 | T_{\lambda\mu} | a_1 J_1 M_1 \rangle$ of a tensor operator between any states with certain angular momentum quantum numbers (and arbitrary additional quantum numbers a_1, a_2) can be different from zero if and only if the conditions (B.38) and (B.39) are fulfilled. For example, multipole transitions of multipolarity λ are strictly forbidden if $\Delta J = |J_2 - J_1| > \lambda$ or $\lambda > J_1 + J_2$.

As a particular case, the angular momentum selection rules restrict multipole moments that are allowed to have nonvanishing *expectation values* in a state with angular momentum J. Here we are interested in the *diagonal* elements $J_1 = J_2 = J$. The rule (B.38) shows that the allowed multipoles are those of rank λ satisfying

$$0 \leq \lambda \leq 2J. \tag{B.40}$$

As follows from (B.40), a system with angular momentum $J = 0$ accepts $\lambda = 0$ only, and therefore can have nonzero charge (2.9) but none of higher multipoles. A system with spin $\frac{1}{2}$, such as the nucleon or electron, can have $\lambda = 0$ or 1, that is, charge and dipole moments, electric (2.11) or magnetic (2.13). A nonzero quadrupole moment, $\lambda = 2$, appears only for systems with $J \geq 1$.

B.2.4 Vector coupling of angular momenta

We have found that in quantum mechanics two subsystems with rotational quantum numbers j_1, m_1 and j_2, m_2 being coupled together can form systems with various quantum numbers j_3, m_3 with respect to their rotation *as a whole*. The probability amplitudes of different possible outputs j_3, m_3 of the vector coupling are given by the *Clebsch-Gordan coefficients* (CGC) $\langle j_3 m_3 | j_1 m_1; j_2 m_2 \rangle$:

$$|j_1 m_1; j_2 m_2\rangle = \sum_{j_3 m_3} \langle j_3 m_3 | j_1 m_1; j_2 m_2 \rangle |(j_1 j_2) j_3 m_3\rangle, \tag{B.41}$$

where the last notation reminds us of the angular momenta (j_1, j_2) of the constituents. The allowed values of j_3, m_3 in (B.41) are given by the same selection rules (B.38), (B.39).

The CGC perform a transformation between two possible sets of basis states (two separate subsystems and the combined system). Both sets are complete, orthonormalized, and equally good, albeit either could be more or less convenient in a given physical situation. The transformation from one set to another one is *unitary*, so that the coefficients $\langle j_1 m_1; j_2 m_2, | j_3 m_3 \rangle$, which perform the inverse transformation,

$$|(j_1 j_2) j_3 m_3\rangle = \sum_{m_1 m_2} \langle j_1 m_1; j_2 m_2 | j_3 m_3 \rangle |j_1 m_1; j_2 m_2\rangle, \tag{B.42}$$

are complex conjugate with respect to those in (B.41). At the standard choice of phases for the matrix elements of the angular momentum, the CGC are *real*, and we will use the notation

$$C^{j_3 m_3}_{j_1 m_1 j_2 m_2} = \langle j_1 m_1; j_2 m_2 | j_3 m_3 \rangle = \langle j_3 m_3 | j_1 m_1; j_2 m_2 \rangle. \tag{B.43}$$

The orthonormalization conditions for both sets of states give

$$\sum_{m_1 m_2} C^{j_3 m_3}_{j_1 m_1 j_2 m_2} C^{j'_3 m'_3}_{j_1 m_1 j_2 m_2} = \delta_{j_3 j'_3} \delta_{m_3 m'_3} \tag{B.44}$$

and

$$\sum_{j_3 m_3} C^{j_3 m_3}_{j_1 m_1 j_2 m_2} C^{j_3 m_3}_{j_1 m'_1 j_2 m'_2} = \delta_{m_1 m'_1} \delta_{m_2 m'_2}. \tag{B.45}$$

B.2.5 Wigner-Eckart theorem

We have worked out the selection rules for the tensor operator $T_{\lambda\mu}$ related to rotational invariance. For the angular momenta J_2, J_1, and λ satisfying (B.38) and (B.39), in general there are many nonzero matrix elements (we indicate explicitly other quantum numbers of the states which are fixed for a given set of matrix elements)

$$\langle a_2 J_2 M_2 | T_{\lambda\mu} | a_1 J_1 M_1 \rangle. \tag{B.46}$$

All matrix elements with different combinations of projections contain the same physics, differing in the mutual orientation of the states $|a_1 J_1 M_1\rangle$, $|a_2 J_2 M_2\rangle$ and of the probe $T_{\lambda\mu}$. That is why, for instance, in tables of physical quantities one can find only one number for a magnetic moment of a particle or of a nucleus instead of the set of numbers corresponding to various matrix elements $\langle JM' | \mu_\mu | JM \rangle$. It is possible to separate the universal geometric information from the specific characteristics of a system under study.

Let us consider the action of the tensor operator $T_{\lambda\mu}$ onto the initial state $|a_1 J_1 M_1\rangle$. As a result of the vector coupling of the angular momenta (B.37), one can obtain for the intermediate state only the angular momentum projection $M' = M + \mu$ and the magnitude of angular momentum J' allowed by the triangle conditions $\mathbf{J'} = \mathbf{J} + \vec{\mu}$. The relative amplitudes of possible intermediate states $|J'M'\rangle$ are given by the CGC as in (B.41),

$$T_{\lambda\mu} |a_1 J_1 M_1\rangle = \sum_{J'M'} C^{J'M'}_{\lambda\mu J_1 M_1} |a_1 (T_\lambda J_1) J'M'\rangle, \tag{B.47}$$

Now we have to project the state $|J'M'\rangle$ onto the final state $|a_2 J_2 M_2\rangle$. Because of the orthogonality of eigenfunctions corresponding to different eigenvalues of Hermitian operators, only the term $J' = J_2, M' = M_2$ in the sum (B.47) survives. Moreover, the matrix element (B.46) cannot change if the initial state, final state, and operator are undergoing a *common* rotation. Therefore the result of the last projection $\langle a_2 J_2 M_2 | J'M'\rangle$ does not depend on the specific value of $M' = M_2$.

We came to the important conclusion: in any matrix element (B.46) of a tensor operator between the states with a certain angular momentum and its projection, the entire dependence on the magnetic quantum numbers M_1, μ and M_2 enters through the CGC only. The remaining factor does not carry any M-dependence and characterizes the physical ampli-tude of the process, regardless of the orientation of the system. Note that all rotational selection rules are already included in this CGC. This is the essence of the *Wigner-Eckart theorem*.

Using the $3j$ symbol instead of the CGC, we write the result as

$$\langle a_2 J_2 M_2 | T_{\lambda\mu} | a_1 J_1 M_1 \rangle = (-)^{J_2 - M_2} \begin{pmatrix} J_2 & \lambda & J_1 \\ -M_2 & \mu & M_1 \end{pmatrix} \langle a_2 J_2 \| T_\lambda \| a_1 J_1 \rangle. \tag{B.48}$$

Here the M-independent factor is introduced as a double-barred (*reduced*) matrix element. The phase factor for the final state in (B.48) is in accordance with the arguments related to time conjugation: the final state ($M_2 = \mu + M_1$) has to be reversed to make the situation symmetric. We see that the geometric part of information is factored into the $3j$-symbol while the intrinsic orientation-independent physics is concentrated in the reduced matrix element. As we declared above, only one number is sufficient to describe the whole set of matrix elements (B.46) if the rotational quantum numbers of the states and the operators are known.

The number shown in physical tables for the expectation values of multipole operators in a state with angular momentum J is, by convention, taken for the substate with the maximum projection $M = J$. Then $\mu = 0$, and the tabular value is

$$T_\lambda(a, J) \equiv \langle aJJ | T_{\lambda 0} | aJJ \rangle. \tag{B.49}$$

For example, the vector ($\lambda = 1$) component needed in (B.49) is $V_0 = V_z$; recall (B.23). The tabular magnetic moment therefore is the expectation value of its projection onto the quantization axis z in the state with the maximum alignment along the z-axis.

B.2.6 Vector model

The Wigner-Eckart theorem provides us with the justification of a simple procedure used from the very early days of atomic physics for calculating the expectation values, as for example,

$$\langle aJM' | \mathbf{V} | aJM \rangle \tag{B.50}$$

where the initial and final states belong to the same multiplet but may differ by the projection of the angular momentum, and $\mathbf{V}$ is an arbitrary vector operator.

The naive although correct way of reasoning is following. The semiclassical image of the state $|JM\rangle$ is that of precession. The angular momentum vector $\mathbf{J}$ of the length $\sqrt{J(J+1)}$ has a projection M onto the quantization axis and it is precessing around this axis forming a cone with the fixed polar angle θ, $\cos\theta = M/\sqrt{J(J+1)}$. The transverse components $J_{x,y}$ are averaged out and have zero expectation values $\langle J_x \rangle$ and $\langle J_y \rangle$ but nonzero mean square values $\langle J_x^2 \rangle$ and $\langle J_y^2 \rangle$. The sum $\langle J_x^2 + J_y^2 \rangle$ supplements M^2 to the total magnitude $J(J+1)$ of the angular momentum squared. In this situation any vector $\mathbf{V}$ related to the system can be in average aligned along the only available preferential direction, namely that of the angular momentum **J**. This proportionality of the two vectors can be written as the *vector model*

$$\mathbf{V} = v(a, J)\mathbf{J}, \tag{B.51}$$

where the coefficient of proportionality is a scalar $v(a, J)$ which can depend on the length of the angular momentum and on other characteristics of the state (a), and the equality has to be understood as the equivalence of the two operators for any matrix element *within* the multiplet. We find this factor by taking the projection on **J** in both parts of (B.51):

$$v(a, J) = \frac{\langle (\mathbf{V} \cdot \mathbf{J}) \rangle}{\mathbf{J}^2} = \frac{\langle (\mathbf{V} \cdot \mathbf{J}) \rangle}{J(J+1)}. \tag{B.52}$$

Instead of this loose derivation, we can use the Wigner-Eckart theorem (B.48). Recalling that any vector is a tensor operator of rank 1 and introducing its spherical components V_μ according to (B.7), we can write the matrix element (B.50) between the states of the same multiplet as

$$\langle aJM'|V_\mu|aJM \rangle = (-)^{J-M'} \begin{pmatrix} J & 1 & J \\ -M' & \mu & M \end{pmatrix} \langle aJ\|V\|aJ \rangle. \tag{B.53}$$

Exactly in the same way, we find for the angular momentum **J**

$$\langle aJM'|J_\mu|aJM \rangle = (-)^{J-M'} \begin{pmatrix} J & 1 & J \\ -M' & \mu & M \end{pmatrix} \langle aJ\|J\|aJ \rangle. \tag{B.54}$$

Eliminating the 3j-symbol, we find that matrix elements of any vector **V** and of the angular momentum **J** are proportional as in the vector model (B.51) with the coefficient

$$v(a, J) = \frac{\langle aJ\|V\|aJ \rangle}{\langle aJ\|J\|aJ \rangle}. \tag{B.55}$$

It is worthwhile to stress again that the whole procedure makes sense only for the transitions *within* the multiplet $|aJM\rangle$. While **J** acts only inside the multiplet, an arbitrary vector **V** can have also off-diagonal elements in J and a matrix elements (B.50) that are unrelated to the matrix elements of **J**.

To establish the final correspondence of (B.52) and (B.55), we calculate the expectation values of scalar quantities $\mathbf{J}^2$ and $(\mathbf{J} \cdot \mathbf{V})$. The calculation is straightforward: write down the scalar product in spherical components (B.9); express the matrix element sought as a product of matrix elements of individual vectors with the summation over the intermediate projection (since at least one of the vectors is **J**, all intermediate states have the same quantum numbers aJ); apply the Wigner-Eckart theorem (B.48) to each of the factors; and sum over intermediate projections. The result is

$$\langle aJM''|(\mathbf{J} \cdot \mathbf{V})|aJM \rangle = \frac{\delta_{M''M}}{2J+1} \langle aJ\|J\|aJ \rangle \langle aJ\|V\|aJ \rangle. \tag{B.56}$$

Here **V** is an arbitrary vector operator. As it should be, the matrix elements of a scalar quantity do not depend on the orientation ($M = M_2$). In the particular case $\mathbf{V} \Rightarrow \mathbf{J}$, the left-hand side of (B.56) is equal to $\delta_{M''M}J(J+1)$. It defines the reduced matrix element for the angular momentum,

$$\langle aJ\|J\|aJ\rangle^2 = J(J+1)(2J+1). \tag{B.57}$$

Finally, combining these results, we obtain

$$\langle aJM'|V_\mu|aJM\rangle = \frac{\langle aJ|(\mathbf{J}\cdot\mathbf{V})|aJ\rangle}{J(J+1)}\langle aJM'|J_\mu|aJM\rangle, \tag{B.58}$$

which is nothing but the vector model (B.51), (B.52). Since the matrix element of $(\mathbf{J}\cdot\mathbf{V})$ in (B.58) does not depend on M, we do not need to indicate the projections explicitly.

Appendix C | Symmetries

C.1 Time Reversal

The equations of quantum as well as classical mechanics are time-reversible if there is no variable in time or magnetic external fields. In contrast to other discrete symmetries, this symmetry does not correspond to any conserved Hermitian operator. Under time reversal, not only do operators of linear and angular momenta have to change sign but the direction of processes has to be reversed as well. A *final* state of a particle with momentum $\mathbf{p}$ and spin $\mathbf{s}$ is to be transformed into the *initial* state of the time-reversed process with momentum $-\mathbf{p}$ and spin $-\mathbf{s}$. Therefore the time reversal operation $\mathcal{T}$ includes transposition (or complex conjugation $\mathcal{K}$) of observables.

Let us define the time reversal operation as

$$\mathcal{T} = U_T \mathcal{K} O_T, \tag{C.1}$$

where O_T changes the time direction in the quantities which are explicitly time-dependent, and U_T is a unitary operator which has to ensure correct transformations of physical quantities. Due to the complex conjugation $\mathcal{K}$, the operator (C.1) is not linear in usual sense; it acts on the coefficients of the linear superposition

$$\mathcal{T}(a\Psi_1 + b\Psi_2) = a^* \mathcal{T}\Psi_1 + b^* \mathcal{T}\Psi_2 \tag{C.2}$$

(sometimes it is called *antilinear*).

When applied to the Schrödinger equation

$$i h \frac{\partial}{\partial t} \Psi(t) = H \Psi(t), \tag{C.3}$$

the operation (C.1) gives

$$-i h \frac{\partial}{\partial(-t)} \left(U_T \Psi^*(-t) \right) = \mathcal{T} H \mathcal{T}^{-1} \left(U_T \Psi^*(-t) \right). \tag{C.4}$$

Let us designate the time reversed quantities with a tilde,

$$A \Rightarrow \tilde{A} \equiv \mathcal{T} A \mathcal{T}^{-1}. \tag{C.5}$$

The reversed dynamics (C.4) are governed by the reversed Hamiltonian

$$\tilde{H} = \mathcal{T} H \mathcal{T}^{-1} = U_T H^*(-t) U_T^{-1}. \tag{C.6}$$

As seen from (C.4), the new state vector

$$\tilde{\Psi}(t) = U_T \Psi^*(-t) \tag{C.7}$$

satisfies the same Schrödinger equation (C.3) as the original vector $\Psi(t)$ if the Hamiltonian is $\mathcal{T}$-invariant,

$$\tilde{H} = H. \tag{C.8}$$

If the Hamiltonian is time-independent, it has stationary eigenfunctions,

$$\Psi(t) = \Psi(0) e^{-(i/\hbar)Et}. \tag{C.9}$$

If the Hamiltonian is also $\mathcal{T}$-invariant, (C.9) shows that the time-reversed ("time conjugate") function with the amplitude $U_T \Psi^*(0)$ is also an eigenfunction with the same energy E. For a nondegenerate eigenvalue E, there is only one eigenfunction corresponding to this energy so that the time reversed function can differ from the original one only by a constant phase. However, in the degenerate case, one can have two mutually time-reversed linearly independent states Ψ and $\tilde{\Psi}$ with the same energy.

The specific form of U_T depends on the representation used for the description of a specific system. For spinless particles described in the coordinate representation by their coordinates $\mathbf{r}$ and momenta $\mathbf{p} = -i\hbar\nabla$ only, time reversal should give $\mathbf{r} \Rightarrow \mathbf{r}$, $\mathbf{p} \Rightarrow -\mathbf{p}$, or

$$\tilde{\mathbf{r}} = U_T \mathbf{r}^* U_T^{-1} = U_T \mathbf{r} U_T^{-1} = \mathbf{r}, \tag{C.10}$$

$$\tilde{\mathbf{p}} = U_T \mathbf{p}^* U_T^{-1} = -U_T \mathbf{p} U_T^{-1} = -\mathbf{p}. \tag{C.11}$$

In order to satisfy these conditions we do not need additional operators U_T, so it is sufficient to put $U_T = 1$. The $\mathcal{T}$-invariant Hamiltonian has to be an even function of momenta $\mathbf{p}$. For free motion the stationary solution is a plane wave $\exp(i\mathbf{k} \cdot \mathbf{r})$ with (c-number rather than operator) momentum $\mathbf{p} = \hbar\mathbf{k}$. The time reversal operation gives, for $U_T = 1$, the conjugate function $\exp(-i\mathbf{k} \cdot \mathbf{r})$, and this is what we expect for the wave propagating in the reverse direction. The solutions with momenta $\mathbf{p}$ and $-\mathbf{p}$ are degenerate.

C.2 Spin Transformation and Kramer's Theorem

For particles with intrinsic degrees of freedom as spin, it is necessary to specify the unitary matrix U_T that would ensure a correct transformation of these variables. Any angular momentum operator $\mathbf{J}$ is $\mathcal{T}$-odd,

$$\tilde{\mathbf{J}} = \mathcal{T} \mathbf{J} \mathcal{T}^{-1} = U_T \mathbf{J}^* U_T^{-1} = -\mathbf{J}. \tag{C.12}$$

For the orbital part $\mathbf{l}$ this follows from the transformation of the momentum $\mathbf{p}$, (C.11). But we need an extra operator U_T to transform the spin variables.

In the standard representation of the Pauli matrices (A.82), only one of them, σ_y, is imaginary, while σ_x and σ_z are real. This corresponds to the usual choice of phases of the matrix elements of the angular momentum (A.72)–(A.74) when the lowering $J_x - iJ_y$ and raising $J_x + iJ_y$ combinations have real matrix elements (A.71). In this representation one can take

$$U_T = \eta_T \sigma_y \tag{C.13}$$

with an arbitrary phase factor η_T, $|\eta_T|^2 = 1$, as the unitary operator performing time reversal. Using the identity (A.83) accumulating the whole algebra of the Pauli matrices, it is easy to check that

$$\tilde{\mathbf{s}} = U_T \mathbf{s}^* U_T^{-1} = -\mathbf{s}, \tag{C.14}$$

as it should be under time reversal (C.12).

Consider a system of A particles with spin $\frac{1}{2}$. The natural generalization of (C.13) should be

$$U_T = (\eta_T)^A \sigma_y(1) \cdots \sigma_y(A), \tag{C.15}$$

since the spin variables of all particles are to be reversed. Taking into account that the matrices σ_y are imaginary and $\sigma_y^2 = 1$, we find for this system

$$T^2 = U_T \mathcal{K} U_T \mathcal{K} = (-)^A. \tag{C.16}$$

Let a system with a $\mathcal{T}$-invariant Hamiltonian be in a stationary state Ψ. If this state is not degenerate, it can be changed under time reversal by not more than a phase factor, $T\Psi = \exp(i\alpha)\Psi$. But then

$$T^2 \Psi = T(e^{i\alpha}\Psi) = e^{-i\alpha} T\psi = e^{-i\alpha} e^{i\alpha}\Psi = \Psi. \tag{C.17}$$

Hence, for a nondegenerate state $T^2 = 1$ regardless of a number of particles. According to (C.16), this means that a system with an odd number of particles with spin $\frac{1}{2}$ cannot have a nondegenerate stationary state. We came to the *Kramers theorem*: stationary states of a $\mathcal{T}$-invariant system of an odd number of particles with spin $\frac{1}{2}$ are *degenerate*, at least twofold.

In the simplest case of a single particle with no spin-orbit coupling or other spin-dependent forces, this is merely a degeneracy of spin states $\chi_\pm$. For a particle in a central field which includes a spin-orbit potential, a stationary single-particle state $|jm\rangle$ with total angular momentum $j = l \pm 1/2$ is $(2j + 1)$-degenerate since the rotationally invariant Hamiltonian cannot change its eigenvalue if the orientation is changed. An external *electric* field can split this degenerate multiplet. However, the *electric* field, like any other field keeping time reversal invariance, does not distinguish between the time-conjugate orbits $|jm\rangle$ and $|j - m\rangle$, they stay degenerate. An actual splitting depends on m^2. In cases with no axial symmetry, as in crystals, m is not a constant of motion anymore, but the degeneracy of time conjugate orbits still holds.

Contrary to that, the *magnetic* field changes sign under time reversal. A system with an external magnetic field $\mathbf{B}$ is not time reversal invariant; the level splitting $E_m(\mathbf{B})$ depends on m and the degeneracy is lifted. If a source (current) generating magnetic field **B** is a part of the system under consideration, so that the total time reversal operation includes $\mathbf{B} \to -\mathbf{B}$, the entire system becomes again $\mathcal{T}$-invariant. Then the degeneracy is restored because for each state $|m; \mathbf{B}\rangle$ there is a conjugate state $|-m; -\mathbf{B}\rangle$ with equal energy. If a system is externally cranked, the angular velocity $\boldsymbol{\Omega}$ also changes sign under time reversal, and the situation is the same as for magnetic field.

C.3 Time-conjugate Orbits

As we saw in the preceding sections, the behavior of the wavefunction under time reversal depends on the spin of the state and on the representation. We will use the representation where the spinors are transformed with the matrix U_T of (C.13) with the phase factor $\eta_T = -i$. Thus, at our choice of η_T, the time reversal operator coincides with the rotation around the y-axis by an angle $180°$,

$$U_T = R_y(\pi). \tag{C.18}$$

Acting on the spinor χ_m with $s_z = m = \pm(\frac{1}{2})$, the operator U_T changes $m \to -m$, and the phase factor gives

$$U_T \chi_+ = \chi_-, \quad U_T \chi_- = -\chi_+, \tag{C.19}$$

which can be expressed as

$$U_T \chi_m = (-)^{1/2-m} \chi_{-m}. \tag{C.20}$$

We know that, with respect to rotations, a system with angular momentum J can be thought of as constructed of $2J$ spins $\frac{1}{2}$. Looking at the time reversal behavior, we have, as in the proof of the Kramers theorem, to perform the transformation (C.20) for each spin. As a result, the state $|JM\rangle$ changes the sign of M and acquires the phase factor with the exponent $\sum (\frac{1}{2} - m) = J - M$. Thus, the definition of the time conjugate state, consistent with (C.20), is

$$|\tilde{J}\tilde{M}\rangle - U_1 |JM\rangle = (-)^{J-M}|J-M\rangle. \tag{C.21}$$

Note that the second time reversal would restore the original state $|JM\rangle$ with the phase factor $(-)^{2J}$ which equals 1 for an integer J and -1 for a half-integer J.

We already witnessed the appearance of the phase (C.21) in the vector coupling of angular momenta when it was related to reversed motion, $\mathbf{J} \to -\mathbf{J}$. The definition (C.21) is consistent with the phase choices of matrix elements of angular momenta and $3j$-symbols. Unfortunately, the traditional definition of spherical functions Y_{lm} differs from the one suggested by (C.21). Since Y_{lm} are functions of the coordinates, they undergo complex conjugation under time reversal, and the related phase factor is $(-)^m$,

$$Y_{lm}(\mathbf{n}) \Rightarrow Y_{lm}^*(\mathbf{n}) = (-)^m Y_{l-m}(\mathbf{n}), \tag{C.22}$$

instead of $(-)^{l-m}$ as it would be in accordance with the rule (C.21), This was the reason for the modified definition of spherical harmonics used by many authors, for example in

Landau and Lifshits [LL65], where the extra factor i^l is added to the normal expression of Y_{lm}. Then the complex conjugation agrees with (C.21) since $(i^l)^* = (-)^l i^l$. One should be careful in using the phase conventions of various authors.

C.4 Two-component Neutrino and Fundamental Symmetries

In the limit of zero mass, the neutrino reveals remarkable properties. For a massless particle, there is no rest frame. In any frame it is moving with speed of light. It has spin $\frac{1}{2}$, and two independent spin states can be classified by taking the momentum axis as that of the quantization. Actually this is the only physical direction associated with a massless particle. Then the spin projection defines the helicity (2.22). A particle with $h = +1(h = -1)$ is similar to a right (left) screwdriver. These states are analogous to the circularly polarized states of the photon (spin 1 with projections ± 1 onto the direction of the wave propagation).

The combination of definite helicity and absence of mass produces new important consequences. Helicity is a scalar with respect to rotations, but in general it is not a Lorentz scalar. For example, it has no meaning at all in the rest frame. However, for a massless particle, helicity is Lorentz-invariant, the momentum of a particle is transformed together with its spin.

The experiments show that the neutrinos ν ("particles") are always left-polarized, whereas the antineutrinos $\bar{\nu}$ ("antiparticles") are always right-polarized. This statement would be exact in the limit of zero mass; for nonzero mass, the degree of longitudinal polarization is v/c but in the majority of physical situations the neutrino velocity is close to the speed of light. The unique correlation of the *lepton number,* which distinguishes particles from antiparticles, with helicity manifests that some fundamental symmetries are violated in nature.

First of all, parity is not conserved any more. The space inversion changes the sign of the helicity but does not convert the neutrino into the antineutrino. Thus, applying the inversion $\mathcal{P}$ to the left-polarized neutrino, we obtain the nonexisting right-polarized antineutrino—the symmetry with respect to the $\mathcal{P}$ operation is lost. Since the neutrinos are produced in weak interactions only, we conclude that, in contrast to strong and electromagnetic interactions, the weak interactions do not conserve parity. If so, the exact stationary states in the nuclear world, where all interactions are present simultaneously, do not have in general certain parity Π. They are superpositions $\alpha|\Pi\rangle + \beta|-\Pi\rangle$. However, because the admixtures of the opposite parity states are due to the weak interactions, typically one of the coefficients in this combination is very small, except for the cases where some nuclear enhancement mechanisms significantly increase parity mixing.

One of the signatures of the parity nonconservation is the mixed character of the electromagnetic transitions between the two states. Assume, for example, that an unperturbed excited state has quantum numbers $J^\Pi = 1^+$, and we observe the magnetic dipole radiation (M1) to a lower state $J^\Pi = 0^+$, in agreement with the selection rules for M1-operators:

$\Delta J = 1$, no parity change. The electric dipole (E1) transition between these states, permitted by angular momentum, is forbidden by parity. It becomes allowed because of the admixtures of opposite parity to initial and final states. The corresponding amplitude is proportional to the interference of two components of the wavefunctions, $\alpha_1 \beta_2^*$ or $\beta_1 \alpha_2^*$.

Another manifestation can be seen in nuclear reactions, when the cross sections of processes interconnected by the inversion transformation turn out to be different. Sometimes it is formulated as the statement that "the results of identical experiments in mirror-reflected laboratories are not mirror-reflected." Thus, in the first experiment where the parity nonconservation was discovered Wu et al. [Wu57], β-decay of polarized nuclei ^{60}Co (here the β electron is accompanied by the electronic antineutrino $\bar{\nu}_e$),

$$^{60}\text{Co} \rightarrow {}^{60}\text{Ni} + e^- + \bar{\nu}_e, \tag{C.23}$$

the angular distribution of the decay electrons, $\sim(1 + a\cos\theta)$, where the angle θ is the one between the electron momentum and the spin direction of the initial nucleus, displayed the preference for the electrons moving opposite to the nuclear spin. This is essentially the quantity of the same type as the helicity (B.40), pseudoscalar that shows that in the mirror-reflected laboratory the angular distribution would be different, $\sim(1 - a\cos\theta)$. An impressive example was seen in scattering of longitudinally polarized slow neutrons off unpolarized heavy nuclei, when the cross sections for different neutron helicities were different (here a very large nuclear enhancement of the effect was observed).

C.5 Charge Conjugation

Another important discrete symmetry is related to the existence of particles and antiparticles. The corresponding transformation C converts all particles into their antiparticles, changing the signs of all charges (electric, baryonic, leptonic, strangeness, etc.) to the opposite. Neutral particles such as π^0 or photons (they have all charges equal to zero) are transformed into themselves, and, because $C^2 = 1$, we can distinguish the neutral particles with definite *charge parity* $C = \pm 1$. The strong and electromagnetic interactions are invariant under the C operation. An electron in an electromagnetic field behaves similarly to a positron in the field of the opposite direction. In order to reveal the symmetry with respect to charge conjugation, we have to invert the sign of the field, that is, to assign to the quantum of the electromagnetic field, the photon, the charge parity $C_\gamma = -1$.

Among many other consequences of invariance under charge conjugation, one can mention the so-called *Furry theorem* in QED: the processes whose only result is the change of an even (odd) photon number to an odd (even) number, are forbidden. Thus, for example, photon-photon interaction processes $2\gamma \rightarrow 3\gamma$ are impossible. The neutral pion decays into two photons, and therefore has positive charge parity.

The neutrino properties discussed earlier show that in weak interactions the charge conjugation symmetry is also broken. There is no longer a full symmetry between particle and antiparticle worlds because the C operation transforms the left neutrino into the nonexistent (in the limit of zero mass) left antineutrino. Both $\mathcal{P}$ and $\mathcal{C}$ symmetries are destroyed simultaneously.

The picture is invariant with respect to the *combined inversion* $\mathcal{CP}$. It means the conversion of the left neutrino into the right antineutrino. The transition to the antiworld should be accompanied by the mirror reflection; then the results of the corresponding experiments would be the same. $\mathcal{CP}$-invariance is nearly exact. As far as we know, it is violated in rare decays of K-mesons (kaons), and similar B-mesons. The probability of $\mathcal{CP}$-violating decay of neutral kaons is only 0.2% of the probability of normal $\mathcal{CP}$-conserving decay.

The production of neutrinos in the weak interactions shows that $\mathcal{P}$- and $\mathcal{C}$-symmetries separately are completely violated. However, this is just the most distinct manifestation. As we mentioned, parity nonconservation takes place in the nuclear weak processes without a neutrino as well. The direct observation of the corresponding $\mathcal{C}$-violation unfortunately would require experiments with antinuclei.

C.6 Electric Dipole Moment

After we have discussed that the parity conservation is not the universal rule of particle and nuclear interactions, we can return to the question of allowed and forbidden multipoles. The existence of the electric dipole moment in a system with spin $\geq \frac{1}{2}$ is permitted if the restrictions related to the parity conservation are lifted. Of course we still keep the restrictions imposed by the requirements of the rotational invariance.

However, the problem is more complicated. The dipole operator $\mathbf{d}$ is a polar vector. Its expectation value can be calculated with the aid of the vector model. This gives for the effective dipole operator for a particle of spin $\frac{1}{2}$:

$$\mathbf{d} = \frac{\langle(\mathbf{d} \cdot \mathbf{s})\rangle}{s^2}\mathbf{s} = \frac{4}{3}\langle(\mathbf{d} \cdot \mathbf{s})\rangle\mathbf{s}. \tag{C.24}$$

The result is determined by the expectation value of the pseudoscalar quantity $(\mathbf{d} \cdot \mathbf{s})$. Since, as a result of the weak interactions, the stationary states have no certain parity, this expectation value can differ from zero. However, a nonzero value of this quantity would contradict time reversal invariance.

Indeed, the spinors $|\frac{1}{2}, m\rangle$ with spin projection m are transformed according to (C.20) under time reversal. The dipole moment $\mathbf{d}$, like the coordinate vector $\mathbf{r}$, is invariant under $\mathcal{T}$-transformation ("$\mathcal{T}$-even") while the spin vector $\mathbf{s}$, like any angular momentum, is $\mathcal{T}$-odd. Therefore the scalar product $(\mathbf{d} \cdot \mathbf{s})$ is $\mathcal{T}$-odd. If $\mathcal{T}$-invariance holds, the expectation value of a time-reversed operator in a time-reversed state should be the same as before the $\mathcal{T}$-transformation,

$$\langle\tfrac{1}{2}, m|(\mathbf{d} \cdot \mathbf{s})|\tfrac{1}{2}, m\rangle = \langle\tfrac{1}{2}, -m| - (\mathbf{d} \cdot \mathbf{s})|\tfrac{1}{2}, -m\rangle^* = -\langle\tfrac{1}{2}, -m|(\mathbf{d} \cdot \mathbf{s})|1/2, -m\rangle \tag{C.25}$$

(expectation values of any Hermitian operator are real). At the same time, the quantity $(\mathbf{d} \cdot \mathbf{s})$ is a rotational scalar, and its expectation value is the same in all substates of the multiplet. Thus, it is equal to zero. This derivation holds for any angular momentum J of the state (not necessarily spin $\frac{1}{2}$). The nonzero helicity $\propto (\mathbf{p} \cdot \mathbf{s})$, in contrast to (C.25), can exist being the product of the two $\mathcal{T}$-odd vectors.

We have shown that a nonzero electric dipole moment of a particle in a stationary state would be a signature of the combination of the parity nonconservation along with the violation of the time reversal invariance. The experiment up to now was unable to discover a dipole moment of a particle. Current data are compatible with zero at the uncertainty level of 10^{-23} e·cm for the proton, 10^{-25} e·cm for the neutron, and 10^{-26} e·cm for the electron.

Recently, the *anapole moment* was discovered experimentally in the nucleus ^{133}Cs. This is a quantity characteristic to the current in the toroidal coil; the main contribution to the anapole moment is due to the operator $\mathbf{a} = [\mathbf{r} \times \mathbf{s}]$. We see that $\mathbf{a}$ is a polar $\mathcal{T}$-odd vector which can exist in nuclear states with nonzero spin J, the corresponding effective operator being

$$\mathbf{a} = \frac{\langle (\mathbf{a} \cdot \mathbf{J}) \rangle}{J(J+1)} \mathbf{J}. \tag{C.26}$$

The quantity $(\mathbf{a} \cdot \mathbf{J})$ is a $\mathcal{T}$-even pseudoscalar, and requires only parity nonconservation but not $\mathcal{T}$-violation. The anapole moment was discovered by the parity violation in atomic radiative transitions induced by the weak interactions between atomic electrons and the nucleus.

C.7 $\mathcal{CPT}$-Invariance

We conclude this appendix with brief mention of the famous CPT-theorem (R. Lüders and W. Pauli). According to this theorem, any theory preserving the fundamental principles as Lorentz invariance, unitarity (conservation of probability), and the proper relation between the spin value and statistics of particles (integer values correspond to Bose-Einstein statistics, while half-integer spins correspond to Fermi-Dirac statistics) is invariant with respect to the combined application of charge conjugation $\mathcal{C}$, spatial inversion $\mathcal{P}$ and time reversal $\mathcal{T}$. Basically this means that antiparticles in the world obtained by the inversion of all four Minkowski coordinates behave in the same way as particles in the original world.

It follows from the $\mathcal{CPT}$-theorem that particles and their antiparticles have exactly equal masses. If they are unstable, their full lifetimes are also exactly equal (in general, this is not correct for partial lifetimes into specific decay channels). There is no experimental evidence for a violation of the $\mathcal{CPT}$-theorem.

Validity of the $\mathcal{CPT}$-theorem allows one to test fundamental symmetries in an indirect way. For instance, the violation of the combined inversion $\mathcal{CP}$ in the decays of neutral kaons is, according to the $\mathcal{CPT}$-theorem, at the same time a signal of $\mathcal{T}$-noninvariance. Presence in an experiment of $\mathcal{T}$-invariant effects violating parity, as in the case of the anapole moment, proves that the charge symmetry $\mathcal{C}$ is violated as well. This statement can be made with no direct measurements of similar processes with antimatter.

Appendix D | Relativistic Quantum Mechanics

D.1 Lagrangians

For discrete systems the *Hamilton principle* yields the *Lagrangian equations*

$$\delta \int_{t_1}^{t_2} L(q_i, \dot{q}_i) \, dt = 0 \rightarrow \frac{d}{dt}\left(\frac{\partial L}{\partial \dot{q}_i}\right) - \frac{\partial L}{\partial q_i} = 0, \tag{D.1}$$

where the *Lagrangian* is the difference between the kinetic energy and the potential energy, i.e. $L = T - V$. The *Hamiltonian* is given by $H = \sum_i p_i \dot{q}_i - L$, where $p_i = \partial L/\partial \dot{q}_i$.

If η_i is the displacement of particle i from its equilibrium position (see figure D.1), then

$$L = \frac{1}{2}\sum_i^N \left[m\dot{\eta}_i^2 - k(\eta_{i+1} - \eta_i)^2\right] = \sum_i^N a\frac{1}{2}\left[\frac{m}{a}\dot{\eta}_i^2 - ka\left(\frac{\eta_{i+1} - \eta_i}{a}\right)^2\right] = \sum_i^N a\mathcal{L}_i, \tag{D.2}$$

where a is the separation distance between the equilibrium positions of two neighboring particles and $\mathcal{L}_i$ is the linear Lagrangian density.

In continuum systems we can make the substitutions

$$a \rightarrow dx, \qquad \frac{m}{a} \rightarrow \mu = \text{linear mass density,}$$

$$\frac{\eta_{i+1} - \eta_i}{a} \rightarrow \frac{\partial \eta}{\partial x}, \qquad ka \rightarrow Y = \text{Young modulus.} \tag{D.3}$$

Now η is a function of x; $\eta(x)$, and $L = \int \mathcal{L} \, dx$ is the *Lagrangian density* given by

$$\mathcal{L} = \frac{1}{2}\left[\mu\dot{\eta}_i^2 - Y\left(\frac{\partial \eta}{\partial x}\right)^2\right]. \tag{D.4}$$

The variational principle, $\delta \int_{t_1}^{t_2} L \, dt = 0$, (using $\eta(t_2) = \eta(t_1) = 0$) leads to the Euler-Lagrange equations

$$\frac{\partial}{\partial x}\frac{\partial \mathcal{L}}{\partial(\partial \eta/\partial x)} + \frac{\partial}{\partial t}\frac{\partial \mathcal{L}}{\partial(\partial \eta/\partial t)} - \frac{\partial \mathcal{L}}{\partial \eta} = 0. \tag{D.5}$$

Figure D.1. Particles connected by identical springs.

For the example given above, the Euler-Lagrange equations become

$$Y\frac{\partial^2 \eta}{\partial x^2} - \mu\frac{\partial^2 \eta}{\partial t^2} = 0, \tag{D.6}$$

which is the wave equation with velocity $\sqrt{Y/\mu}$.

The *Hamiltonian density* is defined by

$$\mathcal{H} = \dot{\eta}\frac{\partial \mathcal{L}}{\partial \dot{\eta}} - \mathcal{L} = \frac{1}{2}\mu\dot{\eta}^2 + \frac{1}{2}Y\left(\frac{\partial \eta}{\partial x}\right)^2 = T + V. \tag{D.7}$$

The quantity $\partial\mathcal{L}/\partial\dot{\eta}$ is known as the canonical momentum.

D.1.1 *Covariance*

Generalizing (D.5) to three dimensions, $\mathcal{L}$ depends on ϕ, $\partial\phi/\partial x_k$ ($k = 1,\ 2,\ 3$), and $\partial\phi/\partial t$. The Euler-Lagrange equations become

$$\sum_{k=1}^{3} \frac{\partial}{\partial x_k}\frac{\partial \mathcal{L}}{\partial (\partial\phi/\partial x_k)} + \frac{\partial}{\partial t}\frac{\partial \mathcal{L}}{\partial (\partial\phi/\partial t)} - \frac{\partial \mathcal{L}}{\partial \phi} = 0. \tag{D.8}$$

A 4-*vector* is defined by

$$b_\mu = (b_0,\ b_1,\ b_2,\ b_3) = (b_0,\ \mathbf{b}) \tag{D.9}$$

with the convention that the Greek letters μ, ν, λ, etc. vary from 0 to 3, and the roman letters i, j, k, etc. vary from 1 to 3.

The coordinate vector x_μ is defined by

$$x_\mu = (x_0,\ x_1,\ x_2,\ x_3) = (ct,\ \mathbf{x}), \tag{D.10}$$

whereas the coordinate vector x^μ is defined by

$$x^\mu = (-x_0,\ x_1,\ x_2,\ x_3) = (-ct,\ \mathbf{x}). \tag{D.11}$$

A Lorentz transformation is given by

$$x'_\mu = a_\mu^\nu x_\nu, \tag{D.12}$$

where a repeated superscript and subscript has the meaning of a sum. Since a Lorentz transformation preserves the length of a vector (that is, $x^\mu x_\mu = x'^\mu x'_\mu$), we have

$$a_\nu^\mu a_\lambda^\nu = \delta_\lambda^\mu, \quad (a^{-1})_\nu^\mu = a_\nu^\mu, \tag{D.13}$$

where $\delta_{\nu\lambda}$ is a Kronecker delta. Thus,

$$x_\mu = (a^{-1})_\mu^\nu x'_\nu = a_\mu^\nu x'_\nu. \tag{D.14}$$

By definition, a 4-vector transforms like $x_\mu.$, and

$$\frac{\partial}{\partial x'_\mu} = \frac{\partial x_\nu}{\partial x'_\mu}\frac{\partial}{\partial x_\nu} = a^\nu_\mu \frac{\partial}{\partial x_\nu}. \tag{D.15}$$

Thus, $\partial/\partial x_\mu$ is also a 4-vector.

A *scalar product* is defined by

$$b \cdot c = b^\mu c_\mu = \mathbf{b} \cdot \mathbf{c} - b_0 c_0 \tag{D.16}$$

and does not change in a Lorentz transformation

$$b' \cdot c' = a^\nu_\mu b_\nu a^\lambda_\mu c_\lambda = \delta^{\nu\lambda} b_\nu c_\lambda = b \cdot c. \tag{D.17}$$

A *second degree tensor* transforms like

$$t'_{\mu\nu} = a^\lambda_\mu a^\sigma_\nu t_{\lambda\sigma}, \tag{D.18}$$

and similarly for tensors of higher dimension.

Equation (D.8) can be written as

$$\frac{\partial}{\partial x_\mu}\left[\frac{\partial \mathcal{L}}{\partial\left(\partial\phi/\partial x_\mu\right)}\right] - \frac{\partial \mathcal{L}}{\partial\phi} = 0. \tag{D.19}$$

This equation is covariant, that is, it has the same form in all systems of reference.

D.2 Electromagnetic Field

The *Maxwell equations*, in Heaviside-Lorentz units, are given by

$$\nabla \cdot \mathbf{E} = \rho, \qquad \nabla \times \mathbf{B} - \frac{1}{c}\frac{\partial \mathbf{E}}{\partial t} = \frac{\mathbf{j}}{c}, \tag{D.20}$$

$$\nabla \cdot \mathbf{B} = 0, \qquad \nabla \times \mathbf{E} + \frac{1}{c}\frac{\partial \mathbf{B}}{\partial t} = 0. \tag{D.21}$$

In these units the fine-structure constant is $e^2/4\pi\hbar c \simeq 1/137.04$, which is equal to $e^2/\hbar c$ in the Gaussian system (cgs) and $e^2/\left(4\pi\hbar c\epsilon_0\right)$ in MKS units. The fields and potentials in these units are related to the corresponding quantities in the Gaussian system by $1/\sqrt{4\pi}$; for example, $\frac{1}{2}\left(|\mathbf{E}|^2 + |\mathbf{B}|^2\right)$ in these units must read $(1/8\pi)\left(|\mathbf{E}|^2 + |\mathbf{B}|^2\right)$ in Gaussian units. However, expressions like $\mathbf{p} - e\mathbf{A}/c$ are the same in both units, because

$$\left(\sqrt{4\pi}e\right)\left(\mathbf{A}/\sqrt{4\pi}\right) = e\mathbf{A}. \tag{D.22}$$

Introducing the *antisymmetric* tensor

$$F_{\mu\nu} = \begin{pmatrix} 0 & B_3 & -B_2 & -iE_1 \\ -B_3 & 0 & B_1 & -iE_2 \\ B_2 & -B_1 & 0 & -iE_3 \\ iE_1 & iE_2 & iE_3 & 0 \end{pmatrix}, \tag{D.23}$$

and the current

$$j_\mu = (c\rho, \mathbf{j}), \tag{D.24}$$

(D.21) can be written in the compact form

$$\frac{\partial F\mu\nu}{\partial x_\nu} = \frac{j_\mu}{c}. \tag{D.25}$$

Because $F_{\mu\nu}$ is antisymmetric, it obeys the relation $\partial F_{\mu\nu}/\partial x_\mu \partial x_\nu = 0$, which means that

$$\frac{\partial j_\mu}{\partial x_\mu} = 0. \tag{D.26}$$

The *vector potential* is introduced by

$$\frac{\partial A_\nu}{\partial x_\mu} - \frac{\partial A_\mu}{\partial x_\nu} = F_{\mu\nu}, \tag{D.27}$$

and (D.20) can be written as

$$t_{\lambda\mu,\nu} + t_{\mu\nu,\lambda} + t_{\nu\lambda,\mu} = 0, \tag{D.28}$$

where the third degree tensor $t_{\lambda\mu,\nu}$ is defined by

$$t_{\lambda\mu,\nu} = \frac{\partial F_{\mu\nu}}{\partial x_\nu} = \frac{\partial}{\partial x_\nu}\left(\frac{\partial A_\mu}{\partial x_\lambda} - \frac{\partial A_\lambda}{\partial x_\mu}\right). \tag{D.29}$$

Using the Euler-Lagrange equations, it is straightforward to show that the Lagrangian

$$\mathcal{L} = -\frac{1}{4}F_{\mu\nu}F_{\mu\nu} + (j_\mu A_\mu)/c, \tag{D.30}$$

reproduces the Maxwell equations (D.20) and (D.21).

We can rewrite (D.25) as

$$\Box A_\mu - \frac{\partial}{\partial x_\mu}\left(\frac{\partial A_\nu}{\partial x_\nu}\right) = -\frac{j_\mu}{c}. \tag{D.31}$$

We can redefine A_μ, without changing $F_{\mu\nu}$, as

$$A_\mu^{\text{new}} = A_\mu^{\text{old}} + \frac{\partial \chi}{\partial x_\mu}, \quad \text{and} \tag{D.32}$$

where

$$\Box\chi = -\frac{\partial A_\mu^{\text{old}}}{\partial x_\mu}, \tag{D.33}$$

and thus,

$$\frac{\partial A_\mu^{\text{new}}}{\partial x_\mu} = \frac{\partial A_\mu^{\text{old}}}{\partial x_\mu} + \Box\chi = 0. \tag{D.34}$$

Since the vector potential is used to simplify the calculations, we can therefore use a simpler equation

$$\Box A_\mu = -\frac{j_\mu}{c}, \tag{D.35}$$

where A_μ obeys

$$\frac{\partial A_\mu}{\partial x_\mu} = 0. \tag{D.36}$$

Equation (D.36) is known as *Lorentz condition*.

But, even by using (D.36), the potential A_μ is not univocally determined. We can still make an additional transformation of the form

$$A_\mu \to A'_\mu = A_\mu + \frac{\partial \Lambda}{\partial x_\mu}, \tag{D.37}$$

where Λ obeys the equation

$$\Box \Lambda = 0. \tag{D.38}$$

The transformation (D.37) is known as the *gauge transformation*.

D.3 Relativistic Equations

The energy momentum relation

$$E^2 = p^2 + m^2 \implies \widehat{H}_0 \psi(\mathbf{x}, t) = \left[\widehat{\mathbf{p}}^2 + m^2\right] \psi(\mathbf{x}, t), \tag{D.39}$$

together with the quantization rules $\widehat{\mathbf{p}} = -i\nabla$ and $\widehat{H}_0 = i\partial_t$, leads to the wave equation

$$\left[\frac{\partial^2}{\partial t^2} - \nabla^2 + m^2\right] \psi(\mathbf{x}, t) = 0. \tag{D.40}$$

This is known as the *Klein-Gordon equation*. It is considered as the appropriate equation for spin-zero particles, for example, the π-mesons.

Another relativistic equation, proposed by Dirac, is linear in the space-time derivatives:

$$i\frac{\partial \Psi}{\partial t}(\mathbf{x}, t) = H_0 \Psi(\mathbf{x}, t). \tag{D.41}$$

where

$$H_0 = \frac{1}{i}\left[\alpha_2 \frac{\partial}{\partial x^1} + \alpha_2 \frac{\partial}{\partial x^2} + \alpha_3 \frac{\partial}{\partial x^3}\right] + \beta m. \tag{D.42}$$

Above, α_i and β are dimensionless constants, commuting with $\mathbf{r}$ and $\mathbf{p}$. Defining

$$\alpha = (\alpha_1, \alpha_2, \alpha_3), \tag{D.43}$$

we get

$$H_0 = \alpha \cdot \mathbf{p} + \beta m. \tag{D.44}$$

Applying the operator $\partial/\partial t$ in (D.41) gives

$$\frac{\partial}{\partial t}\left(i\frac{\partial \Psi}{\partial t}\right) = (\alpha \cdot \mathbf{p} + \beta m)\frac{\partial \Psi}{\partial t}$$

$$\implies i\frac{\partial^2 \Psi}{\partial t^2} = (\alpha \cdot \mathbf{p} + \beta m)\left(\alpha \cdot \frac{\mathbf{p}}{i}\Psi + \frac{\beta}{i}m\Psi\right).$$

$$= i\sum_{k,j} \frac{\alpha_j \alpha_k + \alpha_k \alpha_j}{2}\frac{\partial^2 \Psi}{\partial x^k \partial x^j} - m\sum_k (\alpha_k \beta + \beta \alpha_k)\frac{\partial \Psi}{\partial x^k} - i\beta^2 m^2 \Psi. \tag{D.45}$$

To reduce this equation to the Klein-Gordon equation (for which $E^2 = p^2 + m^2$), it is necessary that

$$\alpha_j\alpha_k + \alpha_k\alpha_j \equiv \{\alpha_k, \alpha_j\} = 2\delta_{kj},$$
$$\alpha_k\beta + \beta\alpha_k \equiv \{\alpha_k, \beta\} = 0,$$
$$\implies \alpha_i^2 = \beta^2 = 1. \tag{D.46}$$

These equations can only be satisfied if α_i and β are matrices. Thus Ψ must be a vector with N components:

$$\Psi = \begin{pmatrix} \psi_1 \\ \psi_2 \\ \psi_3 \\ \vdots \\ \psi_N \end{pmatrix}. \tag{D.47}$$

The matrices α_i and β must have the properties

1. Hermiticity:

   $$\alpha_i^\dagger = \alpha_i, \quad \beta^\dagger = \beta. \tag{D.48}$$

   because H_0 is Hermitian.. This, together with (D.46), implies that the eigenvectors of α_i and β must be ± 1.
2. $\mathrm{Tr}\alpha = \mathrm{Tr}\beta = 0$.
3. N must have even dimension.
   Both properties 3 and 4 follow directly by using (D.46).
4. $N \geq 4$.
   $N = 2$ is not possible because the Pauli matrices (σ, I) form a complete set of 2×2 matrices. However, the matrix I always commutes and (D.46) cannot be satisfied. Thus, $N = 4$ is the smallest possibility, and

   $$\Psi = \begin{pmatrix} \psi_1 \\ \psi_2 \\ \psi_3 \\ \psi_4 \end{pmatrix}, \quad \Psi^\dagger = \left(\psi_1^* \psi_2^* \psi_3^* \psi_4^*\right). \tag{D.49}$$

5. Representation of α_i and β:
   The above conditions lead to many possible representations of α_i and β. The most popular representation is

   $$\alpha_i = \begin{pmatrix} 0 & \sigma_i \\ \sigma_i & 0 \end{pmatrix}, \quad \beta = \begin{pmatrix} 1 & 0 \\ 0 & -1 \end{pmatrix}, \tag{D.50}$$

   where σ_i, 1, and 0 in (D.50) are 2×2 matrices.

D.3.1 Particle at rest

For a particle at rest $\mathbf{p} = 0$ so that $\nabla \Psi = 0$ and

$$i\frac{\partial \Phi}{\partial t}(\mathbf{x}, t) = \beta m \Phi(\mathbf{x}, t). \tag{D.51}$$

In terms of the wave function components

$$i\begin{pmatrix} \partial\phi_1/\partial t \\ \partial\phi_2/\partial t \\ \partial\phi_3/\partial t \\ \partial\phi_4/\partial t \end{pmatrix} = m \begin{pmatrix} 1 & 0 & 0 & 0 \\ 0 & 1 & 0 & 0 \\ 0 & 0 & -1 & 0 \\ 0 & 0 & 0 & -1 \end{pmatrix} \begin{pmatrix} \phi_1 \\ \phi_2 \\ \phi_3 \\ \phi_4 \end{pmatrix} = m \begin{pmatrix} \phi_1 \\ \phi_2 \\ -\phi_3 \\ -\phi_4 \end{pmatrix}. \tag{D.52}$$

These equations have 4 solutions given by

$$\Phi_1 = e^{-imt}\begin{pmatrix} 1 \\ 0 \\ 0 \\ 0 \end{pmatrix}, \quad \Phi_2 = e^{-imt}\begin{pmatrix} 0 \\ 1 \\ 0 \\ 0 \end{pmatrix}, \tag{D.53}$$

$$\Phi_3 = e^{+imt}\begin{pmatrix} 0 \\ 0 \\ 1 \\ 0 \end{pmatrix}, \quad \Phi_4 = e^{+imt}\begin{pmatrix} 0 \\ 0 \\ 0 \\ 1 \end{pmatrix}. \tag{D.54}$$

We identify Φ_1 and Φ_2 as positive energy solutions for up and down spin states, respectively. The other solutions, Φ_3 and Φ_4, are negative energy solutions, or antiparticle solutions, with spin up and down, respectively. We therefore see the necessity of 4 components.

D.3.2 Covariant form: γ matrices

Defining the matrices

$$\gamma^0 \equiv \beta, \quad \gamma^i \equiv \beta\alpha_i, \quad \gamma \equiv \beta\alpha, \quad \gamma^\mu \equiv (\gamma^0, \gamma), \tag{D.55}$$

multiplying the Dirac equation by γ^0, and using $\beta^2 = 1$, we obtain

$$\left(i\gamma^0 \frac{\partial}{\partial t} - \gamma \cdot \frac{\nabla}{i}\right)\Psi(\mathbf{x}, t) = m\Psi(\mathbf{x}, t),$$

$$\left(\gamma^0 p_0 - \gamma \cdot \mathbf{p}\right)\Psi(\mathbf{x}, t) = m\Psi(\mathbf{x}, t). \tag{D.56}$$

or

$$\gamma^\mu p_\mu \Psi(\mathbf{x}, t) = m\Psi(\mathbf{x}, t),$$

where $p_\mu = i\partial_\mu$.

Table D.1: Bilinear covariants built with Dirac γ matrices.

$\Gamma_{4\times4}$	Definition	Transformation	Number
γ	$\beta\alpha$	space vector	3
γ_0	β	time vector	1
$1 \equiv I$	identity	scalar	1
$\sigma^{\mu\nu}$	$\frac{i}{2}\left[\gamma^\mu, \gamma^\nu\right]$	traceless tensor	6
γ_5	$i\gamma^0\gamma^1\gamma^2\gamma^3$	pseudoscalar	1
$\gamma_5\gamma^\mu$	$\gamma_5\gamma^\mu$	pseudovector	4

Using (D.46) we can prove that

$$\gamma^\nu\gamma^\mu + \gamma^\mu\gamma^\nu \equiv \{\gamma^\nu, \gamma^\mu\} = 2g^{\mu\nu},$$
$$\left(\gamma^i\right)^2 = -1 \quad (i = 1, 2, 3), \quad \left(\gamma^0\right)^2 = 1,$$
$$\gamma^\dagger = -\gamma, \quad \gamma_0^\dagger = \gamma_0. \tag{D.57}$$

The most popular representations of γ are

$$\gamma^i = \begin{pmatrix} 0 & \sigma_i \\ -\sigma_i & 0 \end{pmatrix}, \quad \gamma^0 = \begin{pmatrix} 1 & 0 \\ 0 & -1 \end{pmatrix}, \quad \gamma^5 = \begin{pmatrix} 0 & 1 \\ 1 & 0 \end{pmatrix}. \tag{D.58}$$

It is also useful to generalize the matrices σ as

$$\sigma^{ij} = \begin{pmatrix} \sigma_k & 0 \\ 0 & \sigma_k \end{pmatrix}, \quad \sigma^{0i} = i\alpha_i = i\begin{pmatrix} 0 & \sigma_i \\ \sigma_i & 0 \end{pmatrix}. \tag{D.59}$$

The indices (i, j, k) take values 1, 2, and 3 (or x, y, z), and can be cyclically permuted. The σ's are Pauli matrices

$$\sigma_1 = \begin{pmatrix} 0 & 1 \\ 1 & 0 \end{pmatrix}, \quad \sigma_2 = \begin{pmatrix} 0 & -i \\ i & 0 \end{pmatrix}, \quad \sigma_3 = \begin{pmatrix} 1 & 0 \\ 0 & -1 \end{pmatrix}. \tag{D.60}$$

There are only 16 4×4 independent matrices. Using the notation Γ_i for these matrices, one can show that they can all be built from the γ matrices, as shown in table D.1.

The table also shows the transformation properties of $\overline{\Psi}\Gamma\Psi$. The spatial parts σ^{ij} $(i, j = 1, 2, 3)$ are related to spin, while the mixed space-time parts $\sigma^{0\mu}, \sigma^{\mu 0}$ are related to the velocities (proportional to α). To emphasize these relations, we write

$$\Sigma_k = \sigma^{ij}_{4\times4}, \tag{D.61}$$

where Σ is seen as a four-dimensional generalization of the Pauli matrices.

D.4 Probability and Current

Multiplying the Dirac equation to the left by $\Psi^\dagger$ and subtracting the result by the Hermitian conjugate of this operation we get

$$\frac{\partial \left(\Psi^\dagger \Psi\right)}{\partial t} + \nabla \cdot \left(\Psi^\dagger \alpha \Psi\right) = 0. \tag{D.62}$$

This equation has the form of a continuity equation, where the probability is given by

$$\rho = \Psi^\dagger \Psi = |\psi_1|^2 + |\psi_2|^2 + |\psi_3|^2 + |\psi_4|^2\,, \tag{D.63}$$

and the current is given by

$$\mathbf{j} = \Psi^\dagger \alpha \Psi \equiv \Psi^\dagger \beta \beta \alpha \Psi \equiv \overline{\Psi} \gamma \Psi. \tag{D.64}$$

Note that $\mathbf{j}$ is given in terms of the *Dirac adjoint* $\overline{\Psi}$ defined by

$$\overline{\Psi} = \Psi^\dagger \beta \equiv \Psi^\dagger \gamma_0. \tag{D.65}$$

The continuity equation can be expressed as $\partial_\mu j^\mu = 0$, where

$$j^\mu = \left(\rho, \mathbf{j}\right) = \overline{\Psi} \gamma^\mu \Psi. \tag{D.66}$$

D.5 Wavefunction Transformation

The Lorentz transformations are given by

$$x'^\nu = a_\mu^\nu x^\mu. \tag{D.67}$$

In a new referential O', $\Psi'\left(x'\right)$, is related to $\Psi\left(x\right)$ in the referential O, by

$$\Psi'\left(x'\right) = L_\nu\left(a\right) \Psi\left(x\right), \tag{D.68}$$

where $L_\nu\left(a\right)$ changes only the components of $\Psi\left(x\right)$. The Dirac equation must be invariant under this transformation. That is,

$$i\gamma^\mu \frac{\partial \Psi\left(x\right)}{\partial x^\mu} = m\Psi\left(x\right),$$
$$i\gamma^\mu \frac{\partial \Psi'\left(x'\right)}{\partial x'^\mu} = m\Psi'\left(x'\right). \tag{D.69}$$

Using

$$\Psi\left(x\right) = L_\nu^{-1}\left(a\right) \Psi'\left(x'\right), \qquad \frac{\partial}{\partial x^\mu} = a_\mu^\nu \frac{\partial}{\partial x'^\nu}, \tag{D.70}$$

in (D.69), we get

$$i\gamma^\mu a_\mu^\nu \frac{\partial}{\partial x'^\nu} L_\nu^{-1}\left(a\right) \Psi'\left(x'\right) = mL_\nu^{-1}\left(a\right) \Psi'\left(x'\right). \tag{D.71}$$

Multiplying the equation above by $L_v(a)$, and because $L_v(a)$ commutes with the derivatives and with a_μ^v, one gets

$$\left[a_\mu^v L_v(a) \gamma^\mu L_v^{-1}(a)\right] i \frac{\partial}{\partial x'^v} \Psi'(x') = m\Psi'(x'). \tag{D.72}$$

Therefore, Lorentz invariance implies

$$\left[a_\mu^v L_v(a) \gamma^\mu L_v^{-1}(a)\right] = \gamma^\mu,$$
$$\implies a_\mu^v \gamma^\mu - L_v^{-1}(a) \gamma^\mu I_v(a). \tag{D.73}$$

For a spin $\frac{1}{2}$ particle, the rotation by an angle θ implies that

$$\Psi'(x') = U_R(\theta)\Psi(x), \quad \text{where} \quad U_R(\theta) = e^{-i\mathbf{J}\cdot\theta} = e^{-i\theta\cdot\sigma/2}, \tag{D.74}$$

where $\sigma/2$ is the generator of infinitesimal rotations. Since the generalization of the Pauli matrices σ_k is given by equation (D.61), we assume that the relativistic operator for rotations in 3-dimensions, around the axis k, is in the form

$$U_R(\theta) = e^{-i\Sigma_k\theta_k/2}. \tag{D.75}$$

To generalize (D.75) for Lorentz transformations, we imagine these transformations as rotation in space-time and we replace σ^{ij}, the operator of infinitesimal rotations around the axis k, by σ^{0k}, the generator of infinitesimal velocity transformations along an axis k. We see that the rotation "angle" is imaginary. For transformations along an axis, there is a *rapidity parameter* λ that determines the velocity $u = u(\lambda)$ and is additive for successive transformations:

$$\cosh\lambda = \gamma_u = \left(1 - \beta_u^2\right)^{-1/2} = \left(1 - u^2\right)^{-1/2}, \quad \sinh\lambda = \gamma_u\beta_u. \tag{D.76}$$

In terms of this quantity, the Lorentz transformation along the x-axis, can be expressed in terms of an imaginary angle:

$$x' = x\cosh\lambda - t\sinh\lambda, \quad \text{and} \quad t' = t\cosh\lambda - x\sinh\lambda. \tag{D.77}$$

An observer O who sees a particle moving with velocity v along x uses (D.77) to determine the *rapidity* of the particle and denotes it by λ_0. Similarly, the observer O' sees the same particle moving with velocity v' and uses (D.77) to determine the *rapidity*, denoting it by λ_0'. We can show that the relation between the two velocities is

$$v'(\lambda_0') = \frac{u(\lambda) + v(\lambda_0)}{1 + uv'}, \quad \text{that is,} \quad \lambda_0' = \lambda_0 + \lambda. \tag{D.78}$$

Thus, the parameter λ is additive for transformations along an axis (as with rotations around an axis). The transformation is "active," and since the conventional Lorentz transformation (D.67), (D.68) is passive, we have a sign to worry about. We assume that the generalization of (D.75) is

$$L_v(a) \equiv L_v(\lambda) = \exp\left[\mp\frac{1}{2}i\lambda_k\sigma^{0k}\right], \tag{D.79}$$

where k is the axis for the velocity increase and the $+(-)$ sign is for active (passive) transformations. Expanding the potential in a series and using the relation

$$\left(\sigma^{0k}\right)^2 = \left(i\alpha_k\right)^2 = -I_{4\times4}, \tag{D.80}$$

we obtain the relation

$$L_v = \exp\left[\mp\frac{1}{2}i\lambda_k\sigma^{0k}\right] = I\cosh\frac{\lambda_k}{2} \mp i\sigma^{0k}\sinh\frac{\lambda_k}{2}. \tag{D.81}$$

One can show that (D.73) works for $v = 0, k$, from the relations

$$\left\{\gamma^0, \sigma^{0k}\right\} = \left\{\gamma^k, \sigma^{0k}\right\} = 0,$$

$$L_v^{-1}\gamma^0 L_v = e^{-i\lambda_k\sigma^{0k}}\gamma^0 = \cosh\lambda_k\gamma^0 - \sinh\lambda_k\gamma^k$$

$$= a_\mu^0\gamma^\mu = \gamma_v\gamma^0 - \gamma_v\beta_v\gamma^k.$$

$$L_v^{-1}\gamma^k L_v = \gamma_v\gamma^k - \gamma_v\beta_v\gamma^0. \tag{D.82}$$

Note that L_v *is not unitary*, $L_v \neq L_v^{-1}$. This means that the normalization of Ψ changes under the transformation. However, $L_v^\dagger = L_v$. This change in Ψ is necessary to keep the total probability constant, since a volume element also changes udder this transformation. Although $L_v^\dagger$ is not equal to L_v^{-1}, there is a simple relation between them:

$$L_v^{-1} = \gamma_0 L_v^\dagger \gamma_0. \tag{D.83}$$

D.5.1 Bilinear covariants

In table D.1 we have 16 matrices that, when sandwiched between the Dirac spinor Ψ and its adjoint $\overline{\Psi}$, transform as indicated. For example, the current

$$j^\mu(x) = \overline{\Psi}(x)\gamma^\mu\Psi(x) = \Psi^\dagger(x)\gamma^0\gamma^\mu\Psi(x) \tag{D.84}$$

is a 4-vector. For example, relating $j'^\mu(x')$ to $j^\mu(x)$, we have

$$j'^\mu(x') = \Psi'^\dagger(x')\gamma^0\gamma^\mu\Psi'(x') = \Psi^\dagger(x)L_v^\dagger\gamma^0\gamma^\mu L_v\Psi(x)$$

$$= \Psi^\dagger(x)\gamma^0\left(\gamma^0 L_v^\dagger\gamma^0\right)\gamma^\mu L_v\Psi(x) = \Psi^\dagger(x)\gamma^0\left[L_v^{-1}\gamma^\mu L_v\right]\Psi(x)$$

$$= \Psi^\dagger(x)\gamma^0\left[a_v^\mu\gamma^v\right]\Psi(x) = a_v^\mu\overline{\Psi}(x)\gamma^v\Psi(x), \tag{D.85}$$

or

$$j'^\mu(x') = a_v^\mu j^v(x). \tag{D.86}$$

Similarly, $\overline{\Psi}(x)I\Psi(x)$ is a scalar

$$\overline{\Psi}I\Psi = \overline{\Psi}'I\Psi', \tag{D.87}$$

and $\overline{\Psi}\sigma^{\mu v}\Psi$ is second degree tensor

$$\overline{\Psi}'(x')\sigma^{\mu v}\Psi'(x') = a_\alpha^\mu a_\beta^v\overline{\Psi}(x)\sigma^{\alpha\beta}\Psi(x). \tag{D.88}$$

D.5.2 Parity

The parity transform is defined by

$$P\Psi(x) = \Psi'(x') = \Psi'(-\mathbf{x}, t).$$
(D.89)

To include parity as part of the Lorentz group operators, we have to find a 4×4 P matrix with the property

$$P^{-1}\gamma^\nu P = a\,(P)^\nu_\mu\,\gamma^\mu, \qquad \text{where}\, a\,(P) = \begin{pmatrix} 1 & 0 & 0 & 0 \\ 0 & -1 & 0 & 0 \\ 0 & 0 & -1 & 0 \\ 0 & 0 & 0 & -1 \end{pmatrix}.$$
(D.90)

Note that this is an improper Lorentz transformation, since $\det(a) = -1$, while the previous transformations were proper, with $\det(a) = 1$. When a matrix P satisfying (D.89) acts on a spinor Ψ, the new $P\Psi$ has opposite *intrinsic parity*, that is, P changes the internal parts of the wavefunctions. The parity operator acting on the external part is different and can have its own eigenvalues. For example, for the eigenstates of angular momentum, the parity eigenvalues are the familiar $(-1)^l$.

The choice $P = \gamma^0 = \beta$ for the parity operator has the desired effect:

$$P \begin{pmatrix} \psi_1 \\ \psi_2 \\ \psi_3 \\ \psi_4 \end{pmatrix} = \beta \begin{pmatrix} \psi_1 \\ \psi_2 \\ \psi_3 \\ \psi_4 \end{pmatrix} = \begin{pmatrix} \psi_1 \\ \psi_2 \\ -\psi_3 \\ -\psi_4 \end{pmatrix}.$$
(D.91)

This relation is the microscopic explanation for why fermions and antifermions have *opposite intrinsic parity* in the Dirac theory.

The *pseudoscalar object* γ^5 and the *pseudovector* (or *axial vector*) $\gamma^5\gamma^\mu$ of table D.1 behave as a scalar a pseudoscalar under proper Lorentz transformations $[\det(a) = 1]$, but gain a negative sign for parity transformations.

$$P^{-1}\gamma^5 P = -\gamma^5, \qquad P^{-1}\gamma^5\gamma^\mu P = -\gamma^5\gamma^\mu.$$
(D.92)

D.6 Plane Waves

In the Dirac theory, the absence of a positive energy solution is identified as an antiparticle. Thus, it is conventional to revert to the spin in the negative energy eigenvectors so that it refers to real antiparticles. For the particle at rest, this means

$$\Phi_1 = \Phi_{0\uparrow}^{(+)}(0) = e^{-imt}u_\uparrow^{(+)}(0), \qquad \Phi_2 = \Phi_{0\downarrow}^{(+)} = e^{-imt}u_\downarrow^{(+)}(0),$$
$$\Phi_3 = \Phi_{0\downarrow}^{(-)} = e^{+imt}u_\downarrow^{(-)}(0), \qquad \Phi_4 = \Phi_{0\uparrow}^{(-)} = e^{+imt}u_\uparrow^{(-)}(0),$$
(D.93)

with

$$u_\uparrow^{(+)}(0) = \begin{pmatrix} 1 \\ 0 \\ 0 \\ 0 \end{pmatrix}, \quad u_\downarrow^{(+)}(0) = \begin{pmatrix} 0 \\ 1 \\ 0 \\ 0 \end{pmatrix}, \quad u_\downarrow^{(-)}(0) = \begin{pmatrix} 0 \\ 0 \\ 1 \\ 0 \end{pmatrix}, \quad u_\uparrow^{(-)}(0) = \begin{pmatrix} 0 \\ 0 \\ 0 \\ 1 \end{pmatrix}. \tag{D.94}$$

The index 0 refers to the momentum, m_s to spin, and the argument (0) to the position **x**. It is important to remember that the spinors for a particle at rest are base vectors in which a general 4×1 Dirac spinor can be expanded.

For $p \neq 0$ the Dirac equation is

$$\left(i\gamma^0 \frac{\partial}{\partial t} + i\gamma.\nabla \right) \Phi(\mathbf{x}, t) = m\Phi(\mathbf{x}, t). \tag{D.95}$$

The factor mt in the exponent of (D.93) is the scalar product $p^\mu x_\mu$ in the referential where the particle is at rest. Using the covariance property, in another referential one has

$$\Phi_{ps}^{(\pm)}(\mathbf{x}, t) = e^{\mp ip^\mu x_\mu} u_s^{(\pm)}(p). \tag{D.96}$$

We obtain the plane wave spinors $u_s^{(\pm)}(p)$ doing a Lorentz transformation along the z-axis, with the operator L_v:

$$u_s^{(\pm)}(p) = L_v(\lambda) u_s^{(\pm)}(0),$$

$$L_v(\lambda) = e^{\mp i\lambda\sigma^{0k}/2} = I\cosh\frac{\lambda}{2} \mp i\sigma^{0k}\sinh\frac{\lambda}{2}. \tag{D.97}$$

One can show that for a particle with momentum **p**,

$$\tanh\frac{\lambda}{2} = \frac{p}{E_p + m}, \quad \cosh\frac{\lambda}{2} = \sqrt{\frac{E_p + m}{2m}}, \tag{D.98}$$

and $L_v(\lambda)$ can be rewritten as

$$L_v(\lambda) = \sqrt{\frac{E_p + m}{2m}} \left(1 + \frac{\alpha \cdot \mathbf{p}}{E_p + m} \right). \tag{D.99}$$

Replacing the explicit representation of the matrix α in eq. D.97, we get

$$u_\uparrow^{(+)}(p) = N \begin{pmatrix} 1 \\ 0 \\ \frac{p_z}{E_p+m} \\ \frac{p_+}{E_p+m} \end{pmatrix}, \quad u_\downarrow^{(+)}(p) = N \begin{pmatrix} 0 \\ 1 \\ \frac{p_-}{E_p+m} \\ -\frac{p_z}{E_p+m} \end{pmatrix}, \tag{D.100}$$

$$u_\downarrow^{(-)}(p) = N \begin{pmatrix} \frac{p_z}{E_p+m} \\ \frac{p_+}{E_p+m} \\ 1 \\ 0 \end{pmatrix}, \quad u_\uparrow^{(-)}(p) = N \begin{pmatrix} \frac{p_-}{E_p+m} \\ -\frac{p_z}{E_p+m} \\ 0 \\ 1 \end{pmatrix}, \tag{D.101}$$

where

$$p_\pm = p_x \pm ip_y, \quad N = \sqrt{\frac{E_p + m}{2m}}. \tag{D.102}$$

These spinors $u_s^{(\pm)}(p)$ describe free particles with spin $s = \pm\frac{1}{2}$ (in their rest frame), energy $\pm E_p$, and 4-momentum p. (Note that the negative energy solutions have their 3-momenta reversed, according to the interpretation for the antiparticle solutions).

D.6.1 Summary of plane wave spinor properties

1. The normalized wavefunction is

$$\Phi_{ps}^{(\pm)}(\mathbf{x}, t) = e^{\mp i p^\mu x_\mu} u_s^{(\pm)}(p). \tag{D.103}$$

   The space-time dependence is in the exponential, the spin dependence is in the spinor, and *p is not* an operator.

2. The spin is a *good quantum number* only in the particle's rest frame or for a motion along z. If $p_x = p_y = 0$, while $p_z \neq 0$, the u's are eigenstates of Σ_x, but with contributions from the small components.

3. The heliticity $\sigma \cdot \hat{\mathbf{p}}$ is a good quantum number for these u's.

4. When (D.103) is inserted in the Dirac equation one obtains the equation for the free spinors $u^{(\pm)}(p)$:

$$\left(\gamma^\mu p_\mu \mp m\right) u^{(\pm)}(p) = 0. \tag{D.104}$$

   Note again that *p is not* an operator, and that the equation above is a matricial equation.

5. The explicit Dirac equations for the positive and enervative energies are:

$$\left[\gamma^0 p_0 - \gamma \cdot \mathbf{p} - m\right] u^{(+)}(p) = 0,$$
$$\left[-\gamma^0 p_0 - \gamma \cdot (-\mathbf{p}) - m\right] u^{(-)}(p) = 0. \tag{D.105}$$

   We see that the spinors of positive energy have energy and momentum opposite to the spinors of negative energy.

6. The adjoint Dirac spinor $\bar{u}(p) = u^\dagger(p)\gamma^0$ satisfies the transposed Dirac equation:

$$\bar{u}^{(\pm)}(p) \left(\gamma^\mu p_\mu \mp m\right) = 0. \tag{D.106}$$

7. The u's satisfy the Lorentz invariant orthogonality relations:

$$\bar{u}_s^{(b)}(p) u_{s'}^{(b')}(p) = b \delta_{bb'} \delta_{ss'}, \tag{D.107}$$

   where $b = \pm$ for the positive and negative energy solutions.

8. The probability density ρ for the plane waves is

$$\rho = \Phi^\dagger \Phi = u_s^{(b)}(p)^\dagger u_{s'}^{(b')}(p) = \frac{E_p}{m} \delta_{bb'} \delta_{ss'}. \tag{D.108}$$

   Since ρ is proportional to the energy, it is not a Lorentz invariant.

9. The *completeness relation* is

$$\sum_{b=\pm,\, s=\uparrow\downarrow} b u_s^{(b)}(p) \bar{u}_s^{(b)}(p) = I_{4\times4}. \tag{D.109}$$

   Thus, the mathematical completeness requires negative energy degrees of freedom even for low-energy processes.

10. The *current*

$$\mathbf{j} = \bar{u}_s^{(b)}(p)\gamma u_{s'}^{(b')}(p) = \frac{\mathbf{P}}{m}\delta_{bb'}\delta_{ss'} = \frac{\mathbf{P}}{E_p}\rho\delta_{bb'}\delta_{ss'},$$ (D.110)

has the expected plane wave property, that is, $j = \mathbf{v}_g\rho$, a group velocity times a density.

D.6.2 Projection operators

The operators

$$\Lambda_{\pm}(p) = \frac{\pm\gamma^{\mu}p_{\mu} + m}{2m}$$ (D.111)

have the properties

$$\Lambda_+u(p) = u(p), \qquad \Lambda_-u(p) = 0,$$

$$\Lambda_-v(p) = v(p), \qquad \Lambda_+v(p) = 0,$$

$$\Lambda_{\pm}^2 = \Lambda_{\pm}, \qquad \Lambda_+\Lambda_- = \Lambda_-\Lambda_+ = 0,$$

$$\Lambda_+ + \Lambda_- = I,$$ (D.112)

where we define the particle and antiparticle spinors as

$$u_s(p) = u_s^{(+)}(p), \qquad v_s(p) = u_s^{(-)}(p).$$ (D.113)

Using the u's and the v's as base vectors, we see the Λ_+ (Λ_-) is a *projection operator* that eliminates the antiparticle (particle) part of the wavefunction, filtering the particle (antiparticle) part. Since the basis is complete, we have

$$\sum_i |i\rangle\langle i| = I, \Rightarrow \sum_i [u_s(p)\bar{u}_s(p) - v_s(p)\bar{v}_s(p)] = I,$$ (D.114)

which can be easily verified by substitution.

From the equations above, we get

$$\Lambda_+(p) = \sum_{s=1}^{2} u_s(p)\bar{u}_s(p) = \frac{\gamma^{\mu}p_{\mu} + m}{2m},$$ (D.115)

$$\Lambda_-(p) = \sum_{s=1}^{2} v_s(p)\bar{v}_s(p) = \frac{-\gamma^{\mu}p_{\mu} + m}{2m}.$$ (D.116)

D.7 Plane Wave Expansion

One can show that $\sqrt{m/(VE_p)}\Phi_s^{(b)}$ is normalized to unity, showing that the plane wave satisfies the orthogonality conditions:

$$\int d^3x \left[\Phi_s^{(b)}(x)\right]^{\dagger} \Phi_{s'}^{(b')}(x) = \delta_{bb'}\delta_{ss'}\frac{E_p V}{m}.$$ (D.117)

Thus, a general solution of the general Dirac equation has the expansion

$$
\begin{aligned}
\Psi(\mathbf{x}, t) &= \sum_{ps} \sqrt{\frac{m}{VE_p}} \left[e^{i(\mathbf{p}\cdot\mathbf{x} - p_0 t)} u_s^{(+)}(p) b_{ps}^{(+)}(\mathbf{p}) + e^{-i(\mathbf{p}\cdot\mathbf{x} - p_0 t)} u_s^{(-)}(p) b_{ps}^{(-)}(\mathbf{p}) \right] \\
&= \sum_{ps} \sqrt{\frac{m}{VE_p}} \left[e^{ipx} u_s^{(+)}(p) b_{ps}^{(+)}(\mathbf{p}) + e^{ipx} u_s^{(-)}(p) b_{ps}^{(-)}(\mathbf{p}) \right],
\end{aligned}
\tag{D.118}
$$

where the $b's$ are expansion coefficients.

D.8 Electromagnetic Interaction

The interaction with the electromagnetic field is obtained by using the minimal coupling; $p^\mu \rightarrow p^\mu - qA^\mu$. The Dirac Hamiltonian becomes

$$
\left(i \frac{\partial}{\partial t} - q\phi \right) \Psi(\mathbf{x}, t) = \left[\alpha \cdot (\mathbf{p} - q\mathbf{A}) + \beta m \right] \Psi(\mathbf{x}, t).
\tag{D.119}
$$

This equation incorporates the charge q with an external field. It is invariant by a gauge transformation:

$$
A^\mu \rightarrow A^\mu - \partial^\mu \chi(x), \quad \Psi(x) \rightarrow e^{iq\chi(x)} \Psi(x).
\tag{D.120}
$$

We can remove the temporal dependence associated with a rest mass m, and splitting Ψ in upper and lower components.

$$
\Psi(\mathbf{x}, t) = e^{-imt} \begin{pmatrix} \psi_U(\mathbf{x}, t) \\ \psi_L(\mathbf{x}, t) \end{pmatrix}.
\tag{D.121}
$$

In this equation, ψ_U and ψ_L are two-dimensional spinors (that is, 2×1, or Pauli) still containing the temporal dependence of the kinetic energy. Replacing the above equation in the Dirac equation (D.119) we obtain that the terms with α couple ψ_U and ψ_L, yielding

$$
i \frac{\partial \psi_U}{\partial t} = \sigma \cdot (\mathbf{p} - q\mathbf{A}) \psi_L + q\phi \psi_U,
\tag{D.122}
$$

$$
i \frac{\partial \psi_L}{\partial t} = \sigma \cdot (\mathbf{p} - q\mathbf{A}) \psi_U + q\phi \psi_L - 2m\psi_L.
\tag{D.123}
$$

D.9 Pauli Equation

The assumption of time dependence (D.121) produces an asymmetric term proportional to $-2m$ in (D.123). This has two important consequences. The first is that in the nonrelativistic limit, $m \rightarrow \infty$, and $\partial \psi_L / \partial t$ becomes exceptionally large. This leads to the image of a charge jumping back and forth between the positive and negative energy components (known as *Zitterbewegung*). The second is that, if the kinetic energy is small compared to its rest mass, the lower components ψ_L are small compared to the upper component. This is formally seen solving (D.123) for ψ_L:

$$\psi_L = \frac{\sigma \cdot (\mathbf{p} - q\mathbf{A})}{2m} \psi_U - \frac{i \, (\partial/\partial t) - q\phi}{2m} \psi_L. \tag{D.124}$$

Solving for ψ_L and replacing it in (D.122) for ψ_U we get the Klein-Gordon equation.

An approximation for the lower component is obtained expanding the equation above in a power series on ratios between momentum, energy, and the rest mass m:

$$\psi_L \simeq \psi_{L0} \left[O(v)^0 \right] + \psi_{L1} \left[O(v) \right] + \cdots,$$

$$\psi_{L0} \simeq 0, \qquad \psi_{L1} \simeq \frac{\sigma \cdot (\mathbf{p} - q\mathbf{A})}{2m} \psi_U. \tag{D.125}$$

We can approximately decouple ψ_L and ψ_U, inserting ψ_{L1} in (D.120):

$$i \frac{\partial \psi_{U_0}}{\partial t} \simeq q\phi_{U_0} + \frac{\left[\sigma \cdot (\mathbf{p} - q\mathbf{A}) \right] \left[\sigma \cdot (\mathbf{p} - q\mathbf{A}) \right]}{2m} \psi_{U_0}. \tag{D.126}$$

Using the relation

$$\sigma \cdot \mathbf{A} \sigma \cdot \mathbf{B} = \mathbf{A} \cdot \mathbf{B} + i\sigma \cdot (\mathbf{A} \times \mathbf{B}), \tag{D.127}$$

we can rewrite (D.126) as

$$i \frac{\partial \psi_{U_0}}{\partial t} \simeq \frac{(\mathbf{p} - q\mathbf{A})^2}{2m} \psi_{U_0} - \frac{q}{2m} \left[\sigma \cdot (\nabla \times \mathbf{A} + \mathbf{A} \times \nabla) \right] \psi_{U_0} + q\phi_{U_0}. \tag{D.128}$$

Since $\mathbf{p}$ is the operator $-i\nabla$ acting to the right, the terms $\nabla \times \mathbf{A}$ and $\mathbf{A} \times \nabla$ *do not* cancel. To calculate the gradient terms we use the vector identity

$$(\nabla \times \mathbf{A} + \mathbf{A} \times \nabla) \psi_U = \nabla \times (\mathbf{A}\psi_U) + \mathbf{A} \times \nabla\psi_U$$

$$= \psi_U \nabla \times \mathbf{A} + (\nabla\psi_U) \times \mathbf{A} + \mathbf{A} \times \nabla\psi_U$$

$$= \psi_U (\nabla \times \mathbf{A}) = \psi_U \mathbf{B}. \tag{D.129}$$

We thus obtain the *Pauli equation*,

$$i \frac{\partial \psi_{U_0}}{\partial t} \simeq H_P \psi_{U_0}, \tag{D.130}$$

which has the form of the Schrödinger equation, but with the Pauli Hamiltonian

$$H_P = \frac{(\mathbf{p} - q\mathbf{A})^2}{2m} - \frac{q}{2m} \sigma \cdot \mathbf{B} + q\phi. \tag{D.131}$$

Although the Dirac equation automatically includes higher order terms than the Pauli equation, we see that even in a lower approximation, the Dirac theory predicts a gyromagnetic factor $g = 2$ for fermions. Specifically, knowing that the magnetic dipole interaction with an external magnetic field is given by

$$H' = -\mu \cdot \mathbf{B}, \tag{D.132}$$

we can identify this with the second term of the Pauli Hamiltonian, eq. D.131, and we deduce the magnetic dipole moment μ and the g–factor as

$$\mu = -\mu_B g\mathbf{S} \equiv -\mu_B \sigma \qquad \left(\mu_B = \frac{q}{2m} \right)$$

$$\Rightarrow g = g_D = 2, \tag{D.133}$$

where μ_B is the Bohr magneton. Thus, the Dirac equation predicts that any particle with spin $\frac{1}{2}$ has $g = 2$.

The prediction (D.133) agrees perfectly with the experiments for electrons and muons in which case $q = e$ and m is the particle mass (e.g., $m_\mu \simeq 205 m_e$). The small deviation from $g = 2$ for the electron, or the muon, is due to radiative corrections. In contrast, the magnetic moments of strongly interaction particles, like protons and neutrons, are very different from the Dirac predictions of 2 and 0 respectively:

$$\frac{g_p}{2} = 2.7928474, \quad \frac{g_n}{2} = -1.9130427. \tag{D.134}$$

The deviations from the Dirac value μ_D are characterized by the *anomalous momentum* κ,

$$\mu = \mu_D + \kappa \mu_B. \tag{D.135}$$

This anomalous magnetic interaction is included phenomenologically in the Dirac theory by the addition of an explicit term $-\kappa F_{\mu\nu}\sigma^{\mu\nu}/2$ to the Hamiltonian. The nucleon momenta can be explained, in principle, using the quark model.

D.9.1 Spin-orbit and Darwin terms

Higher order corrections beyond the Pauli Hamiltonian can be obtained by replacing ψ_{L1} for ψ_L in (D.122). One gets

$$\Psi_{L_2} \simeq \frac{\sigma \cdot (\mathbf{p} - q\mathbf{A})}{2m}\psi_U - \frac{[i(\partial/\partial t - q\phi)][\sigma \cdot (\mathbf{p} - q\mathbf{A})]}{2m}\psi_U. \tag{D.136}$$

The next step is to insert (D.136) in the equation for ψ_U, (D.122), and to identify the right side as $H\psi_U$. But, before that, the wave equation must be renormalized to make the Hamiltonian Hermitian and therefore produce a correct nonrelativistic limit. After resolving this complication, we get

$$H_0 \simeq \left[\frac{(\mathbf{p} - q\mathbf{A})^2}{2m} - \frac{p^4}{8m^3} \right] + q\phi - \frac{q}{2m}\frac{\sigma \cdot \mathbf{B}}{2m} - \frac{iq}{8m^2}\mathbf{p} \cdot \mathbf{E}$$
$$- \left[\frac{iq}{8m^2}\sigma \cdot (\nabla \times \mathbf{E}) + \frac{q}{4m^2}\sigma \cdot (\mathbf{E} \times \mathbf{p}) \right], \tag{D.137}$$

We recognize the second term inside the first parenthesis as a relativistic correction to the kinetic energy

$$\sqrt{p^2 + m^2} - m \simeq \frac{p^2}{2m} - \frac{p^4}{8m^3} + \cdots. \tag{D.138}$$

This is familiar in atomic physics, where it slightly lowers energy levels compared to the nonrelativistic values. The term $\sigma \cdot (\nabla \times \mathbf{E})$ in (D.137) is identically zero for a spherically symmetric potential. But the second term is zero for a spherically symmetric potential. But the term $\sigma \cdot (\mathbf{E} \times \mathbf{p})$ contains the *spin-orbit interaction* responsible for the *fine-structure effect* of the atomic levels. To show this we rewrite this term as

$$H_{so} = -\frac{q}{4m^2}\sigma \cdot (\mathbf{E} \times \mathbf{p}) = \frac{q}{4m^2}\sigma \cdot \frac{\partial V(r)}{\partial r}\frac{\mathbf{r}}{r} \times \mathbf{p}$$
$$= \frac{q}{4m^2}\frac{1}{r}\frac{\partial V(r)}{\partial r}\sigma \cdot \mathbf{l}. \tag{D.139}$$

The term $\mathbf{p} \cdot \mathbf{E}$ in (D.137), known as the Darwin term, is related to the Laplacian of the central potential

$$V_{dar} = -\frac{iq}{8m^2}\mathbf{p} \cdot \mathbf{E} = \frac{1}{8m^2}\nabla \cdot \nabla V(r) = \frac{1}{8m^2}\nabla^2 V(r). \tag{D.140}$$

Because $\nabla^2 (1/r) \propto \delta (r)$, this is a contact interaction and thus only affects the S states in atoms. Further insight on the nature of this term can be obtained by considering a charge that, when confined to a bound state, can oscillate between states of positive and negative energies To take this assumption further, imagine that this *Zitterbewegung* leads the charge to select a region in space of the size of its Compton wavelength $\Delta r \simeq 1/m$ around a point r. As a consequence, the Hamiltonian contains an extra term to account for this fluctuation:

$$H' \simeq \langle V (r + \Delta r)\rangle - \langle V(r)\rangle$$
$$\simeq \left\langle V(r) + \frac{\partial V}{\partial r}\Delta r + \frac{1}{2}\sum_{i,j}\Delta r_i \Delta r_j \frac{\partial^2 V}{\partial r_i \partial r_j}\right\rangle - \langle V(r)\rangle$$
$$\simeq \frac{1}{6}(\Delta r)^2 \nabla^2 V \simeq \frac{1}{6m^2}\nabla^2 V. \tag{D.141}$$

This indeed resembles (D.140).

Appendix E | Useful Constants and Conversion Factors

The values inside parentheses indicate the standard deviation. For example, in the first line the error in the electron charge can be understood to be $0.000000063 \times 10^{-19}$ C.

E.1 Constants

Electric charge

$$e = 1.602176462(63) \times 10^{-19} \text{ C} = 1.200\sqrt{\text{MeV} \cdot \text{fm}}$$

$$e^2 = 1.440 \text{ MeV·fm}$$

Planck constant

$$h = 6.62606876(52) \times 10^{-27} \text{ erg·s}$$

$$= 4.13566727(16) \times 10^{-21} \text{ MeV·s}$$

$$\hbar = h/2\pi = 1.054571596(82) \times 10^{-27} \text{ erg·s}$$

$$= 6.58211889(26) \times 10^{-22} \text{ MeV·s}$$

Speed of light

$$c = 299792458 \text{ m/s}$$

$$\hbar c = 1.9733 \times 10^{-11} \text{ Mev·cm} = 197.327 \text{ MeV·fm}$$

Gravitational constant

$$G = 6.673(10) \times 10^{-11} \text{ m}^3 \cdot \text{kg}^{-1} \cdot \text{s}^{-2}$$

Boltzmann constant

$$k = 1.3806503(24) \times 10^{-16} \text{ erg/K}$$

Avogadro number

$$N_A = 6.02214199(47) \times 10^{23} \text{ mol}^{-1}$$

Molar volume

$$V_m = 22.413996(39) \text{ l/mol } (273.15 \text{ K}; 101325 \text{ Pa})$$

Faraday constant

$$F = 96485.3415(39) \text{ C/mol}$$

Compton wavelength

$$\chi_e = \hbar/m_e c = 386.1592642(28) \text{ fm (electron)}$$

$$\chi_p = \hbar/m_p c = 02103089089(16) \text{ fm (proton)}$$

Nuclear magneton	$\mu_N = 3.152451238(24) \times 10^{-14}$ MeV/T
Bohr magneton	$\mu_B = 5.788381749(43) \times 10^{-11}$ MeV/T
Fine structure constant	$\alpha = e^2/\hbar c = 1/137.03599976(50)$
Electron classical radius	$r_e = e^2/m_e c^2 = 2.817940285(31)$ fm
Bohr radius	$a_0 = \hbar^2/m_e e^2 = 0.5291772083(19) \times 10^{-8}$ cm

E.2 Masses

Electron	$m_e = 9.10938188(72) \times 10^{-28}$ g $= 5.485799110(12) \times 10^{-4}$ u
	$\quad = 0.510998902(21)$ MeV/c^2
Muon	$m_\mu = 0.1134289168(34)$ u $= 105.6583568(52)$ MeV/c^2
Pions	$m_{\pi^0} = 134.9764(6)$ MeV/c^2
	$m_{\pi^\pm} = 139.56995(35)$ MeV/c^2
Proton	$m_p = 1.67262158(13) \times 10^{-24}$ g $= 1.00727646688(13)$ u
	$\quad = 938.271998(38)$ MeV/c^2
Neutron	$m_n = 1.00866491578(55)$ u $= 939.565330(38)$ MeV/c^2
Hydrogen atom	$m_H = 1.007825036(11)$ u $= 938.791$ MeV/c^2

E.3 Conversion Factors

Length	1 fermi $= 1$ fm $= 10^{-15}$ m $= 10^{-13}$ cm
Area	1 barn $= 1$b $= 10^{-24}$ cm^2 $= 10^2$ fm^2
Mass	1 unit of atomic mass $= 1$ u $= (1/12)$ m$(^{12}_{6}$C$)$
	$= 1.66053873(13) \times 10^{-24}$ g $= 931.494013(37)$ MeV/c^2
	$= 1822.872 \; m_e$
Energy	1 eV $= 1.602176462(63) \times 10^{-12}$ erg
	$= 1.073544206(43) \times 10^{-9}$ u·c^2
	1 erg $= 10^{-7}$ J
Temperature ($k = 1$)	1 MeV $= 1.16 \times 10^{10}$ K $= 1.78 \times 10^{-30}$ kg

References

[Ab94] F. Abe et al., *Phys. Rev.* **D50** (1994) 2966.

[Ad50] R. K. Adair, *Rev. Mod. Phys.* **22** (1950) 249.

[Ald56] K. Alder, A. Bohr, T. Huus, B. Mottelson and A. Winther, *Rev. Mod. Phys.* **28** (1956) 432.

[An32] C. D. Anderson, *Science* **76** (1932) 238.

[AN36] C. D. Anderson and S. H. Neddermeyer, *Phys. Rev.* **50** (1936) 263.

[Ar85] P. Armbruster, *Ann. Rev. Nucl. Sci.* **35** (1985) 135.

[Ar96] R. Arnold et al., *Z. Phys.* **C72** (1996) 239.

[AS72] M. Abramowitz and I. A. Stegun (eds.), *Handbook of Mathematical Functions*, Dover, New York, 1972.

[AW95] G. Audi and A. H. Wapstra, *Nucl. Phys.* **A595** (1995) 409.

[Az95] R. E. Azuma et al., *Proc. ENAM 95,* M. de Saint Simon and O. Sorlin (eds), Gifsur-Yvette: Editions Frontières, (1995).

[Aj60] F. Ajzenberg-Selove, *Nuclear Spectroscopy* Part B, Academic Press, New York, 1960.

[Ax62] P. Axel, *Phys. Rev.* **126** (1962) 671.

[Ba10] H. Bateman, *Proc. Cambridge Phil. Soc.* **15** (1910) 423.

[Ba57] J. Bardeen, L. N. Cooper, and J. R. Schrieffer, *Phys. Rev.* **108** (1957) 1175.

[Ba74] R. Bass, *Nucl. Phys.* **A231** (1974) 45.

[Ba89] J, N. Bahcall, *Neutrino Astrophysics*, Cambridge University Press, Cambridge, 1989.

[Ba96] R. M. Barnett et al., *Phys. Rev.* **D54** (1996) 1.

[Bah69] J. N. Bahcall and R. M. May, *Ap. J.* **155** (1969) 501.

[Bah69b] J. N. Bahcall and C. P. Moeller, *Ap. J.* **155** (1969) 511.

[Bah89] J. N. Bahcall, *Neutrino Astrophysics*, Cambridge University Press, Cambridge, 1989.

[Bar94] F. C. Barker, *Nucl. Phys.* **A575** 361 (1994).

[Bau86] G. Baur, *Phys. Lett.* **B 178**, 135 (1986).

[BB30] W. Bothe and H. Becker, *Z. Phys.* **66** (1930) 289

[BB52] H. A. Bethe and S. T. Butler, *Phys. Rev.* **85** (1952) 1045

[BB86] G. F. Bertsch and R. A. Broglia, *Physics Today*, August 1986, p. 44.

[BB88] C. A. Bertulani and G. Baur, *Phys. Rep.*, **163** (1988) 299.

[BBFH57] E. M. Burbidge et al., *Rev. Mod. Phys.* **29** 547 (1957).

[BBM02] C.A. Bertulani, M.S. Hussein and G. Muenzenberg, *Physics of Radioactive Nuclear Beams*, Nova Science Publishers, Hauppage, NY, 2002.

[BBR86] G. Baur, C. A. Bertulani and H. Rebel, *Nucl. Phys.* **A 458** (1986) 188.

[BD64] J. D. Bjorken and S. D. Drell, *Relativistic Quantum Mechanics*, McGraw-Hill, New York, 1964.

[BD04] C.A. Bertulani and P. Danielewicz, *Introduction to Nuclear Reactions*, IOP Publishing (Taylor and Francis), London, 2004.

[Be49] L. A. Beach, C. L. Peacock and R. G. Wilkinson, *Phys. Rev.* **76** (1949) 1624.

[Be72] K. E. Bergkvist, *Nucl. Phys.* **B39** (1972) 317.

[Be79] H. A. Bethe, G. Brown, J. Applegate and J. M. Lattimer, *Nucl. Phys.* **A324** (1979) 487.

[Bet00] S. Bethke, *J. Phys.* **G26** (2000) R27.

[Bet37] H.A. Bethe, *Rev. Mod. Phys.* **9** (1937) 69.

[Bet38] H. A. Bethe, and C. L. Critchfield, *Phys. Rev.* **54** (1938) 248.

[Be39] H. A. Bethe, *Phys. Rev.* **55** (1939) 103, 434.

[Bet49] H. A. Bethe, *Phys. Rev.* **76** (1949) 38.

[BG37] W. Bothe and W. Gentner, *Z. Phys.* **106** (1937) 236.

[BG77] P. J. Brussaard and P. W. M. Glaudemans, *Shell-Model Applications in Nuclear Spectroscopy*, North-Holland, Amsterdam, 1977.

[Bh52] A. B. Bhatia, Kun Huang, R. Huby and H. C. Newns, *Phil. Mag.* **43** (1952) 485.

[BK47] G. C. Baldwin and G. S. Klaiber, *Phys. Rev.* **71** (1947) 3; *Phys. Rev.* **73** (1948) 1156.

[Bl96] Y. Blumenfeld, *Nucl. Phys.* **A599** (1996) 289c.

[BM69] A. Bohr and B. Mottelson *Nuclear Structure*, vols. I and II Benjamin, New York, 1969.

[Bo13] N. Bohr, *Phil. Mag.* **26** (1913) 1, 476, 857.

[Boh36] N. Bohr, *Nature* **137** (1936) 344.

[BP62] B. Buck and F.G. Perey, *Phys. Rev. Lett.* **8** (1962) 444.

[BS82] S. Björnholm and W. J. Swiatecki, *Nucl. Phys.* **A391** (1982) 471.

[BT98] A. B. Balantekin and N. Takigawa, *Rev. Mod. Phys.* **70** (1998) 77.

[Bu57] E. M. Burbidge, G. R. Burbidge, W. A. Fowler, and F. Hoyle, *Rev. Mod. Phys.* **29** (1957) 547.

[BW39] N. Bohr and J. A. Wheeler, *Phys. Rev.* **56** (1939) 426.

[BW52] J. M. Blatt and V. F. Weisskopf, *Theoretical Nuclear Physics*, Wiley, New York, 1952.

[Can05] L.F. Canto, P. R. S. Gomes, R. Donangelo, and M. S. Hussein, *Phys. Reports* **424** (2006) 1.

[CG28] E. U. Condon and R. W. Gurney, *Nature* **122** (1928) 439; *Phys. Rev.* **33** (1929) 127.

[Ch31] S. Chandrasekhar, *Astrophys. J.* **74** (1931) 81; *Rev. Mod. Phys.* **56** (1984) 137.

[Ch32] J. Chadwick, *Proc. Roy. Soc. (London)* **A136** (1932) 692, *Nature* **129** (1932) 312.

[Ch55] O. Chamberlain, E. Segrè, C. Wiegand and T. Ypsilantis, *Phys. Rev.* **100** (1955) 947.

[Ch58] D. M. Chase, L. Wilets, and A. R. Edmonds, *Phys. Rev.* **110** 1080, 1958.

[Ch64] J. H. Christenson, J. W. Cronin, V. L. Fitch, and R. Turlay, *Phys. Rev. Lett.* **13** (1964) 138.

[Ch80] L. M. Chirovsky, W. P. Lee, A. M. Sabbas, J. L. Groves and C. S. Wu, *Phys. Lett.* **B94** (1980) 127.

[Ch84] S. Chandrasekhar, *Rev. Mod. Phys.* **56** (1984) 137.

[ChR79] M. Chemtob and M. Rho (eds.), *Mesons and Nuclei*, vol. I, North Holland, Amsterdam, 1979.

[CJ32] I. Curie and F. Joliot, *Compt. Rend. Acad. Sci.* **194** (1932) 273, 708, 876, 2208.

[Cla84] D. D. Clayton, *Principles of Stellar Evolution and Nucleosynthesis*, University of Chicago Press, Chicago, 1984.

[Co56] B. Cork, G. R. Lambertson, O. Piccioni and W. A. Wenzel, *Phys. Rev.* **104** (1956) 1193.

[Co74] S. Cohen, F. Plasil and W. J. Swiatecki, *Ann. Phys.* **82** (1974) 557.

[Coo57] C. W. Cook, W. A. Fowler, C. C. Lauritsen, and T. Lauritsen, *Phys. Rev.* **107** (1957) 508.

[CS51] E.U. Condon and G. Shortley, *The Theory of Atomic Spectra*, Cambridge University Press, Cambridge, 1951.

[Da52] R. Davis Jr., *Phys. Rev.* **86** (1952) 976.

[Da96] B. F. Davis et al., *Nucl. Phys.* **A599** (1996) 277c.

[DB88] S. S. Dietrich and B. L. Berman, *At. Data and Nucl. Data Tables* **38** (1988) 199.

[Des93a] P. Descouvemont, *Phys. Rev.* **C47** 210 (1993).

[Des98] P. Descouvemont, *Tours Symposium on Nuclear Physics III (AIP Conf. Proc. 425)*, American Institute of Physics, New York, 1998.

[Ed60] A. R. Edmonds, *Angular Momentum in Quantum Mechanics*, Princeton University Press, Princeton, (1960).

[En66] H. A. Enge, *Introduction to Nuclear Physics*, Addison-Wesley, Reading, MA, 1966.

[En74] R. Engfer et al., *At. Data and Nucl. Data Tables* **14** (1974) 509.

[ESW96] R. K. Ellis, W. J. Stirling, and B. R. Webber, *QCD and Collider Physics*, Cambridge University Press, Cambridge, 1996.

[FCZ67-11] Fowler W A, Caughlan G E and Zimmermann B A 1967 *Ann. Rev. Astron. Astrophys.* **5** 525.

[Fe34] E. Fermi, *Z. Phys.* **88** (1934) 161.

[Fe53] H. Feshbach, C. E. Porter, and V. F. Weisskopf, *Phys. Rev.* **90** (1953) 166.

[FFN85-11] G. M. Fuller, W. A. Fowler, and M. Newman *Ap. J.* **293** (1985).

[Fie89] R. D. Field, *Applications of Perturbative QCD*, Addison-Wesley, New York, 1989.

[For03] C. Forssén, N. B. Shul'gina and M. V. Zhukov, *Phys. Rev.* **C67** 045801 (2003).

[FR47] J. L. Fowler and L. Rosen, *Phys. Rev.* **72** (1947) 926.

[Fri51] E. Frieman and L. Motz, *Phys. Rev.* **89** (1951) 648.

[Fr91] J. I. Friedman, *Rev. Mod. Phys.* **63** (1991) 615.

[FT50] E. Feenberg and G. Trigg, *Rev. Mod. Phys.* **22** (1950) 399.

[FTV93] W. H. Press, B. P. Flannery, S. A. Teukolsky, and W. T. Vetterling, *Numerical Recipes*, Cambridge University Press, Cambridge, 1993.

[Ga28] G. Gamow, *Z. Phys.* **51** (1928) 204.

[Ga57] R. L. Garwin, L. M. Lederman and M. Weinrich, *Phys. Rev.* **105** (1957) 1415.

[Ge53] M. Gell-Mann, *Phys. Rev.* **92** (1953) 833.

[Ge64] M. Gell-Mann, *Phys. Lett.* **8** (1964) 214.

[Gh50] S. N. Ghoshal, *Phys. Rev.* **80** (1950) 939.

[GL48] E. Gardner and C. M. G. Lattes, *Science* **107** (1948) 270.

[Gl70] S. L. Glashow, J. Iliopoulos, and L. Maiani, *Phys. Rev.* **D2** (1970) 1285.

[GN11] H. Geiger and J. M. Nuttall, *Phil. Mag.* **22** (1911) 613.

[Go50] H. Goldstein, *Classical Mechanics*, Addison-Wesley, Reading, MA, 1950.

[GS51] M. Goldhaber and A. W. Sunyar, *Phys. Rev.* **83** (1951) 906.

[GT48] M. Goldhaber and E. Teller, *Phys. Rev.* **74** (1948) 1046.

[GW73] D. J. Gross and F. Wilczek, *Phys. Rev. Lett.* **30** (1973) 1343.

[Ha49] O. Haxel, J. H. D. Jensen, and H. E. Suess, *Phys. Rev.* **75** (1949) 1766.

[Ha56] B. Hahn, D. G. Ravenhall, and R. Hofstadter, *Phys. Rev.* **101** (1956) 1131.

[Ha65] S. M. Harris, *Phys. Rev.* **B138** (1965) 509.

[He25] W. Heisenberg, *Z. Phys.* **33** (1925) 879.

[He32] W. Heisenberg, *Z. Phys.* **77** (1932) 1.

[HF52] W. Hauser and H. Feshbach, *Phys. Rev.* **A87** (1952) 366.

[HH29] W. Heitler and G. Herzberg, *Naturwiss.* **17** (1929) 673.

[HM84] F. Halzen and A. D. Martin, *Quarks and Leptons: An Introductory Course in Modern Particle Physics*, Wiley, New York, 1984.

[Ho46] F. Hoyle, *M. N. R. A. S.* **106** (1946) 343.

[Ho57] R. Hofstadter, *Ann. Rev. Nucl. Sci.* **7** (1957) 231.

[Ho71] T. L. Houk, *Phys. Rev.* **C3** (1971) 1886.

[Ho96] S. Hofmann, *Nucl. Phys. News* **6** (1996) 26; S. Hofmann et al., *Z. Phys.* **A354** (1996) 229.

[Hoy53] F. Hoyle, D. N. F. Dunbar, W. A. Wenzel, and W. Whaling, *Phys. Rev.* **92** (1953) 1095.

[Hoy54] F. Hoyle, *Ap. J. Sup.* **1** (1954) 121.

[HS39] O. Hahn and F. Strassmann, *Naturwiss.* **27** (1939) 11, 89.

[HT03] P.G. Hansen and J. A. Tostevin, *Annu. Rev. Nucl. Part. Sci.* **53** 219 (2003).

[Hy71] E. K. Hyde, *The Nuclear Properties of the Heavy Elements*, vol. 3, *Fission Phenomena*, Dover, New York, 1971.

[HW53] D. L. Hill and J. A. Wheeler, *Phys. Rev.* **89** (1953) 1102.

[Ib65] I. Iben Jr., *Astrophys. J.* **141** (1965) 993.

[In54] D. R. Inglis, *Phys. Rev.* **96** (1954) 1059.

[Ja75] J. D. Jackson, *Classical Electrodynamics*, Wiley, New York, 1975.

[Jac99] J. D. Jackson, *Classical Electrodynamics* (3rd edition) Wiley, New York, 1999.

[JB50] J. D. Jackson and J. M. Blatt, *Rev. Mod. Phys.* **22** (1950) 77.

[JF58] R. C. Jensen and A. W. Fairhall, *Phys. Rev.* **109** (1958) 942.

[Jo72] A. Johnson, H. Ryde and S. A. Hjorth, *Nucl. Phys.* **A179** (1972) 753.

[Ke73] G. A. Keyworth et al., Third International Symposium on Physics and Chemisty of Fission, vol. 1 (1973) 85; *Phys. Rev.* **C8** (1973) 2352.

[Kn66] D. J. Knecht, P. F. Dahl, and S. Messelt, *Phys. Rev.* **148** (1966) 1031.

[Ko68] L. Kowalsky, J. C. Jodogne and J. M. Miller, *Phys. Rev.* **169** (1968) 894.

[Ko86] S. E. Koonin, *Computational Physics*, Addison-Wesley, Reading, MA, 1986.

[Ko92] M. Koshiba, *Phys. Rep.* **220** (1992) 229.

[Ku86] W. Kündig et al., *Weak and Eletromagnetic Interactions in Nuclei*, H. V. Klapor (ed.), Springer, Berlin, 1986.

[La47] C. M. G. Lattes, H. Muirhead, G. P. S. Occhialini and C. F. Powell, *Nature* **159** (1947) 694.

[La80] R. D. Lawson, *Theory of the Nuclear Shell Model* Oxford: Clarendon Press, 1980.

[Lac80] M. Lacombe et al., Phys. Rev. **C21**, 861 (1980).

[La96] Yu. A. Lazarev et al., *Phys. Rev.* **C54** (1996) 620.

[Lan96] K. Langanke and C. A. Barnes, *Advances in Nuclear Physics* **22** 173, J. W. Negele and Vogt E. (eds.), New York: Plenum Press, 1996.

[Le68] M. Lefort, *Nuclear Chemistry*, Van Nostrand, Amsterdam, 1968.

[Le74] A. Leprêtre et al., *Nucl. Phys.* **A219** (1974) 39.

[Lep97] G.P. Lepage, "How to Renormalize the Schroedinger Equation," *Proceedings of the A. Swieca Summer School*, World Scientific, 1997, C.A Bertulani et al. (eds), nucl-th/9706029.

[LeC54] J.M.B. Lang and K.J. Le Coutuer *Proc. Phys. Soc.* **67A** 586, 1954.

[Lei95] J. R. Leigh et al. *Phys. Rev.* **C52** (1995) 3151.

[Lib55] W. F. Libby, *Radiocarbon Dating*, 2nd ed., University of Chicago Press, Chicago, 1955.

[Lin79] R. A. Lindgren, W. L. Bendel, L. W. Fagg, and E. C. Jones, Jr. *Phys. Rev. Lett.* **35**, 1423 (1975).

[LL65] L.D. Landau and E.M. Lifshitz, Quantum Mechanics: Nonrelativistic Theory, vol. 3, Pergamon Press, Oxford, 1965.

[LMA80] S. M. Lee, T. Matsuse and A. Arima A *Phys. Rev. Lett.* **45** (1980) 165.

[LS78] C. M. Lederer and V. S. Shirley (eds.), *Table of Isotopes*, Wiley, New York, 1978.

[Lu80] V. A. Lubimov, E. G. Novikov, V. Z. Nozik, E. F. Tretyakov, and V. S. Kosic, *Phys. Lett.* **B94** (1980) 266.

[LY56] T. D. Lee and C. N. Yang, *Phys. Rev.* **104** (1956) 254.

[MA40] E. McMillan and P. H. Abelson, *Phys. Rev.* **57** (1940) 1185.

[Ma49] M. G. Mayer, *Phys. Rev.* **75** (1949) 1969; **78** (1950) 16.

[Ma68] H. Marshak, A. Langsford, C. Y. Wong and T. Tamura, *Phys. Rev. Lett.* **20** (1968) 554.

[Ma75] N. Marty, M. Morlet, A. Willis, V. Comparat and R. Frascaria, *Nucl. Phys.* **A 238** (1975) 93.

[Mac89] R. Machleidt, *Adv. Nucl. Phys.* **89** (1989) 189.

[Mac01] R. Machleidt, *Phys. Rev.* **C 63** (2001) 024001.

[MF39] L. Meitner and O. R. Frisch, *Nature* **143** (1939) 239, 471.

[MG69] U. Mosel and W. Greiner, *Z. Phys.* **222** (1969) 261.

[Mi67] A. Michalowicz, *Kinematics of Nuclear Reactions*, Iliffe Books, London, 1967.

[Mi87] K. Miura et al., *Nucl Phys.* **A467** (1987) 79.

[MJ55] M. G. Mayer and J. H. D. Jensen, *Elementary Theory of Nuclear Shell Structure*, Wiley, New York and Chapman Hall, London, 1955.

[MN55] B. R. Mottelson and S. G. Nilsson, *Phys. Rev.* **99** (1955) 1615.

[Mo30] N. F. Mott, *Proc. Roy. Soc. (London)* **A126** (1930) 259.

[Mo58] R. Mössbauer, *Z. Phys.* **151** (1958) 124.

[Mo81] U. Mosel *Comm. Nucl. Part. Phys.* **9** (1981) 213.

[Mo88] S. Mordechai et al., *Phys. Rev. Lett.* **61** (1988) 531.

[MS94] H. Metcalf and P. van der Straten, *Phys. Rep.* **244** (1994) 203.

[MT68] E. Migneco and J. P. Theobald, *Nucl. Phys.* **A112** (1968) 603.

[Mu67] M. L. Muga, C. R. Rice and W. A. Sedlacek, *Phys. Rev. Lett.* **18** (1967) 404.

[Mu87] T. Muta, *Foundations of Quantum Chromodynamics*, World Scientific, Singapore, Singapore, 1987.

[Mv67] K. W. McVoy, *Ann. Phys.* **43** (1967) 91.

[MV94] M. Moe and P. Vogel, *Annu. Rev. Nucl. Part. Sci.* **44** (1994) 247.

[My77] W. D. Myers, W. J. Swiatecki, T. Kodama, L. J. El-Jaick and E. R. Hilf, *Phys. Rev.* **C15** (1977) 2032.

[MW79] C. Mahaux, H. A. Weidenmüller, *Ann. Rev. Part. Nucl. Sci.* **29** 1 (1979).

[Na70] J. B. Natowitz, *Phys. Rev.* **C1** (1970) 623.

[Ne60] W. E. Nervik, *Phys. Rev.* **119** (1960) 1685.

[New82] R. Newton, *Scattering Theory of Waves and Particles* (Berlin: Springer Verlag, Berlin), 1982.

[Nie81] N.K. Nielsen, *Am. J. Phys.* **49** (1981) 1171.

[Ni55] S. G. Nilsson, *Mat. Fys. Medd. Dan. Vid. Selsk.* **29** n 16 (1955).

[NN53] T. Nakano and K. Nishijima, *Prog. Theor. Phys.* **10** (1953) 581.

[No50] L. W. Nordheim, *Phys. Rev.* **78** (1950) 294.

[Og75] Yu. Ts. Oganessian, A. S. Iljinov, A. G. Demin, and S. P. Tretyakova, *Nucl. Phys.* **A239** (1975) 353.

[Og99] Yu. Ts. Oganessian et al., *Phys. Rev. Lett.* **83** (1999) 3154.

[Opi51] E. J. Öpik, *Proc. Roy. Irish Acad.* **A54** (1951) 49.

[Pa34] W. Pauli, *Rapports du Septième Conseil de Physique Solvay, Bruxelas, 1933*, Gauthiers-Villars Cie, Paris, 1934.

[Pa52] A. Pais, *Phys. Rev.* **86** (1952) 663.

[Pa68] D. Paya et al., *Nuclear Data for Reactors*, vol. II, IAEA, Vienna, 1968, 128.

[PB62] F. G. Perey and B. Buck *Nucl. Phys.* **32** (1962) 353.

[PC61] J. L. Powell and B. Crasemann, *Quantum Mechanics*, Addison-Wesley, Reading, MA, 1961, 140.

[Pe50] I. Perlman, A. Ghiorso and G. T. Seaborg, *Phys. Rev.* **77** (1950) 26.

[Pe63] F. G. Perey, *Phys. Rev.* **131** (1963) 745.

[Po62] S. M. Polikanov et al., *Soviet Phys. JETP* **15** (1962) 1016.

[Po73] H.D. Politzer, *Phys. Rev. Lett.* **30** (1973) 1346.

[PR60] R. V. Pound and G. A. Rebka Jr., *Phys. Rev. Lett.* **4** (1960) 337.

[Pr62] M. A. Preston, *Physics of the Nucleus*, Addison-Wesley, Reading, MA, 1962.

[Pr89] P. B. Price, *Annu. Rev. Nucl. Part. Sci.* **39** (1989) 19.

[PS90] Š. Piskoř and W. Schäferlingová, *Nucl. Phys.* **A510** (1990) 301.

[PS95] M.E. Peskin, D.V. Schroeder, *An Introduction to Quantum Field Theory*, Westview Press, Boulder, CO, 1995.

[PW71] R. Pitthan and Th. Walcher, *Phys. Lett.* **36B** (1971) 563.

[Ra33] I. I. Rabi et al., several articles in *Physical Review* between 1933 and 1936.

[Ra879] Lord Rayleigh, *Theory of Sound*, vol. II, Macmillan, London, 1877; *Proc. Roy. Soc. (London)* **A29** (1879) 71.

[RC53] F. Reines and C. L. Cowan Jr., *Phys. Rev.* **92** (1953) 830.

[Re65] F. Reif, *Fundamentals of Statistical and Thermal Physics*, McGraw-Hill, Tokyo, 1965.

[Re68] R. V. Reid, Jr., *Ann. Phys.* **50** (1968) 411.

[Ree59] H. Reeves and E. E. Salpeter, *Phys. Rev.* **116** (1959) 1505.

[RH50] L. Rosen and A. M. Hudson, *Phys. Rev.* **78** (1950) 533.

[Ro51] J. M. Robson, *Phys. Rev.* **83** (1951) 349.

[Rol74] C. Rolfs and R. E. Azuma, *Nucl. Phys.* **A227** 291 (1974).

[RR88] C. Rolfs and W. S. Rodney, *Cauldrons in the Cosmos*, University of Chicago Press, Chicago, 1988.

[RT64] B. W. Ridley and J. F. Turner, *Nucl. Phys.* **58** (1964) 497.

[Ru11] E. Rutherford, *Phil. Mag.* **21** (1911) 669.

[Ru19] E. Rutherford, *Phil. Mag.* **37** (1919) 537.

[Sa80] A. Salam, *Rev. Mod. Phys.* **52** (1980) 525.

[Sa92] H. Sagawa, *Phys. Lett.* **B286** (1990) 7.

[Sal52] E. E. Salpeter, *Phys. Rev.* **88** (1952) 547; see also *Ap. J.* **115** (1952) 326.

[Sal54] E. E. Salpeter, *Australian J. Phys*, **7** (1954) 373.

[Sal57] E. E. Salpeter, *Phys. Rev.* **107** (1957) 516.

[Sc26] E. Schrödinger, *Ann. der Phys.* **79** (1926) 361, 489.

[Sc54] L. I. Schiff, *Prog. Theor. Phys.* **11** (1954) 288.

[Sch98] H. Schatz et al., *Phys. Rep.* **294** 167 (1998).

[See65] P. A. Seeger, W. A. Fowler, and D. D. Clayton, *Astrophys. J. Suppl.* **11** 121 (1965).

[Si97] A. De Silva, M. K. Moe, M. A. Nelson, and M. A. Vient, *Phys. Rev.* **C56** (1997) 2451.

[Si98] B. Singh, J. L. Rodriguez, S. S. M. Wong, and J. K. Tuli, *Nucl. Data. Sheets* **84** (1998) 487.

[SJ50] H. Steinwedel and J. H. D. Jensen, *Z. Naturforsch.* **52** (1950) 413.

[Sk57] T. H. R. Skyrme, *Proc. Phys. Soc.* (London) **A70** (1957) 433.

[Sk59] T. H. R. Skyrme, *Nucl. Phys.* **9** (1959) 615.

[Sk66] S. J. Skorka, J. Hertel, and T. W. Retz-Schmidt, *Nucl. Data* **2** (1966) 347.

[Sm95] R. Smolańczuk, J. Skalski, and A. Sobiczewsky, *Phys. Rev.* **C52** (1995) 1871.

[So70] A. J. Soinsky, R. B. Frankel, Q. O. Navarro, and D. A. Shirley, *Phys. Rev.* **C2** (1970) 2379.

[Stok78] R. G. Stokstad et al., *Phys. Rev. Lett.* **41** (1978) 465.

[SS55] G. L. Squires and A. T. Stewart, *Proc. Roy. Soc.* **A230** (1955) 19.

[ST37] J. Schwinger and E. Teller, *Phys. Rev.* **52** (1937) 286.

[St67] V. M. Strutinsky, *Nucl. Phys.* **A95** (1967) 420.

[St70] E. P. Steinberg, B. D. Wilkins, S. B. Kaufman, and M. J. Fluss, *Phys. Rev.* **C1** (1970) 2046.

[St96] M. Steiner et al., *Phys. Rev. Lett.* **76**, 26 (1996).

[Sta99] D.M. Stamper-Kurn et al., *Phys. Rev. Lett.* **80** (1999) 2072.

[Sue56] H. E. Suess and H. C. Urey, *Rev. Mod. Phys.* **28** (1956) 53.

[Sue00] K. Suemmerer and B. Blank, *Phys. Rev.* **C61** (2000) 034607.

[SvH69] E. E. Salpeter, H. M. van Horn *Ap. J.* **155** (1969) 183.

[SW86] B. D. Serot and J. D. Walecka, *Adv. Nucl. Phys.* **16** (1986) 1.

[Ta85] I. Tanihata et al., *Phys. Rev. Lett.* **55** (1985) 2676.

[Ta92] I. Tanihata et al., *Phys. Lett.* **B289** (1992) 261.

[Ta96] N. Takaoka, Y. Motomura and K. Nagao, *Phys. Rev.* **C53** (1996) 1557.

[Th84] C. E. Thorn, J. W. Olness, E. K. Warburton and S. Raman, *Phys. Rev.* **C30** (1984) 1442.

[Tv72] A. Tveter, *Nucl. Phys.* **A185** (1972) 433.

[Van94] L. Van Wormer et al., *Astrophys. J.* **432** 326 (1994).

[VB72] D. Vautherin and D. M. Brink, *Phys. Rev.* **C5** (1972) 626.

[Wag68] R. V. Wagoner *Ap. J. Suppl.* **18** 247 (1969).

[Wa52] B. E. Watt, *Phys. Rev.* **87** (1952) 1037.

[Wa58] A. H. Wapstra, *Handbuch der Physik* **38** (1958) 1.

[Wa93] L. Wang et al., *Phys. Rev.* **C47** (1993) 2123.

[We35] C. F. von Weizsäcker, *Z. Phys.* **96** (1935) 431.

[We51] V. F. Weisskopf, *Phys. Rev.* **83** (1951) 1073.

[We67] S. Weinberg, *Phys. Rev. Lett.* **19** (1967) 1264.

[We75] G. W. Wetherill, *Ann. Rev. Nucl. Sci.* **25** (1975) 283.

[Wei99] S. Weinberg, *The Quantum Theory of Fields*, 3 vols., Cambridge University Press, Cambridge, 1996–2000.

[Wi73] J. Wilczyński, *Nucl. Phys.* **A216** (1973) 386.

[Wir95] R. B. Wiringa, V. G. J. Stoks, and R. Schiavilla, *Phys. Rev.* **C51** (1995) 38.

[Wo73] C. Y. Wong, *Phys. Rev. Lett.* **31** (1973) 766.

[Wo87] A. Van der Woude, *Prog. Part. Nucl. Phys.* **18** (1987) 217.

[Woo86] S. E. Woosley and T. A. Weaver, *Ann. Rev. Astr. Astrophys.* **24** (1986) 205.

[Wu57] C. S. Wu, E. Ambler, R. W. Hayward, D. D. Hoppes, and R. P. Hudson, *Phys. Rev.* **105** (1957) 1413.

[Xu04] H. M. Xu, C. A. Gagliardi, R. E. Tribble, A. M. Mukhamedzhanov, and N. K. Timofeyuk, *Phys. Rev. Lett.* **73** (1994) 2027.

[Yu35] H. Yukawa, *Proc. Phys. Math. Soc. Japan* **17** (1935) 48.

[Zh93] M. V. Zhukov, B. V. Danilin, D. V. Fedorov, J. M. Bang, I. J. Thompson, and J. S. Vaagen, *Phys. Rep.* **231** (1993) 231.

[ZM73] A. M. Zebelman and J. M. Miller, *Phys. Rev. Lett.* **30** (1973) 27.

[Zw64] G. Zweig, CERN Report N° 8182/TH401, 1964 (unpublished).

Index